TASTE CHEMISTRY

TASTE CHEMISTRY

by

Robert S. Shallenberger

New York Agricultural Experiment Station
Cornell University
Geneva, New York (USA)

BLACKIE ACADEMIC & PROFESSIONAL
An Imprint of Chapman & Hall
London · Glasgow · New York · Tokyo · Melbourne · Madras

Published by Blackie Academic & Professional, an imprint of Chapman & Hall, Wester Cleddens Road, Bishopbriggs, Glasgow G64 2NZ

Chapman & Hall, 2–6 Boundary Row, London SE1 8HN, UK

Blackie Academic & Professional, Wester Cleddens Road, Bishopbriggs, Glasgow G64 2NZ, UK

Chapman & Hall Inc., 29 West 35th Street, New York NY10001, USA

Chapman & Hall Japan, Thomson Publishing Japan, Hirakawacho Nemoto Building, 6F, 1–7–11 Hirakawa-cho, Chiyoda-ku, Tokyo 102, Japan

DA Book (Aust.) Pty Ltd, 648 Whitehorse Road, Mitcham 3132, Victoria, Australia

Chapman & Hall India, R. Seshadri, 32 Second Main Road, CIT East, Madras 600 035, India

First edition 1993

Typeset in 10/12pt Times, by Alden Multimedia, Northampton
Printed in Great Britain at the University Press, Cambridge

ISBN 0 7514 01501

A catalogue record for this book is available from the British Library

Library of Congress Cataloging-in-Publication data

Shallenberger, R. S.
Taste chemistry / by Robert S. Shallenberger.
p. cm.
Includes bibliographical references and indexes.
ISBN 0 7514 0150 1
1. Taste—Physiological aspects. I. Title.
QP456S53 1992 92-11271
612.8′7—dc20 CIP

Dedication

To those who have, and to those who do
To those who would, and to those who will
Pursue the science of 'Sweetness and Light.'

Preface

The object of this text is to examine, and elaborate on the meaning of the established premise that 'taste is a chemical sense.' In particular, the major effort is directed toward the degree to which chemical principles apply to phenomena associated with the inductive (recognition) phase of taste. A second objective is to describe the structure and properties of compounds with varying taste that allow decisions to be made with respect to the probable nature of the recognition chemistry for the different tastes, and the probable nature of the receptor(s) for those tastes. A final objective is to include appropriate interdisciplinary observations that have application to solving problems related to the chemical nature of taste.

Taste is the most easily accessible chemical structure-biological activity relationship, and taste chemistry studies, i.e. the chemistry of sweetness, saltiness, sourness, and bitterness, have application to general biology, physiology, and pharmacology. Because it involves sensory perception, taste is also of interest to psychologists, and has application to the food and agricultural industries. The largest portion of the text is directed toward sweetness as, due to economic and other factors, the majority of the scientific studies are concerned with sweetness.

The text begins with a prologue to describe the problems associated with the study of taste chemistry. Then, there is an introductory chapter to serve as an overview of the general interdisciplinary knowledge of the subject. It is followed by a chapter on the fundamental chemical principles that apply to taste induction chemistry.

A series of chapters then present and discuss the structure and properties of compounds that elicit various tastes. Their primary purpose is to demonstrate possible relations between structure and taste activity, particularly sweetness. These relations seem relatively straightforward in some cases, but lead to ambiguity, uncertainty, and contradiction in others. An example is the often confusing structural identity and taste of the *syn*- and *anti*-oximes, amino acid enantiomers, and peptides. A second purpose is that, on perusing the text, further possible insight into structure-taste activity relations may occur to the reader, and he/she will have available sufficient structure-taste relations to afford preliminary examination of those thoughts.

These are followed by a chapter that elaborates upon the common chemical features associated with each taste, and the next chapter presents the probable chemical mechanism(s) for them. A penultimate chapter introduces the role that symmetry, chirality, and topology seem to have in taste chemistry. The final chapter presents the probable chemical nature of taste receptors, and an epilogue contains a discussion of the extent to which problems associated with comprehension of taste seem to be resolved by application of chemical principles.

A glossary with definitions of stereochemical and other terms is included, and the definitions are cross-referenced to the text. The need for the glossary arose with the decision not to digress too far from the taste chemistry theme because an abstract symmetry or chirality principle needed to be introduced. It is hoped that the glossary treatment, plus the context with which the terms are used in the main body of the text are sufficient to convey their meaning.

All sections in the text are identified with a number code for cross-reference purposes. The page number for each section can be found in the text contents at the front of the book. There is also an author index and a subject index.

ACKNOWLEDGEMENTS

The author thanks the associations, companies, and organizations that granted permission to reproduce copyright material. They are the Academic Press, The American Chemical Society and its Division of Chemical Education, the American Pharmaceutical Association, Walter de Gruyter & Company, The MacMillan Publishing Company, McGraw-Hill, Inc., Oxford University Press, Pergamon Press, Inc., The Psychonomic Society, The Royal Society of Chemistry, Springer-Verlag, and Verlagsgesellschaft mbH.

The author is grateful to colleagues who submitted copies of reprints, thoughts, caveats, and copies of manuscripts not yet published. They are

Hans-Dieter Belitz, James N. BeMiller, Gordon G. Birch, James R. Daniel, Grant E. DuBois, Jan R. Frijters, Dieter Glaser, Murry Goodman, Leslie Hough, Michael G. Lindley, Frieder Lichtenthaler, J.A. Maga, Mohamed Mathlouthi, Lucio Merlini, Tetsou Suami, P.A. Temussi, Arnold van der Heijden, Henk van der Wel, and Thomas J. Venanzi. Hopefully, their valued contributions have neither been misrepresented nor misinterpreted. My long-time colleague, Terry E. Acree has been most helpful, and I gratefully acknowledge his encouragement and interest, but in particular I valued our many discussions.

The help and encouragement from Grace Fox Parsons (Gracie), another long time colleague, is fondly acknowledged. I thank Stuart G. Reeves for diligent perusal of the manuscript, and constructive comments on the chemistry and general organization. In particular, help with the significance of solution recognition reactions versus binding reactions, from the biochemist's viewpoint, proved to be valuable. Last, but not least, I thank Mary van Buren, and Jeanne Samimy, of the New York State Agricultural Experiment Station Library for their help.

Robert S. Shallenberger
Geneva, NY (USA)

Table of Contents

Prologue

Taste is a subject of interest and study for a number of different academic disciplines. Some of them are chemistry, biochemistry (molecular biology), pharmacology, psychology, and neurophysiology. As a consequence of valuable interdisciplinary interaction among taste investigators, knowledge of taste has advanced markedly in recent years. At the same time, knowledge of taste has become replete with enigmas and apparent contradictions, both within and across the scientific disciplines. For chemistry, and even though taste is widely considered to be a *chemical sense*, what is meant by that term is fuzzy. It can mean that taste is caused by the general physical properties of chemicals. It can also mean that taste results from chemical reactions. A chemist might presume that different tastes are elicited by a chemical probe that reacts in different ways with substances that taste sweet, salty, bitter, or sour. If so, it would seem logical to assign the initiation of the sweetness, sourness, saltiness, and bitterness sensations to different chemical reactions.

However, there is a paucity of criteria usually employed to characterize the presumably different chemical reactions. There is no change in temperature, volume, or pH. There is no measurable increase or decrease in oxidation number. No reaction products result that either precipitate, evolve as a gas, or change the color of an indicator. There is not even a chemical equation to balance, and if there are chemical reactions responsible for different tastes, they must belong to that nebulous group of weak, easily reversible reactions known as recognition-chemistry reactions. Those weak reactions are ill-defined, and have common features. As a consequence of the tentative chemical nature of taste, taste theory has proponents to support any one of several tenets. Among them, it is proposed that different tastes are due to chemicals of different size, shape, or volume. The distinction among the sweet, salt, sour, and bitter tastes is then assigned to different receptors for the different tastes, and even for different substances with a given taste modality, such as sweetness.

A particularly difficult problem encountered by assigning taste to chemical reactions is the need to account for the fact that the affinity constants for some sweet substances (sugars) are so low, they seem superficial, while the sweetness potency of other substances is now known to vary by five log units. On the other hand, the sour and salt tastes are represented by substances that vary in potency by factors of only two to three. Substances are known that block the sweet taste of substances, but not the other tastes. Still other compounds taste exceedingly bitter to some people, and not bitter at all to others.

Louis Pasteur concluded that there is a handedness (chiral) feature contained in the structure of the receptor for sweetness, as the right-handed form of asparagine tastes sweet while the mirror-image left-handed form does not. An exception to the conclusion is that only some antipodal amino acids exhibit the phenomenon. Moreover, although different tastes for enantiomers was logically believed to also apply to the right- and left-handed sugars, it is now known that this is not true either. A receptor for sweetness that recognizes handedness in some cases, but not for others, serves as the basis for a chiral anomaly of highest order. Explanations to account for the handedness phenomena are as varied as the number of scientific disciplines devoted to the study of taste, and include proliferation of receptor numbers. Among them are right-handed receptors for sweet D-amino acids, left-handed receptors for sweet L-amino acids, and left-handed receptors for bitter L-amino acids, and no receptors for tasteless amino acids. By the same token, it is proposed that there are both right- and left-handed sweet receptors for the sugars, or that the sugars elicit sweetness by some physical process, and other sweeteners do not. This author believes that if taste is indeed a chemical sense, and caused by chemical reactions, the only acceptable resolution of the above chiral phenomena must be derived by application of stereochemical principles. Such resolution of the stereochemical anomaly would serve to confirm the chemical-principle basis of taste, and perhaps shed light on the fundamental stereochemical nature of taste, including the stereochemical nature of the receptor(s). Otherwise, it is the nonchemical taste-principles that are most valid, and not the premise that taste is caused by chemical reactions.

With special reference to taste induction and chemical recognition phenomena, and as stated in the preface, the purpose of this text is to determine the extent to which application of chemical principles can account for the aforementioned, and other taste phenomena. After consideration of the data available, and after the deliberations presented in the text of this treatise, the degree to which the endeavor seems to have been successful is summarized in the epilogue.

CHAPTER 1

Chemical Nature, Psychology, and Physiology of Taste

Two of the five fundamental human senses, recognized since antiquity, are chemical senses and three are physical senses. Taste, smell, and the common chemical sense are caused by chemical substances. Sight, sound, and touch are caused by physical forces (light waves, pressure waves, and physical pressure, respectively).

Taste is elicited by water soluble molecules that interact with receptors on the tongue and in the oral cavity. The compounds have been variously described as sapid, saporific, or saporous substances meaning, simply, that they elicit taste. Presumably, the compounds possess saporous groups responsible for their taste. Smell is elicited by volatile chemical substances that reach the olfactory mucous membrane receptors in gaseous form. They have been described as odiferous substances, or simply, *odorants*. By analogy, saporous substances are *tastants*. The common chemical sense is considered by many to be a component of taste. Taste is caused by chemicals, but knowledge of the chemical nature of taste requires multidisciplinary treatment. This treatise begins with an introduction to the general chemistry of taste, along with knowledge derived from psychology and physiology.

1.1 GENERAL NATURE OF TASTE

1.1.1 The Basic Tastes

The number of recognized primary tastes has varied over the years. Applying the 'Doctrine of Opposites,' Aristotle considered sweet and bitter taste to be examples of the doctrine, and believed that all other tastes lay between the two extremes. Linnaeus expanded the number to include sweet, sour, sharp, salty, bitter, fatty, insipid, astringent, viscous, aqueous, and nauseous. Wund, the founder of experimental psychology, first reduced the number to six (sweet, salt, bitter, sour, metallic, and alkaline), and then to only sweet, salt, sour, and bitter. The notion of only four basic tastes received support from a number of 19th century investigators (cf. Pfaffmann *et al.*, 1971), and McBurney (1974) lists eleven reasons for them.

It has been proposed by Henning (1915), somewhat along the line of thought espoused by the ancient Greeks, that the four tastes constitute a psychological continuum, and mixed tastes can be placed upon the surface of a psychological continuum in which sweet, salt, bitter, and sour are the apexes of a regular tetrahedron (Fig. 1.1). More recently, a spherical modification of the tetrahedron was proposed by Haefeli & Glaser (1985).

In contrast, Frings (1954) believed that all tastes form a psychological continuum that has points of familiarity in a continuous taste spectrum. The psychological scale of taste qualities of Schiffman & Erickson (1971), and Erickson & Schiffman (1975), uses a multidimensional space rating technique employing two types of psychological data, judgment of similarity and semantic differential rating. When these parameters are plotted in three dimensions, compounds that elicit sweet, salt, sour, and bitter tastes group together on the surface of a tetrahedral structure similar to that of Henning's. Compounds that elicit an alkaline taste (e.g., NaOH) fall outside the tetrahedral structure, i.e. 'Off the sweet, salty, sour, bitter surface.' From a more pragmatic viewpoint, Erickson (1985) pointed out that in some cases, rather than being inherent in the (chemical sensory) data, such classifications simply reflect a human need to organize and understand sensory phenomena.

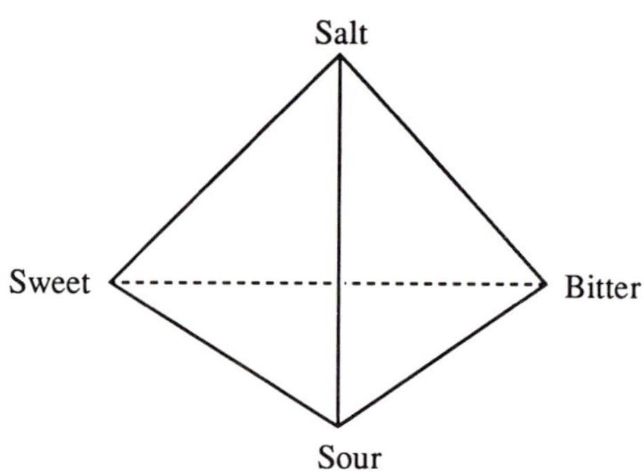

FIG. 1.1. Henning's (1915) tetrahedral taste continuum.

Evidence to indicate that the notion of four basic tastes is an over-simplification can be found in several psychological studies. Assignment of names to taste categories using the traditional four basic taste systems may underestimate the number of taste categories that exist (Ishii & O'Mahony, 1987). This seemed to be particularly true when a taste panel was asked to assign a number of different substances to the four tastes. Among them were sucrose and aspartame, strong and weak sodium chloride, lithium chloride, citric acid, caffeine and several 'umami' substances. To characterize the taste of magnesium sulfate, a taste panel may require six or more descriptive terms (Kelling & Halpern, 1988).

Part of the problem of assigning tastes to four modalities is that many compounds have mixed tastes. In addition, from the psychological view, taste has both an analytic and a synthetic component. When compounds that have different tastes are mixed, the resulting tastes are always of the same quality as the components, and from an analytic point of view, no new tastes are generated (McBurney, 1974). Erickson (1985) believes that taste also has a synthetic component, and certain mixtures of substances (e.g., glucose and saccharine) create a new taste in the same sense that mixing primary colors leads to a new color. According to Schifferstein & Frijters (1990), whether taste is analytic or synthetic is not necessarily a perceptual problem, but is due to the task and the instructions given to taste panel subjects.

If taste is a chemical sense that is caused by chemical reactions, the problem of the number of basic tastes, and whether taste is psychologically analytic or synthetic, or some combination of both, may be simplified. This would be particularly true if different initial reaction mechanisms are partly responsible for the different tastes. If the reaction time to sodium chloride saltiness, for example, is less than the reaction time to the bitterness of caffeine, their single tastes should readily be distinguishable in a mixture. This seems to be the case (Kuzniciki & Turner, 1988), and the taste of NaCl and caffeine in a mixture can readily be detected, i.e. the perception of each is analytic. If the reaction time and intensity to one compound is about the same as that for a second, then a new taste may have indeed been psychologically synthesized. An example may be the multiple tastes of monosodium glutamate (Matsuo & Yamamoto, 1989), and other compounds with the same property (5′-ribonucleotides). The taste of these substances is known as 'umami,' or, 'tasty substance,' and believed by some to be a taste primary.

1.1.2 Definition of Sweetness, Saltiness, Sourness, and Bitterness

The four primary tastes have established chemical reference substances that serve to demonstrate them. Thus, different tastes can be defined through reference to these substances. By definition, then sweetness, saltiness, sourness, and bitterness are demonstrable *fundamental propositions*. As no two chemical substances have the same chemical properties, the definitions of sweet, salt, sour, and bitter will vary with the chemical identity of the reference substance. The definition also considers the possibility that different tastes may be due to different types of chemical reactions. If so, no two compounds can possibly have identical taste recognition chemistry attributes.

1.1.3 Taste Profiles and Taste Spectra

A corollary to the aforementioned taste definition is that the taste of a substance depends on its chemical activity and the nature of the chemical activity. If a chemical has potential for more than one type of chemical activity, it will then have potential for mixed taste attributes. Taken together, the taste profile and taste spectrum then need to be considered in order to characterize the taste of substances.

1.1.3.1 *Taste Profiles*

When the taste of substances is described, the reaction time, the taste intensity, and the duration of the taste yield a time-intensity *profile* (Lee & Pangborn, 1986; Janusz *et al.*, 1991). An idealized plot of a taste profile is described in Fig. 1.2 for equal amounts of the crystalline forms of two carbohydrate sweeteners, sucrose and fructose.

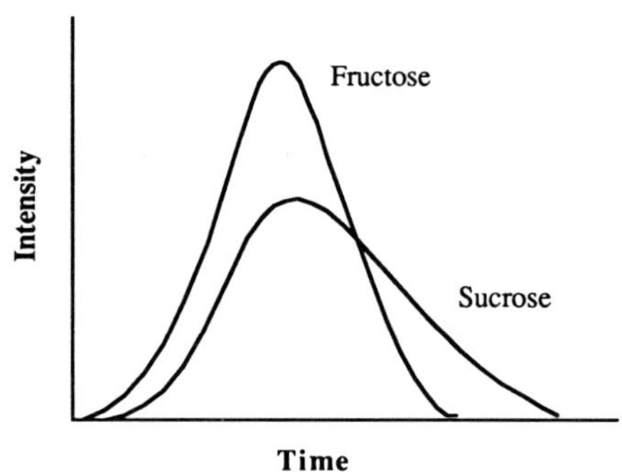

FIG. 1.2. Diagrammatic plot of the sweetness intensity of crystalline fructose and sucrose versus the duration of the sensation.

Glucose, glycine, chloroform, cyclamate, aspartame, and dihydrochalcones taste sweet, but these substances have different sweetness profiles, particularly when concentration effects are included. The sweetness of chloroform is perceived instantly, but is fleeting. Aspartame yields a distinct sweetness time-intensity profile at a concentration far less than that for the sugars. It takes a little while for the sensation of sweetness to develop to full intensity when solutions of glucose are placed on the tongue, and the taste persists for only a short while. On the other hand, the sweetness of dihydrochalcones is persistent, to a fault. Most high-potency sweet substances have pronounced taste attributes in addition to sweetness. Clearly, different substances with the same overall taste quality may not only have different taste profiles, they may also have different taste quality spectra.

1.1.3.2 *Taste Spectra*

A taste spectrum includes the presence or absence of all primary tastes, secondary (alkaline, metallic) tastes, and elements of the common chemical sense (coolness, hotness, astringency). The problem of defining a specific taste is analogous to attempts to define a specific color, such as red. The basic color primaries are red, blue, and yellow, and they can be described by a particular wavelength of the electromagnetic spectrum, or by reference to x, y, z coordinate numbers. Tastes probably could also be described through reference to the fundamental nature of the chemical reactions that elicit them, if these were known, or do indeed exist.

With respect to sensitivity to the primary tastes, it has been stated that 'people live in entirely different worlds' (Blakeslee & Fox, 1932). Similarly, each person has a different view of the nature of the color at a specific absorption wavelength. Consequently, a number of mathematical approaches has been devised to specify colors, but not with complete success. The difficulty of defining taste, even within a given primary, has its color counterpart in the plethora of subjective measurement techniques that have been developed. Not only do mathematical approaches to color specification overlap, but so do subjective methods for taste specification (Section 1.2). Finally, and in contrast to taste, color perception is purely synthetic. The mixing of two color primaries yields a different color. While there also seems to be a synthetic component to taste, which varies with the concentration of each taste substance, taste is largely analytic, particularly when two different tastes are present in a mixture.

The distinction and importance of the notion of taste profiles and spectra is well-known in the sugar industry. Crystalline sucrose may have a purity greater than 99·93%, and the composition of the remaining 0·07% is glucose and fructose (0·02%), moisture (0·03%), and ash (0·02%). The concentration of impurity is not particularly significant from the compositional viewpoint (99·93% purity is quite high), but it may affect the sugar's taste spectrum. At one time, the relative sweetness scores assigned to sucrose obtained from the sugar beet and sugar cane were 97·9 and 100, respectively. Upon examining the composition of sugar ash, Iida (1955) found that beet sugar was lower in calcium, sulfur, and phosphorous, but contained considerably more potassium.

When calcium sulfate and calcium hydrogen phosphate were added in the proper quantity to crystals of beet sugar, the cane and beet sugars were then judged to be equally sweet. Thus, with respect to the fundamental proposition that 'sugar (sucrose) tastes sweet,' it is clear that the perceived taste is not due to sucrose alone. It is also due to the nature and concentration of (cocrystallized) salts. Beet sugar solutions may also have an earthy odor due to the presence of geosmin (Monte & Maga, 1981/82). The relative sweetness score (97·9) assigned to beet sugar was not a 'relative' sweetness score *per se*, but a taste (or flavor) spectrum specification, instead. With respect to the taste spectrum of sucrose, Richter & Campbell (1939–1940) and Neuman (1944) found sucrose solutions, at or near the recognition threshold for sweetness taste bitter and/or sour. Whether the assessment is a record of the taste of the salts is not clear. What is clear is that the taste attributes of any substance can be recorded as a sweet, salt, sour, and bitter spectrum. The significance of taste spectra specifications is most apparent in the psychological assessment of compounds traditionally used to demonstrate the sweet, salt, sour, and bitter tastes.

1.1.3.3 *Taste Spectra Specification for Basic Tastes*

In an evaluation of substances used as standards for taste assessment,

TABLE 1.1
Percent of taste panel personnel assigning taste attributes to various standard tastants. After Meiselman & Dzendolet (1967)

Substance	*Sour*	*Salt*	*Bitter*	*Sweet*	*None*
0·002 M HCl	63·4	1·7	32·5	2·5	0·0
0·04 M NaCl	5·9	80·0	5·9	4·2	4·2
$8{\cdot}0 \times 10^{-6}$ M Quinine	20·8	0·0	45·0	9·2	25·0
0·025 M Sucrose	2·6	0·9	5·0	90·8	0·9

Meiselman & Dzendolet (1967) found that 90·8% of the subjects tested thought that 0·025 M sucrose tasted sweet, while only 80% thought that 0·04 M NaCl was truly salty. In addition, only 63% felt that 0·002 M HCl was truly sour, but on the other hand, 25% thought that quinine was tasteless, 21% thought that it was sour, and 45% thought it was bitter. All concentrations used were above the recognition thresholds (Table 1.2). For subsequent comparisons, these data are shown in Table 1.1.

Using persons who could reliably describe the taste of substances with respect to the four primary tastes, McBurney & Schick (1971) found that all standard taste substances are usually capable of eliciting the four primary tastes. Their results are shown in Fig. 1.3. The taste spectrum

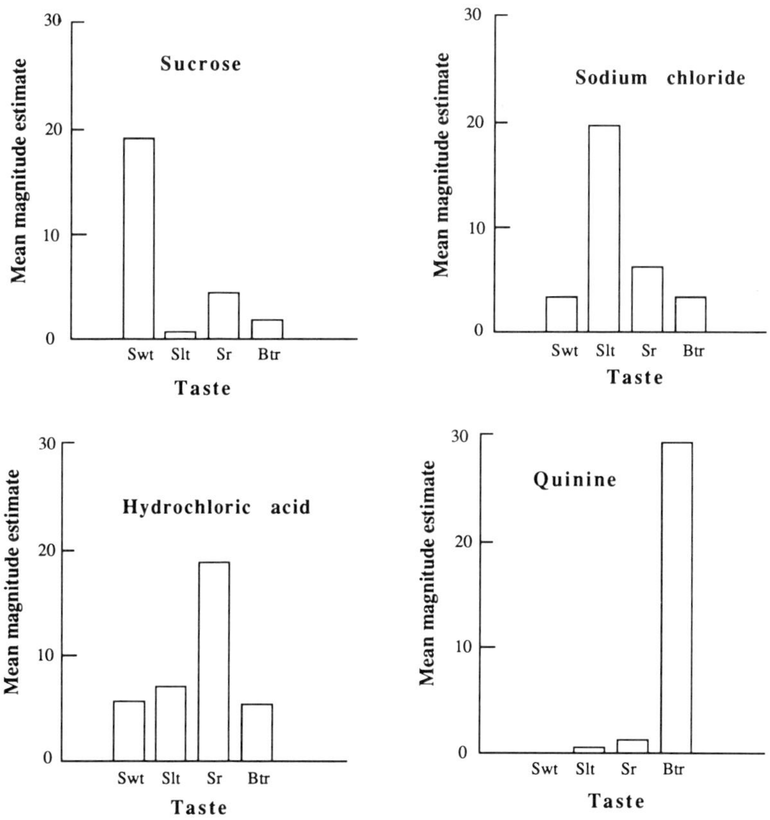

FIG. 1.3. The taste spectrum for various primary taste substances (McBurney & Schick, 1971).

shown for sucrose is in accord with the near-threshold sour/bitter evaluation by Neuman (1944), and in agreement with the taste distribution data of Table 1.1.

No other compound seems capable of duplicating the saltiness character of sodium chloride, but it also has a mixed taste spectrum (Fig. 1.3). It also has a percentage distribution of 'tasters' (Table 1.1) that agree with the taste spectrum. Hydrochloric acid, another very pure chemical substance, seems to have a disproportionately large share of its taste spectrum assigned to the three other tastes, and the spectrum again resembles the distribution of other tastes among tasters. In contrast to the taste distribution data given in Table 1.1, the taste spectrum of quinine is predominantly bitter. The extent to which the taste spectrum or distribution of tastes for a given chemical is due to impurities, or due to ability of the compounds to elicit other tastes, is not easily determined.

1.1.4 Morphological Receptors for the Basic Tastes

1.1.4.1 *Distribution of Tastes*

The mid-dorsal (center) surface of the tongue is insensitive to all taste qualities, and different areas of the tongue are not uniformly receptive to the different tastes (Pfaffmann *et al.*, 1971). Hänig (1901) stated that sensitivity to sweetness is greatest at the tip of the tongue, and while salt sensitivity is more general, it too is greatest at the tip. Hänig (1901) also stated that the tongue is more sensitive for bitter at the back, but McBurney (1974) reported that the tip is sensitive for bitter also. Henkin & Christiansen (1967) report that the soft palate is even more sensitive to bitter-tasting compounds than any tongue area.

1.1.4.2 *Distribution of Receptors*

There are several types of taste papillae (protuberances) on the tongue. Fungiform papillae are the smallest and are located on the front two-thirds of the tongue (Fig. 1.4A). Anatomically, they resemble a button mushroom, and each contains one to eight onion-like (taste bud) units, containing taste cells, on the top of the papillae (Figs 1.4C, D). The papillae are also connected with nerves devoid of a myelin (lipoprotein) sheath, and do not terminate on specialized receptors, but terminate in the papilla's epithelium (upper surface) itself (Henkin *et al.*, 1969). Circumvallate papillae (Fig. 1.4B) are located at the rear (posterior) one-third of the tongue. They are fewer in number, but larger, and contain more taste cell units. They are also surrounded by a moat. The number of

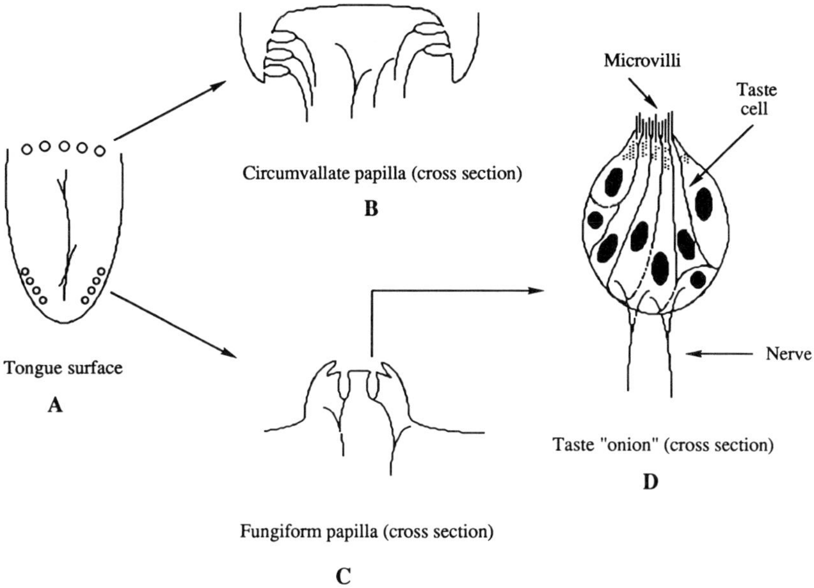

FIG. 1.4. Location of circumvallate papillae (B), fungiform papillae (C), on the surface of the tongue (A), and a diagram (D) of a single taste bud (onion).

unmyelinated nerve fibers is also greater than for the fungiform papillae. Foliate papillae are found on each side of the back of the tongue, and filiform papillae are located on the dorsal surface. The latter do not contain taste buds. Taste buds are also found on the palate, the pharynx, and the larynx (Pfaffmann, 1951).

The taste buds, or onions, are massively penetrated by unmyelinated nerves that reach the taste cells, and continue upward to reach a level only a few microns distant from the taste pit (the opening to the onion). The junction complexes formed by the taste cells provide such a tight seal just below the pit as to probably make it difficult for taste substances to contact directly the nerve endings (Murray, 1971). The microvillae themselves are believed to be extensions of the taste cells.

1.1.4.3 *Response of Taste Buds to Different Stimuli*

With taste made up of four primaries, presumably with different chemical mechanisms to elicit each taste, it would seem that different taste buds might respond preferentially to compounds that elicit specific tastes. This is not always the case. Kiesow (1896), in confirmation of an earlier study by Öhrwall (1891), reported that some individual taste papillae may react exclusively to salt, sweet, and sour. Others react to combinations of two, three, or all four of the basic taste stimuli. Later, Pfaffmann (1941, 1955), Kimura & Beidler (1961), Beidler (1966), Harper *et al.* (1966), and Bealer & Smith (1975) reported that only small numbers of papillae have qualitative specificity when stimulated by chemicals. This indicates that single taste papillae, or single receptor cells, react to all chemical stimuli. More recently, Plattig & Innitzer (1976), and Arvidson & Friberg (1980) again reported multiple sensitivity in single fungiform papillae and in single taste buds.

Data obtained by Pfaffmann (1955), for the neurological response of single rat taste fibers to solutions of sucrose, NaCl, HCl, and quinine, are shown in Fig. 1.5. When compared with the data obtained for humans in Table 1.1, and with the four-taste spectrum for the same substances in Fig. 1.3, it is found that the data for sweet, salt, and sour tastes are again strikingly similar. The data for the neurological response to quinine corresponds even better with human taste variability to quinine than to the assessment of the taste spectrum. However, the latter data were obtained by excluding persons insensitive to the bitter taste of phenylthiocarbamide (PTC).

The neurological responses to all four tastes in Fig. 1.5 are believed to be 'across-nerve' phenomena. It is also possible that the total response, for both the rat and the human, is due to single nerve response. This would be particularly true if given taste substances are capable of eliciting more than one taste chemical reaction (Section 1.1.3.2). On the other hand, von Békésy (1964, 1966) found that electrical stimulation of a papilla elicited a sensation corresponding to one and only one of the primary tastes. Dzendolet & Murphy (1974) confirmed von Békésy's findings. Still, a number of other experiments show only a small portion of papillae exhibit qualitative specificity (Price & DeSimone, 1977). In view of the mixed sensations that can be obtained with electrical stimulation (Ajdukoviċ, 1990), the conclusions of von Békésy and Dzendolet & Murphy seem questionable.

It is not uncommon to find some taste fibers that respond more to one taste than to another. Although there is evidence for sweet, acid, salt, and bitter sensitive taste fibers (Frank, 1985), multiple response seems to be the rule. Possibly, there is a base line reaction for a given taste in selected fibers, and the noise observed is due to partial response to other tastes. To reconcile these observations, various theories of neural coding have been proposed, particularly the *labeled line* system of coding. The notion of a

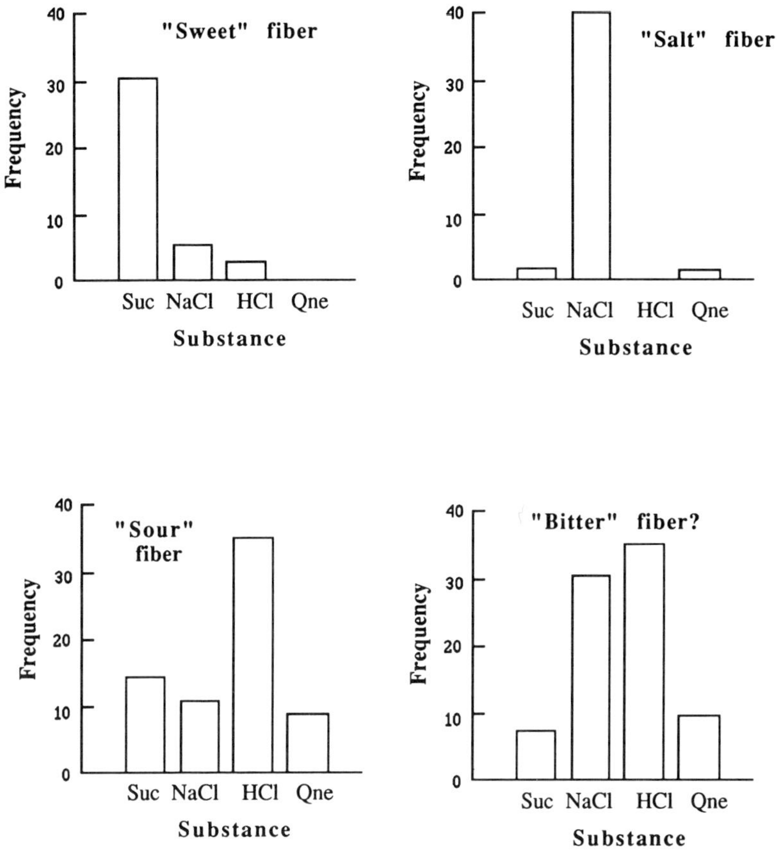

FIG. 1.5. Neurological response from single taste fibers (rat) to solutions of sucrose, NaCl, HCl, and quinine (Pfaffmann, 1955).

labeled line to the central nervous system implies activation of a modality-specific cell. According to Bartoshuk (1987), if there were only one type of receptor for sweetness, sweeteners would yet stimulate other receptor mechanisms as well. Those other receptor mechanisms then cause nonsweet sensations. Hellekant & Ninomiya (1991) have provided strong electrophysiological evidence for the labeled line notion in chimpanzees. Smith (1985) concluded that the distinction between labeled line and other notions of neural coding (e.g., across-fiber coding) is a matter of belief in whether activity in a single neuron can, by itself, represent a single taste quality. Admittedly, neural coding mechanisms proposed to account for the distinctions between the basic tastes are rather complex. The picture is even more complicated by the possibility (Section 1.1.3) that the total taste spectrum of a given chemical may be due to an ability of a given substance to elicit one or more of the primary tastes. The only safe conclusion seems to be that there are both single and multiple response fibers.

In addition to the general psychological and neurophysiological features of taste, there are several rather intriguing chemical features. They include taste blindness, blocking of specific tastes, and the modification of taste by certain substances.

1.1.5 Taste 'Blindness'

There is wide variation in recognition thresholds for sucrose sweetness (3-fold), sodium chloride saltiness (80-fold), hydrochloric acid sourness (200-fold), and quinine bitterness (200-fold). The range in thresholds for the bitter taste of phenylthiourea (phenyl thiocarbamide, PTC) is about 9000-fold because there are two basic types of response to the bitterness of the compound. Some people can taste the bitterness of the substance and others cannot. The inability to taste PTC is called taste 'blindness,' and has received much attention (Amerine *et al.*, 1965). The phenomenon is a simple Mendelian recessive trait that occurs in about one out of four people. There is also wide racial variation, and varied response for given individuals at different times. The differences in individuals may not carry over to the bitter taste of all other bitter compounds, but Hall *et al.* (1975) report that PTC nontasters assign lower bitterness scores to caffeine. Mela (1989) found that the bitterness of quinine and urea is also lower for nontasters.

A second example of taste 'blindness' is the inability of persons with congenital idiopathic hypoparathyroidism to recognize sweetness, although they give a normal response for salty, sour, and bitter substances (Henkin & Shallenberger, 1970). These persons do respond to the subsidiary tastes of compounds. They consider sucrose and fructose to taste sour. Glucose, galactose, and cyclamate are said to taste bitter. The general loss of taste, with increased sensitivity to sourness and bitterness, is associated with a variety of ailments and treatment with drugs.

1.1.6 Taste Blocking and Modification

General anesthetics such as pantocain and novocain suppress all tastes (Skramlik, 1963). Examination of over twenty anesthetics revealed that their action, as might be expected, varies with structure. As indicated by elevation of the threshold for quinine hydrochloride, sodium chloride, tartaric acid and sucrose, the general order of effectiveness on the four tastes was found to be bitter > salt > sour > sweet. When a high-potency sweetening agent was included in the series, the order became bitter > sweet > salt > sour. The most effective substance was pantocain. Even 10 min after application to the tongue, crystalline quinine possessed only textural attributes.

1.1.6.1 *Taste Blocking*

Three triterpene saponin glycosides are known that suppress or 'block' sweetness. They are the gymnemic acids, ziziphens, and holodulcin (Kolodny & Kennedy, 1988). The gymnemic acids are found in the leaves of *Gymnema sylvestre*, a plant native to India, and known since antiquity for ability to block the sweet taste of sugar. When the tongue is treated with a solution of the gymnemic acids, crystals of sucrose have only a gritty textural attribute. To the present writer, the solution of gymnemic acids itself was perceived as being overwhelmingly sweet, so that the subsequent taste of the sucrose crystal was only gritty. Dilute solutions of gymnemic acids are reported to suppress only sweetness, and to have little effect on sourness, saltiness, and bitterness, whereas higher concentrations suppress all tastes (Kurihara, 1971).

The gymnemic acids (Fig. 1.6) are a glucuronic acid glycoside of a hexahydroxy triterpene, with the aglycone hydroxyl groups esterified by various combinations of at least five acids (e.g., formic, acetic, and propionic acid). The compounds have varying sweetness-suppressant activity, but leaf extracts, and a compound designated 'A1,' at a concentration of 5×10^{-3} M, can block the sweetness of either crystalline sucrose, or a 1 M (34%) solution of sucrose. Compound A1 also suppresses the sweetness of cyclamate, D-amino acids, lead acetate, and salts of beryllium, but not chloroform (Kurihara, 1969). Parenthetically, chloroform also has a sweet odor (Henkin & Shallenberger, 1971), which may lead to sweet taste. Because gymnemic acids preferentially suppress sweetness, the mode of action would seem to be competitive in binding with a receptor specific for sweetness. A second possibility is that it destroys the conditions necessary for the recognition chemistry unique to sweetness (Section 10.8). The other two triterpene saponin glycosides may act in a similar fashion, but the time-course of inhibition and recovery differ (Kolodny & Kennedy, 1988).

The compounds cyclohexyl acetic acid, indole acetic acid, α-naphthol sulfate, and methyl 4,6 dichloro-4,6-dideoxy-α-D-galactopyranoside also

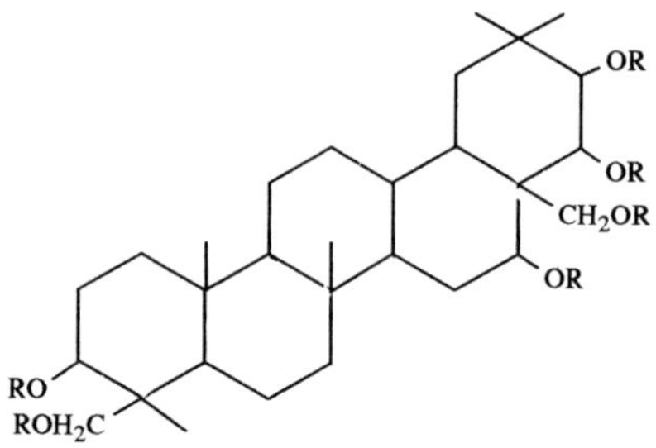

Gymnemic acid. R = various acids + glucuronic acid.

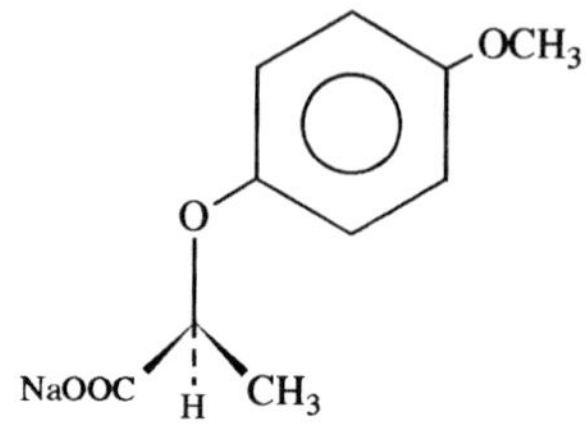

Na(+) 2-(4-Methoxyphenoxy)phenoxypropionate (Lactisol)

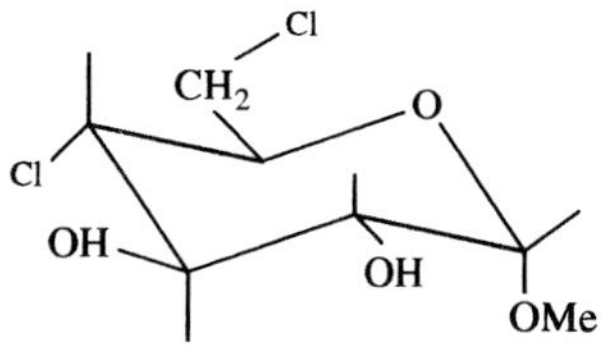

Methyl-4,6-dideoxy-4,6-dichloro-α-D-galactoside

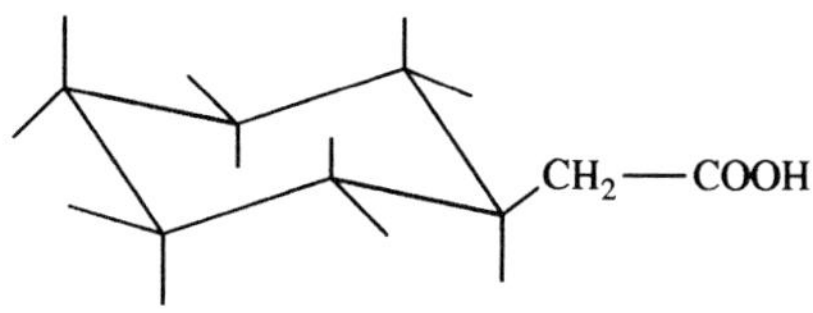

Cyclohexyl acetic acid

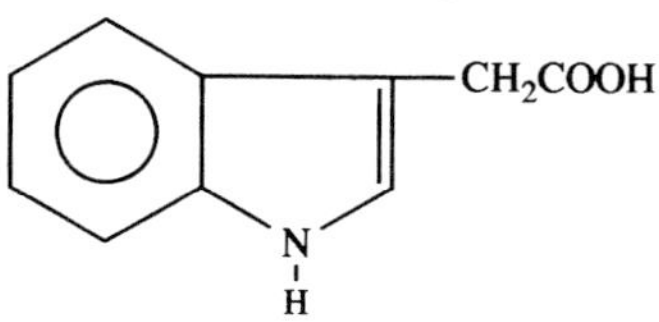

Indoleacetic acid

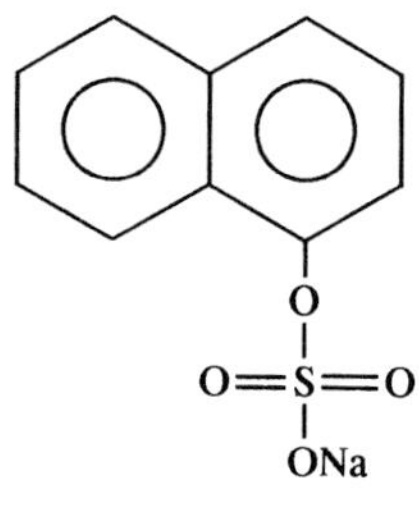

α-Naphthyl sulfate

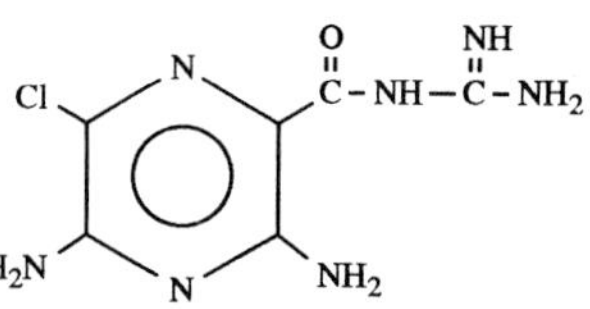

1-(3,5-diamino-6-chloropyrazinoyl) guanidine (Amiloride)

FIG. 1.6. Structure of sweetness suppressants.

inhibit sweetness (Lindley, 1991). Their structures are shown in Fig. 1.6. The latter compound needs to be employed at a concentration of about 1% (0·05 M) to inhibit the sweetness of sucrose and other sweeteners (Schiffman *et al.*, 1987). As it is also intrinsically bitter, it is not quite clear to what extent sweetness is suppressed, or merely masked.

Of particular importance among compounds known to suppress

sweetness is Na (+)-2-(methylphenoxy)phenoxypropioniate (lactisol, PMP). This substance shows action when accompanied by a sweetener, i.e. the tongue does not have to be pre-treated with the inhibitor (Lindley, 1991). In addition, PMP and other phenoxyalkanoic acid inhibitors are pure chemical substances, suitable for use in kinetic studies. While the mode of action of other sweetness inhibitors is uncertain, that of PMP appears to be purely chemical as only the (+)-enantiomer is effective. Michaelas-Menten plots suggest that the inhibition is competitive (see, however, Section 10.8). Increasing the lipophilicity of the phenyl ring increases sweetness-inhibition potential. PMP alone tastes slightly bitter and metallic, but it blocks the sweetness of 10% sucrose at a concentration of only 0·005%. The effect is reversible and first-order with respect to inhibitor concentration. It too does not block the salt, sour, and bitter tastes.

The substance known as amiloride (Fig. 1.6) is reported to block the saltiness of sodium and lithium chlorides (Schiffman *et al.*, 1983), but has no effect on the response to potassium and ammonium chlorides (Hellekant *et al.*, 1988). In addition, 0·5 mM amiloride was reported to block 44–73% of the sweetness of sugars and high-potency sweeteners.

1.1.6.2 *Taste Modification*

Miraculin is a base-reacting glycoprotein with a molecular weight of about 44 000, and is obtained from the berries of the West African plant *Synespalum dulcifum*. It has the unique ability (Kurihara, 1971; Inglett, 1971) to convert the taste of acids from sour to sweet. Treatment of the tongue with 7×10^{-7} M concentrations of protein can render 0·02 M citric acid as sweet as 0·4 M (14%) sucrose, and the effect is persistent. Treatment of the tongue with $2{\cdot}0 \times 10^{-8}$ M concentrations of protein causes acids to taste sweet for up to 20 min, and the action of $2{\cdot}3 \times 10^{-6}$ M solutions last for up to 3 h.

Still other substances are purported to enhance other tastes, but they also seem to have unique intrinsic tastes. They include monosodium glutamate (MSG), and two 5′-ribonucleotides, disodium 5′-inosate (IMP) and disodium 5′-guanylate (GMP). Monosodium glutamate is used at concentrations of 0·2–0·5%, in the presence of NaCl to enhance food flavors. At concentrations of 0·05–1% it tastes slightly sweet and salty, much on the order of NaCl (Section 3.4.1). At higher concentrations the taste is thought to be unique, and is described as ‘umami’ or ‘tasty substance’ in the Japanese language. The umami taste may be a ‘meaty’ taste. This recognition threshold seems to be at

$8{\cdot}5 \times 10^{-4}$ M (1·4%). The unique taste is found only in the L-form. For reviews of its purported taste enhancing properties (see Amerine *et al.*, 1965; Birch, 1986).

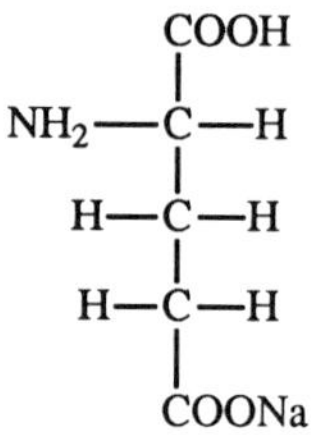

Monosodium glutamate (MSG)

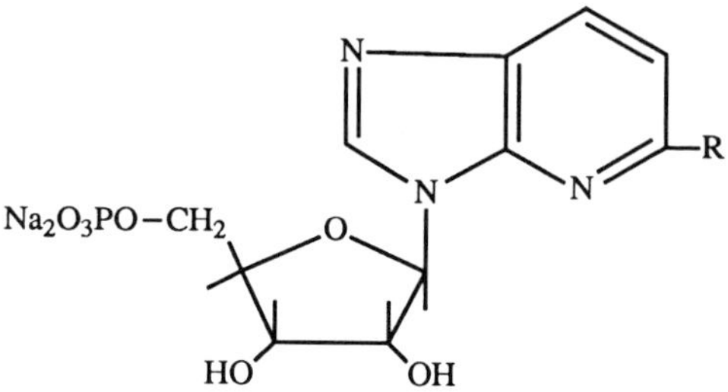

Disodium inosine-5'-monophosphate (R = H)
Disodium guanosine-5'monophosphate (R = NH_2)

Other substances purported to have the umami tastes are disodium 5′-inosate (IMP) and disodium 5′ guanylate (GMP). When a dilute solution (0·02%) of MSG is tasted, its salty, unusual taste is experienced. Then when an 0·02% solution of a nucleotide is subsequently tasted, the same salty and unusual taste is experienced, but the intensity is increased several times. The same result is experienced beginning with the nucleotide solution, and then the MSG solution. The test serves as an example of taste synergism, and it has been proposed that umami is a basic taste (Yamaguchi, 1987).

1.2 TASTE MEASUREMENT

A number of methods are available to measure the intensity of tastes. They include determination of the threshold concentration of a substance, paired, and multiple comparison techniques, and sensory magnitude scaling techniques. All methods require application of statistical procedures to establish the significance of the results. The taste of substances recorded in Beilstein or the Merck index is often the opinion of only one or two tasters. Those assessments rely on the same principle that is applied to a strip of litmus paper. If the litmus turns red to a test solution, the pH is less than 7 and the solution is acid. If the litmus turns blue, the pH is greater than 7 and the solution is basic. The 'dye' for sweet and other tastes may turn blue at a pH of 8 for some persons, at pH 6 for others, and at pH 5 for still others.

1.2.1 Threshold Determinations

One of the oldest techniques for demonstrating differences in taste potency among substances is determination of the *limen*, or threshold concentration, of a substance. This is defined as the minimum concentration of a substance for which a particular taste can be recognized by a significant number of tasters, and is called the *recognition threshold*. Recognition thresholds for compounds may be determined several different ways. A method of 'limits' and of 'constant stimuli' (Pfaffmann *et al.*, 1971), and a method of successive approximations or comparisons are most favored (Amerine *et al.*, 1965). Marin *et al.* (1991) evaluated three methods for determining recognition thresholds; those calculated from interval estimates, a two-phase linear regression model, and a new logistic regression model where the threshold is defined as the point of maximum initial curvature of a typical sigmoidal dose-response curve.

Care must be exercised in the application of threshold data. The values cannot be used for determining relative taste intensity when the relative taste intensity and taste spectrum of a substance change with concentration, which is sometimes the case. Wide variation in results is found with different procedures, and there is an element of uncertainty whether the threshold value obtained belongs to the taste attribute in question. Nevertheless, recognition thresholds have important application as indexes to the magnitude of a compound's chemical activity toward the taste receptor.

Recognition thresholds for substances usually associated with eliciting the sweet, salt, sour and bitter tastes are given in Table 1.2. A number of other tables are also available (e.g., Amerine *et al.*, 1965). The recognition threshold value for sucrose sweetness, 0·017 M, shown in Table 1.2, corresponds with the value (0·011 M) used extensively by Wieser *et al.* (1977). The recognition threshold for acid taste is relatively constant at 0·001–0·003 N for all acids (Pfaffmann, 1959).

Activity constants for taste are in agreement with the contention that recognition thresholds are an index of chemical activity toward the receptor site. Shamil *et al.* (1988*b*) found that activity constants (K_m), determined from maximum sweetness-intensity and time to maximum-intensity ratios, correlate well with thresholds. The observation affords an overview of chemical principle application to threshold data. For

TABLE 1.2
Median molar recognition thresholds for taste (Pfaffmann *et al.*, 1971)

Taste	*Substance*	*Molecular weight*	*Median*	*Range*
Sweet	Sucrose	342·2	0·017	0·012–0·037
	Saccharine (Na)	241·1	$2{\cdot}3 \times 10^{-5}$	$2{\cdot}0–4{\cdot}0 \times 10^{-5}$
	Beryllium chloride	80·0	0·000 3	
	Sodium chloride	58·5	0·020	0·001–0·10
Salt	Sodium chloride	58·5	0·030	0·003–0·085
	Potassium chloride	74·6	0·017	0·001–0·070
Sour	Hydrochloric acid	36·5	0·000 9	5×10^{-5}–0·01
	Acetic acid	60·1	0·001 8	0·000 1–0·005 8
Bitter	Quinine sulfate	746·9	$8{\cdot}0 \times 10^{-6}$	$4 \times 10^{-6}–1{\cdot}1 \times 10^{-5}$
	Quinine HCl	360·88	$3{\cdot}0 \times 10^{-5}$	$2 \times 10^{-6}–4 \times 10^{-4}$
	Caffeine	194·1	$7{\cdot}0 \times 10^{-4}$	$3 \times 10^{-4}–1 \times 10^{-3}$
	Phenyl thiourea[a]	152·21	$2{\cdot}0 \times 10^{-5}$	
			0·008	$2 \times 10^{-7} > 0{\cdot}017$
	Urea	60·1	0·120	0·116–0·130
	$MgSO_4 \cdot 7H_2O$[b]	246·49	0·004 6	0·004 2–0·005

[a]Low value applies to 'tasters' and high value applies to 'nontasters.'
[b]Epsom salt.

example, if the activity of a given molecule toward the receptor is 10 times that of another, then the recognition thresholds, as a mass action phenomenon, should be 1 and 10, respectively. By selecting examples from Table 1.2 (median threshold molar concentrations), relative activity for sweet, salt, sour, and bitter can be listed as follows:

Sweet (0·017 M sucrose), relative activity = 1 × (unity)

Salt (0·03 M NaCl), relative activity = 0·57×

Sour (0.0009 M HCl), relative activity = 18·8×

Bitter (0·0007 M caffeine), relative activity = 24·3×

The spread between the relative strength of the purported interactions for sour and bitter, as compared with salt and sweet, are too large to reflect simple differences among weak chemical bonding phenomena. They are greater yet when other bitter or sour tasting substances are included in the list, but decrease for sweetness when the intensely sweet substances are included. This is one of the reasons why, in Section 1.1, it seems that the degree of overlap between the chemical sense of taste, and the common chemical sense, is uncertain. It is quite possible that receptors for the common chemical sense are not taste receptors at all, but a general array of free nerve endings. These then respond to irritation, or thc type of sensation (pain, hotness) that might be expected from the action of a strong acid or strong base.

'Detection' thresholds are lower than recognition thresholds, and are seldom caused by the sensory quality in question, but by another taste modality or sensory attribute (tactile, common chemical, odor, etc.) instead. Although the relative sweetness of glucose is only about two-thirds that of sucrose, its detection threshold is much lower, probably because of residual hydrochloric acid used in preparing it from starch. The detection threshold of sodium chloride is about 0·01 M, but it tastes sweet to most people at that concentration. The recognition threshold for the saltiness note of sodium chloride is 0·03 M. The detection and recognition thresholds for some compounds are open to question on the basis that they have potential to elicit more than one taste, e.g. quinine hydrochloride and the acid salts of crystalline amino acids.

1.2.2 Comparison Tests

Comparison procedures have value for demonstrating differences in taste intensity among samples. They are also used to determine thresholds for substances, either as a pure compound in solution, or in mixtures. Among them are paired-comparison techniques, duo-trio, and triangle-intensity tests. A review of the attributes of each procedure has been compiled by Amerine *et al.* (1965).

1.2.3 Scaling Procedures

Magnitude estimation procedures are generally accepted as the favored method for demonstrating the effect of concentration on the intensity of a sensory attribute (Pfaffman *et al.*, 1971). These procedures provide meaningful ratio scales of sensory magnitude (Moskowitz & Sidel, 1971), and more recently have been used to judge intensities of sensation of two or more modalities on the same scale (Marks *et al.*, 1988).

1.3 PSYCHOPHYSICAL PRINCIPLES

When a cup of coffee containing one teaspoon of sugar is not sweet enough, and the person drinking it wishes to make it about twice as sweet, does that person add just one more teaspoon of sugar? The procedure employed to answer the question requires application of the psychophysical laws. These 'laws' relate and quantify the magnitude of a perceived sensation to the magnitude of the stimulus.

1.3.1 Psychophysical Laws

According to Weber (cf. Amerine *et al.*, 1965), the increase in the stimulus intensity (S) required to produce a *just noticeable* sensory *difference* (j.n.d.) is a constant (k) where $\Delta S/S = k$ (the Weber fraction). Holway & Hurvich (1937) found that this fraction decreased with increasing salt concentration, while Schutz & Pilgrim (1957*b*) found that it first decreased, and then increased for both saltiness and sweetness. The Weber fractions over several concentrations were, on the average: salty (NaCl) 0·153; sweet (sucrose) 0·172; sour (citric acid) 0·224; bitter (caffeine) 0·302. This suggests that lower concentrations of bitter and sour substances are needed to elicit a j.n.d. than for salt and sweet. Weber fractions seem to be too variable to be meaningful when derived from relative sensory values. Although they seem to represent the effect of some type of psychological increment, it is not linear, as the equation seems to imply (Amerine *et al.*, 1965).

Fechner (cf. Trout & Sharp, 1937) added the response parameter (R) to the Weber ratio and derived the relation to be $\Delta S/S = k\,\Delta R$. In logarithmic form, the expression is $R = \log S$. Unlike the Weber ratio, the Fechner expression is not restricted in application to j.n.d.'s. Thus, increases in the magnitude of the sensory response (R) are proportionate to the logarithm of the stimulus intensity (S). In order for a sensory response to proceed in an arithmetic manner with increasing S, S needs to increase in geometric fashion (Fig. 1.7). The Fechner law was applied by Shallenberger & Acree (1971) to establish the relative sweetness of

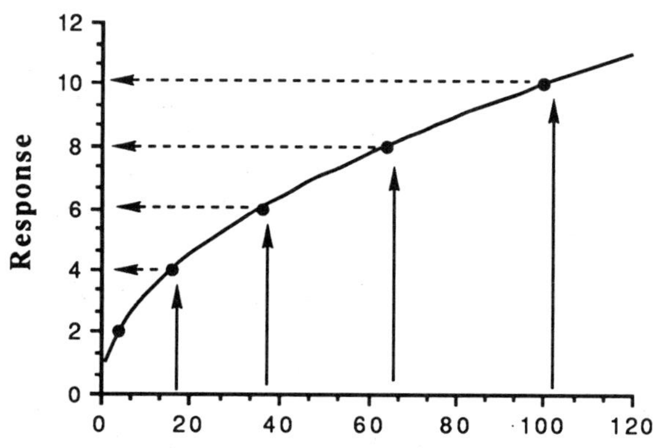

FIG. 1.7. Relation between sensory response and increasing stimulus intensity (Trout & Sharp, 1937).

crystalline sugars. Presuming that the crystalline sugars were tasted as saturated salivary solutions, and that $R = S^{0\cdot5}$, as suggested by Fig. 1.7, sweetness scores (R) were averaged, then squared, and relative sweetness calculated using the sweetness of crystalline sucrose as unity. Such scores, were found to have stoichiometric significance with respect to sugar relative sweetness (Section 6.5).

For quantitative sensory measurements Stevens (1957) proposed that 'equal stimulus ratios produce equal subjective ratios,' or $R = kS^n$. Stevens' law is believed to be, among other things, relatively unencumbered by obscure assumptions about j.n.d.'s. In logarithmic form, Stevens' exponential law is

$$\log R = n \log S + \log k \tag{1.1}$$

where n in the exponential equation is the slope of the logarithmic plot. Stevens' law is particularly applicable to magnitude estimation methods. Ultimately, ratios *per se*, are not determined. Instead, subjects assign numbers to stimuli proportionate to the perceived magnitude of the sensation. When the sensory scores and stimulus concentration are plotted on a log–log basis, if n exceeds 1·0, it is stated (Moskowitz, 1971*a*) that the response *accelerates* with concentration. If n is less than 1·0, the sensory response *decelerates* with concentration. The plot shown in Fig. 1.7 is a deceleration curve with $n = 0{\cdot}5$. Its mirror-image would be an example of an inductive or accelerating curve with $n > 1$.

For the sweetness of substances, relative sweetness scores are not the

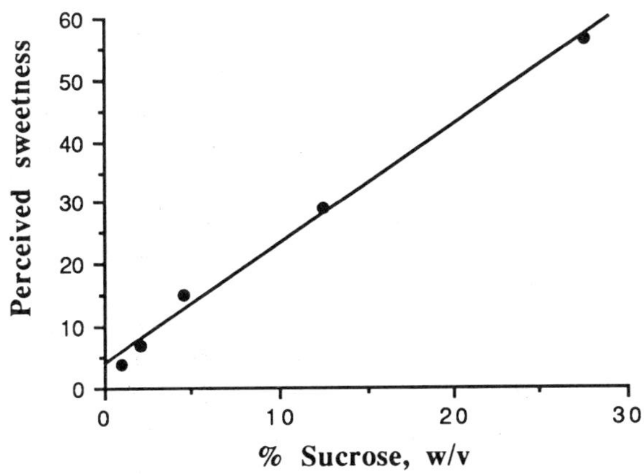

FIG. 1.8. Relation between perceived sweetness and sucrose concentration (Cameron, 1947).

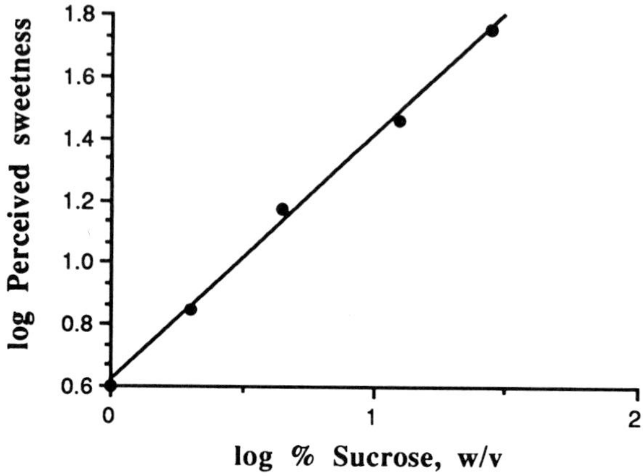

FIG. 1.9. Relation between perceived sweetness and sucrose concentration plotted on log–log coordinates (Cameron, 1947).

same as perceived sweetness scores assigned to different concentrations of a single substance. The former are intrinsically adjusted for whatever exponential relations there may be, while the latter are not. For the perceived sweetness of sucrose, over the range of 1–27% (w/v), the data of Cameron (1947) are shown in Fig. 1.8. This simple arithmetic relation is quite good ($r = 0{\cdot}994$). The exponential function for this data is therefore 1, as also found by Beebe-Center & Waddell (1948) and DuBois *et al.* (1991). Plotted on log–log coordinates the exponent becomes 0·793, and the correlation ($r = 0{\cdot}997$) is slightly improved (Fig. 1.9).

Moskowitz (1970) explored the sweetness response to increasing sucrose concentrations over a wider range of sucrose concentrations (up to 64% w/v). The first four values neatly fit a log–log plot with an exponent of 1·4, as shown in Fig. 1.10. For comparative purposes, the exponent for saltiness and sweetness is 1·3–1·5 (Stevens, 1960; Moskowitz, 1970, 1971*b*), and for bitterness and sourness it is 1·0–1·1 (Moskowitz, 1971*b*).

The relation between perceived sweetness intensity for a number of carbohydrate and other sweeteners was studied by DuBois *et al.* (1991). The results showed that, over the range 2–16% sucrose, scores were linear, and in agreement with the data shown in Fig. 1.8. A linear relation was also found for other sugars. However, the panel employed by those investigators had been trained to equate the percentage sucrose with an equivalent sweetness score. Two percent sucrose was assigned a sweetness

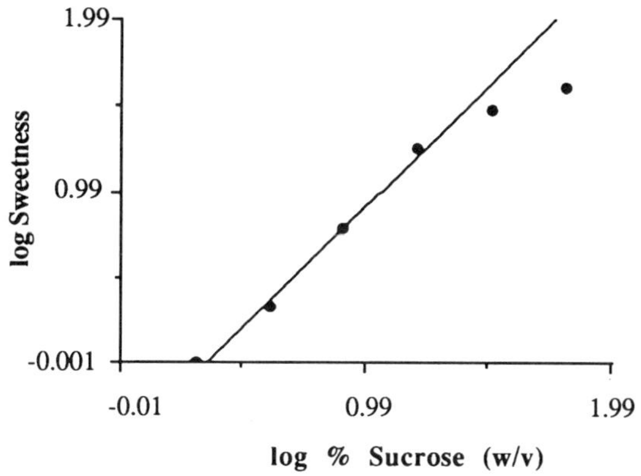

FIG. 1.10. Increase in the perceived sweetness of sucrose with increasing concentrations of sucrose (Moskowitz, 1970).

score of 2, 5% sucrose a score of 5, etc. Although the panel was trained to assign a score of 15 to a sucrose concentration of 16%, it was assumed that the panel had been trained to 'linearize a hyperbolic function.' On submitting the high-potency sweeteners to the panel, plots of the sweetness scores versus concentration were invariably hyperbolic. Results for sucrose and the high-potency sweetener aspartame are summarized in Fig. 1.11. From Fig. 1.11, it is clear that the terms intensity and potency have different meanings. All compounds with the

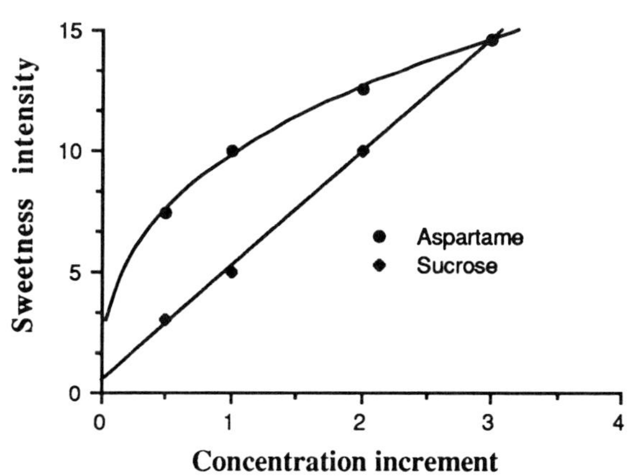

FIG. 1.11. Increase in sweetness intensity of sucrose and aspartame with increasing concentration. Sucrose increments are 2·5, 5, 10, and 15%. Aspartame increments are 0·0005, 0·001, 0·002, and 0·003% (DuBois *et al.*, 1991).

same taste quality are capable of eliciting the maximum taste *intensity* for that specific taste. The concentrations needed to do so are not the same, i.e. their *potency* may differ markedly.

From the kinetic viewpoint, it was found that most of the high-potency sweeteners and the sugars (as the assumed hyperbolic nature was rendered linear by panel training), best fit the Beidler equation (Section 10.7.1). The finding suggests that the receptor-stimulus interaction is on a one-to-one basis. Several compounds (saccharine, cyclamate, and sucralose) fit an expanded Beidler equation somewhat better, indicating reaction kinetics of higher order. This study is an example of the type of significant chemical data that may be obtained, provided the psychophysical laws are considered, by employing the only measure of perceived taste intensity available (panel scores).

When all taste parameters that will subsequently be discussed are considered, a picture somewhat different from that described above emerges. The parameters are subliminal concentrations, psychophysical laws, the law of mass action, chemical equilibria, lipophilicity/hydrophobicity, and number of hetero- and homophoric saporous groups per molecule. The net sensory response to the taste of substances must describe a sigmoid function when taste intensity is plotted against concentration. An inductive effect, or accelerating segment of the curve will be observed as the concentration approaches the recognition threshold. It is followed by a near-linear function. As mass action effects take hold at an inflection point in the sigmoid curve, the plot will demonstrate *decelerating* character. Unless a highly detailed and wide range of taste substance concentrations are employed, the results obtained will describe various segments of the sigmoid relationship.

To answer the problem posed at the beginning of this section, the simplest linear equation $R = 3{\cdot}89 + 1{\cdot}94S$ (Fig. 1.8), where R is the sweetness score and S is the percentage sugar (w/v), can be used. If one teaspoon (5 g) of sugar is added to one cup of coffee (237 ml), the sugar concentration is 2·1% w/v, and the sweetness score is 7·86. To double the sweetness to 15·72, the sucrose needed is 6·10%, which requires two additional teaspoons. Fechner's law states that this is the approximate amount needed, and the reader is invited to derive the answer to the problem using the exponential equations. Probably, no one would ever formulate or approach the problem in the fashion described, and few persons, if any, probably ever think in terms of 'halving' or 'doubling' a sensory attribute. As for the psychophysical principles themselves,

Stevens (1957) stated that, 'Many authors have screamed that (the psychophysical laws) are nonsensical, meaningless, and impossible, but those who follow these methods go ahead and do it anyhow.'

1.3.2 Adaptation and Cross-Adaptation

With continuous exposure the taste of solutions of a saporous compound will diminish, over time, and nearly disappear. The experience is called *adaptation* and is a form of fatigue. When applied to a single substance it is known as *self-adaptation* (Bartoshuk, 1987). Continuous and uninterrupted exposure, unlike the conventional procedure for ingesting a substance, is caused by running a solution of tastant over the tongue, or simply by retaining the solution in the mouth without swallowing it. If adaptation to the sweetness of sucrose diminishes the sweetness intensity of a second sweetener, *cross-adaptation* is said to occur. These procedures have been applied to the problem of the types and number of receptors needed for the different 'primary' tastes. They have also been used to confirm or disavow that taste is a chemical sense, and as an index to the number and types of receptors within a given taste quality.

A plot of adaptation and recovery for the taste of sodium chloride at different concentrations is shown in Fig. 1.12 using the data of Hahn (1934). The data given for the ordinate are threshold concentrations (cf. Pfaffmann, 1951).

A kinetic plot of the same data is shown in Fig. 1.13. Both adaptation to, and recovery from, the taste of table salt neatly fit requirements

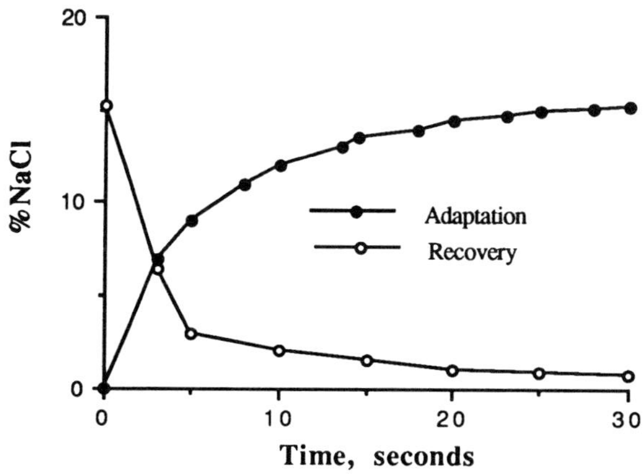

FIG. 1.12. Adaptation-recovery data for sodium chloride. Data of Hahn (1934).

required of a first-order (unimolecular) equilibrium chemical reaction (Section 2.5.2). Adaptation to NaCl proceeds at a rate commensurate with interaction with the receptor sites by a constant percentage of stimulant at any given time. Recovery, in turn, proceeds at a rate proportionate to a constant percentage of the undissociated tastant–receptor complex remaining at a given time. Both processes can be described as a reversible equation:

$$\text{Tastant} + \text{Receptor} \underset{k_2,\ \text{recovery}}{\overset{k_1,\ \text{adaptation}}{\rightleftharpoons}} \text{(Tastant-Receptor interaction)} \qquad (1.2)$$

As judged from the slopes of kinetic plots for the adaptation process ($-0{\cdot}054x$) and the recovery process ($-0{\cdot}0612x$) for salt, the rate of recovery is about 20% more rapid than the rate of adaptation. If the effect of other factors, such as rate of salivary flow, are excluded, the kinetic data indicate that the tastant–receptor interaction is at a higher energy level than the reactants (tastant + receptor). The resulting propensity to dissociate would account, in part, for the more rapid recovery rate. From the viewpoint of mass action principles adaptation occurs when recognition dissociation is prevented by excess tastant (Section 2.4.2.2). In support of the description of adaptation equilibria, loss of neural discharge accompanies adaptation (Zotterman, 1971). Whatever the mechanism of adaptation might be, the course of events

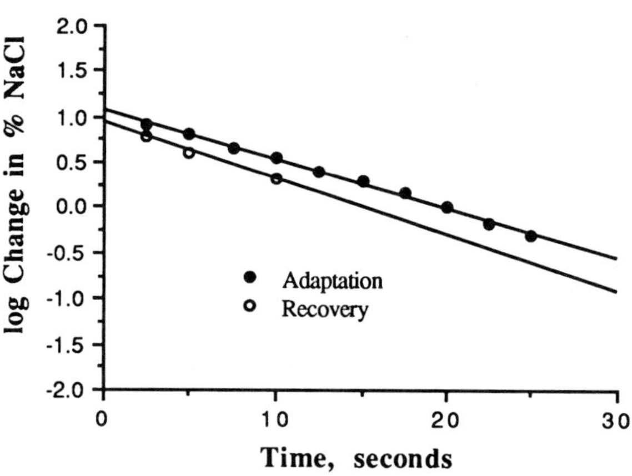

FIG. 1.13. Kinetic plot of the adaptation-recovery. Data of Hahn (1934).

responsible do not seem to occur at the outset of the taste recognition reaction, but at some later point in another process.

Not all adaptation data fit the first-order (unimolecular) kinetic pattern over the complete course of events, although all recovery data show ideal unimolecular kinetics. There is an induction phase for adaptation to sucrose (Pfaffmann, 1951; Bartoshuk, 1987) that appears to be related to a lag period for orientation of the compound with the sweetness receptor. From then on the course of events fit requirements for first-order kinetics for adaptation and recovery.

The kinetic nature of adaptation-recovery data supports the assumption (cf. Lazareff, 1922) that there is facile combination between the stimulus and a receptor. The interaction is necessary for a taste response to occur. The data further suggest that all stimuli that elicit the same taste quality are mediated by the same receptor. McBurney & Bartoshuk (1973) observed that there is little, if any, cross-adaptation between substances with different tastes. From this observation they concluded that different taste qualities are probably initiated by different receptor mechanisms. Also, there is cross-adaptation among sweeteners (McBurney, 1972), acids (McBurney *et al.*, 1972), and among salts (McBurney & Lucas, 1966), but cross-adaptation has been known to fail for some bitter substances (McBurney *et al.*, 1972). The general inability for substances with different taste quality to exhibit cross-adaptation is taken to be strong perception evidence for the four basic taste modalities (McBurney, 1974).

Results of other adaptation studies are not particularly clear-cut. If the premise is accepted that sweet, salt, sour and bitter can, at best, only be defined as fundamental propositions (Section 1.1.2), then it is not possible for two different chemicals to have the same complete set of sensory attributes. Thus, when the results of chemical cross-adaptation studies are recorded, it can be expected that they will vary as widely as the chemical properties of the compounds examined.

1.3.2.1 *Cross-Adaptation to Saltiness*

According to Hahn *et al.* (1940), adaptation to any salt, such as NaCl, NaBr, KCl, or $CaCl_2$, has no effect on the threshold for any other salt. On the other hand, McBurney & Lucas (1966) and Smith & McBurney (1969) observed that there is indeed cross-adaptation among salts, and it is most pronounced among salts with common cations. Smith & Frank (1972) also observed cross-adaptation to saltiness using an electrophysiological technique.

1.3.2.2 *Cross-Adaptation to Sourness*
According to Pfaffmann *et al.* (1971) and McBurney (1974) adaptation to an acid adapts the sour taste for all other acids. Contrary to that generalization, Ganzevles & Kroeze (1987) found that HCl does not exhibit either self- or cross-adaptation for sourness, but carboxylic acids showed both self- and cross-adaptation. However, the latter observation is based on comparing solutions of equal molarity. Adaptation to citric acid does not reduce the sour taste of magnesium sulfate and urea (McBurney, 1972), but these substances do not have a proton to begin with. Furthermore, bitterness and sourness are psychologically confusing (Pfaffmann *et al.*, 1971; O'Mahony *et al.*, 1979), and represent two different mechanisms for induction of unpleasant taste. Sodium sulfate also does not have a hydrogen atom, but is reported to taste both sour and bitter (Smith & McBurney, 1969).

1.3.2.3 *Cross-Adaptation to Bitterness*
McBurney (1972) observed that after adaptation to quinine·HCl and H_2SO_4, the compounds were rendered tasteless, the bitterness of caffeine was reduced, and the bitterness of magnesium sulfate, urea, and phenylthiocarbamide was not affected. Subsequently (McBurney, 1974) it was found that after adaptation to quinine hydrochloride, nearly all other bitter tasting substances exhibit reduced bitterness intensity except KNO_3. Cross-adaptation apparently occurs for some bitter substances, but not for others. The observation leads to the notion that bitterness can be caused by different chemical mechanisms.

1.3.2.4 *Cross-Adaptation to Sweetness*
McBurney (1972) found that sugars and high-potency sweetening agents exhibit cross-adaptation. Schiffman *et al.* (1981*b*), found that sugars and sugar alcohols at equi-sweet concentrations (glucose, 1·1 M; fructose, 0·6 M; galactose, 0·1 M; maltose, 1·2 M; xylose, 1·3 M; sorbitol, 1·2 M; xylitol, 1·0 M) seem to cross-adapt, as adaptation to any one of them diminished the sweetness of another. The authors also found that after adaptation to equi-sweet concentrations of calcium cyclamate (0·6%) or sodium saccharine (0·045%), there is little effect on the sweetness of the sugars. On the other hand, adaptation to acetosulfame K (0·35%) enhanced the sweetness of the sugars, and adaptation to the sweetness of sodium saccharine enhanced the sweetness of glucose by a highly significant 13%. This occurred even though adaptation to glucose reduced the sweetness intensity of nonsugar (or sugar alcohol) sweeteners.

Lawless & Stevens (1983) also found various cross-adaptation abilities among the intense sweeteners dihydrochalcone, aspartame, and saccharine. All adapted themselves and were cross-adapted by sucrose. Either partial cross-adaptation or no cross-adaptation occurred with the intense sweeteners when the latter were used as the preadaptation rinse. In addition, 'asymmetric' cross-adaptation occurred where one compound adapted another, but not *vice versa.*

1.3.2.5 *Effect of Adaptation on Other Taste Qualities*

According to Dallenbach & Dallenbach (1943), and as might be expected from the change in thresholds for mixtures of substances (Section 1.3.3), adaptation to bitter taste lowers the threshold for other tastes. On the other hand, adaptation to citric acid, and bitter compounds, did not reduce sweetness, but enhanced (potentiated) it, e.g. the sweetness of sucrose, but adaptation to HCl did not potentiate sweetness. If taste is a chemical sense, and the mechanisms for different tastes are different, it is difficult to conceive of consistent cross-adaptation phenomena between taste qualities. The subjects of adaptation and cross-adaptation are complex. The latter are affected not only by the taste quality of the adapting substance, but also by its chemical activity. Correspondingly, McBurney (1974) observed that while bitter adaptation does not reduce saltiness, especially if the bitter substance is quinine·HCl, sour adaptation will reduce bitterness.

1.3.2.6 *Mechanism of Adaptation*

In view of the kinetic nature of adaptation data, the initial mechanism of adaptation would seem to be chemical kinetic and equilibria phenomena. In addition, a compound's activity toward its particular receptor site governs the nature of data obtained. As mentioned in Section 1.3.2, whatever the mechanism of adaptation may be, from the viewpoint of taste as a chemical recognition phenomenon, clearly some substance essential for completion of the transduction cascade (Section 1.4) becomes limiting. The extent to which it becomes limiting is somehow proportionate to the concentration and taste-inducing activity of the adapting solution. From the physiological viewpoint, Pfaffmann (1959) assumed that adaptation reflects the exhaustion of some receptor substance in a manner analogous to the bleaching of visual purple by light.

In contrast to the aforementioned conclusion, Hellekant (1969) believed that cross-adaptation is the result of competition for taste receptors. Upon discovering that there is no cross-adaptation between oral and intravascular stimuli, Hellekant *et al.* (1986) further suggested that, in addition to competition, cross-adaptation is due to limitation of available receptors on the membrane of the sensory cells. It does not seem to be the result of a generator potential or similar events after the binding of the stimulus to the receptors. Other mechanism proposals (cf. Pfaffmann, 1959) include: (1) change in permeability of the receptor membrane for the adapting stimulus only; and (2) adaptation is an example of the 'central and peripheral (nerve) process working in opposition to each other.' A particularly significant characteristic of adaptation is the observation (DuBois *et al.*, 1993) that at equi-sweet concentrations, high-potency sweeteners exhibit adaptation at a much more rapid rate than do the sugars. The observation is best explained in terms of chemical activity toward the receptor as proposed at the outset of this section, and is given further consideration in Sections 10.4.1 and 12.4.2.3.

1.3.3 Taste of Mixtures

In assessing the taste of mixtures, the notions of taste suppression and taste synergism arise. In mixtures of substances with different tastes, suppression of each taste is the rule but whether the total taste intensity is less than the sum of the perceived taste intensities of the components is not easily determined. In mixtures of compounds that elicit similar tastes (sugars), synergism or addition is said to occur (Pfaffmann *et al.*, 1971; Frijters & DeGraaf, 1989). The mechanism of taste synergism need not necessarily be true chemical synergism ($1 + 1 = 3$), as noted for the nucleotides and monosodium glutamate, but supplemental action due to superimpositioning of different taste profiles and spectra. The sweetness of the sugars shows sweetness synergism that has been traced to superimpositioning of sweetness profiles (Dahlberg & Penczek, 1941). Curtis *et al.* (1984) report that sugar mixtures show synergism at low concentrations, and suppression at high concentrations. It would seem that this type of synergism is of the profile superimposition type. At higher sugar concentrations the effects become obscured. DeGraaf & Frijters (1988) observed that the taste interaction between two sugars in a mixture, is somewhat synergistic at low sweetness levels, additive at intermediate levels, and suppressive at high sweetness levels. Another form of taste synergism has been described as psychological addition along an expanding function (Bartoshuk & Gent, 1985), but how the principle might apply to taste chemistry considerations is not clear.

Ultimately, assessment of the taste of mixtures of substances is difficult, and again brings up the argument whether mixed taste phenomena are synthetic or analytic (Section 1.1.1). Other factors, such as water taste, and psychological expressions that cover notions of *compensation, simultaneous compensation, contrast,* and *successive contrast* need to be invoked (Bartoshuk, 1987).

The most comprehensive early compilation of taste interactions was made by Heymans (1899) and was directed toward the effect of one tastant on the recognition threshold of another. In essence, thresholds for the taste of a given substance are elevated in the presence of a substance that has a different taste (Table 1.3), which is in accord with the observation that mixtures of tastants have a mutually suppressive action. The interrelation receives additional attention in Chapter 12. The data of Heymans also has application to the suprathreshold taste of mixtures, but discussion is reserved for Chapter 12. There, the taste of mixtures also has application to the problem of the types of taste receptors.

1.4 TRANSDUCTION

Whatever the initial recognition chemistry of taste might be, or whether receptors are specific or nonspecific (cf. Bryant *et al.*, 1989), an electrical impulse is generated. It is transmitted to the brain along the *chorda tympani* nerve from the front two-thirds of the tongue, and the *nervus glosspharyngeus* for the back one-third of the tongue. A sequence of intermediate chemical events is involved in transduction. An earlier thought on the nature of transduction (Beidler, 1966) drew an analogy between a taste cell and a battery. The source of energy was believed to be an uneven flow of potassium and sodium ions across the cell membrane. The inside of the cell was electrically negative with respect to the outside, and when a tastant reacted with the receptor, Na^+ and K^+ ions leaked through the lipid–protein membrane. Exchange then took place with similar ions in the thin layer of extracellular solution. The increase in ion permeability was thought to result in decreased electrical voltage across the membrane. The magnitude of the voltage (30–40 mV), was then presumed to be proportionate to the chemical intensity of the initial interaction, or to the number of receptor sites filled. The voltage change stimulated the nerve endings encircled by the cell membrane, and a series of brief (1–3 ms) impulses were then carried to the brain. Certain tastes can also be stimulated electrically, but the taste qualities obtained are not easily explained (Bujas, 1971). Anodal electrical stimulation clearly elicits a sour taste, while cathodal stimulation effects are ambiguous, although bitter and sweet sensations have been reported (Ajduković, 1990).

Beidler (1954) proposed that the initial chemistry of taste occurs at the surface of the taste cell membrane, but transduction studies can be interpreted to indicate that the salt taste occurs only after ions enter the taste cell (DeSimone *et al.*, 1981; Schiffman *et al.*, 1983; Schiffman, 1990). Possibly, some salt diffuses into the cell, and some of the ions also interact with the taste receptors before diffusion. Nevertheless, a change in membrane potential results, and several different transduction mechanisms may be responsible for it. Different transduction mechanisms are then believed to be responsible for the four basic tastes. The present author does not believe that different types of induction chemistry for the different tastes precludes the probability of different transduction mechanisms for the different tastes. On the contrary, the nature of the chemical recognition reaction probably affects the nature of the transduction chemistry.

For taste initiation, Teeter & Gold (1988) subscribe to receptor cell depolarization due to entry of salts through ungated channels in the membrane. Acids are thought to depolarize cells by hydrogen-ion blocking of potassium channels. Avenet *et al.* (1988) state that transduction of the sour and salty tastes occurs by proton-modulated K^+ channels and by Na^+ channels at the apical taste cell. In addition to the direct effects of taste stimuli on ion channels, intracellular messengers are also believed to contribute to transduction. Bitter stimuli induce a rise in intracellular calcium concentration, possibly through release of inositol triphosphate. Cyclic nucleotides may be intracellular messengers for sweetness induced by sugars. Tonosaki & Funakoshi (1988) found that cyclic guanidine monophosphate (cGMP) decreased potassium conductance, and that sucrose also depolarizes cells by decreasing potassium conductance.

For sweetness, Simon (1991) considered two basic models with which sugars bind to receptors on the apical membrane. In the first, a channel is opened permitting small cations to enter and depolarize taste cells. In the second, the binding starts a messenger cascade of events that either close potassium channels, and/or open chloride channels. Sweetness in particular seems to be mediated by a chain of events characterized as a cascade. A sequence for the stimulus-induced depolarization of receptor cells was outlined by van der Heijden (1993). It combines the observations of Kinnamon (1988) and Teeter & Gold (1988). The

TABLE 1.3
Effect of increasing concentrations of sucrose, NaCl, HCl, and quinine · HCl on the thresholds for the sweet, salt, sour, and bitter tastes (Heymans, 1899)

% Substance	*Sweet threshold (% Sucrose)*	*Salt threshold (% NaCl)*	*Sour threshold (% HCl)*	*Bitter threshold (% Quinine · HCl)*
Sucrose				
0·0	—	0·25	0·003 8	0·000 6
1·25	—	0·56	0·003 5	0·001 0
2·5	—	0·66	0·004 5	0·001 7
5·0	—	0·98	0·006 0	0·002 5
10·0	—	1·23	0·006 8	0·004 6
15·0	—	1·83	(0·004 5)	0·007 0
20·0	—	2·08	(0·003 8)	0·009 9
25·0	—	2·55	(0·004 5)	0·014 0
NaCl				
0·0	0·40	—	0·003 4	0·000 3
0·625	0·55	—	0·003 1	0·001 1
1·25	0·85	—	0·004 1	0·001 0
2·5	1·68	—	0·005 1	0·002 5
5·0	2·93	—	0·006 5	0·005 7
7·5	4·75	—	0·008 3	0·008 3
10·0	6·30	—	0·011 8	0·011 6
12·5	(9·40)	—	0·012 8	0·014 6
HCl				
0·0	(0·54)	0·24	—	0·000 5
0·012 5	(0·41)	0·31	—	0·000 8
0·025	0·39	0·34	—	0·001 1
0·05	0·63	0·38	—	0·001 6
0·10	0·90	0·46	—	0·003 1
0·15	1·20	0·54	—	0·004 3
0·20	1·58	0·64	—	0·005 5
0·25	1·86	0·70	—	0·006 9
Quinine				
0·0	0·80	0·25	0·003 3	—
0·05	1·18	0·35	0·007 0	—
0·1	1·43	0·41	0·010 3	—
0·2	2·13	0·58	0·016 3	—
0·4	3·00	0·98	0·024 8	—
0·6	3·95	1·30	0·027 5	—
0·8	4·95	1·60	0·041 0	—
1·0	(6·50)	(2·25)	0·046 5	—

sequence is (1) Chemical stimulus ⟶ (2) Activation of a guanidine triphosphate-binding protein (G protein) ⟶ (3) Activation of adenylate cyclase ⟶ (4) Production of cyclic nucleotides (messengers), such as cyclic adenosine monophosphate (cAMP) or guanadine monophosphate (cGMP) ⟶ (5) Closing of an apical K^+ channel ⟶ (6) Opening of a voltage-dependent Ca^{2+} channel ⟶ (7) Influx of Ca^{2+} near synapses ⟶ (8) Neurotransmitter release at synapses. The latter events are thought to occur at the basolateral end of the taste cell.

The cascade of events postulated by Lancet & Ben-Arie (1991) parallels the above cascade through step (4). Step (5) then becomes the activation (by cAMP) of the enzyme protein kinase (PK-A), which catalyzes protein phosphorylation. Then in (6), a phosphorylation-gated potassium channel is closed, when acted upon by PK-A, leading to sensory cell depolarization and neurotransmitter release. A diagram of a sweetness cascade is shown in Fig. 1.14. Although not indicated in the diagram, the G-protein and the enzyme adenyl cyclase are believed to be bound to the membrane (Lancet & Ben-Arie, 1991).

In summary, in transduction, the four tastes seem to be mediated, in part, by modulation of K^+ conductance, and it seems unlikely that a single transduction mechanism can explain such diverse taste qualities as salt, sweet, bitter, and sour (Bigiani & Roper, 1991). The latter conclusion also can be applied, of course, to taste induction chemistry.

1.5 OVERVIEW OF THE TASTE OF CHEMICALS

The basic tastes of substances are essentially food and/or wholesomeness criteria. However, there are several tastes that, while probably due to chemical reactions, do not fit the basic taste classification. These are the 'alkaline' taste of base-reacting substances, and the metallic and/or 'medicinal' tastes. As stated in the Prologue, taste is believed to be a chemical sense, but the significance of the definition is uncertain. The following sections, taken in large part from Moncrieff (1967), present an overview of known chemical substance properties and reactions that affect taste. The first purpose of the overview is to demonstrate how the meaning of the term chemical sense is fuzzy. The second purpose is to present the series of perplexing features associated with taste chemistry. The third purpose is to establish the need for Chapter 2, enroute to resolving those taste chemistry enigmas in the remainder of the text.

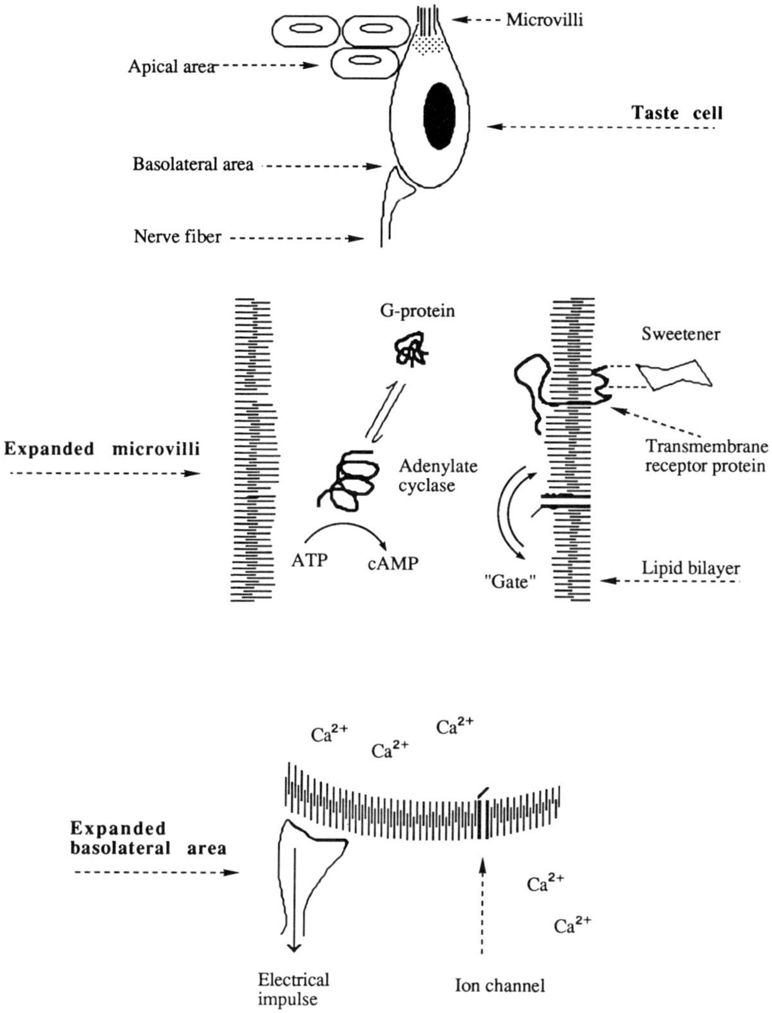

FIG. 1.14. Diagrammatic representation of the transduction events that take place upon interaction of a tastant with a taste receptor.

1.5.1 Chemicals that Taste Sweet, Salty, Sour, and Bitter

Substances that taste sweet, salt, bitter and sour are found in the inorganic and organic classes of compounds. Substances that taste salty and sour invariably dissociate (ionize) in solution, while substances that taste sweet and bitter generally do not. Inorganic substances that do ionize, but nevertheless taste sweet, are dilute solutions of sodium and potassium chloride, sodium and potassium hydroxides, lead acetate and carbonate, and salts of beryllium. Bitterness is an additional attribute of inorganic salts other than sodium chloride, such as potassium chloride and magnesium sulfate. Organic substances that taste sweet and/or bitter encompass the sugars, amino acids, peptides, proteins, and a large number of compounds that are derivatives of cyclohexane, aniline, benzene, etc. High-potency sweetness is usually associated with many of these substances, and high-potency bitterness with many more organic substances. The salt taste occurs in many inorganic substances, but especially the halide salts. Saltiness is also an attribute of the metal salts of common organic compounds such as acetic, citric, malic, and tartaric acids. Sourness is also a general attribute of the inorganic (hydrochloric, sulfuric, phosphoric) and organic (acetic, citric, etc.) acids.

The sourness or tartness of acids is largely related to the hydrogen ion concentration of their solutions. It is also related to total acid concentration in those cases where the acid does not dissociate completely. On a molar basis, sulfuric acid (H_2SO_4) is twice as sour as hydrochloric acid (HCl) as it has two protons that dissociate. Both of the latter mineral acids completely dissociate in solution. As acetic acid (CH_3COOH), which does not dissociate completely, nevertheless tastes just as sour, on a molar basis, as does HCl, both the dissociated hydrogen ions, and the hydrogen ions with potential to dissociate, play a role in the total sourness of acids.

It is generally conceded that only sodium chloride elicits a purely salty taste. Attempts to relate the salt taste to anions and cations have a long history, not only with respect to the salt taste quality, but also to the intensity of saltiness. Substances that are predominantly sweet are found in all chemical classes, but are usually associated with multiple hydroxyl groups, and α- but not β- or γ-amino acids. Other compounds containing the NH_2 group that taste sweet are amides, ureas and hydrazides. On ascending a homologous series, the sweetness of compounds tends to either disappear (sugars > oligosaccharides > polysaccharides), or become even more bitter (L-amino acids > peptides > proteins). Benzene compounds (Fig. 1.15) that contain a single NO_2 group are often sweet (nitrobenzene), but those containing two or more NO_2 groups (*m*-nitrobenzene, picric acid) are bitter. Salicylic acid is an example of a compound that is predominantly sweet, but also sour. The transitional relation between sweetness and bitterness is exemplified by the ureas. Urea itself is bitter, but α-dimethyl urea is sweet.

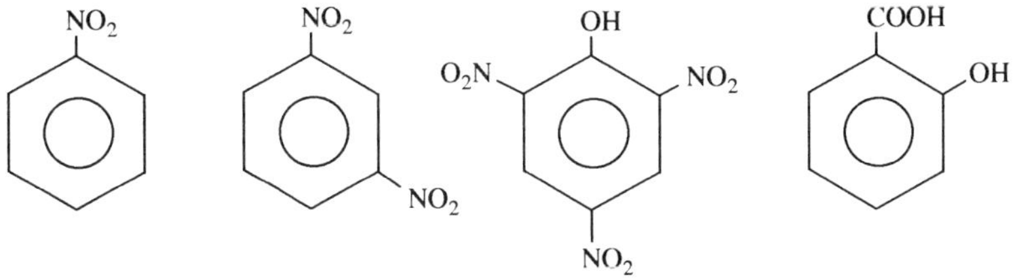

Nitrobenzene (Sweet) | *m*-Nitrobenzene (Bitter) | Picric acid (Bitter-sour) | Salicylic acid (Sweet-sour)

FIG. 1.15. Structure and taste of substituted benzene substances.

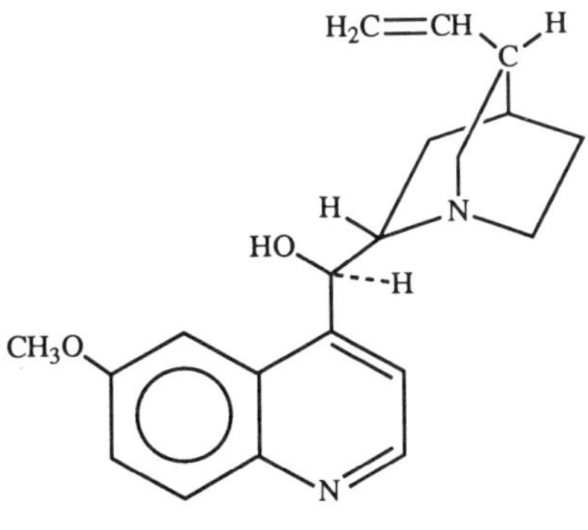

Quinine

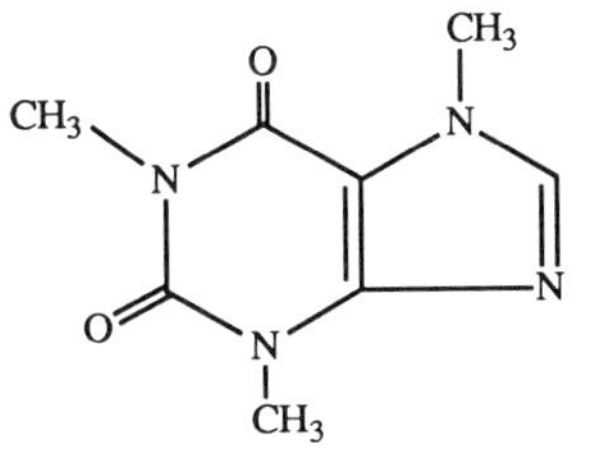

Caffeine

By itself, bitterness is an unpleasant taste, and like sweetness, bitterness is elicited by chemicals of all classes, but alkaloids (quinine, caffeine) are usually employed to characterize (define) the sensation. Chemically, alkaloids are free bases. Other substances that taste bitter are caesium chloride, potassium iodide, and magnesium sulfate. Generally, salts that contain the caesium or rubidium cation, or the iodide anion, are bitter. The transition from sweetness to bitterness occurs with seemingly subtle changes in constitution. The presence of one nitro group may impart sweetness, but two or more, as stated above, impart bitterness. Nitrogen compounds that are basic are usually bitter, and those that are acidic are sour and/or bitter. Those that are neutral to litmus are often sweet, such as the amino acids, hydrazides, and ureas.

1.5.2 Taste Transformations

The salt and sour tastes are simply related chemically, since neutralization of a sour tasting acid with an appropriate base converts the taste from sour to salt.

$$\text{HCl (sour)} + \text{NaOH} \longrightarrow \text{NaCl (salt)} + \text{HOH}$$

Among the aromatic series of compounds, the introduction of a halogen atom may transpose the taste from sweet (*m*-aminobenzoic acid) to bitter (4-chloro-*m*-aminobenzoic acid). A more striking case is the transposition of ethoxyphenylthiourea from bitter to sweet by simply replacing the sulfoxide group with a carbonyl group (dulcin), as shown in Fig. 1.16. This takes place even though sulfur and oxygen are closely related members of the Group VI elements of the periodic table. The naturally occurring amino acid L-phenylalanine is bitter. Transposition of the spatial position of the —NH_2 group on the lone asymmetric carbon atom from 'left' to 'right' forms the mirror-image compound (enantiomer) D-phenylalanine, which tastes sweet.

Crystals of the sugar α-D-mannose taste sweet, but when they are dissolved in water, mutarotation occurs. In mutarotation, α-D-mannose establishes an equilibrium with β-D-mannose, which is bitter. The taste of the mixture is both sweet and bitter. The only difference between the two substances is that the position of a hydroxyl group about one asymmetric carbon atom is transposed (Fig. 1.16). The compound formed is not an enantiomer, as was the case for phenylalanine, but a 'partial' enantiomer (diastereoisomer) instead. Lastly, when two of the hydrogen atoms of one of the amine groups of urea are substituted by methyl groups, the taste is transformed from bitter to sweet.

The compounds shown in Fig. 1.16, and used as examples of how structure modification alters taste, were selected for specific purposes. The phenylthiourea and dulcin compounds demonstrate that a relatively minor change in the nature (electronic configuration) of a single atom is sufficient to transpose tastes. The taste is transposed for the two amino acids that are chemically equivalent, atom for atom, but bear a mirror-image right–left relation to each other. This indicates that there may be a stereochemical requirement for the sweet and bitter tastes of organic substances, but the nature of the steric requirement is difficult to determine. Not all enantiomers have opposite tastes. Indeed, some have identical taste. The two sugars, α-D- and β-D-mannose are not enantiomers, as the configuration about only one carbon atom has been transposed. Nevertheless, one tastes sweet, and the other tastes bitter. Substitution of two polar hydrogen atoms of urea with two nonpolar methyl groups also transposes the tastes of these substances,

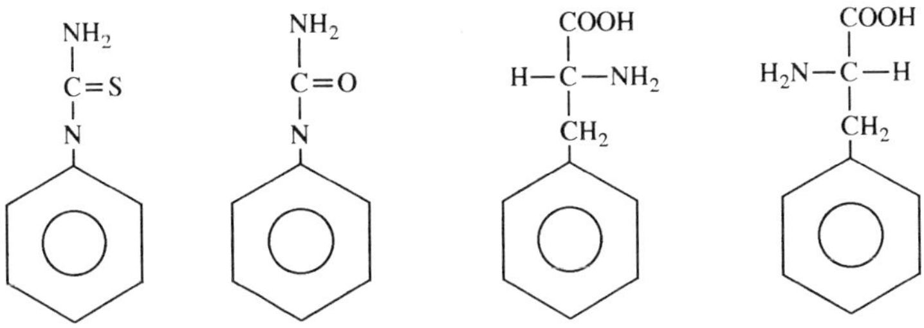

Phenylthiourea (Bitter) Dulcin (Sweet) D-Phenylalanine (Sweet) L-Phenylalanine (Bitter)

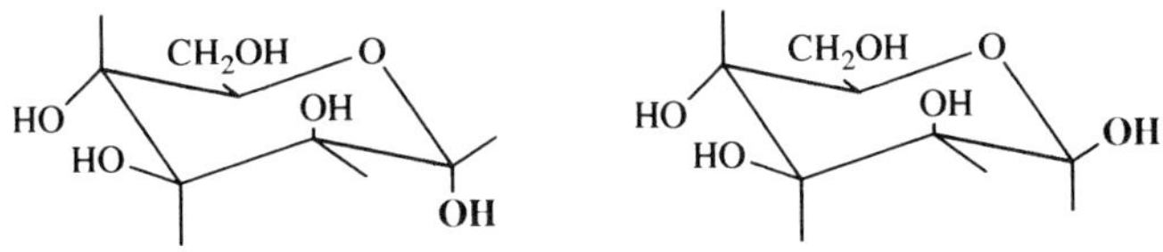

α-D-Mannose (Sweet) β-D-Mannose (Bitter)

Dimethyl urea (Sweet) Urea (Bitter)

FIG. 1.16. Effect of structure alteration on taste.

suggesting that in some cases the chemistry of taste has an apolar (hydrophobic) feature.

1.5.3 Common Chemical Sense Taste Attributes

There are several common chemical sensations allied with the sense of taste. A sensation of 'hotness' can be experienced with the ingestion of capsaicinoid oils, such as occur in *Capsicum frutescens L. Solanaceae*, or Cayenne pepper. Capsaicin causes a hot sensation at a concentration of 1 ppm, but the substance itself is reported to be tasteless and odorless. The taste of nitroglycerin is both sweet and burning. Endothermic substances that dissolve with absorption of heat, such as menthol, have a cooling sensation. Sugar alcohols possess this property, in addition to tasting sweet. The astringency of tannins (polyhydroxyphenolic substances) is probably related to their ability to precipitate proteins. Wilkins (1964) has shown that the pungency of onions and garlic is due to thiopropional sulfoxide ($CH_3CH_2CH{=}S{=}O$). The volatile compound is lachramatory (produces tears). The latter property is not intrinsic, but is due to hydrolysis and formation of sulfuric acid when the compound is dissolved in water. The structures of substances that elicit the common chemical sense, and an example of the lachramal reaction is shown in Fig. 1.17.

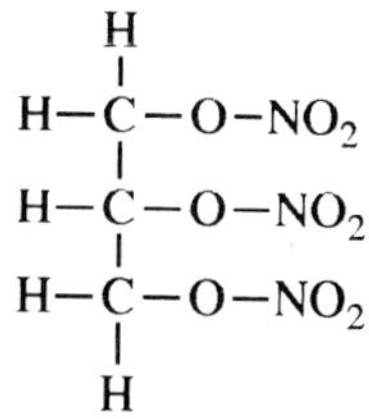

Nitroglycerin

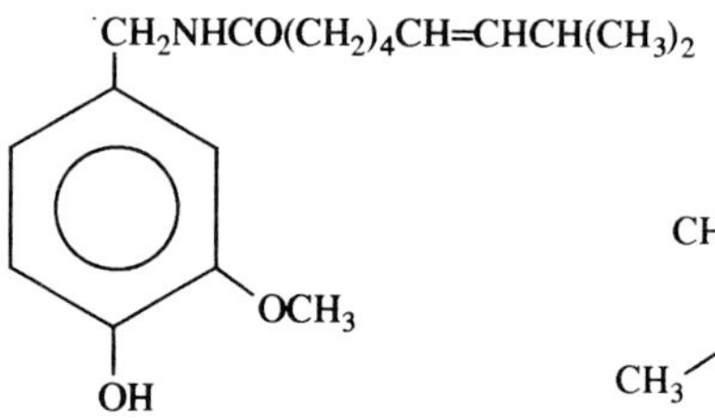

Capsaicin (Hot, burning)

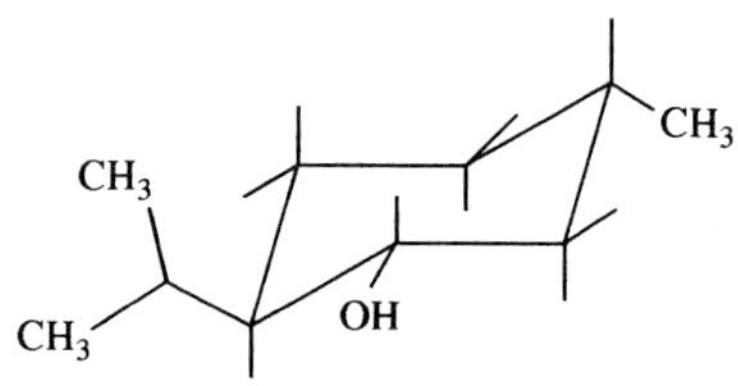

(+)-Menthol (Cooling)

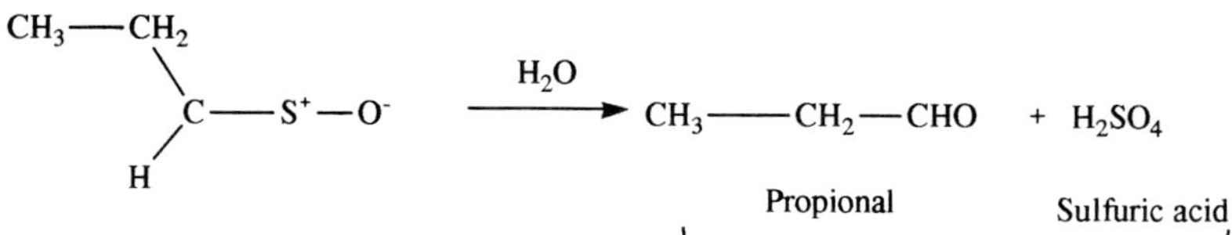

Thiopropional sulfoxide

Propional

Sulfuric acid

Lachramators

FIG. 1.17. Substances that elicit the common chemical sensations of hotness and coldness and the approximate reaction that produces tears.

1.6 CHEMICAL INTERRELATIONS AMONG THE FOUR TASTES

A diagrammatic chemical relation among the four primary tastes is shown in Fig. 1.18. The boldest arrowheads indicate the major reversible relations, while the finer arrowheads indicate subsidiary reversible interrelations. Neutralization of an acid is a prime example of reversible interaction 1. Solvation of α-D-mannose (sweet) with water, and formation β-D-mannose (bitter) is an example of reaction 2. Reactions 1 and 2 are chemical transformations, but subsidiary or secondary interrelations (3 and 4) are caused even more subtly. The bitterness and sourness of salts (3) are changed merely by changing the concentration. Potassium chloride tastes sweet at low concentration, and bitter at higher concentrations. Sodium chloride transforms from sweet at low concentration to salty at higher concentrations, and is an example of (4). Tertiary interrelations (5) and (6) are due to changes in concentration, and to intrinsic multiple tastes. Some salts, such as beryllium chloride and lead acetate, are sweet at low concentrations, but sour at higher concentrations (6), while magnesium ammonium sulfate and salicylic acid are both sweet and sour. Still others (magnesium sulfate) are both salty and bitter (5), with the predominate attribute again depending on concentration.

The relations described in Fig. 1.18 alone seem to be justification for taste quantification procedures that assign scores to substances for all four taste attributes (cf. Fig. 1.3). While many secondary tastes are nearly subliminal, several substances elicit distinct multiple tastes (Shamil *et al.*, 1987), such as sodium benzoate (Fox, 1932) and monosodium glutamate. Such interrelations would also seem to justify the construction of

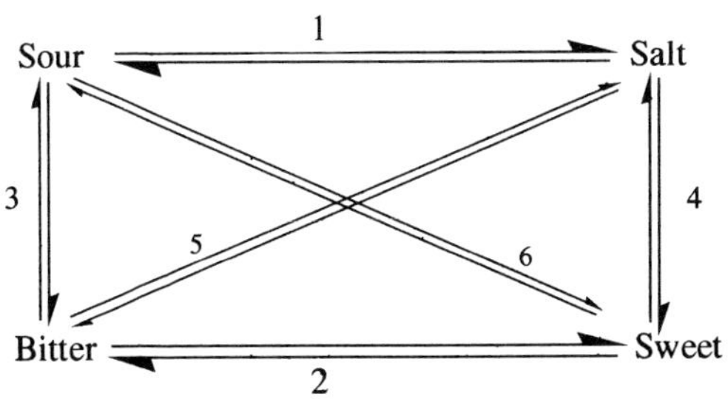

FIG. 1.18. Diagrammatic representation of the chemical relations among the four basic tastes.

psychological taste continua, such as Henning's taste tetrahedron. Nevertheless, Fig. 1.18 is not to be construed to imply either a psychological or a chemical continuum among the primary tastes. It serves only to emphasize chemical interrelations that affect taste, such as changing concentration, substitution, acid-base neutralization, and isomerization reactions.

CHAPTER 2

Taste Chemistry Principles

The principles that serve as the foundation for the discipline of chemistry have application to taste chemistry. They include atomic theory, molecular orbital symmetry and stereochemistry. This chapter begins with the notion of chemical equivalents because if taste is a chemical sense, elicited by chemical reactions, the principles of chemical equivalency must apply. The concept of chemical equivalents is simple and straightforward. Nevertheless, the significance of it, even today, seems to have become lost somewhere in the complexity of taste chemistry phenomena.

2.1 CHEMICAL EQUIVALENTS

The atomic theory that accounts, in part, for the formation of compounds from the elements is guided by the laws of constant composition, and proper and reciprocal proportions. One gram mole of sodium and 1-g mol of chlorine are chemically equivalent as they form 1-g mol of sodium chloride. Two gram moles of hydrogen are chemically equivalent to 1-g mol of oxygen, and when combined they form 1-g mol of water. Carried over to solutions, chemical equivalency is expressed in terms of *normality*. This forms the basis for balancing chemical equations as an electronic statement of the nature of a chemical reaction. To express the hydrogen ion concentration of acid solutions, it can be stated that a 1-g M solution of HCl is 1N with respect to H^+ ($HCl \rightarrow H^+ + Cl^-$). A 1-g molar solution of H_2SO_4 is 2N with respect to H^+ ($H_2SO_4 \rightarrow 2H^+ + SO_4^{2-}$). Therefore a 1M solution of sodium hydroxide will neutralize an equal volume of a 1M solution of HCl ($HCl + NaOH \rightarrow NaCl + H_2O$). However, it takes a 2M solution of

NaOH to neutralize an equal volume of a 1M solution of H_2SO_4 ($H_2SO_4 + 2NaOH \rightarrow Na_2SO_4 + 2H_2O$). The NaOH solution is 1N with respect to OH^-, but the sulfuric acid solution is two normal with respect to H^+.

In oxidation-reduction ('redox') reactions the gram equivalent weight of an oxidizing or reducing agent is the amount of compound that will produce a 1-mol change in oxidation number (i.e. Avogadro's number of electrons). In the oxidation of sodium to the sodium ion ($Na \rightarrow Na^+ + e^-$) the equivalent weight is equal to the atomic weight. For magnesium ($Mg \rightarrow Mg^{2+} + 2e^-$) the equivalent weight is one-half of the atomic weight. In taste chemistry, as in other areas of biological chemistry, chemical equivalency takes on subtle forms compounded by other chemical events. A single compound has varying equivalency for all four tastes (Section 1.1.3.2), and subsidiary chemical reactivity and properties.

2.2 WATER SOLUBILITY AND REACTIONS IN WATER

To possess taste, and to gain access to the receptor site, a substance must be soluble in water. Monomolecular water is a dipolar compound with a slight negative charge on the oxygen atom and a slight positive charge on the two hydrogen atoms. Substances soluble in water either contain polar groups (OH, NH_2), or they ionize in solution. In either case, the substance interacts strongly enough with water to break *its* structure. In the former case, substances hydrogen bond to the water, and in the latter case substance solubility is a fine balance between the crystalline lattice energy of ions, the hydration energy, and entropy. Polar tastant groups are hydrated, and anions are surrounded by molecules of water with the $H^{\delta+}$ atoms pointed toward the negative ion. Cations are surrounded by water molecules with the oxygen atoms ($O^{\delta-}$) pointed toward the positive ion. Nonhydrating water molecules then stabilize the solution by keeping the ions separated.

Reactions in water solution are not well understood because there is a huge perturbing influence by the solvent, and there is little clear idea as to the true nature of reacting species. In some cases the reacting species may have one or more molecules of water bound to it, some so strongly that the water's chemical and physical properties are drastically changed. This is the case for high concentrations of sugar in water, and for the water hydration shell of metallic and nonmetallic ions, and even for particular chemical groups in some substances, such as the imino and the oxime groups.

2.3 pH, p*K*, AND TOTAL ACIDITY

The concepts of an acid, the normality of an acid, the pH of solutions, the p*K* of acids and bases, the action of buffers, and the notion of total acidity have application in several areas of taste chemistry.

2.3.1 Definition of an Acid and Base

The simplest definition of an acid is that traditional acid properties, including sourness, are the result of dissociation of substances with the formula HX, to yield H^+ and X^- ions. As X^- might also contain a proton capable of dissociation, it too can be an acid. Examples are HCl, H_2SO_4 and HSO_4^-.

$$HCl = H^+ + Cl^-$$

$$H_2SO_4 = H^+ + HSO_4^-,$$

$$HSO_4^- = H^+ + SO_4^{2-}$$

A base is a metal hydroxide (MOH) that dissociates in water to yield a positively charged metal ion (M^+) and the negatively charged hydroxyl ion (OH^-). Examples that cover this case are NaOH, KOH, and $Ca(OH)_2$.

A second case arises when the substance itself does not possess either H^+ or OH^-, but induces their formation by reacting with water. Carbon dioxide and aluminum salts are examples of the former ($CO_2 + H_2O \rightarrow H^+ + HCO_3$), and ammonia is an example of the latter ($NH_3 + H_2O \rightarrow NH_4^+ + OH^-$). This is the *general solvent system* definition, and it has application to the general sourness of acids.

Another way to treat the distinction between an acid and a base is to define an acid as any substance that can donate a hydrogen ion, and a base as any substance that can accept a hydrogen ion. All acid-base reactions by this definition are *proton transfer reactions*, and is known as the Brønsted–Lowry definition. The dissociation of HX by this definition is not simply breakup of HX to H^+ and X^-, but proton transfer from HX to a molecule of solvent that functions, in turn, as the base.

$$HX + H_2O \rightarrow H_3O^+ + X^-$$

By this definition, HX is the acid, and water is the base. For the products, H_3O^+ is also an acid, and the hydroxyl ion a base. On each side of the equation, every Brønsted–Lowry acid is paired with a Brønsted–Lowry base, and both sides are said to have a *conjugate* acid-base pair. The significant feature of this definition is that the role of a particular substance is changeable, and can be described in terms of conjugate acids and bases. HCl is the conjugate acid of Cl^-, and acetic acid is the conjugate acid of the acetate ion. H^+ is the conjugate acid of the base H_2O, and H_2O in turn is the conjugate acid of the base OH^-. The Brønsted–Lowry definition of an acid and a base, and the notion of conjugate acids and bases, is particularly appropriate for application to the taste properties of amino acids.

The third way to define an acid and a base is known as the Lewis definition, and has application to the interaction of tastants with receptor sites, including hydrogen bonding and charge transfer reactions. According to the Lewis definition, an acid is any molecular or atomic species that acts as an electron pair acceptor, and is *electrophilic*. A base, by this definition, is an electron pair donor, and is *nucleophilic*.

2.3.2 pH

The pH of a solution is the negative common logarithm of the hydrogen ion concentration. Water ionizes to a slight extent, and the ions are hydrated.

$$H_2O_{(\text{liquid})} = H^+{}_{(\text{aq})} + OH^-{}_{(\text{aq})}$$

The equilibrium constant for the disssociation of water is

$$K_d = \frac{[H^+][OH^-]}{[H_2O]}$$

In a chemical reaction, including taste reactions, the concentration of water, regardless of its involvement in the reaction is

$$1000\,\text{g liter}^{-1}/18{\cdot}0\,\text{g mol}^{-1} = 55{\cdot}5\,\text{mol liter}^{-1}$$

or $K_w = 55{\cdot}5\,K_{eq} = [H^+][OH^-]$. The equilibrium constant is the *ion product constant* for water. At temperatures from 0 to 60°C, it is nearly constant at a value of 10^{-14}, although like all equilibrium constants, it increases with temperature due to increased dissociation. As the hydrogen ion concentration $[H^+]$ equals the hydroxyl ion concentration $[OH^-]$ in pure water, $[H^+] = [OH^-] = 10^{-7}\ \mathrm{mol}^{-1}$. As the definition of pH states that $\mathrm{pH} = -\log_{10}[H^+]$, the hydrogen ion concentration of water is $\mathrm{pH} = -\log_{10}[10^{-7}] = 7$. Similarly, the pOH is 7, and the $\mathrm{p}K_w = -\log_{10} K_w = 14$.

When a strong acid, i.e. one that dissociates completely in solution (HCl, HI, H_2SO_4), is added to water, the H^+ concentration is greatly increased. For a 0·1M solution of HCl dissolved in water ($H^+ = 1 \times 10^{-1}$M),

$$\mathrm{pH} = H^+ = -\log(1 \times 10^{-1}) = 1$$

and the pH of a 1M solution of HCl = 0. The process is not simply dissociation. It is the result of the stronger interaction of H_2O molecules with H^+, as opposed to the weaker interaction of H^+, with Cl^- ions. HCl is an acid in the Brønsted–Lowry sense (proton donor) and Cl^-, with little affinity for a proton, is a weak conjugate base. In the Brønsted–Lowry sense, NaOH is a base because in its dissociation in water solution to form Na^+ and OH^- ions, the OH^- ions interact most strongly with water. By accepting the water protons, the water dissociation equilibrium is perturbed. Thus, water molecules interact more strongly with the OH^- ion than with the sodium ion. Here, OH^- is the strong base and Na^+ is the weak conjugate acid.

2.3.3 p*K*

The p*K* of an acid is defined as the negative common logarithm of a compound's dissociation constant. A weak acid or base is one that does not completely dissociate in water, and a strong acid is one that does. The p*K* for the strong acid HCl is 1, and equal to pH. For an acid that does not dissociate completely (acetic, citric, lactic), only a portion of the protons associate with water.

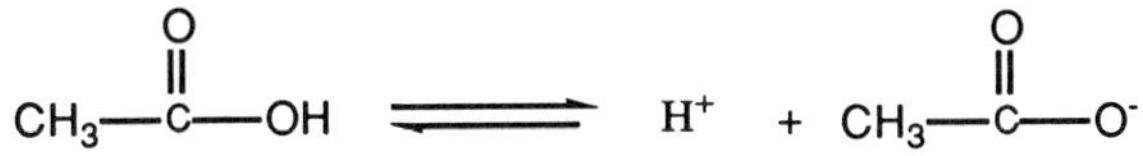

and the dissociation constant for the acid is

$$K_a = \frac{[H^+][CH_3COO^-]}{[CH_3COOH]}$$

For acetic acid, the value of K_a is 0·0000176 ($1{\cdot}76 \times 10^{-5}$) molar. The p$K_a$ is $-\log_{10} K_a$, or 4·75. Not all acids have a single pK value. They may be di- or tri-basic. These acids have more than one group that dissociates to yield hydrogen ions. The pK_a's of some common acids important to taste chemistry are: acetic, 4·75; citric, 3·14, 4·77, 6·39; malic, 3·4, 5·11; lactic, 3·08; phosphoric, 2·12, 12·67; picric, 0·38. In solvents other than water, i.e. relatively apolar solvents, HCl is a weak acid because it does not completely dissociate. This property of acids has application to their sourness. The lack of dissociation of amino acids in apolar solvents has general membrane and other biological (taste) consequences.

2.3.4 Total Acidity

The notion of total acidity is an application of chemical equivalency principles. Acids with different equivalency have, on a normality basis, equal acidity. Although the pK_a of HCl and acetic acid are different, resulting in different pH for the same concentration, they nevertheless have the same acid equivalency. A 0·1N solution of either acid, which is also 0·1M, requires a 0·1N solution of base to effect neutralization. On the other hand, molar solutions of inorganic and organic acids that are dibasic, require twice as much normal base to neutralize them. Molar solutions of tricarboxylic acids, of course, require three times as much normal base solution.

2.4 MASS ACTION AND CHEMICAL EQUILIBRIA

A second fundamental chemical principle important to taste chemistry is mass action and accompanying equilibria phenomena.

2.4.1 Mass Action

If different tastes are due to chemical reactions, they are subject to the laws of chemical equilibria, and the law of mass action is the touchstone of chemical dynamics and equilibria. The law of mass action developed, over a period of many years, to improve upon the notion of chemical affinity, which had eluded definition and attempts to measure it. At one time, the nature of affinity was even believed to be allied to Newton's theory of gravity, but chemical attraction forces decrease too rapidly with increasing distance for this to be true. In time, affinity became recognized as a consequence of chemical dynamics, and related to reaction rate probability theory, i.e. reactivity.

According to Jones (1914), during studies of the rate of sucrose hydrolysis in 1850, Wilhelmy found that the amount of sucrose hydolyzed to glucose and fructose, for a given unit of time, was proportionate to the amount of sucrose present. In other words, as the change in sucrose concentration per unit of time was a constant ($d\text{Sucrose}/dt = K$), the velocity of the reaction was governed solely by the amount of sucrose present at any given time.

Again, according to Jones (1914), the phenomenon observed by Wilhelmy was formally developed as an expression of mass action by Guldberg & Wagge in 1867. In their own words (Jones, 1914), it is stated that 'Suppose two substances, A and B react, and are transformed into two new substances A′ and B′. However, under the same conditions A′ and B′ can also transform themselves into A and B. Neither the formation of A′ and B′ nor A and B is complete, and at the end of the reaction, we have the four substances present, A, B, A′, and B′. The force that causes the formation of A′ and B′ is in equilibrium with that which causes the formation of A and B. The force that causes the formation of A′ and B′ increases proportionally to the affinity coefficients of the reaction $A + B = A' + B'$, but it also depends on the masses of A and B.'

If the active masses of A and B are designated p and q, and the affinity coefficient by K, the force is equal to $(K)(p)(q)$. Disregarding extraneous forces, and considering only the forces between A′ and B′, which are p' and q', then the affinity coefficient of the reaction $A' + B' = A + B$ is K' and the force for the reformation of A and B is $K'p'q'$. This force is in equilibrium with the first force, consequently

$$Kpq = K'p'q' \tag{2.1}$$

By determining the active masses of p, q, p', and q' experimentally, the affinity coefficients of K and K' can be derived. With the ratio K'/K, the result of the reaction for 'any original condition' of the four substances can be found. It is derived in the following manner:

Let P, Q, P', and Q' be the absolute masses of the four substances A, B, A′, and B′ and let x be the number of atoms of A and B that are transformed into A′ and B′. Then, at constant volume (V),

$$p = \frac{P - x}{V}, \quad q = \frac{Q - x}{V}, \quad p' = \frac{P' + x}{V}, \quad q' = \frac{Q' + x}{V} \tag{2.2}$$

Substitution in eqn (2.1) yields

$$K\left(\frac{P-x}{V}\right)\left(\frac{Q-x}{V}\right) = K'\left(\frac{P'+x}{V}\right)\left(\frac{Q'+x}{V}\right) \tag{2.3}$$

Multiplying eqn (2.3) by V^2, leads to

$$(P-x)(Q-x) = \frac{K'}{K}(P'+x)(Q'+x) \tag{2.4}$$

If the two substances do not change in active mass during reaction, and have equal (stoichiometric) value, eqn (2.4) becomes

$$(Q-x) = \frac{K'}{K}(Q'+x) \tag{2.5}$$

and

$$x = Q - \frac{K'}{K}(Q) \div 1 + \frac{K'}{K} \tag{2.6}$$

This in essence is the *law of mass action*. As an equilibrium expression, in modern terms, it is

$$n\text{A} + m\text{B} \rightleftharpoons p\text{A}' + q\text{B}' \tag{2.7}$$

and

$$\frac{[A']^{p}\,[\text{B}']^{q}}{[A]^{n}\,[\text{B}]^{m}} = K \tag{2.8}$$

The brackets indicate concentration in moles/liter. The exponents are the powers to which the concentrations are to be raised. The ratio expression for eqn (2.8) is the mass action expression of eqn (2.7). At equilibrium, the ratio is equal to the equilibrium constant K, and the condition is the *law of chemical equilibrium*.

2.4.2 Chemical Equilibria

A hypothetical plot of the formation of A′ and B′ from A and B in the equilibrium reaction $\text{A} + \text{B} = \text{A}' + \text{B}'$ (eqn (2.7)) is shown in Fig. 2.1. Initially, due to mass action effects, the decrease in concentration of A and B is rapid, and the increase in concentration of A′ and B′ is also rapid. After time has elapsed, the concentration of the four substances becomes static at t_e. Although A and B still react to form A′ and B′, B′ and A′ also react to form A and B; the forward and reverse reactions take place at the same rate. The equilibrium will remain indefinitely at this position, unless additional slow reactions occur, and/or one or more of the reactants are somehow removed from the system.

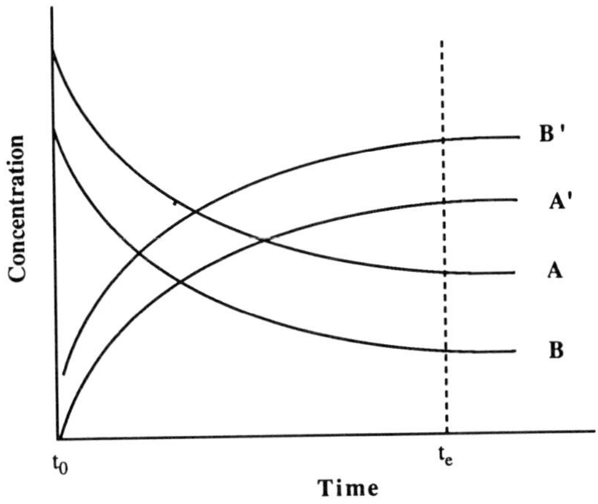

FIG. 2.1. Plot of a hypothetical mass action equilibrium phenomenon for the formation of A′ and B′ from A and B, and also the formation of A and B from A′ and B′.

2.4.2.1 *Chemical Activity*

In an equilibrium expression the concentration expression K, can be replaced with a dimensionless term. An example is chemical *activity* where K is used to describe the propensity for a reaction to occur. The older chemical expression for activity is affinity. That term now has special connotation in structure-activity relationships that is significantly different from the notion of activity (Section 2.10.1). The strict chemical definition of affinity is that it is the attractive force exerted in different degrees between substances of particles that causes them to enter into and remain in chemical combination. In taste chemistry, a case can be made for the activity of a recognition interaction, but the substances do not necessarily enter into strict combination (affinity).

The mass action equation leads (Section 2.6.3) to $\Delta G = -RT \ln K$ (see eqn (2.18)). The final equilibrium state, as characterized by K, is related to the free-energy change from reactants to products in the standard (G°) state. Usually, if ΔG° is zero, $K = 1$. When ΔG° is greater than zero, $K < 1$, indicating that the reaction does not proceed very far from left to right. If ΔG° is less than zero, K is greater than one, and reaction to the right is favored.

2.4.2.2 *Taste Chemical Equilibria*
Taste chemistry phenomena are examples of simple chemical equilibria and mass action. For a chemical compound to elicit a specific taste quality, the tastant (T) must interact with a receptor (R), or

$$\text{Tastant } (T) + \text{Receptor } (R)$$
$$= \text{Tastant } (T) : \text{Receptor } (R) = \text{sensory response}$$

Then, $[T : R] \div [T][R]$ must equal a constant, K. If more tastant is added to the system, the number of receptors interacting with T is increased, and there is corresponding increase in T : R, but the value of K is not altered. Put another way, increasing the concentration of T increases the reaction rate between T and R to yield an increase in the rate of the T : R interaction.

At the outset the T : R 'complex' will form faster than it decomposes. From then on, T : R decomposition will speed up to reestablish the value for the constant K. If T is continually added to the system to maintain the amount of T : R, secondary events in the transduction process are not able to respond accordingly, and adaptation occurs. On the other hand, if T is constantly removed from the system (washed away), T : R will decrease. Taste intensity will also decrease until $K = 0$, or falls below the recognition threshold.

From psychophysics and physiology (Sections 1.1.3.2 and 1.1.4.3) it was found that a given chemical compound may contain potential to elicit all four taste attributes. The taste of compound T, in terms of mass action equilibria phenomenon, is then a composite of $T_{sweet} + T_{sour} + T_{salt} + T_{bitter}$. The taste attributes experienced will be some mixed response to the interaction of the tastant with the receptor(s). Should the mass of one component of the composite be much greater than the other three, a single taste should predominate. If two or more of the components are significant, mixed taste will result. If at equilibrium one of the components is removed from the system more rapidly than another, there is potential for an 'after-taste'.

2.5 CHEMICAL KINETICS

The rate (kinetics) of chemical reactions, also has application to taste. The course of events, over time, determines whether a reaction is mono-, bi-, or of higher molecular order (Frost & Pearson, 1961). An indication of chemical activity can be obtained from calculated rate (velocity) and equilibrium constants. From a kinetic point of view, chemical reactions are classified as zero-, first-, and higher-order reactions.

2.5.1 Zero-Order Reactions

If a reaction proceeds at a constant rate, with time, it is a zero-order reaction. The zero indicates that the rate is independent of the concentration of one of the reactants, which is usually present in great excess. These are the recognition reaction conditions established for adaptation studies, although actual adaptation kinetics are first order (Section 1.3.2). An example of zero-order kinetics, taken from enzymology, is the rate with which the sugar glucose is produced from a large excess of starch by action of the enzyme glucoamylase. The rate of the reaction depends only on the concentration of the enzyme. The amount of substrate transformed with time, per unit of enzyme, is a constant. Such plots use only arithmetic coordinates, and are employed to assign activity coefficients to enzyme isolates or preparations.

2.5.2 First-Order Reactions

First-order (mono- or unimolecular) reactions proceed at a rate proportionate to the concentration of *one* of the reactants, and are 'first-order with respect to it', e.g. A or B. The simplest form of a first-order reaction is A → B. Compound A is then transformed to B at a rate proportionate to a given amount (percentage) of A being transformed per unit time. The change (d) in the concentration [A] of A at time (t) is contained in the expression

$$-\frac{d[A]}{dt} = K[A] \tag{2.9}$$

where K is a constant. The integrated form of eqn (2.9) is

$$\log \frac{[A_0]}{[A]} = \frac{K_t}{2{\cdot}3026} \tag{2.10}$$

where $[A_0]$ is the molar concentration of A at zero time and [A] is the concentration of A at time t. This form is useful for solving kinetic data graphically. In the age of the ubiquitous computer, it is just as easy to derive the natural logarithm of $[A_0]$ minus [A] and calculate the regression equation for the effect of time. The slope of the resulting equation, which is linear for first-order data, is then the rate constant K in

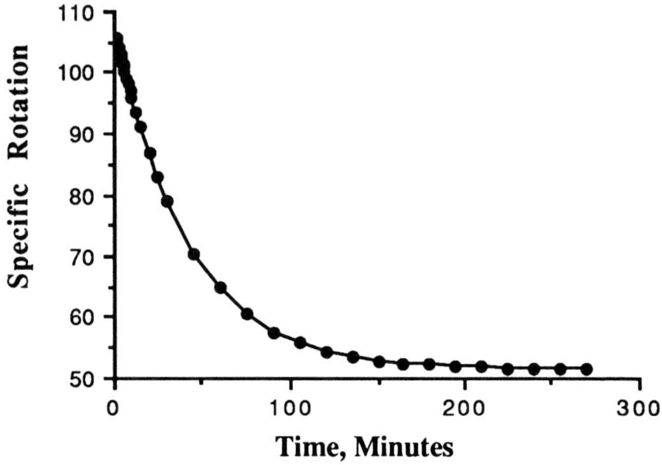

FIG. 2.2. Change (mutarotation) in the specific rotation of α-D-glucopyranose after dissolving the crystalline form of the sugar in water at 23°C.

either the initial equation, or in the exponential form of the initial equation

$$[A] = [A_0]_e^{-Kt} \tag{2.11}$$

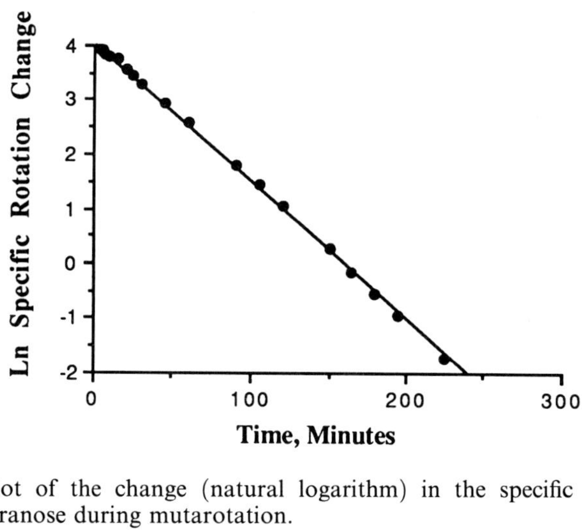

FIG. 2.3. Plot of the change (natural logarithm) in the specific rotation of α-D-glucopyranose during mutarotation.

2.5.3 Calculation of First-Order Velocity Constants

An example of the arithmetic nature of a set of first-order kinetic data, using the change in optical rotation upon dissolving crystalline α-D-glucose in water is shown in Fig. 2.2. Beginning with an initial specific optical rotation of $+112°$, the rotation changes, over a period of time, until equilibrium is attained at $+52{\cdot}5°$. The plot of the logarithm of the specific rotation change for α-D-glucopyranose, with time, is shown in Fig. 2.3. The plot confirms that mutarotation of α-D-glucopyranose is on the order of 'A reversibly goes to B'. In this case, **B** is the isomer β-D-glucopyranose, and the specific rotation is at equilibrium at $+52{\cdot}5°$

$$\mathrm{A} \underset{k_2}{\overset{k_1}{\rightleftharpoons}} \mathrm{B}$$

The regression equation for the plot shown in Fig. 2.3 is $y = 4{\cdot}0863 - 0{\cdot}0266x$. The slope of the equation $(-0{\cdot}0266)$ is the velocity constant K for eqn (2.10) and is also the exponent K for eqn (2.11). However, the reaction is reversible, and the velocity constant K is really the sum of the specific rate constants $k_1 + k_2$ for the forward and reverse reactions at 23°. Together, the two sugars make up $> 99\%$ of the equilibrium mixture.

The velocity constants k_1 and k_2, obtained from the natural logarithmic form of the first-order equation for α-D- and β-D-glucose mutarotations at 20°, are 0·0092 and 0·0053, respectively (Bates *et al.*,

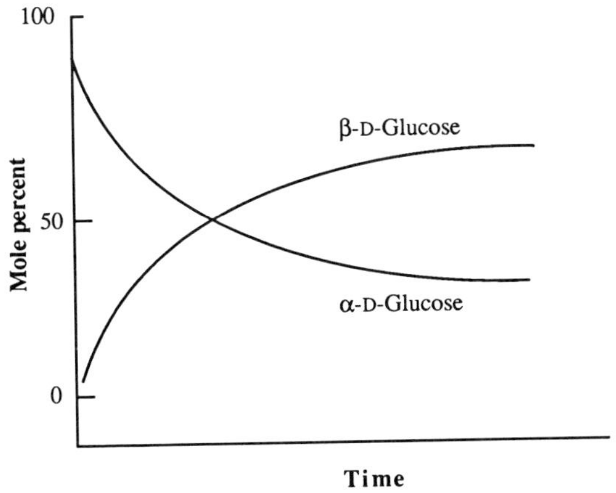

FIG. 2.4. Mass action equilibrium plot for the mutarotation of either α-D-glycopyranose or β-D-glucopyranose.

1942). The position of the equilibrium is obtained from $0{\cdot}0092 \div 0{\cdot}0092 + 0{\cdot}0053 = 63{\cdot}5\%$ β-D-glucopyranose and $0{\cdot}0053 \div 0{\cdot}0092 + 0{\cdot}0053 = 36{\cdot}5\%$ α-D-glucopyranose. The course of first-order kinetic events that characterizes the glucose mutarotation reaction in terms of mass action is shown in diagrammatic form in Fig. 2.4.

2.5.4 Calculation of First-Order Equilibrium Constants

Equilibrium constants, using optical rotation as a measure of concentration, can also be calculated from kinetic data. The significance of this sample calculation is that equilibrium constants can also be calculated using sensory intensity data as an index of concentration. Care is needed to ensure that the sensory data employed are a linear function of the stimulus concentration.

The equilibrium constant k_1/k_2 for the simple first-order glucose mutarotation reaction is calculated from the expression

$$\frac{k_1}{k_2} = \frac{r_\alpha - r_\infty}{r_\infty - r_\beta}$$

where r_α = the initial specific rotation of α-D-glucopyranose (+112°), r_β is the initial specific rotation of β-D-glucopyranose (+19°), and r_∞ is the equilibrium specific rotation (+52°). The ratio k_1/k_2 is then $112 - 52{\cdot}5/52{\cdot}5 - 19 = 59{\cdot}5/33{\cdot}5$, or 1·78. The percentage β-anomer is $59{\cdot}5/93{\cdot}0 = 64\%$, in good agreement with the result calculated from the reaction velocity constants. Thus, both rate constants and equilibrium constants from mass action data can be employed to estimate the concentration of A and B in simple equilibrium phenomena.

In certain cases it may only be necessary to plot the relation between two variables, such as sensory magnitude (as an estimate of concentration) versus time, on semi-logarithmic coordinates to ascertain if a relation is first-order. There are many reactions that are first-order with respect to one of the reactants, but not the others. There are also cases where the concentration of one of the reactants is so high the decrease in concentration is nearly immeasurable, and the reaction appears to be first-order, although it is second-order, or bimolecular. This is the case for the role of water in the hydrolysis of sucrose (Section 5.3.4). If the reaction is truly of measurable second-order nature, the rate is proportional to the concentration of two reactants. For the reaction $A + B \rightarrow C$, the plot of $1/C$ versus time is linear.

2.6 THERMODYNAMICS

A source of energy is needed for the tastant : receptor (T : R) recognition interaction to occur, but it does not have to be large. The ability of NaCl to dissolve in water is due to a chemical reaction, even though the starting materials, as in taste phenomena, can be recovered unchanged. For the solvation of sodium chloride, the source of energy required to disrupt the crystalline salt lattice structure is derived from the structure of water. An increase in entropy occurs, and as a small amount of heat is evolved during the solvation of the salt, the reaction is weakly exothermic. Presumably, the initial reactions of taste chemistry are not very different energetically, from the solvation of substances, as water solubility is prerequisite for taste.

Although the *first law of thermodynamics* states that energy can neither be created nor destroyed, energy can be transformed. The energy consumed by interaction of a substance with the taste receptor is accompanied by energy liberation. The increase in energy (ΔE) must equal the heat absorbed by the system minus the work performed

$$\Delta E = \text{Heat} - \text{Work} \tag{2.12}$$

For taste, the work performed by the system is transmission of a nerve impulse, followed by dissociation of the complex.

The *second law of thermodynamics* states that conversion of heat into work is accompanied by change in the temperature of the system

$$\Delta H = \Delta G + T\Delta S \tag{2.13}$$

In eqn (2.13), H is the enthalpy (the total energy of the system), G is the Gibbs' free energy (often 'F'), or the portion of the total energy that a system can use to perform work, T is the absolute temperature, and S is the entropy. It is not possible to determine the absolute values for a given system, but it is possible to determine their change (Δ).

2.6.1 Enthalpy

When a taste interaction is initiated, it occurs at constant pressure (P), and if there is a change in the volume (ΔV) of the system, there will be work done equal to $P\Delta V$. Enthalpy (H) is the thermodynamic property that simultaneously takes into account the energy of the system and the work done on the immediate surroundings, or $\Delta H = \Delta E + \Delta(PV)$. Thus, the increase in enthalpy of a system is equal to the increase in total energy plus the increase in the product $P \times V$. As taste occurs at constant pressure, $\Delta(PV)$ is $P\Delta V$, and the work done on constant pressure surroundings, by a change in volume, ΔV is

$$\Delta H = \Delta E + P\Delta V \tag{2.14}$$

The above equation is equal to $\Delta E = \Delta H - P\Delta V$, for comparison to the statement of the first law. Thus, if the only work done by the system is $P\Delta V$, then at constant pressure, $\Delta H =$ the heat absorbed. Put another way, when heat is added to a system at constant pressure, some of the heat is used to increase the internal energy of the system, and the rest does the work. Any physical system leans toward a state of lower energy. In addition, any system leans, at the same time, to a state of higher entropy. The two forces oppose each other, and circumstances determine the direction of the final state.

2.6.2 Entropy

The second law of thermodynamics can be restated to say that any system, left alone, will change toward a state of higher disorder. The term entropy (S) is used to describe relatively disordered states of matter as opposed to an ordered state, as found in a crystalline lattice structure. A liquid has a higher state of atomic or molecular disorder than a crystal, and possesses higher entropy. While a physical system leans toward a state of lower energy, probability favors higher disorder (higher entropy). This interplay is best described using the notion of *free energy*.

Total energy (E) is composed of two parts. The first is the aforementioned free energy, or the portion of the total energy available to do work. The second is the unavailable energy. For a mole of material the unavailable energy is TS. It is made up of the entropy (S) times a proportionality factor that governs the weighting to be assigned to the entropy expression, which is the absolute temperature (T). The higher the temperature, the more the disorder, and the less available the total energy becomes. Thus, the free energy term above (G) is related to total energy and entropy as

$$\text{Total energy} = G + TS \tag{2.15}$$

As the total energy is equal to the internal energy plus work (i.e. enthalpy or H), we have

$$H\,(\text{enthalpy}) = E\,(\text{internal energy}) + PV\,(\text{external work}) \tag{2.16}$$

For free energy, the following expression holds

$$G\,(\text{Free energy}) = H\,(\text{enthalpy}) - TS\,(\text{unavailable energy}) \qquad (2.17)$$

2.6.3 Significance of Free Energy in Taste

The salient feature of the change in free energy (ΔG or ΔF) for taste, and other chemical reactions, is that it is the touchstone by which it can be determined if a reaction will proceed spontaneously, and if so, in which direction. Otherwise, it will remain at equilibrium. In an equilibrium system, when no external work is involved, the free energy of a reacting system does not change, i.e. $\Delta G = 0$. In taste, of course, electrical work is done.

For any reaction that is spontaneous, the free energy always decreases, and ΔG is < 0. When the reaction reaches equilibrium $\Delta G = 0$. For the decomposition of products toward an equilibrium, ΔG is positive. Because the equilibrium constant (K) for taste (e.g. the sweetness of sugars) is of the order of 0·001 M, and ΔG is positive, ΔG is taken to be the dissociation constant for the sweetness phenomenon (Lancet & Ben-Arie, 1991).

K	ΔG
0·001	4·077
0·010	2·720
0·100	1·357
1·0	0·0
10·0	−1·357
100·0	−2·720
1000·0	−4·077

ΔG is determined from the relation

$$\Delta G = -RT \ln K \qquad (2.18)$$

where R is the standard gas constant PV/T (0·082057), but expressed as calories (1·98 cal/mol per k). Molar K values are interpreted to be *activity* or *association* constants (K_a). For taste chemistry purposes, their reciprocals (K_d, or dissociation constants) are often used.

The approximate relation between ΔG and molar equilibrium constants at 25°C are tabulated above.

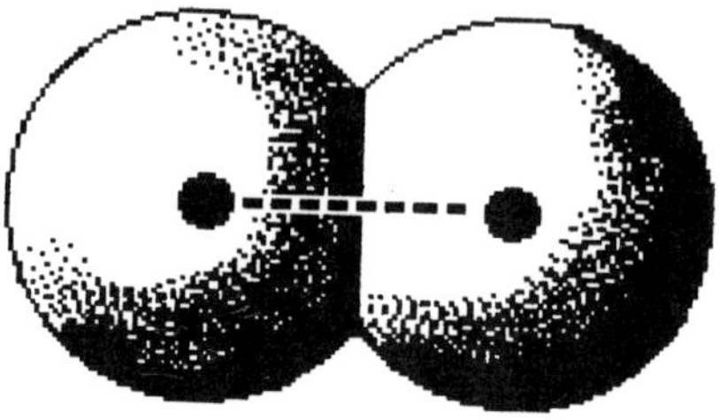

FIG. 2.5. Diagrammatic depiction of the linear single covalent bond between two hydrogen atoms.

2.7 NATURE OF CHEMICAL BONDS AND CHEMICAL REACTIONS

The specific chemical nature of the receptor(s) for taste is not precisely known. However, the number of weak chemical interactions that may occur with it are limited, and can be described in affinity terms, although recognition activity rather than affinity, is probably operative in taste induction chemistry. When two atoms come together, the electrons of one atom influence the electrons and the nucleus of the other, and a 'bond' may be said to form between them leading to a decrease in free energy. The interaction may entail the gain, loss, or sharing of electrons. Chemical bonds have varying strengths, and there are large extremes for a given bond type.

2.7.1 The Covalent Bond

A covalent bond involves sharing of electrons. Sharing of electrons, as in the hydrogen molecule ($H^{\cdot} + {}^{\cdot}H = H : H$) creates an electron pair, which generates a single covalent bond (H–H), as shown diagrammatically in Fig. 2.5.

The covalent bond is highly directional, which in turn governs the shape of the molecule. The hydrogen molecule, which contains only two atoms, is necessarily linear, but the water molecule (H–O–H) is not. Electron sharing by the oxygen atom with two hydrogen atoms follows the direction of two tetrahedrally inclined $2p$ orbitals that form the single covalent bond. Subdivisions of the covalent bond depend upon whether the bond is between two like or unlike atoms. In the former case, the bond is *nonpolar*, and the electric charge distribution between the two atoms, as in gaseous hydrogen, is equally distributed. In water, the bond is between two unlike atoms, and the charge is not equally distributed, resulting in separate positive and negative charges, or a dipole moment. The strength of single covalent bonds is quite high (40–110 kcal mol^{-1}).

2.7.2 The Ionic (Electrostatic) Bond

An ionic bond is formed when an electron is transferred from atom A to B ($A - e \rightarrow B + e$). The attraction between them yields compound (A^+B^-). The attraction is Coulombic, and is called an ionic bond or electrovalent bond. Unlike the covalent bond, it is not highly directional. The Group I elements (Li, Na, K, etc.) or Group II elements (Ca, Mg, etc.) easily give up an electron from the outer valence shell to an element with high electron affinity (paucity of electrons in the outer valence shell). The latter are characterized by the group VII elements (F, Cl, I, etc.), and react with Group I and II elements to form salts. The ionic bond is not particularly strong. As the free energy (ΔG) is only about 2–3 kcal/mol, and ionic bonds are severed upon dissolving the salt in water because hydration of the ions result in an increase in entropy.

Ultimately, there is no sharp distinction between a covalent and an ionic bond, except perhaps relative strength. If the cation and anion have about the same ability to attract electrons, the bond will be nearly nonpolar. If the anion has somewhat greater ability to attract electrons than the cation, the shared electrons will spend more time on the cation. If the electron attracting ability of the anion is much greater than the cation, the electron pair will not be shared at all, and the bond will be purely ionic. This is the case for crystalline NaCl. The distinction has relevance to taste quality, especially the sweetness, bitterness, and the saltiness intensity of ionic substances. For a compound to elicit taste, a number of weak chemical interactions are candidates for the recognition chemistry of the various tastes. In addition to the ionic bond, hydrogen bonds, the hydrophobic bond, van der Waals interactions, and charge transfer interactions may be responsible for the different tastes.

2.7.3 The Hydrogen Bond

Although the hydrogen atom has only one orbiting electron, the anomalous behavior of water requires that the hydrogen atoms of water must be capable of simultaneously participating in two bonds. The 'second' bond that the single orbiting electron can participate in is thought to be largely ionic in character, while the first is largely covalent.

2.7.3.1 *Nature of the Hydrogen Bond*

For a hydrogen bond to occur, a hydrogen atom (H) must be bonded covalently to an electronegative atom (A), such as oxygen or nitrogen. A second electronegative atom (B), must also be close to the first electronegative atom. When this occurs, because the coordination number of hydrogen is two, the hydrogen atom then becomes 'bonded' to the second electronegative atom also.

$$A\text{–}H \rightarrow B$$

Statistically, the single electron of hydrogen spends a portion of the time in partial ionic and even covalent transfer to the second electronegative atom, and 'bonds' to it. Pauling (1960) envisaged the hydrogen bond in ice to involve resonance between structures a through c.

(a) $O\text{–}H : O$ (61%)

(b) $O\text{: }H^{+\,-}\text{:}O$ (34%)

(c) $O\text{: }H \rightarrow O$ (5%)

In the case of hydrogen bonds between two water molecules, the distance between the two, linearly bonded, oxygen atoms is 2·76 Å.

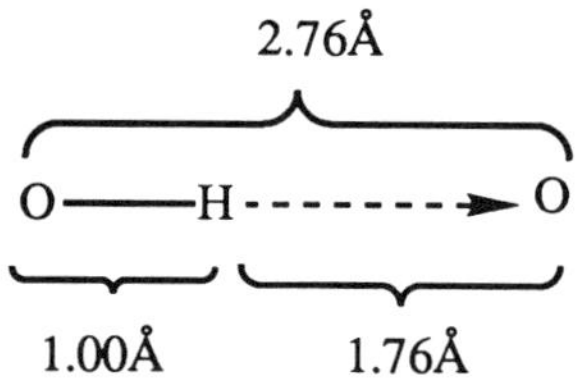

The proton is 1·00 Å from the oxygen atom to which it is covalently bonded, and 1·76 Å from the other. The character of the hydrogen bond

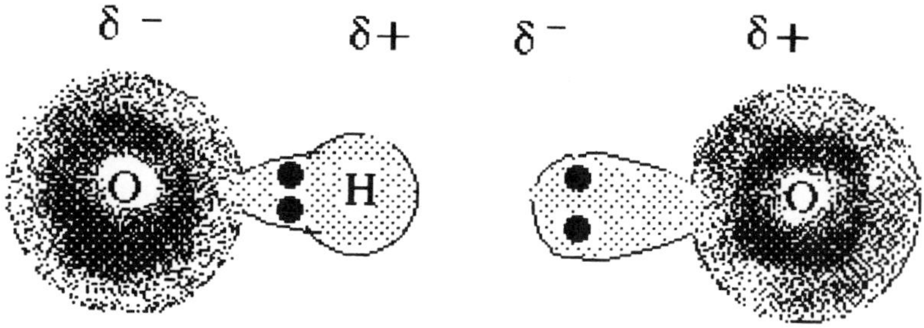

FIG. 2.6. Diagrammatic depiction of the hydrogen bond as a weak covalent attraction between two dipoles.

is believed to be largely ionic. This is because the 1*s* orbital of the hydrogen atom is liberated, to an extent of about 39% (34% ionic bonding and 5% secondary covalent bonding), from participation in covalent bond formation.

Another view is that the bare nucleus (proton) of the covalently bonded hydrogen atom is *delocalized*, and attracted to the electron pair of the second electronegative atom. It is then available for formation of a fractional dipole/dipole covalent bond with the more distant oxygen atom (Fig. 2.6). Jeffrey & Saenger (1991) view hydrogen bonds as *soft bonds* that are very different from covalent bonds in the sense that they are easily deformed by other intermolecular interactions. The stretching and bending force constants are only about 1/15th that for covalent bonds. Hydrogen bonds also display group-pair properties. Thus, the properties of a hydrogen bond depend not only on the first neighbor atoms of both A and B, but also on the sequential nature of the total pattern of bonding.

2.7.3.2 *Strength of the Hydrogen Bond*

The hydrogen bond is not particularly strong (about 3·4 kcal mol^{-1} for liquid water), but an abundance of them markedly alters the properties of substances. An example is water that, with a molecular formula of H_2O ought to be a gas. Hydrogen bonds are generally presumed to be linear, for example, O–H → O, but considerable deviation from linearity (up to 30%) is the more usual case (Donohue, 1968). The strength of the bond varies inversely with the distance between the electronegative atoms, and because nonlinear bonds are longer, they are weaker (Pimentel & McClellan, 1960). The following descending order for the strength of hydrogen bond interactions was offered by Hamilton & Ibers (1968), and is based on observed A to B distances.

O–H → O (2·7 Å)

O–H → N (2·8 Å)

N–H → O (2·9 Å)

N–H → N (3·1 Å)

C–H → O (3·2 Å)

Such criteria have been criticized (Jeffrey & Saenger, 1991) because the A–B separation is a function of the covalent (A–H) bond length, the hydrogen bond length (H· · ·B), and the angle for A–H to B. The distance

TABLE 2.1.
Hydrogen bond distances versus van der Waals radii (Jeffrey & Saenger, 1991).

Type of bond	*H···B distances*	*Sum of van der Waals radii*
O–H···O	1·44–2·10	2·33
O–H···Cl	1·95–2·33	2·63
N–H···O=C	1·58–2·05	2·32
N–H···O	1·60–2·40	2·95
N–H···Cl	2·10–2·20	2·62
C–H···O	2·04–2·65	2·30

between neighboring atoms of the donor and acceptor groups must be substantially less than the sum of their van der Waals radii. Therefore, Jeffrey & Saenger proposed characteristics for hydrogen bonds that indicate considerable overlap in relative strength. These are shown in Table 2.1.

2.7.3.3 *Inter- and Intramolecular Hydrogen Bonds*

The prevailing tenet in the 1960s held that any propensity for intramolecular hydrogen bonding in a substance would be quenched in a polar environment, such as water. One problem with that tenet was that certain polar groups in the structure of substances were intrinsically subject to a hydrophobic environment. A hydroxyl group of a $-CH_2OH$ group in the structure of the sugars is an example. It seemed reasonable to assume that they, at least, would exhibit some intramolecular hydrogen bonding, even in water. Present thinking (cf. Jeffrey & Saenger, 1991) is that of competition, and a net hydrogen bonding energy for the intra- and intermolecular force. For polar, uncharged groups in amino acids and peptides (the peptide bond, tyrosine, serine and threonine –OH groups, and the amide groups of glutamine and asparagine), partitioning studies indicate a hydrogen bonding energy of 1·0–1·4 kcal mol^{-1}. For charged groups (the imido and NH groups of histidine, the carboxyl anion of aspartic and glutamic acids, and the ammonium cation of lysine), it is 1·5–2·8 kcal mol^{-1}.

2.7.3.4 *Cooperative Hydrogen Bonding*

Hydrogen bonds can form continuous chain-like structures, and the binding energy is greater than the sum of the individual bonds. They are presumed to prevail, not only in crystalline structures, but also in water solution. Jeffrey & Saenger recognized two types of such cooperative bonding. The first, applicable to water and carbohydrates, is cooperative σ bonding in those cases where neighboring functional groups, e.g., –OH groups, have donor *and* acceptor capability. Here, polarization occurs through hydroxyl σ bonds.

$$\overset{\delta^-}{\text{O}}\text{—}\overset{\delta^+}{\text{H}}\cdots\overset{\delta^-}{\text{O}}\text{—}\overset{\delta^+}{\text{H}}\cdots\overset{\delta^-}{\text{O}}\text{—}\overset{\delta^+}{\text{H}}\cdots\overset{\delta^-}{\text{O}}\text{—}\overset{\delta^+}{\text{H}}$$

For functional groups that operate as donors only, such as an imino or amino hydrogen atoms, cooperation requires that adjacent hydrogen bonding be linked by bonds with π-electron character.

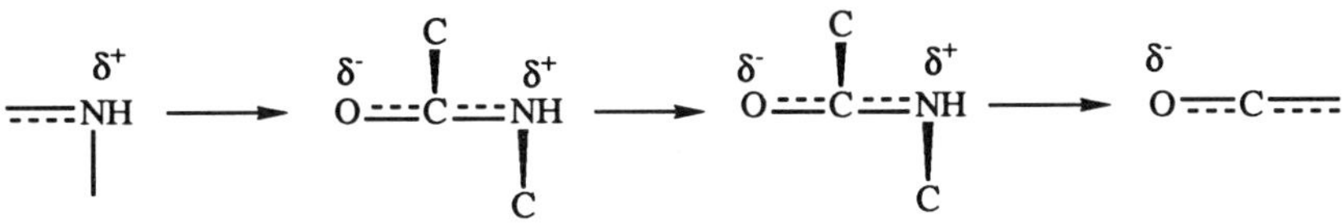

The π-cooperation is important in protein hydrogen bonding as it increases double-bond character, and thus the rigidity of the peptide C–N bond.

2.7.4 The Hydrophobic Bond

The notion of the hydrophobic bond originated with Némethy (1967). When an apolar group, such as a methane (CH_3–R), methylene (RCH_2R'), or methine ($RR'CHR''$), is in a polar environment, the apolar material generates increased order in the polar solvent. Other apolar substituents (e.g., the phenyl radical) or an apolar substance (hydrocarbon) can be expected to behave similarly. The water molecules arrange around the nonpolar solute to form a more extensive number of hydrogen bonds. Such ordering of water molecules is entropically unfavorable, and compensation arises when more than one such apolar group is present. The two apolar groups aggregate, and the ordered water molecules are released. The concomitant decrease in free energy stabilizes the interaction between the two nonpolar groups (substituents, substances). The bonding is probably through van der Waals forces. The substances are attracted to each other, and form a hydrophobic bond (about $-1\,\text{kcal}\,\text{mol}^{-1}$). The hydrophobic interaction between two methylene groups in a polar environment is estimated to be of the order of $-0{\cdot}7$ kcal/mol. The hydrophobic effect can be measured (Jeffrey & Saenger, 1991), by the differential solubility of a substance in an apolar and polar solvent. Here, the data can be directly associated with the

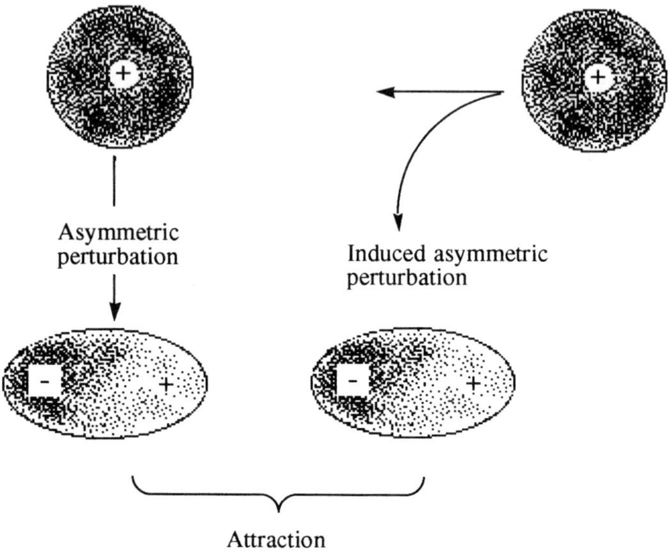

FIG. 2.7. Asymmetric perturbation of a nonpolar atom or molecule, and the induced perturbation and attraction to a second nonpolar atom or molecule.

notion of ordered water molecules around a nonpolar solute, and the loss in energy is related to the surface area of the nonpolar solute.

2.7.5 van der Waals Interactions

There are two van der Waals interactions for either atoms or molecules, one repulsive and the other attractive ($-0{\cdot}5$ to 1 kcal mol^{-1}). At distances greater than about 10 Å, there is little interaction (Fig. 2.7). Probability, with respect to distribution of electrons, affects their behavior. On occasion, electrons about the nucleus of an atom, or an apolar molecule, become asymmetrically dispersed for a fraction of a second. This leads to a slight preponderance of electrons on one side of the atom or molecule, to generate a dipole. As atoms or molecules move closer, the momentary dipole induces a neighboring atom or molecule to assume a dipolar configuration. At a distance of about 5 Å, the two are attracted to each other. The attraction is known as a *dispersion bond.* Although the bonds are relatively weak, they are of sufficient number to hold nonpolar materials together in liquid form. At closer distance (4 Å), the atoms and molecules then begin to repel each other.

The aforementioned example is also classified as an *induced dipole–*

dipole interaction. The magnitude of the interaction depends on the size of the instantaneous dipole of the first atom or molecule, and on the polarizability of the second. An approximate mathematical model, known as the London formula, for the two induced dipoles (Atkins, 1978) is

$$V(R) = \frac{(-3I_1I_2)}{(2I_1 + I_2)} \times \frac{(\alpha_1 + \alpha_2)}{R^6} \qquad (2.19)$$

where I_1 and I_2 are the polarization potentials of the two atoms or molecules, and R is their separation in nm. From the equation, it is estimated that the dispersion energy for two methane molecules is about 1·96 kcal mol^{-1}. Because the attraction varies with the inverse sixth power of R, these forces die off rapidly with increasing distance.

Such an interaction resembles the hydrophobic bond, but the interaction described does not necessarily occur in a polar environment. Salem (1964) takes the view that dispersion forces are always attractive for molecules in their ground states, exist among all groups, and are especially important between nonpolar groups. The hydrophobic bond is the result of the large dispersion attraction between nonpolar groups, and of the large electrostatic attraction between water molecules, together with the weakness of polarization forces.

The London formula also applies to interactions between polarizable molecules (*dipole–dipole interactions*), which are highly directional. As long as the dipole charges are of opposite sign, and properly aligned, the dipoles can be attracted to other dipoles. The attractive forces between two dipoles that are aligned head to tail would seem to be cancelled by the repulsive forces due to dipoles that are aligned head to head or tail to tail (Fig. 2.8). There is a tendency for one dipole to align the other into an

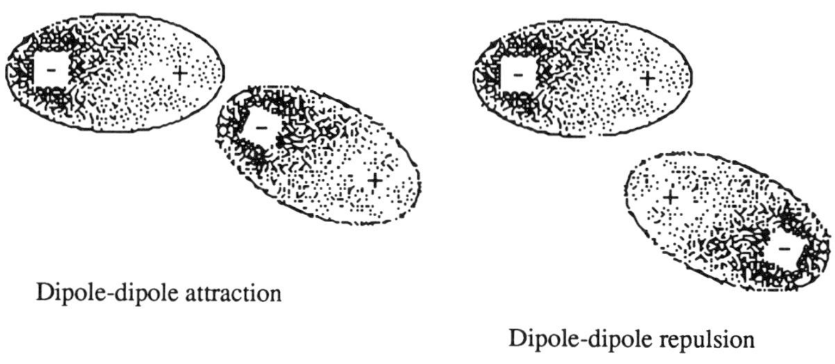

FIG 2.8. Dipole–dipole interaction and repulsion.

arrangement favorable for interaction. The magnitude of the interaction between two dipole moments is about $-0{\cdot}33\,\mathrm{kcal\,mol^{-1}}$ (Atkins, 1978).

For a dipole–induced dipole interaction the magnitude of the effect again depends on the size of the permanent dipole moment, and the polarizability of the second. For HCl in the presence of benzene the magnitude of the interaction energy at a distance of 3 Å (R) is $0{\cdot}33\,\mathrm{kcal\,mol^{-1}}$.

2.7.6 Charge Transfer Reactions

Charge transfer interactions (-1 to $-7\,\mathrm{kcal\,mol^{-1}}$) can be regarded as an interaction between ionic pairs. Donors are rich in π electrons, such as alkenes and aromatic compounds, and have electron donating substituents, such as a hydroxyl or ether group. Weak electron donors are groups that simply have a pair of unshared electrons, such as a carbonyl oxygen atom, a sulfur atom, and a hydroxyl oxygen atom (Baker, 1967). Acceptors are deficient in π electrons, and strongly withdraw them. They can also be regarded as π acids. Acceptors are also the weakly acidic hydrogens of groupings such as chloroform, alkyl-OH, or imidazole-H, the very groups that act as proton donors in hydrogen-bonding systems. There is considerable overlap of characteristics for the various subdivisions of weakly attractive interactions. The hydrogen bond, while largely ionic in character, is also an example of a form of charge transfer phenomenon.

2.7.7 Displacement Reactions

Displacement reactions are bimolecular reactions. The rate of reaction depends on the concentration of two species, where a molecule of one species *collides* with one molecule of another. Displacement reactions are quite general, and occur with a wide range of anions and neutral molecules. The single requirement is that the attacking group must be an atom with a pair of electrons available for bonding (Lewis base). Thus (Streitwieser & Heathcock, 1976), the base is a conjugate base of an acid, and it may be either neutral, or charged.

$$\mathrm{BH} \overset{k_a}{\rightleftharpoons} \mathrm{B{:}^-} + \mathrm{H^+}$$

$$\mathrm{BH^+} \overset{k_a}{\rightleftharpoons} \mathrm{B{:}} + \mathrm{H^+}$$

Conjugate acid Conjugate base

The affinity of a base for a proton is measured by the pK_a of the conjugate acid in water. A weak conjugate acid (more positive pK_a) corresponds to a strong base. A strong conjugate acid (less positive pK_a) corresponds to a weak base. Some pertinent bases and their pK_a are:

Base	*Conjugate acid*	*pK_a*
HO^-	H_2O	15·7
CH_3COO^-	CH_3COOH	4·8
H_2O	H_3O^+	−1·7
Cl^-	HCl	−7
H_2N^-	H_3N	35

The basicity of a base depends on the strength of its bond to a proton in solution. In the displacement reaction the base forms a bond to another atom. The classic displacement atom is carbon, and an example is:

$$HO^- + CH_3Cl \rightarrow CH_3OH + Cl^-$$

Many reactions can be formulated as displacement reactions at atoms other than carbon. Even usual acid-base reactions, such as between water and HCl, can be considered as a displacement on *hydrogen.*

These reactions are extremely rapid, and are considered to be facile equilibria.

It is also common to refer to *leaving groups* in displacement reactions. Any group that is the conjugate base of a strong acid can act as a leaving group, and the facility with which a group functions in this capacity is related to its basicity. A group that is a weak base (conjugate base of a strong acid) will generally be a good leaving group.

2.8 INDUCTIVE AND RESONANCE EFFECTS

Inductive and resonance principles play an important role in taste chemistry. Many organic molecules possess positive and negative electrical charges, leading to a dipole moment if the charges are at different positions in the group, or the molecule. This in turn leads to permanent electron displacement, such as in a pair of unshared electrons of a hydroxyl group, an amino group, or a halogen atom.

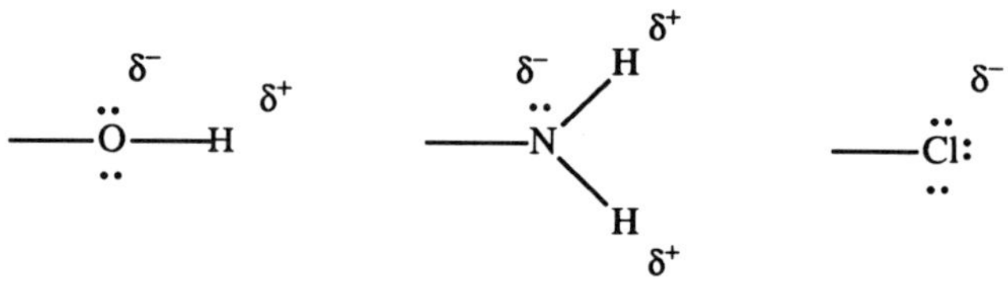

Some of the significant directions and the dipole moment (Debye units) of groups found in tastants are as follows (Fieser & Fieser, 1961)

C–H ($C^+ \rightarrow$ H, 0·4)	H–N ($H^+ \rightarrow$ N, 1·31)
C–N ($C^+ \rightarrow$ N, 1·25)	H–O ($H^+ \rightarrow$ O, 1·51)
C–O ($C^+ \rightarrow$ O, 1·6)	H–S ($H^+ \rightarrow$ S, 0·68)
C–Cl ($C^+ \rightarrow$ Cl, 2·3)	H–Cl ($H^+ \rightarrow$ Cl, 1·08)
C–Br ($C^+ \rightarrow$ Br, 2·2)	H–Br ($H^+ \rightarrow$ Br, 0·78)
C–I ($C^+ \rightarrow$ I, 2·0)	H–I ($H^+ \rightarrow$ I, 0·38)

For the C–H bond, unsaturation may reverse the direction of the dipole, e.g. acetylene. For tastants, certain groups appended to a molecule lead to taste potential, alteration of taste potency, or even alteration of taste *per se*. This is because of their inductive effect on the molecule, or the influence on other groups in the molecule. Groups that attract electrons more strongly than hydrogen (and have negative inductive effects) are NH_3^+, NO_2, CN, COOH, COOR, CH=O, F, Cl, Br, OH, OR, and SH. Groups that attract electrons less strongly than hydrogen (and have positive inductive effects) are CH_3, CH_2R, and COO^-.

The notion of resonance applies to structures that have comparable energy, and interconvert by redistribution of electrons of unsaturated or ionized centers (Fieser & Fieser, 1961). The structure of the resonating acetic acid anion is:

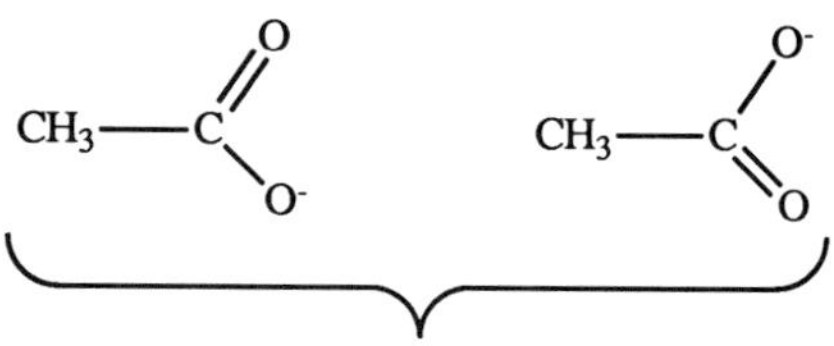

Hybrid anion

Resonance may also be conjugative, resulting from the delocalization and mobility of π electrons in groups that have conjugated double bonds. Some groups increase electron density in conjugated systems (Korolkovas, 1970), and have +R character, and also negative inductive effect. Examples are halogen atoms: $-OH$; $-OR$; $-OOCR$; $-SH$; $-NH_2$, and the amide group ($-NH-COR$). Some that decrease electron density, but also have a negative inductive effect are: $-NO_2$; $-CN$; $-CHO$; $-COOH$; $-CONH_2$ and $-CF_3$. Among those groups that increase electron density, and have positive inductive effects are $-O^-$; $-S^-$; and $-CH_3$.

An $-OH$ group will increase the electron density of an unsaturated system, such as benzene, and the influence is reciprocal. Here, the $-OH$ group is rendered more acidic because the electron density about the oxygen atom is so lowered that it cannot hold the positive hydrogen atom as well. Similarly, an amino group is rendered less basic. According to Fuson (1950), effects are greatest when a group with high electron-releasing power and one of low electron-releasing power occur together. In this case the two tend to neutralize each other. An example pertinent to taste chemistry is the amino group in an amide. Here, the amino group has lost the greater part of its basicity in releasing electrons to the carbonyl group, which is then deprived of its ability to act in a manner typical of a carbonyl group. With particular significance to taste chemistry, inductive effects are known to operate over rather large

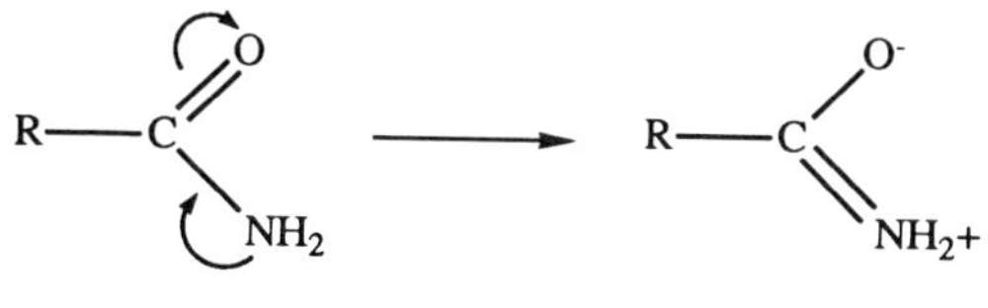

distances. Substitution of benzoic acid esters with a dimethylamino group in the *para* position greatly reduces the rate with which the carboxyl group can be saponified. This example (Fuson, 1950) is shown below:

2.9 ISOSTERIC GROUPS

The term *isosteric* group originally referred to compounds or groups of atoms that have the same number and arrangement of electrons, such as N_2 and CO. In that sense they were truly isosteric (Gr. *isos*, same; *stereos*, solid, space). Subsequently, groups with only similar steric and electronic configurations came to be treated as isosteres, and the notion of *biosteres* was introduced to signify groups with similar biological activity. Some pharmaceutical substances (pharmacophores) have the same general (isosteric) activity (Korolkovas, 1970). Acetylcholine and carbachol have similar muscarinic (muscle smoothing) activity, but one possesses a terminal methyl group and the other a terminal amine group. Another name for such substances are *agonists*.

$$CH_3\text{—}CO\text{—}CH_2\text{—}CH_2\text{—}N(CH_3)_3$$

Acetylcholine

$$H_2N\text{—}CO\text{—}CH_2\text{—}CH_2\text{—}N(CH_3)_3$$

Carbachol

As used in taste chemistry, the term isostere implies that groups or atoms have similar taste functionality, or activity. Different compounds with sweet taste are sweet agonists, and their glycophores (Gr. *glyc*, or *gluc*, sweet; *phoros*, to carry), while chemically dissimilar, are *glycosteric* groups. Substances that elicit the salt taste possess isosteric halophores (Gr. *hals*, salt). Chemically different substances and groups that elicit the bitter taste possess isosteric picrophores (Gr. *picros*, bitter). Compounds that possess sour taste nearly always have a dissociable hydrogen atom. Sour taste may be invoked by a compound with such strong nucleophilic groups, that a proton may be extracted from the receptor for taste, and the sensation is described as sour. The hydrogen ion and the strong nucleophilic group may then be described as isosteric *acidophores*.

In terms of isosteric taste activity, both the $-NH_2$ and $-OH$ groups are isosteric proton donors, and the carbonyl group and an unsaturated center are isosteric proton acceptors. The sulfur atom can act in the same capacity as an oxygen atom as a proton acceptor. Groups that may be isosteric in one sense, can act in opposite fashion in another sense. The $-OH$ group can donate and also accept a proton in hydrogen bonding systems, but the NH_2 group can operate only in donor capacity. The appropriate term for the acceptor groups in this case is saporous *antisteres*. By the same token a sulfur atom may be isosteric with an oxygen atom in hydrogen bonding phenomena, and may also act isosterically as an inductive atom in heterocyclic rings, as there is little effect on taste quality (although taste potency may be altered). When a sulfur atom is substituted for a carbonyl oxygen atom, as in the sweet substance dulcin (Section 1.5.2), taste is transformed to bitter. In this latter sense, sulfur and oxygen are taste antisteres.

2.10 TASTE STRUCTURE-ACTIVITY THEORY

A general expression of the interaction of any biologically active substance with a receptor, which encompasses various applications of the law of mass action (e.g., the Langmuir adsorption isotherm) was proposed by Ariens *et al.* (1979). Applied to saporous substances, it is

$$\mathrm{T} + \mathrm{R} \underset{k_{-1}}{\overset{k_1}{\rightleftharpoons}} \mathrm{TR} \underset{k_{-2}}{\overset{k_2}{\rightleftharpoons}} \mathrm{TR}^* \underset{k_{-3}}{\overset{k_3}{\rightleftharpoons}} \mathrm{T} + \mathrm{R}^* \underset{k_{-4}}{\overset{k_4}{\rightleftharpoons}} \underline{\mathrm{R}} \underset{k_{-5}}{\overset{k_5}{\rightleftharpoons}} R \qquad (2.20)$$

where T = the tastant, R = the receptor, R^* = the activated receptor, and $\underline{\mathrm{R}}$ = the nonreceptive, nonactivated receptor. The $\underline{\mathrm{R}}$ term is employed in special theories to account for fade in dose response curves, and for specific desensitization of receptors. For subsequent application to different theories, the constants $K_1 - K_5$ are equilibrium constants that govern the relative amount of different receptor species. These are defined as $K_i = k_i/k_i$. Receptor species R, TR, and $\underline{\mathrm{R}}$ do not contribute to stimulus (taste) induction, but TR^* and R^* are effective.

2.10.1 Occupancy Theory

The occupancy theory of receptor–tastant interaction states that the intensity of a structure–biological interaction is directly proportional to the number of receptors that interact with the tastant. For the occupation theory of tastant–receptor interaction at equilibrium,

$$\frac{[\mathrm{TR}^*]}{[\mathrm{R}^*]} = K_3[\mathrm{T}] \qquad (2.21)$$

and $K_3[\mathrm{T}] \gg 1$. Upon application the tastant interacts with the receptor and activates it. The affinity is such that the total number of receptors activated by interaction with the tastant is greater than the total number of activated receptors, i.e. $TR^* > R^*$. The intensity of the stimulus is then proportional to the fraction of receptors present in the activated state.

$$SR(:)\frac{(TR^* + R^*)}{R} = \frac{(TR^*)}{R} \tag{2.22}$$

As a consequence (cf. Korolkovas, 1970), the sensory response SR (sweetness, sourness, etc.) will be proportionate to K_3,

$$SR = K_3[TR] \tag{2.23}$$

When the above notion of activated receptors (R^*) is put aside, in the state of equilibrium,

$$\frac{[T][R]}{[TR]} = \frac{k_{-1}}{k_1} = K_1 \tag{2.24}$$

where K_1 is the dissociation constant for the recognition interaction, the total number (t) of receptors is

$$[R_t] = [R] + [RT] \tag{2.25}$$

Therefore

$$K_1 = \frac{[R_t - RT][T]}{[RT]} \quad \text{or} \quad \frac{[RT]}{[R_t]} = \frac{[T]}{[T] + K_1} = \frac{1}{1 + (K_1/[T])} \tag{2.26}$$

When all receptors are 'occupied', the taste intensity will be maximal (m), or

$$SR_m = k_3[R_t] \tag{2.27}$$

By dividing eqn (2.23) by eqn (2.27) we obtain

$$\frac{SR}{SR_m} = \frac{[RT]}{[R_t]} \tag{2.28}$$

When eqn (2.28) is rewritten, the Michaelis–Menten expression of enzymic reaction is obtained.

$$SR = \frac{SR_m[T]}{[T] + K_1} \tag{2.29}$$

In sweetener research, the occupancy theory seems applicable provided all sweeteners have about the same intrinsic activity. Unless restricted by solubility, all sweeteners then can display the same maximum sweetness intensity response. These are logarithmic curves up to an end-point, or a limiting factor beyond which there is no further increases in perceived intensity. Figure 1.7 is an example of a general sensory response plotted against concentration, and DuBois *et al.* (1991) show curves for a number of different sweeteners. Presumably, the limiting factor is the number of receptors available for interaction with the tastant.

2.10.2 Hit–Run Theory

In the hit-activation (hit–run) theory of agonist–receptor interactions, $K_3[\mathrm{T}]$ of eqn (2.21) is much less than 1, and taste intensity is proportional to the fraction of the receptors present in the activated state,

$$\mathrm{S} : \frac{(\mathrm{TR}^* + \mathrm{R}^*)}{r} = \frac{\mathrm{R}^*}{r} \tag{2.30}$$

The sensory response is then proportionate to

$$\mathrm{SR} = K_{\mathrm{t}}\mathrm{R} \tag{2.31}$$

for comparison to eqn (2.23).

2.10.3 Rate Theory

In the rate theory of agonist (tastant) action, activation of the receptors is proportional to the total number of encounters per unit time, and not to the number of occupied receptors. The rate of association is expressed by the equation

$$\mathrm{A} = k_1[\mathrm{T}](1 - p) \tag{2.32}$$

where $[\mathrm{T}]$ is the tastant concentration and $(1 - p)$ is the fraction of free receptors (p is the fraction of occupied receptors). In this example, the rate of association and dissociation is the same,

$$\mathrm{A} = \frac{k_2}{1 + \frac{(k_1/k_2)}{[\mathrm{T}]}} \tag{2.33}$$

For saporous substances, rates of association and dissociation are rapid, but the evidence points to varied taste sensations resulting from unequal association and dissociation rates.

2.10.4 Inhibition

The Michaelis–Menten equation has special application to enzyme kinetic studies. While the expression 'affinity' is used to describe enzyme inhibition and other phenomena, the term has been changed to activity in this section, as the parameters refer to activity, i.e. a sensory response. When the equation is in concentration form (eqn (2.29)), it is suitable for application to taste phenomena, and both sides of the equation can be transformed to the reciprocal and rearranged.

$$\frac{1}{\mathrm{AR}} = \frac{K_m}{\mathrm{SR}_{max}} \frac{1}{[\mathrm{T}]} + \frac{1}{\mathrm{SR}_{max}} \tag{2.34}$$

This form of eqn (2.29) is now a concentration form of the Lineweaver–Burke equation. It has application to derivation of taste kinetics, through analogy to enzyme saturation phenomena, and is particularly useful for deducing the nature of tastant inhibition. A general plot of the equation is given in Figs 2.9–2.11, where it has a slope of K_m/SR_{max}. It also has an intercept of $1/[\mathrm{SR}]_{max}$ on the reciprocal sensory response coordinate, and is an accurate determination of $[\mathrm{T}]_{max}$. The intercept of $-1/K_m$ on the reciprocal tastant concentration coordinate serves as an activity constant. When the constant $-1/K_m$ is small, the activity of the tastant toward the receptor is high.

2.10.4.1 *Competitive Inhibition*

For competitive taste inhibition, the slope of the double reciprocal plot increases with increasing concentrations of inhibitor [I]. The intercept on

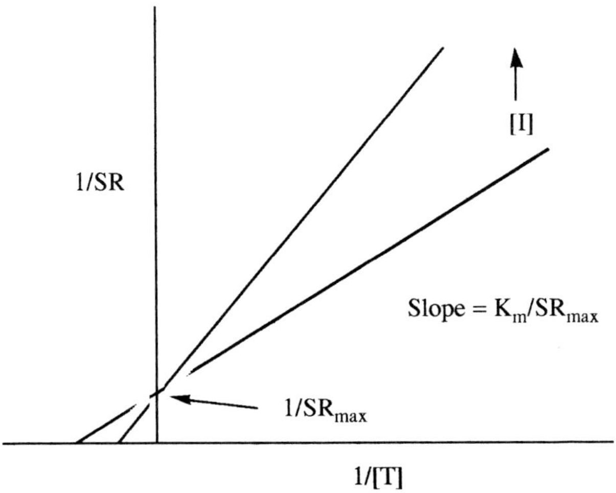

FIG. 2.9. Characteristic plot of the Lineweaver–Burke equation for competitive inhibition.

the abscissa for $-1/K_M$ also increases (Fig. 2.9), and the term $1/SR_{max}$ is not affected. The plot in Fig. 2.9 indicates that the inhibitor competes in some manner with the tastant for the receptor. The inhibition can be confirmed by increasing the concentration of tastant. Here, both the slope and the tastant concentration coordinate's intercept decrease.

2.10.4.2 *Noncompetitive Inhibition*

In noncompetitive inhibition, different concentrations of inhibitor change the slope of the reciprocal plots, but the intercepts for $1/K_m$ remain the same (Fig. 2.10). For enzyme chemistry, the inhibitor is thought to combine with the free enzyme at a site different from that for the enzyme–substrate interaction, and interferes with the enzyme action. Such interaction is reversible, and is not affected by substrate concentration. It depends solely on the tastant concentration.

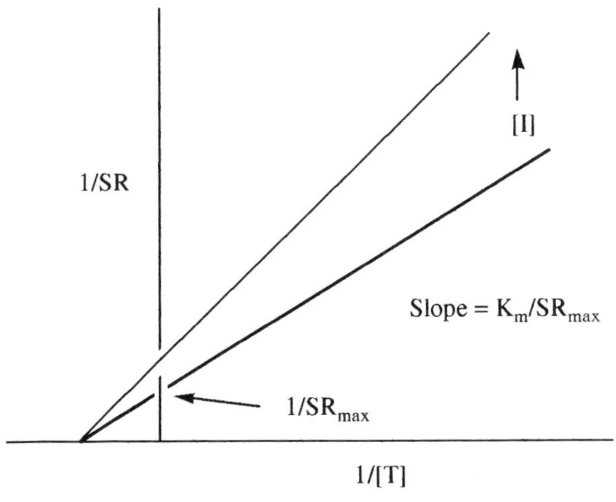

FIG. 2.10. Characteristic plot of the Lineweaver–Burke equation for noncompetitive inhibition.

2.10.4.3 *Uncompetitive Inhibition*

A characteristic plot of uncompetitive inhibition is shown in Fig. 2.11. Here, the slopes for the Lineweaver–Burke equation remain the same, but the intercepts differ. In enzyme chemistry, the inhibitor is thought to bind only with the enzyme at a site (allosteric site) different from that for

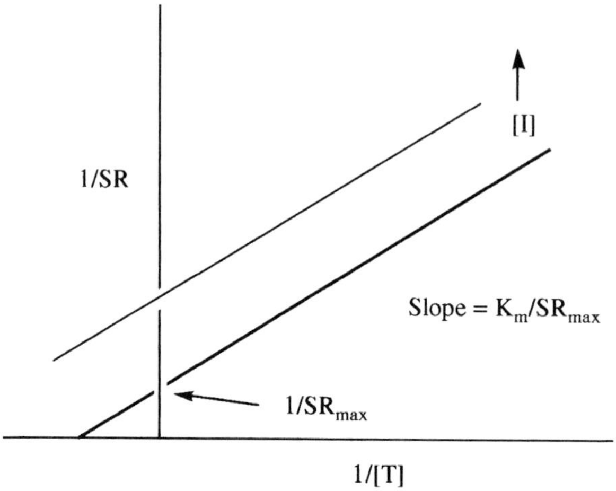

FIG. 2.11. Characteristic plot of the Lineweaver–Burke equation for uncompetitive inhibition.

the enzyme–substrate combination. The degree of inhibition may increase when the substrate concentration is increased, and is common for two-substrate reactions. Another term for uncompetitive inhibition is *coupling inhibition*.

2.11 SYMMETRY AND CHIRALITY

Symmetry and chirality also have application to taste chemistry. Bipolar ions of various sizes and charges are encountered that have different degrees of saltiness, intermixed with other taste attributes. Sourness occurs in bipolar solutions when the hydrogen ion concentration is greater than the hydroxyl ion concentration. The reverse case leads to an unpleasant (alkaline, soapy?) taste. Dipolar sweet substances may or may not have nearly equal but oppositely charged monopolar components. In some cases, only one species of monopole (nucleophile or electrophile) may be present in a chemical structure, and this case may be associated with bitterness. The taste of amino acids depends largely on the configuration (D- or L-) of the α-amino group, but the taste of the sugars is independent of D- or L-configuration. These features of taste chemistry require application of symmetry and chirality principles.

In its broadest sense, symmetry is a universal theme associated with proper proportion and continuity of form. Symmetry is found in nature, music, art, architecture, mathematics, natural law, and abstract physical law. It is believed that all forms of symmetry have a common geometric basis. The symmetry of particular concern here is the symmetry of structure and function (reactivity, taste modality), and it takes two forms (cf. Weyl, 1952). The first is the notion of geometric symmetry, characterized by the operations of translation, rotation, and reflection, and is the basis for analytical symmetry. It is manifested in the geometric repetition of form, as caused by translation along a line in one direction, in a plane in two directions, or in space in three directions. The second form is bilateral symmetry (Section 2.11.4).

All symmetries of the first form are derived from group-theory translational groups and rotational groups. Examples of fundamental linear, planar, and spatial translations are shown using various operations as applied to the figure six. Translation may also be derived

666666	666666 666666 666666	6 6 6 6 6 6 6 6 66 66 666666
Linear translation	Planar translation	Spatial translation

by rotation about an axis, or by reflection. The rotation of the figure six through 90° merely lays the figure on its side, and through 180° inverts it, but it is still the figure 6, unless it is specified as $\underline{9}$. Reflection of the planar numeral six in a mirror line creates a nonsuperposable mirror image six

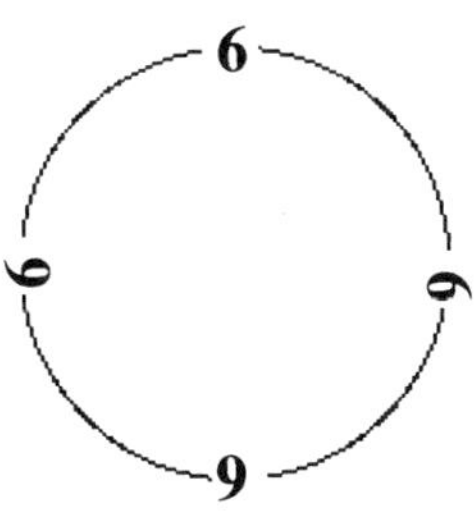

Rotation

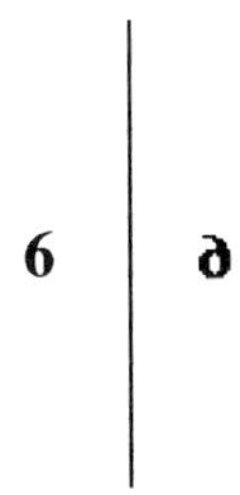

Reflection

'behind the mirror'. The figure six has a rather special symmetry element. It possesses *handedness.* Mirror reflection, because it fundamentally transposes front and back, also transposes one of the two handed forms possible into the other, i.e. from left to right. In the plane of the paper, the reflection cannot be superposed upon the original structure, but it can be if removed from the reflection in the plane of the paper, and rotated through 180°. In symmetry parlance, a sharp distinction is made between the terms superimpose and superpose. To *superimpose* an object is to place one object on another. Hence, the psychological superimposing of taste profiles in Section 1.1.3.1 is, ultimately, the superimposition of two profile time-intensity plots. To *superpose* an object is to place one object upon another and to have them coincide in all geometric respects. All objects have superimposability, but only select objects have superposability. If two sweeteners have superposable time-intensity profiles, they are probably the same sweetener.

One objective of symmetry in chemistry is to establish whether a structure is symmetrical or not. If it is symmetrical, a second objective is to deduce the nature of the symmetry, which yields, in turn, information about chemical properties. For example, the nature of the symmetry elements of a molecule determines whether it will have a dipole moment, or possess optical activity. Application of symmetry principles at the atomic level affords a general picture of the electron distribution for a molecule, and in some cases allows the shape of the molecule to be determined. It was through application of symmetry principles that Dalton was led to formulate the atomic laws. A third objective of symmetry is to facilitate understanding of complex chemical phenomena, which can often be resolved by application of simple symmetry operations (Hochstrasser, 1966).

2.11.1 Principles of Symmetry

Operations that generate symmetry (translation, rotation), are also used to deduce the symmetry elements of a structure. Information about rotation and reflection is derived, in turn, using an algebraic process known as group theory. To derive group theory information, the structure or object under consideration must be placed in a coordinate system, so that symmetry operations can be performed in a consistent manner. This is accomplished by placing the center of gravity of a structure at the origin of Cartesian coordinates. All axes and planes of symmetry will then pass through the origin, and any symmetry operations performed (rotation, reflection) will yield one point that is not changed. The symmetry operations are consequently known as point-symmetry operations. A special form of point-symmetry operations is applied in subsequent chapters (Chapters 9–11) by placing the saporous groups of substances in or at the origin of Cartesian coordinates.

An object or structure can be said to be symmetrical if it remains unchanged after one or more symmetry operations have been applied to it. All told, up to six operations are recognized (Brand *et al.*, 1975; Dmitriev, 1979; Jaffé & Orchin, 1965). The number varies because several are essentially equivalent with respect to an operation, but not with respect to point group theory. One (the identity, E) is not an operation at all, but is necessary for application of group theory. The operations presented here are: (1) rotation about an axis; (2) reflection; (3) the combination of rotation and reflection; and (4) translation, or reflection through a center of symmetry. A symmetry operation is distinct from a symmetry element. The operation of rotation about an axis is not the same as the fact that a structure possesses a rotation axis. Rotation is the operation, and a structure may have a twofold axis of rotation as the element of symmetry. A compound may have a center of symmetry, which is the symmetry element, but translation through that center is a symmetry operation.

2.11.1.1 *Rotation*

For rotation about a symmetry axis, or (C_n), n is defined as $2\pi/n$ or $360°/°r$, where $°r$ is the number of degrees that it is necessary to rotate an object or molecule to obtain a structure identical with the original. The symbol C indicates rotation (Latin *circum*, to rotate). As rotation by 360° will always restore an object to its original position, it is a trivial operation, equal to the identity (E). Rotation operations are only possible in dimensions greater than one.

The proof that a molecule of water is a symmetrical object begins with the observation that the water molecule has a C_2 axis of symmetry.

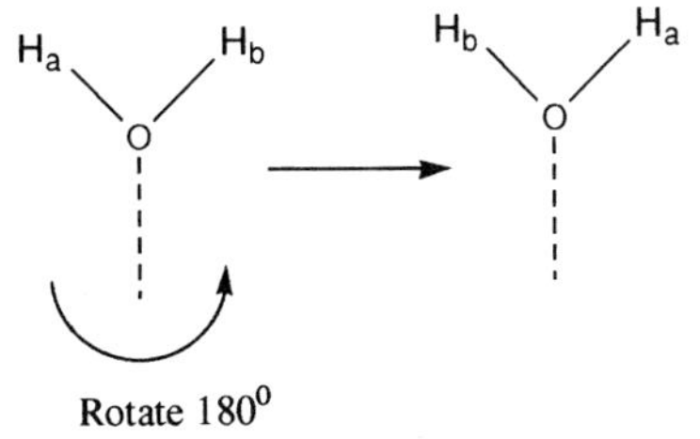

Water

Rotation by 180° results in an orientation indistinguishable from the original, provided that the labeled hydrogen atoms are indistinguishable from each other. Many molecules have a rotational axis of symmetry greater than one. Chloroform has a C_3 axis of symmetry, and benzene has a C_6 axis of symmetry. The nitrogen molecule (N_2), like many other linear molecules, has more than one symmetry axis. The *principal* axis, or Cp axis, is the one that passes through the greatest number of atoms. In a Cartesian coordinate system, with the center of the molecule placed at the origin, the axis of highest order (i.e. the principal axis) is the z axis. The x axis is then normal to the plane for the z axis. The principal axis for the nitrogen molecule is infinite-fold, or C_∞.

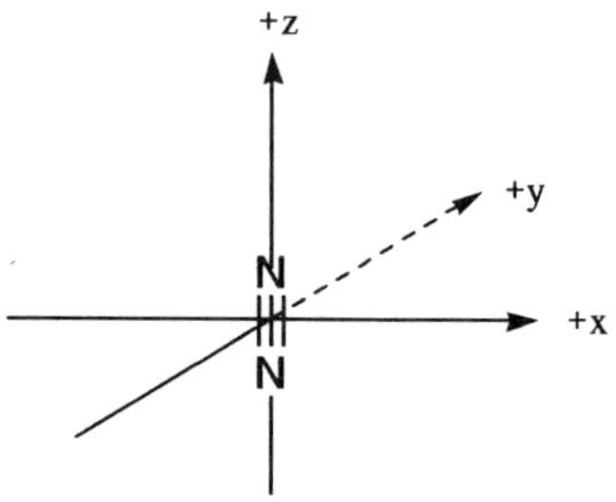

The nitrogen molecule at the origin of Cartesian coordinate axes

A simple rotational axis with a fold greater than one is not by itself a sufficient condition for symmetry. If a simple rotation axis is the only symmetry element that a structure possesses, it can have right and left handed forms (i.e. it is chiral). An example is a skewed conformation for an ethane molecule. Here, a near and a remote hydrogen atom is considered to be a pair, and a skewed ethane molecule can be considered to have a threefold axis of symmetry. The chiral model is but one of many possible conformations and ethane is, in the final analysis, a racemic mixture of chiral conformations. On the other hand, if a compound has

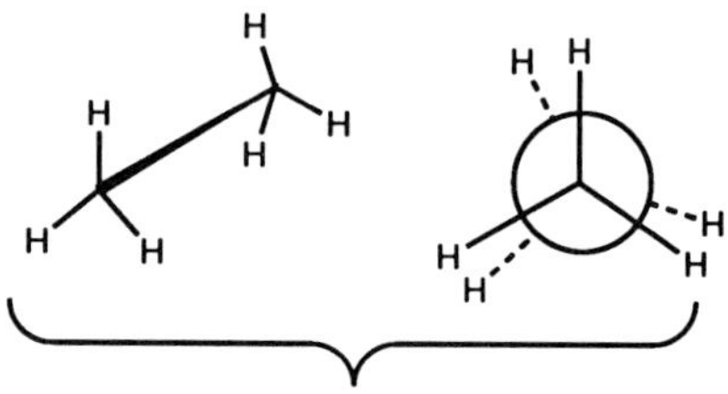

Skewed ethane

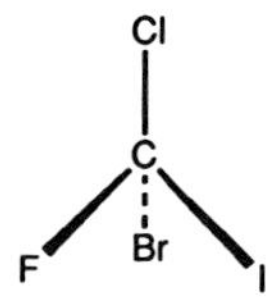

Cl,F,Br,I Methane

no rotational axis of symmetry it is chiral (a few exceptions are known), and possesses optical activity. A tetrahedral carbon atom substituted with four different halogen atoms is an example of a structure without a symmetry axis, and it too is chiral. The operation of rotation about a simple axis has been designated (cf. Brand *et al.*, 1975) as a symmetry operation of the *first kind.*

2.11.1.2 *Reflection*
Reflections in a mirror point in one dimension, a mirror line in two dimensions, or a mirror plane in three dimensions, are also symmetry operations. The reference mirror for these operations always has one less dimensional parameter than the object upon which the symmetry operation is performed.

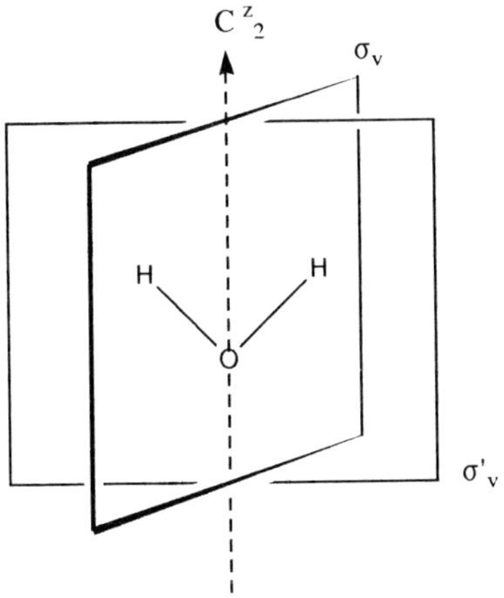

Planes of symmetry in the water molecule

Reflections are made in or at a mirror point to comply with requirements of group theory, as one point must remain unchanged. The use of reflection through a mirror (the object behind the mirror) has translation application, and is used as a criterion of symmetry (or chirality). The water molecule has two planes (σ) of symmetry, one in the xz plane (σ^{xz}), and one in the yz plane (σ^{yz}). Many molecules have an infinite number of such planes, such as the nitrogen molecule.

2.11.1.3 *Rotation–Reflection* (*Proof of Symmetry*)
When an object is rotated about an axis, and the reorientation is next

reflected in a plane perpendicular to that axis, the object has a rotation–reflection axis of symmetry when the resulting structure is identical with the original. When *trans* dichloroethylene is first rotated through 180°, and then reflected in a bisecting two-sided mirror plane, the left side of the molecule 'sees' its right side in reflection, and the right side 'sees' the left side. Assembling the reflections yields the original molecule. The same result is obtained when the reflection operation precedes rotation.

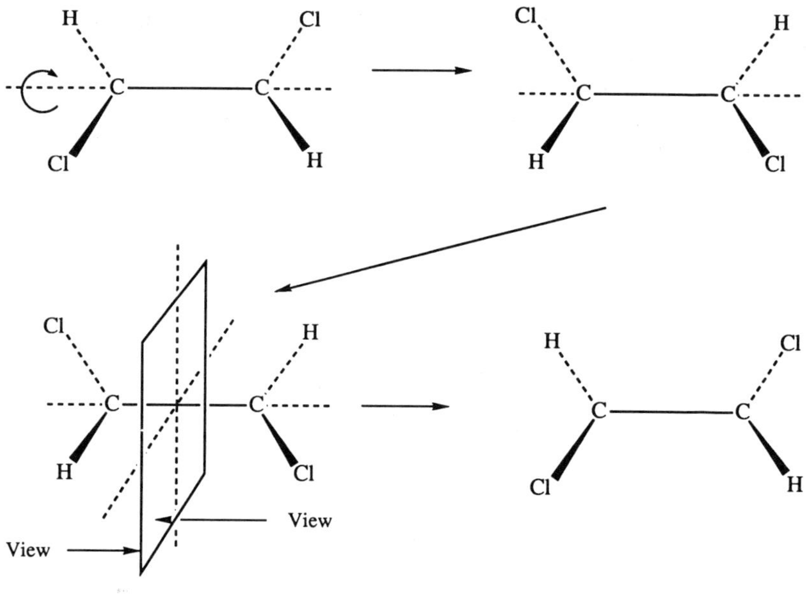

Rotation-reflection (Proof of symmetry)

2.11.1.4 *Rotation–Inversion Axes*

The operation of rotation–inversion is equivalent to rotation–reflection. The order of the rotation is not the same, but inversion is the equivalent of reflection. The tetrahedral methane molecule has (among other elements of symmetry) an inversion center of symmetry, and an inversion–rotation axis. Symmetry is proved or disproved by the operation of rotation–reflection. Inversion through a center followed by rotation is an equivalent operation. This is demonstrated with methane, as it is easier to comprehend. When two operations are required to bring a structure into congruence with the original, they are called symmetry operations of the *second kind*.

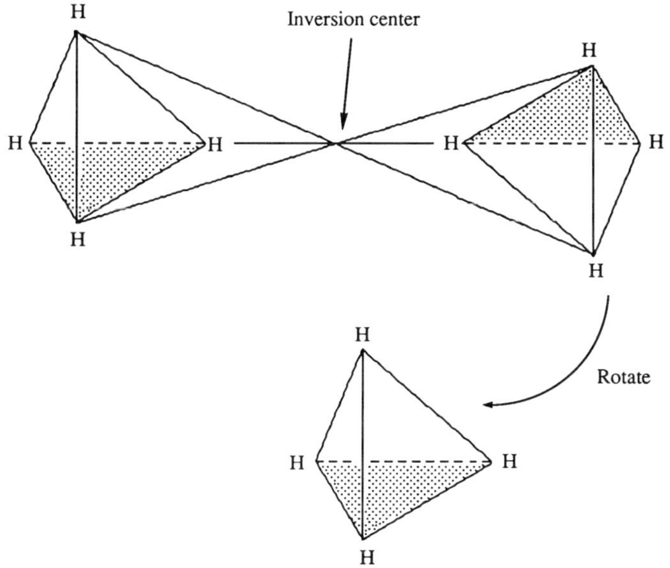

Inversion-rotation of methane

2.11.1.5 *Two-Dimensional Symmetry Point Groups*

Certain symmetry groups in the structure of a compound require that other elements must also be present. At the same time, the presence of certain symmetry elements excludes the possibility of others. For structures in two dimensions, the equilateral triangle (b) in Fig. 2.12 has a reflection (mirror) line of symmetry that bisects the base. Every point on the left side of the base can be translated across the mirror line to an equidistant point on the right, and vice versa. Of the structures shown in Fig. 2.12, structure (b) is the only one with just a single mirror line of symmetry. Two of the structures, the scalene triangle (a), and the parallelogram (c), possesses none. The rectangle (d) and the ellipse (h) have two, the square (e) four, the star five, and the circle an infinite number of mirror lines of symmetry.

Rotation is the second type of symmetry operation possible in two dimensions but it must take place about a symmetry point, as no axes are possible. The parallelogram (c), the rectangle (d), and the ellipse (h) each has a twofold symmetry point-axis. The square (e) has a fourfold point-axis, the Pythagorean star (f) has a fivefold point-axis and the circle (g) has an infinite-fold point-axis.

The last symmetry element possible for a structure in two dimensions is a symmetry center. As in three dimensions, a line drawn through the center from a given point needs to encounter an identical point at an equal distance beyond the center. The structures with a center of symmetry in Fig. 2.12 are the parallelogram (c), rectangle (d), circle (g), and ellipse (h).

Now that the symmetry elements for the structures in Fig. 2.12 are specified, the interrelations associated with them, such as the presence of symmetry elements that exclude the occurrence of others, become apparent. (1) From structures (c, d, e, g, and h), clearly a twofold point-axis, and a center of symmetry always occurs together. This is not true for three dimensions, however. (2) An object with a fourfold symmetry point-axis (e) must also have a twofold point axis. (3) When only two mirror-lines of symmetry are present, they must be perpendicular at the point of intersection. The point of intersection also contains a twofold point-axis (d). The square (e) has two sets of perpendicular symmetry lines, and a fourfold axis of symmetry.

An example of exclusions is that any structure with a fivefold point-axis (f) cannot have a threefold axis, a twofold axis, or a center of symmetry. With respect to structure (a), it can be stated that it does not have a mirror plane of symmetry, nor does it have a rotation axis. The absence of these two elements preclude the presence of a symmetry center. Structure (a) is not symmetrical. It does, of course, have a onefold symmetry axis, as do all the other structures, but that is a trivial symmetry element.

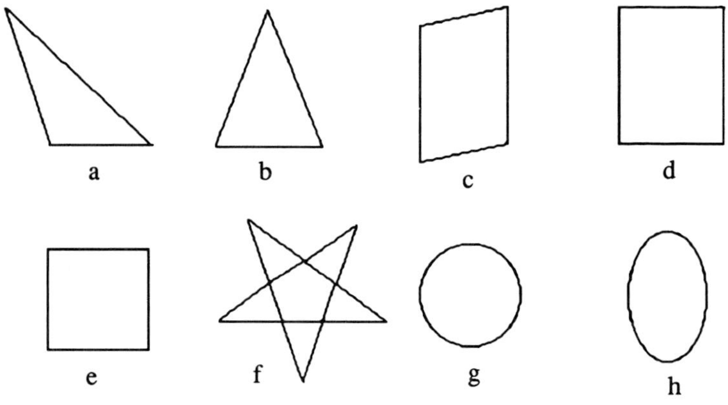

FIG. 2.12. Two-dimensional structures belonging to different symmetry groups.

Six of the structures shown in Fig. 2.12 belong to different two-dimensional symmetry groups, but two, despite their different form (d and h), belong to the same group. Both have a twofold axis of symmetry perpendicular at the point of intersection, and no others. The point of intersection is also a center of symmetry, and the twofold axes are also mirror-lines of symmetry. The group to which (a) belongs is chiral, and has a mirror-line reflection structure (a′) that, in two dimensions, is not superposable on the original. Thus, combined rotation–reflection, followed by superposability (congruence), is proof of symmetry. Lack of superposability, following these operations, i.e. operations of the second kind, are proof of chirality.

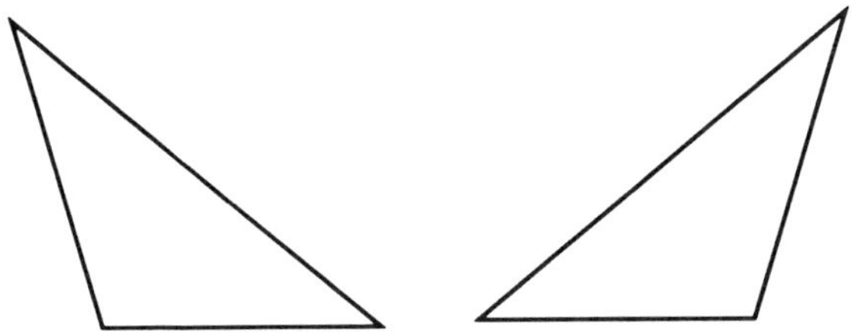

a a'

2.11.1.6 *Three-Dimensional Symmetry Point Groups*
Symmetry operations form a set, as is evident from the symmetry operations performed on two-dimensional objects. In three dimensions, the symmetry operations performed on objects make up sets of elements that comprise a mathematical group. In the group, the product of any two elements affords still another member of the group, and the operation $A \times B = C$ is commutive, i.e. $AB = BA$. For the symmetry multiplication example, $(AB) \times C = A \times (BC)$. In addition, the set also contains the identity (E) and $AE = EA$ for all the elements. A final characteristic of a group is that each element has a reciprocal, and $A^{-1}A = AA^{-1} = E$.

Construction of the group for the symmetry elements in a molecule of water (A C_2 rotation axis and two vertical planes of symmetry) is as follows:

	E	$\mathbf{C_2}$	σ_v	σ_v'
E	E	C_2	σ_v	σ_v'
$\mathbf{C_2}$	C_2	E	σ_v'	σ_v
σ_v	σ_v	σ_v'	E	C_2
σ_v'	σ_v'	σ_v	C_2	E

The product obtained by multiplying any two of the elements contained in the group always results in an element for the same molecule, and σ_v from the horizontal column times C_2 from the vertical column yields σ_v'. It is also seen from the group why E is important. Without the identity element, the collection of operations would not have the properties of the group. The collection is called a *point group* because whatever symmetry operations are performed, at least one group must remain stationary (Dmitriev, 1979). When a character Table for the four symmetry operations is formed for the point group C_{2v} (Jaffé & Orchin, 1965), other properties of the group become apparent, such as the direction of a dipole moment.

While objects in the same two-dimensional group had entirely different shapes, examples of compounds that belong to the C_{2v} three-dimensional point groups have entirely different substituents, reactivities, properties, and shapes. Nevertheless, they also have some properties in common. In this case, it is a dipole moment, and all have taste and/or other sensory attributes. Some of them are shown below. Formaldehyde (a) is a gas with a pungent odor. Hydrogen sulfide (b) is a poisonous gas with the odor of rotten eggs, but with a sweetish taste. Nitrous oxide (c), or laughing gas, also has a sweetish taste and a sweetish odor. Sulfur dioxide (d) is a gas with a suffocating odor, but ethylene dichloride (e) is a heavy liquid with a sweet taste.

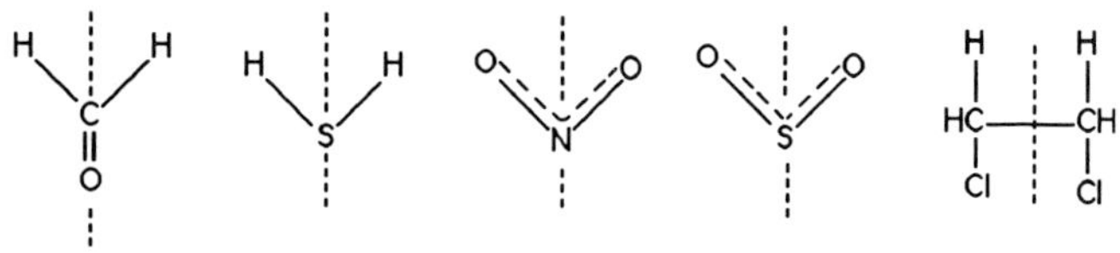

a b c d e

A number of different point groups are recognized. The point group C_n (symmetry axis only), is a special case, as all objects in this group are chiral. These substances have optical rotatory power, unless a racemic mixture of conformations or configurations renders them optically inert. Ethane is an example of the former, and either an L- or D-amino acid is an example of the latter.

2.11.2 Dissymmetry and Asymmetry

When an object falls into the symmetry point group C_n and has only a single axis of symmetry, it fails the ultimate symmetry test of being superposable on its mirror reflection. Thus this symmetry group encompasses objects that are chiral, and right- and left-handed forms are possible. The symmetry group (C_n) is covered by Pasteur's 1860 idea (cf. Pasteur, 1915) of *dissymmetry*. Objects that are obviously symmetrical, but not superposable on their reflected structures, can exist. Pasteur's term 'dissymmetric', is sometimes interpreted (Lowry, 1935) to indicate degraded symmetry. The principal example is the pattern of a regular helix which, while possessing a simple axis of symmetry, is nevertheless not superposable on its reflected form. As a dissymmetric structure in the form of a screw, it has right or left functionality because of the symmetry element. If the right or left helical pattern of the screw were not regular, the structure then would be asymmetric, and nonfunctional. As applied to bilateral symmetry, the notion of dissymmetry implies a disparity between either the size or number of the right and left forms.

An example of the distinction between dissymmetry and asymmetry can be made using the scheme suggested by Ruch (1972). If presented with a random pile of shoes of different size and color, a person can readily separate them into two piles, one containing all the right forms, and the other all the left forms. The task is accomplished by simply comparing the shoes with what is perceived to be a person's own right or left foot. In the final analysis, the perceptive difference has its origin in the form of a partial right- and a left-handed helix. If presented with the opportunity to separate asymmetric objects, such as potatoes or lumps of coal, into two classes, it would soon be found that it is impossible to do so. It is impossible because an asymmetric structure possesses no element of symmetry to serve as a reference point, and an asymmetric structure is characterized more by chaos than by order.

2.11.3 Chirality

The term chiral is derived from the Greek word *cheir*, meaning 'hand', and the study of *chirality* is the study of 'handedness'. In the sense used here, it is the study of three-dimensional objects that have nonsuperposable geometrically opposite structures. The three-dimensional notion of nonsuperposability is necessarily abstract, as no two objects can occupy the same space at the same time. The fact that a glove fashioned for the right hand cannot be fitted over the left hand serves to demonstrate the notion of nonsuperposability. It can, however, be made congruent with the glove for the left hand by everting it (i.e. turn it inside out).

The word chiral was introduced to scientific literature by Lord Kelvin (cf. Whyte, 1958) with the statement, 'I call any geometric figure, or any group of points chiral, and say it has chirality, if its image in a plane mirror, ideally realized, cannot be brought to coincide with itself.' As used by Pasteur, the term dissymmetric, and the term chiral, can be equated on a one-to-one basis (O'Loane, 1980). While the notion of symmetry as a universal theme may be innate, the notion of chirality, which is also a universal theme, must be taught. Ultimately, chirality can be defined as the study, and/or identification and specification of handedness (right or left). A significant distinction can be made between symmetry and chirality. A symmetrical object retains its symmetry during transposition of geometric dimensions, but the geometric features that create a chiral object are extinguished upon ascending geometric continuums. As a case in point, reflection of the two-dimensionally chiral figure 6 in Section 2.11 generates the two-dimensional antipode. The two-dimensional handedness of the figure six is extinguished in three dimensions. Operations of the second kind are allowed, and rotation through 180° brings the reflection back into congruence with the original figure six.

A second distinction between symmetry and chirality (Mislow & Bickart, 1976/77) is that detection of chirality depends on the nature of the measurement. A measurement, such as the melting point, or dipole moment, cannot establish chirality, but other measurements, such as optical rotation, and circular dichroism can. In addition, chirality can be generated by differentially labeling component parts of symmetrical objects, or by skewing a symmetrical object. That feature has application to taste chemistry as like groups in the structure of a compound may have different reactivity.

2.11.3.1 *Demonstration of Chirality*

A demonstration of the abstract principles and notions of chirality is the reflection of the right hand in a mirror. The reflection has the overall structure of the left hand, and bringing the reflected hand up to the mirror will *superimpose* the palm of the right hand upon its reflected image. However, the complete structure of the right hand is not *superposable* on its reflection. Even though the latter statement is also an abstraction, as no two objects can occupy the same space, it is easily demonstrated. A glove made for the right hand cannot be fitted over a left hand as the structures of the right hand and the left hand, are geometric opposites.

In the final analysis, geometric right handedness describes a right-handed screw (helicoid) pattern. A pnemonic for describing the nature of a right-handed screw pattern also serves for derivation of the handedness of objects. The thumb and first two fingers of the right hand, when held erect, describe a partial turn of a right-handed helix. As viewed by the owner of the hand, it has a positive helical slope. The pattern described by the left hand has a negative slope. The infinite symmetry axis of the helix is the symmetry component of chirality. The chiral nature of the helix permits separation of the pile of shoes (Section 2.11.2) into homochiral (Mislow & Bickart, 1976/77) right and left families. As for the potatoes, an astute horticulturist would place them in only one homochiral family, as the helical pattern of the buds is the same. Chemists have long made the distinction between the homochiral D- and homochiral L-families of sugars and amino acids.

In summary, a chiral structure retains just enough symmetry (a simple axis) to allow a left–right distinction. Through lack of a convention, which is right and which is left may not always be certain. On the other hand, when an object cannot be proven to be totally symmetrical, it may be either dissymmetric or asymmetric. Conversely when an object cannot be proven to be chiral, it may be either symmetric, or asymmetric.

2.11.3.2 *Classification of Symmetry and Chirality*

Although chirality is considered to be a special division of symmetry, there is a question on how much can be said about chirality, in a systematic way using symmetry terms and expressions (Mead, 1974). Bilateral symmetry is also a special case of symmetry (Weyl, 1952), and while little can also be said of it using group theory, the notion of bilateral symmetry has important application in biology and to taste chemistry. In summary, symmetry can be said to encompass

I. Symmetry
II. Bilateral symmetry
III. Chirality.

Number II is distinguished from I as it encompasses the combination of antisymmetry elements (handedness, bipoles, dipoles, and electrostatic charges). Number III is distinguished from I in that its primary geometric element is helical, and it is distinguished from bilateral symmetry (II) in that its principal concern is the specification and *distinction* between antipodal structures.

Asymmetry is the complete lack of any element of symmetry. The fact that some structures are not completely symmetrical, but can be distinguished from structures that have no elements of symmetry, separates the notion of dissymmetry from that of asymmetry. There is no distinguishable helicoid pattern of chirality evident in an asymmetric structure.

2.11.3.3 *The Asymmetric Carbon Atom*

The first notion of the importance of chirality in chemistry originated with Pasteur's attempt in about 1860 (cf. O'Loane, 1980; Pasteur, 1915) to describe the geometry of tartaric acids. The effort is also touted to be the origin of the discipline of stereochemistry (cf. Shallenberger & Wienen, 1989). Pasteur discovered that three forms of tartaric acid are possible. Two of them possessed optical activity of equal magnitude, and were chemically identical in all respects. The sign for the optical rotation of one of them was positive, and for the other it was negative. The third form was optically inert. The current structures for these compounds, in Fischer projection, are shown in Fig. 2.13 using current chiral specification procedure (Cahn *et al.*, 1966).

The discovery led Pasteur to ask, 'Are the atoms of the right acid grouped on the spirals of a dextrogyrate helix, or placed at the summits of an irregular tetrahedron, or disposed to some other dissymmetric grouping?' 'We cannot answer these questions.' As creation of new French words is the prerogative of the Académie Française (O'Loane, 1980), the term dissymmetry does not usually appear in dictionaries. When it does, it is used as a synonym for asymmetry. As a result, when Pasteur's article *Researches sur la Dissmétry Moléculaire des Produits Organiques Naturels* was translated, the original term dissymmetry became asymmetry. Pasteur used the term dissymmetric in an apparent attempt to solve a conceptual problem encountered in describing the geometry of a helix, as helical arrangement of atoms served to answer the problem he was trying to solve (Shallenberger & Wienen, 1989). A regular helix is not superposable on its mirror image structure. Nevertheless it has an infinite-fold symmetry axis, and is not without all elements of symmetry, which is what the term *asymmetric* unambiguously implies.

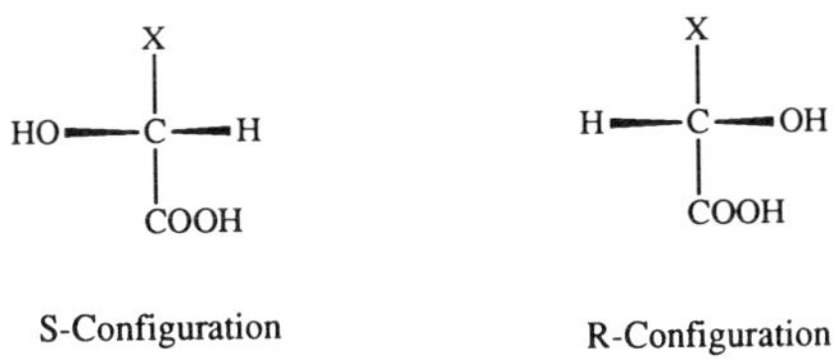

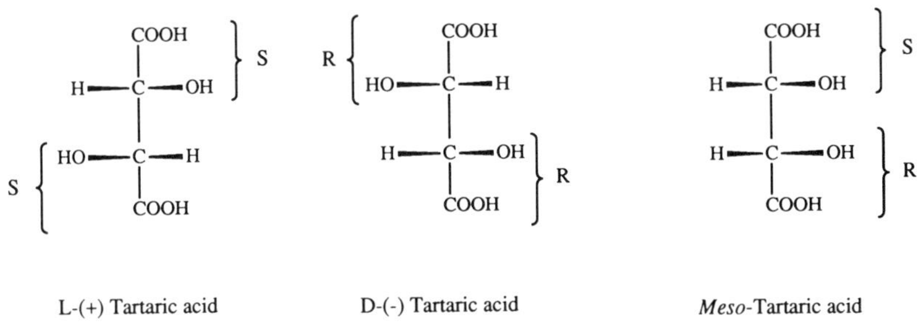

FIG. 2.13. Structure of tartaric acid enantiomers and *meso*-tartaric acid.

Shortly afterwards, van't Hoff (1874) developed the idea of the *asymmetric carbon atom* to explain the occurrence of optical isomers. An asymmetric carbon atom was defined as a carbon atom substituted with four groups of different composition. Although van't Hoff did not use the term dissymmetric (the word was not a Dutch word, either), he stated that, 'In the asymmetric carbon atom we have a condition that distinguishes itself by its helical arrangement' (cf. Riddel & Robinson, 1974). When the four valences of the regular tetrahedral carbon atom are substituted with four different chemical groups, an irregular tetrahedron is obtained. Different groups intrinsically have different size. For the tetrahedron, no two sides are of equal length, and the apices describe a helical pattern, as shown in Fig. 2.14. The irregular tetrahedron, or the helical pattern, can now have a right- and a left-handed form. The handedness of a tetrahedron substituted by four different groups is specified by convention. This requires that the base of the tetrahedron be viewed directly, with (d) remote. If (a) has greater priority (e.g. size, atomic number) over (b) and (b) over (c), etc., the arrangement is plus

(+) when (a) to (b) to (c) proceeds in clockwise fashion. It is minus (−) when the sequence is counterclockwise.

Clockwise (+) Counterclockwise (-)

When (a) has a greater size than (b), etc., the corner of the irregular tetrahedron closest to an observer is the termination of a helical pattern moving from (d) toward the observer. This is shown in Fig. 2.14. Not only is the priority sequence clockwise, but the screw pattern resembles the right-handed thread of a machine screw. It will move away from the operator when turned to the right (clockwise).

In Fig. 2.13, an R, S specification for the chiral nature of the groups in a multiple chiral group of compounds was employed. The chiral family specification for the two enantiomers is D- and L-, and a special convention was applied. The convention is particularly important for specifying the chiral family to which a sugar or an amino acid belongs. For the sugars and amino acids, an asymmetric reference carbon atom is oriented, *after application of a priority sequence*, so that the hydroxyl group or amine group is horizontal. It is also oriented so that those groups project toward the viewer. The compounds will then belong to the D-family when either the hydroxyl or the amine group is toward the viewer's right. Otherwise, the compounds belong to the L-family.

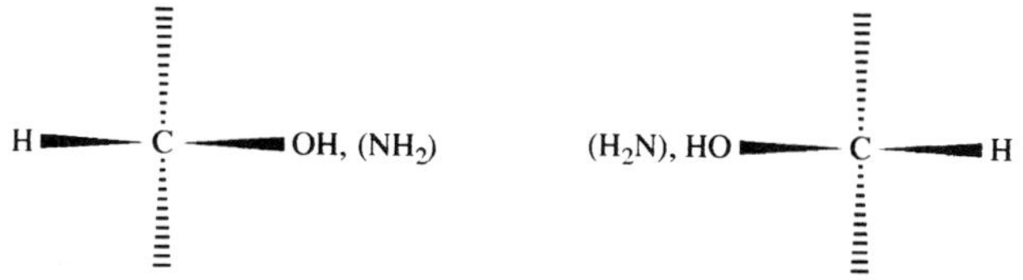

Reference carbon atom configuration for the D-chiral family

Reference carbon atom configuration for the L-chiral family

The special priority sequence differs for the sugars and the amino acids, and they are presented in Sections 5.1.1.2 and 7.1.1, respectively.

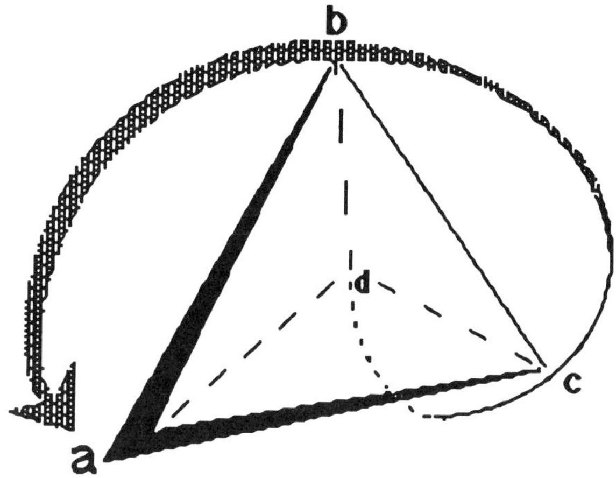

FIG. 2.14. The right-handed helical pattern circumscribing an irregular tetrahedron formed by substituting a carbon atom with four groups of different size (Shallenberger & Wienen, 1989).

Nevertheless, it can be determined that the procedure steadily assigns the S, S configuration for tartaric acid (Fig. 2.13) to the L-series, and the R, R configuration to the D-series, even without a priority sequence convention for those substances. Rotation of the molecules in the plane of the paper leads to the correct chiral family specification in either case. The (+) and (−) signs are merely used to indicate the optical rotatory nature of the compounds, and has no chiral specification significance.

2.11.3.4 *Sources of Chirality in Compounds*

Helical arrays of atoms. Any helical array of atoms, as in the structure of quartz crystals, or hexahelicene, leads to chirality and optical activity (Kirkwood, 1937; Brewster, 1959). A tetrahedral carbon atom, substituted with four groups of different size leads to a helical pattern, and hence to chirality. The dispositions of hydroxyl groups on a five- or six-membered ring describe a helical arrangement, and is the primary source of sugar optical rotatory power. Among the monosaccharides, the most optically active sugar (β-L-arabinopyranose) has the most regular and most steeply pitched helical pattern of hydroxyl substituents.

Differential labeling. The regular tetrahedral carbon atom may also be rendered dissymmetric merely by giving the apices, or faces, four different labels. While such a structure cannot, of course, be optically active, the procedure has important conceptual consequences. Differential labeling (numbering, lettering, coloring) may seem to be a trivial source of chirality, but certain functional groups, seemingly identical in composition and function, may be electronically different. They are differentially labeled by virtue of their location in a compound. Depending upon the nature of substituents, benzene ring protons are not electronically equal. The hydroxyl groups of a ring form of a sugar are another example.

For tetrahedral carbon atoms, mirror-image forms are possible in either the differentially labeled case, or for the four groups of different size. One consequence of this is that one form can be generated from the other, merely by transposing any *two* substituents or labels. A simple linear transposition of any two of the four labels or groups generates an enantiomer. When executed with groups of different size, the net result is to generate a helical pattern that goes the other way. Herein seems to lie a fundamental reason for why simple chemical transformations in the structure of a compound drastically alter taste. The enantiomer generated by transposition of only two substituents is unique to the regular tetrahedron (Mislow & Siegel, 1984). For skeletons other than the regular tetrahedron, transposition of ligands or labels forms a diastereoisomer (Section 5.1.1.1).

Conformational skewing. From the discussion of symmetry it was learned that rotation of substituents about an axis leads to a form (conformation) of ethane that is chiral. The operation is more one of skewing from rotation about single covalent bonds, than the symmetry operation of rotation, *per se*. Anything that will skew an otherwise symmetrical structure is a third source of chemical chirality. The scalene triangle in Fig. 2.12 is chiral because it is skewed.

2.11.3.5 *Chiral Operations and Specification*

The symmetry operation of highest-order is a combination rotation–reflection procedure (or inversion through a center, i.e. $S_n = C_i$), and is known as a symmetry operation of the *second kind*. For practical purposes it can be stated that if an object is superposable upon its mirror-image structure, it is symmetrical. If it cannot be superposed on its mirror-image structure, then it is either chiral or asymmetric. There are no operations that will bring a chiral object into an orientation identical to that of the original orientations. The differential labeling of a regular structure, or the skewing of a regular structure will generate chirality (Shallenberger *et al.*, 1981) and are believed to be chiral operations. Strictly speaking, there is no chiral equivalent for the usual symmetry operations leading to symmetry point groups and the notion of different orders of symmetry. An object is either chiral, or it is not (i.e. achiral).

While there are no chiral operations *per se*, axes, centers, and planes of chirality are recognized and employed, along with a set of sequence rules, to *specify* the nature of molecular chirality (Cahn *et al.*, 1966). The proposed chiral centers, etc., as shown below are operational abstractions. A tetrahedral carbon atom with four different groups appended to it is considered to be a *chiral center*, but only in an abstract sense. The meaning of the term asymmetric carbon atom is also slightly altered from that used previously in this treatise. Cahn *et al.* (1966) set forth the principle that, 'Only when no symmetry operation can convert the molecule (into coincidence with itself), has the molecule no symmetry.' Only then may it be called 'asymmetric in the correct usage of the term'. 'Thus, Cabcd has no symmetry whatever, and therefore the term "asymmetric carbon atom" is entirely correct.' Later on, the authors state, 'However, a model which has no element of symmetry except at most axes of rotation, may be called chiral.'

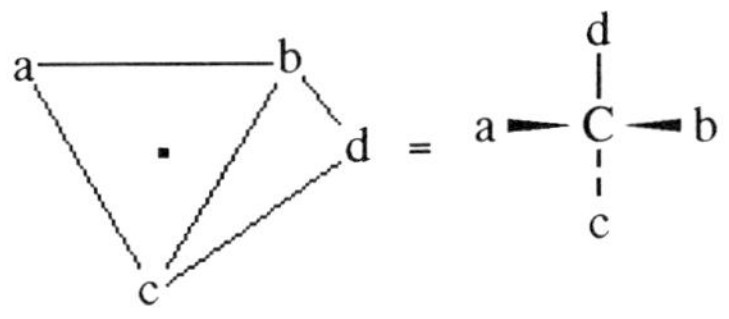

Chiral center

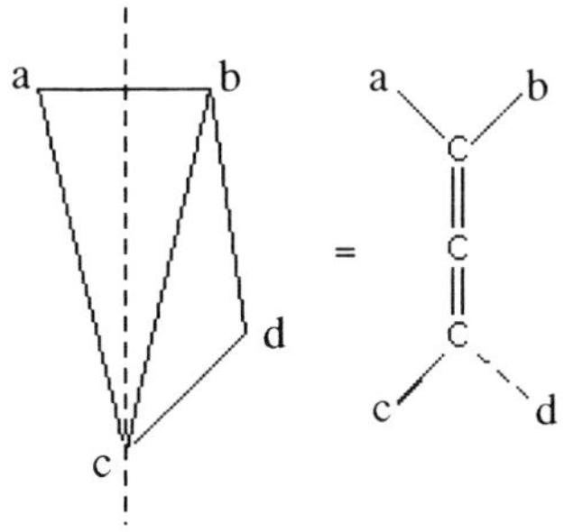

Chiral axis

It would seem that if there is any such thing as a *chiral operation*, it must be a procedure that renders an otherwise symmetrical structure, either handed (dissymmetric) in one or the other *homochiral* senses, or asymmetric. A procedural operation might also be performed to render a topologically homochiral object, such as a regular right-handed screw pattern, asymmetric. Such an operation would merely skew an otherwise symmetrical structure, or in the case of a chiral structure, to skew the screw pattern, and thereby render it asymmetric.

In an attempt to comprehend the double chiral nature of sugar ring structures, Shallenberger *et al.* (1981) pointed out that a sugar ring, made up of asymmetric carbon atoms, is itself chiral. It is rendered chiral by the manner with which carbon atom substituents are appended to it, and because it is puckered. The authors then viewed the differential labeling and skewing of the ring as two separate chiral operations. The chiral operations as performed on a symmetrical hexagonal ring are illustrated in Fig. 2.15. The drawings in Fig. 2.15 demonstrate that two operations are needed to impart handedness to a five- or six-membered ring. If the ring, such as a regular hexagon, is planar, but numbered, then it is not superposable on its mirror-line reflected structure. It is superposable on its mirror-plane reflected structure. If the mirror-image plane structures are skewed in the same direction, they are not superposable in three dimensions, and differential labeling and skewing become chiral operations.

For the hexapyranose sugar rings, the ring is differentially labeled either by the presence of the C-5 hydroxymethylene group, or because all substituent hydroxyl groups are electronically different. Shallenberger *et al.* (1981) also presented a method for algebraically computing the chiral features (conformation, enantiomeric, and anomeric form) displayed by sugar structures. The salient feature of the algebraic procedure is that it portrays the chiral features of the sugars that are now recognized to be absolute. It also has application to specification of the chiral nature of the glycophore for sweetness, and the receptor for sweetness (Section 11.2.2.2).

2.11.3.6 *Prochirality*

The principle of prochirality (Hanson, 1966; Hirshmann & Hanson, 1971), or pseudochirality (Nourse, 1975) originated with the observation

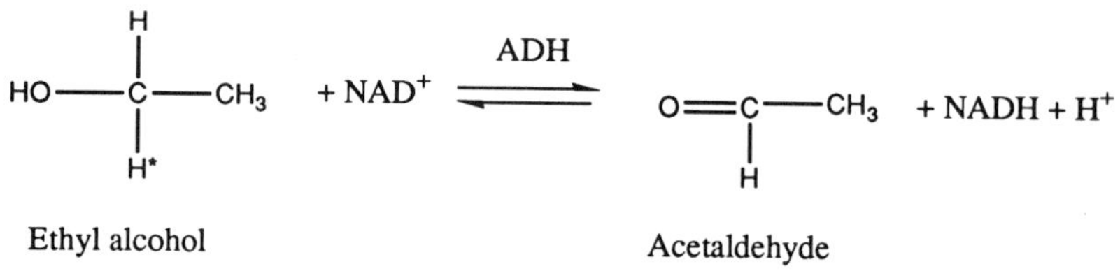

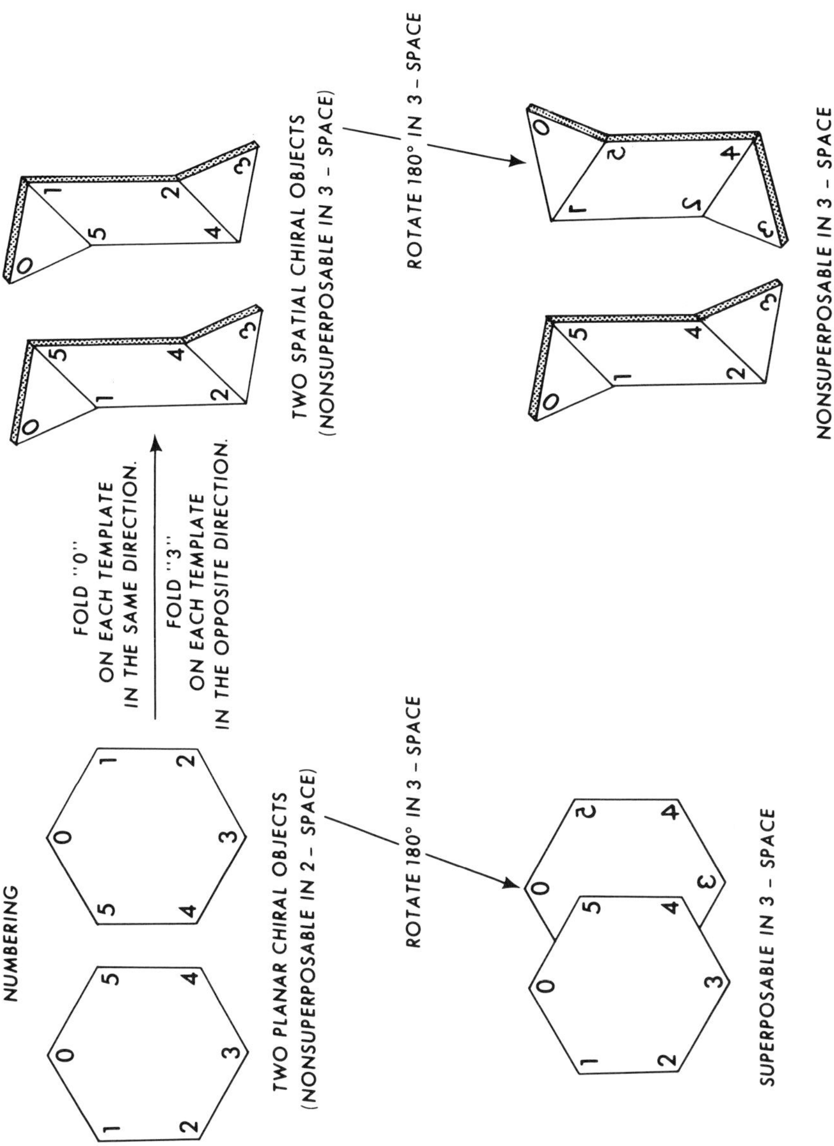

FIG. 2.15. Chiral operations of differential labeling and skewing (Shallenberger *et al.*, 1981).

(Ogston, 1948) that an enzyme, such as alcohol dehydrogenase (cf. Bentley, 1978) is able to distinguish between the two protons of a symmetrical substrate, ethyl alcohol. Alcohol dehydrogenase (ADH) selectively removes a specific methylene ($-CH_2$) proton (asterisked), and a hydroxyl proton from ethyl alcohol to form acetaldehyde, and does so repeatedly when the reaction is reversed and brought forward again (cf. Whitaker, 1972). Thus, with labeled protons, ethanol seems dissymmetric, as two abstract enantiomers are possible. Similarly, the enzyme aconitase readily distinguishes between the two CH_2COOH groups of citric acid, a symmetrical compound. In view of the manner in which it is metabolized, glycerol is not a 'symmetrical three-carbon compound' (Hirschmann, 1960). Other enzyme examples, and nonenzymic distinctions between otherwise symmetrical groups are reviewed by Bentley (1969).

"Plus and minus" ethanol

Some authors considered the phenomenon of prochirality to be an example of *rotational asymmetry*. Ultimately (cf. Prelog, 1976), it was concluded that it is a manifestation of the generation of a type of chirality that results in passing from a three-dimensional continuum to a two-dimensional continuum. There is no inherent three-dimensional configurational dissymmetry in the ethanol structure. The so-called dissymmetry is imposed by the recognition nature of the enzymic reaction (cf. Bentley, 1978). The tetrahedral form of 'dissymmetric' ethanol is shown in Fig. 2.16. Because the tetrahedron is symmetrical (two pairs of sides are equal in length), it is superposable on its mirror image. If the reactive site of the enzyme interacts with the compound in tripartite fashion, then two faces of the tetrahedron, containing the three different groups, are not equal. In one (the back face), the sequence CH_3 to OH to H is counterclockwise, and in the other face (the bottom), the sequence is counterclockwise. If the receptor face, with its reactive sites A, B, C, that correspond to, and interact with CH_3, OH, and H are *counterclockwise*, as in Fig. 2.17, then only the asterisked proton will take part in the reaction.

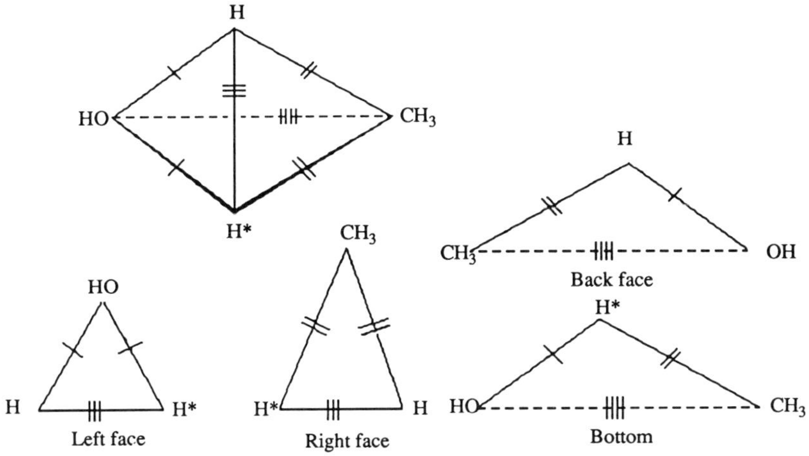

FIG. 2.16. The tetrahedral form of 'dissymmetric' ethanol.

The base of the intact tetrahedron must be viewed with H at the top remote. Then, the sequence for the base CH_3 > OH > H is clockwise. By the same token, if CH_3 was to pair up with receptor site A, OH with B, and H with C, the A, B, C receptor sequence is counterclockwise, because

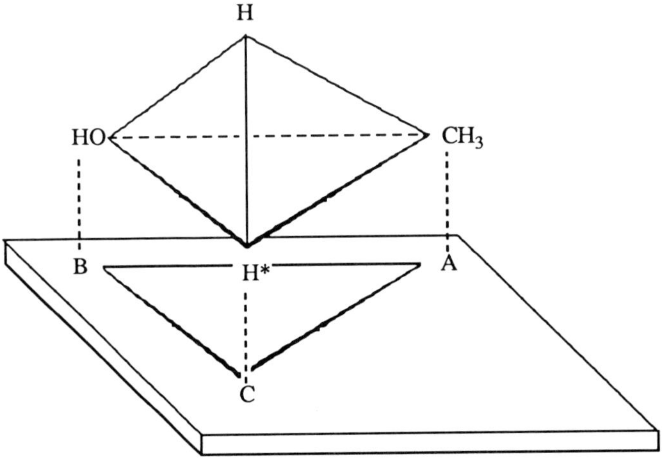

FIG. 2.17. Interaction of 'dissymmetric' ethanol with the alcohol dehydrogenase receptor.

the receptor is *one-sided*. Looking through the tetrahedron to determine the chiral sequence of the base is analogous to trying to tell time by looking at a reflection of a clock in a mirror. It can be done, but the possibility for error increases as a new priority sequence must be established. Looking through the tetrahedron gives the impression that the CH_3, OH, H sequence is clockwise, but it is not. Standard procedure for determining the sequence of substituents for any asymmetric carbon atom must be followed, i.e. by observing the base with the opposite apex remote. Thus the base is *dexter* with respect to CH_3, OH, and H. A clockwise substance orientation is required to interact with a counter-clockwise receptor sequence. Putting the observations above in another context, the catalytic enzyme site 'resides in flatland', and is capable of distinguishing between the two faces of ethanol.

An implication of prochirality is that *either* of the two ethanol protons can be viewed as being prochiral. Substituting either one of them with a fourth ligand will render the molecule chiral. It is perhaps unfortunate that formal application and development of chiral principles is so strongly focused on three-dimensional basic figures with four ligands (cf. Cahn *et al.*, 1966; Prelog, 1976). The emphasis often leads to the intuitive notion that chirality has only three-dimensional significance.

The actual application of the notion of prochirality on alcohol dehydrogenase action has intriguing consequences. For example, it follows from the previous discussion that the biologically active receptor site must also be prochiral! As such, the three-dimensional nature of the receptor (chiral or achiral) is immaterial as long as the *site* is prochiral. Furthermore, if there can be elements of prochirality to otherwise symmetrical three-dimensional structures, there can be elements of prosymmetry in otherwise chiral three-dimensional structures.

2.11.4 Bilateral Symmetry

Bilateral symmetry is the symmetry of left and right. The topology of the human form is the primary example, and this form of bilateral symmetry is strictly geometric (Weyl, 1952). The distinction between left and right is arbitrary and relative. While there is no fundamental difference between left and right to a mathematician, there is clear distinction between left and right in biochemistry. The topic is not covered in symmetry point-group theory. Nevertheless, in the sense that bilateral symmetry results from the combination of geometric opposites, it appears in chemistry in various forms. In abstract form, there are cases of nongeometric bilateral symmetry, but in the final analysis, those abstract forms may also be geometric. For example, there is abstract bilateral symmetry for the pH of a solution at neutrality. The number of positively charged anions and cations is the same. However, their size is different, and a case for bilateral dissymmetry arises with respect to ion size that has application to taste chemistry.

The composition of racemic mixtures is another form of geometric bilateral symmetry. Such mixtures are optically neutral because D- and L-tartaric acid crystals, for example, have equal optical rotatory power, but of opposite sign. The rapid interconversions of conformational forms that are fundamentally chiral (e.g. ethane, ethylene glycol) are examples of conformational bilateral symmetry. A case for bilateral symmetry can also be said to occur in a single structure. These are the *meso* forms of compounds such as *meso*-tartaric acid. Such forms have a bisecting mirror plane of symmetry, as the reflection of the left-hand portion of the molecule generates the right-handed form, and *vice versa*. One of the asymmetric carbon atoms in *meso*-tartaric acid (Fig. 2.13) must be R- and the other S-, as shown below, and *meso*-tartaric acid has an intrinsic bilateral symmetry (cf. Shallenberger & Wienen, 1989).

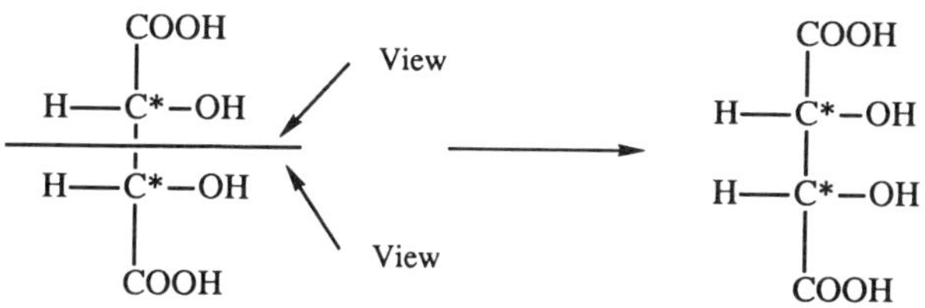

The occurrence of chiral substituents in various structures presents a problem with the conceptualization and definition of an asymmetric carbon atom. The definition problem has a parallel in taste chemistry, as the taste receptor recognizes the chirality of amino acids, but not the chirality of sugars. The traditional definition of an asymmetric carbon atom is that it is a carbon atom substituted by four different groups. Failure to specify that the groups have different size leads to cases where two of the four different groups on a carbon atom can be different only in the sense that they are optical antipodes (cf. Jaeger, 1930). In structure I below, the *rectus* (R) and *sinister* (S) substituents are the same size, but they differ because they are optical antipodes. Number I is the general structure for a *meso* compound such as *meso*-tartaric acid, and has an intrinsic bilateral symmetry. Its reflected structure (S, R) is superposable on the original through a simple rotation operation. On the other hand,

Structure II would also seem to defy the traditional definition of an asymmetric carbon atom because two of the substituent groups are identical, but it is *not* superposable on its mirror image structure. Reflections of the two R groups transform them into two S groups. Number II is not symmetrical, but chiral, as is the enantiomer (S, S). Essentially, neither RR nor SS is bilaterally symmetric, but possess handedness instead. The asymmetric carbon atom definition problem is resolved, either by insisting that the four different groups in the definition need to be of different size, or by application of chiral principles set forth by Mislow & Siegel (1984).

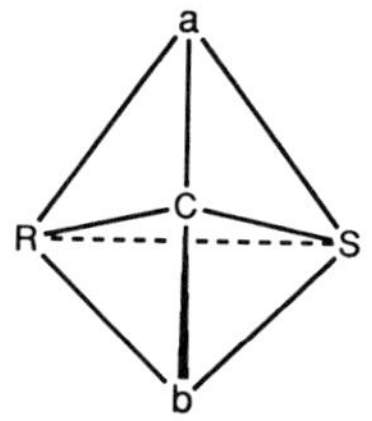

I

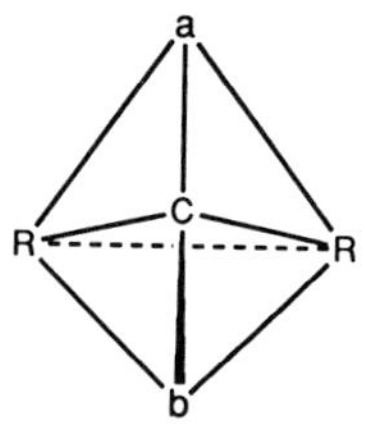

II

2.12 RECOGNITION CHEMISTRY

In addition to the conventional criteria given for the nature of weak chemical interactions (Sections 2.7.3–2.7.7), there is a final notion of chemical interactions. It is known as recognition chemistry, and has application to taste chemistry (cf. Venanzi & Venanzi, 1988*a*, 1991), and to the initial events associated with many biological chemistry phenomena. It has, as its basis, the observation and conviction that chemicals that elicit a biological response do not seem to arrive at the site of interaction through a random walk or through a probability consequence of Brownian motion (Jeffrey, 1993). Molecular electrostatic interactions over a long range, electrostatic topography, hydrophobic topography, and other factors are associated with recognition chemistry.

For the binding of carbohydrate to antibodies, Lemieux (1982) reasoned that the hydrogen bonding of carbohydrates to a combining site is incidental. It only allows a better proximity between the hydrophobic areas on the carbohydrate structure and a complementary hydrophobic region in the combining site. It was further suggested those hydroxyl groups that find their way into the combining site most likely do so in an intramolecularly hydrogen-bonded form. This notion of hydrophobic topology has subsequent application to a model for sweetness recognition chemistry (Chapter 12).

An example from enzyme chemistry for some force other than Brownian movement to guide a substance to the proper site, where full binding and a biochemical transformation take place, is invertase hydrolysis of sucrose (Section 5.3.4). Only a few hydroxyl groups, acting through long-range (cooperative?) hydrogen bondings are required to steer the sucrose molecule to the enzyme binding site. There, full binding occurs, in addition to an appropriate topographic fit. The enzyme then cleaves the sucrose molecule by adding a molecule of water to the glycosidic linkage. In essence, then, the aforementioned discussion of prochirality (Section 2.11.3.6) is an example of a prochiral recognition reaction between the enzyme (alcohol dehydrogenase) and the substrate (ethyl alcohol) whereby the interaction takes place prior to binding and subsequent conversion of the substrate to products, e.g. acetaldehyde.

The kinetics of recognition chemistry reactions is also similar to other reactions. The forces involved also are the same for other weak intermolecular forces, but may even be weaker and/or fewer. Therefore, the topological requirements for recognition chemistry are probably not as rigid as those for enzyme transformations, immunological reactions, and pharmacological responses. Perhaps the most important example of what is meant by taste recognition chemistry is contained in the consideration of the manner with which sugars interact with water (Section 6.6.1). The evidence to indicate that such phenomena are active in taste chemistry is neatly contained in the observations presented by Lancet & Ben-Arie (1991). Those authors proposed that all receptors function by way of multiple elementary interactions (EIs). Based on the assumption of a free energy of 1·4 kcal/EI, 6–9 EIs are required to generate typical pharmacological or enzyme–substrate binding. However, the sugars elicit the sweet taste response with only one standard elementary interaction. Apparently, a salient feature of recognition phenomena is to increase, through orientation, the effective concentration of a tastant, particularly in the case of sweeteners.

CHAPTER 3

Water and Inorganic Compounds

The most general classification of chemical compounds is that of inorganic and organic substances. Inorganic substances generally do not contain a carbon atom, whereas organic substances contain several. Another distinction is that carbon substances containing only a few elements, and have their names derived from those elements, are inorganic compounds (carbon dioxide, hydrogen cyanide, sodium chloride). Water is definitely an inorganic compound (dihydrogen oxide) and methyl alcohol is definitely an organic compound. As with all empirical classification schemes, the distinction is not particularly clear when the lower molecular weight substances are considered. Formaldehyde, CH_2O might be considered to be either an inorganic substance, or the simplest sugar, as it corresponds to the general molecular formula for sugars, or $C_n(H_2O)_n$.

3.1 WATER

The properties of water, and of substances dissolved in water, are so complex it is difficult to distinguish between those important to taste chemistry, and those that are not. Clearly, the hydration characteristics of tastants is significant in molecular recognition (cf. Galema *et al.*, 1990). Water is a liquid at ordinary temperature and pressure, while compounds with comparable molecular weight (O_2, N_2, CO_2, SO_2, H_2S, NH_3) are gaseous. The liquid form has unusually high surface tension, and when acids, bases, or salts are dissolved in it, water becomes a conductor for electric current. Liquid water also has a high dielectric constant, and an uncharacteristically high boiling point (100°C). The boiling point of carbon dioxide is $-78{\cdot}5$°C. The role of water in taste must not be underestimated, even though the full role is not known. To possess taste, chemical substances must be soluble in water, which serves as the transport medium. Water of hydration is also responsible for the solubility of chemical substances. Before the onset of their usual (salt, bitter) taste at higher concentrations, several inorganic substances taste sweet. Change in the structure of water at the chemical receptor for taste is the probable source of the energy (increase in entropy) required for a taste reaction.

3.1.1 Composition and Elemental Structure of Water

The molecular formula of water is H_2O. The two protons, each with a single orbiting electron, combine with an oxygen atom with six orbiting electrons in its outer shell. The electronic configuration of the oxygen atom is (2) $2s^22p^4$. For the Lewis model, the bonding with hydrogen is through covalent bonds, i.e. the sharing of electrons.

$$2\text{H}\cdot + \cdot\underset{\cdot\cdot}{\overset{\cdot\cdot}{\text{O}}}\cdot \rightarrow \text{H}\!:\!\underset{\cdot\cdot}{\overset{\cdot\cdot}{\text{O}}}\!:\!\text{H} \quad \text{or} \quad \text{H–O–H}$$

The Lewis model for water is also drawn illustrating that the sharing of electrons between H and O is often represented as a straight line. The water molecule, however, is not linear. The electron repulsion model for water predicts that the two single covalent oxygen bonds and the two lone electron pairs will be tetrahedral (Fig. 3.1). Although the observed angle is 105°, and not 109°28′ as shown, the tetrahedral orbitals of the

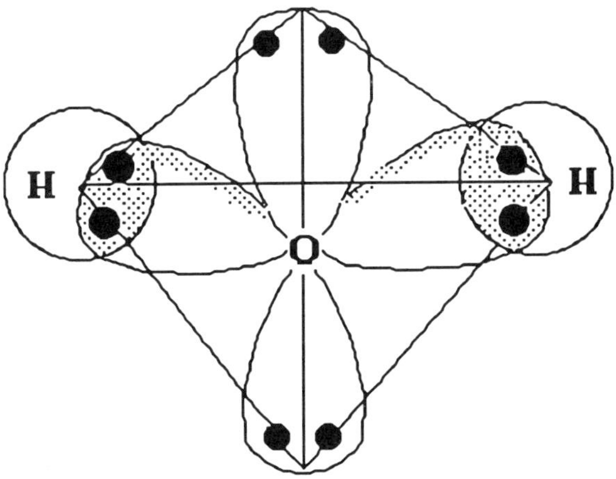

FIG. 3.1. Tetrahedral structure of water.

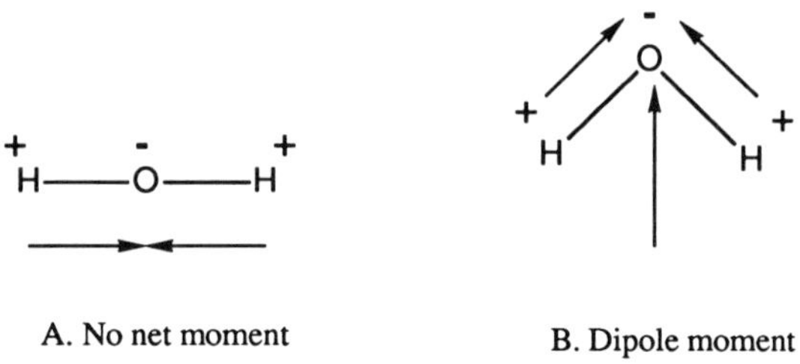

FIG. 3.2. The dipole moment in water.

model are sp^3 hybrids, and the O to H distance is 0·96 Å. Because of the tetrahedral orbitals, the water molecule is nonlinear. The water molecule is also dipolar, and has positive and negative centers that do not coincide. The model that accounts for the separation of charges (dipole moment) in water is the structure shown in B in Fig. 3.2.

At ordinary temperature and pressure a single water molecule, H_2O, with a molecular weight of 18, ought to be a gas. Not only is it not a gas at ordinary temperature and pressure, but it is a liquid with an inordinately high boiling point. The reason is that water has a secondary structure, formed from hydrogen bonds.

3.1.2 Hydrogen-Bonded Structure of Water

The notion that the properties of water are determined by hydrogen bonding is generally accepted. Derivation of a structure for water, however, is fraught with uncertainty. The exercise raises the conceptual problem of what is meant by the term 'structure'. For liquids, any given topology exists only for a very short while, and the topology can be observed only by methods that record the structure at even shorter times (Jeffrey & Saenger, 1991). For water, diffraction studies from the crystalline state support continuum models, such as the tridymite structure of Bernal & Fowler (1933), while thermodynamic evidence favors mixed models with 'instantaneous' views.

3.1.2.1 *Continuum Models*

In continuum models for the structure of water, there is coordination of a water molecule with its neighbors. The coordination, on average, is fourfold, and each oxygen atom is associated with four others, for a total of five. The association of the five molecules of water is also tetrahedral (Bernal & Fowler, 1933), as shown in Fig. 3.3.

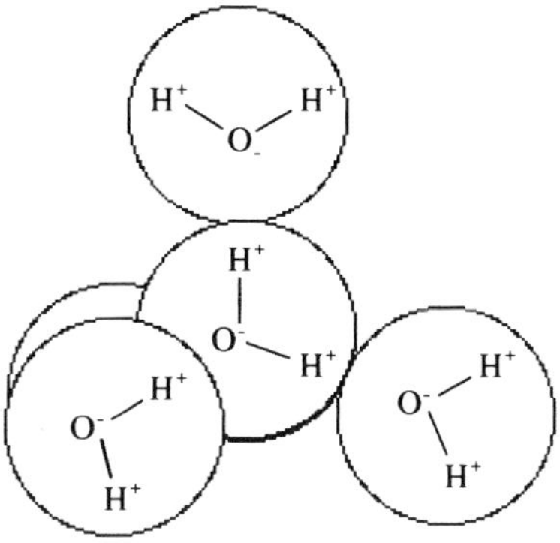

FIG. 3.3. Tetrahedral coordination of water molecules (Bernal & Fowler, 1933).

The tetrahedral coordination of a water molecule with four others serves as the basis for formation of ring forms. The hexagonal ring form serves as the basic unit for even further polymerization, leading to the *tridymite* structure of water. The term tridymite means three pairs of molecules, and the six-membered forms repeat, as shown in Fig. 3.4. Although Bernal & Fowler did not mention hydrogen bonding, they envisaged one conformational form of tridymite water to prevail in an upper layer, and the alternate form to prevail in a lower layer. The repeating units of water in a plane were formed from six-membered ring forms of water with bonding through extracyclic hydrogen atoms that are disposed in equatorial fashion. Vertical stacking was formed through bonding of extracyclic hydrogen atoms disposed in axial fashion. Layering is best caused by drawing molecules together that are made up of only one conformational form, as shown in Fig. 3.4. The latter point, and the notion of two different layers, is probably moot as the two forms are identical.

The polytridymite structure is thought to predominate in ice, and in liquid water. One view has it, that for water, 15–28% of the bonds are broken, and a static representation of this thought is shown for the planar surface of tridymite ice and water in Fig. 3.5.

The drawing depicts the difference between ice and water as a decrease in bonding for water that can collapse and increase density. In the dynamic model, the 15–28% decrease in bonds refers to a general decrease in total bonding strength and greater mobility. The fact that the rings and polymers are formed by hydrogen bonds explains the behavior of water. For example, hydrogen bonding strength is inversely proportional to the temperature. When temperature is increased, ice bonds collapse, volume shrinks, and even more molecules of water

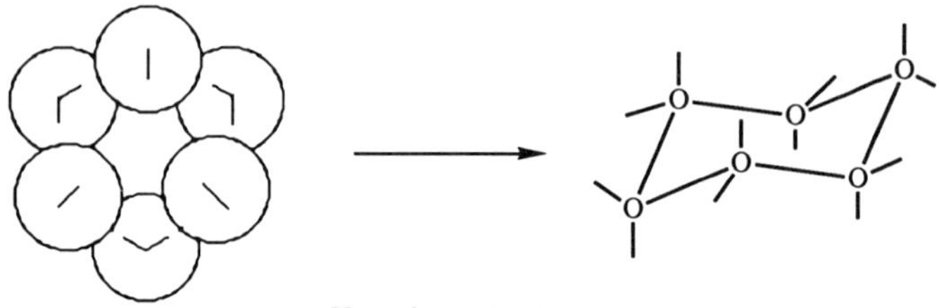

Upper layer structure

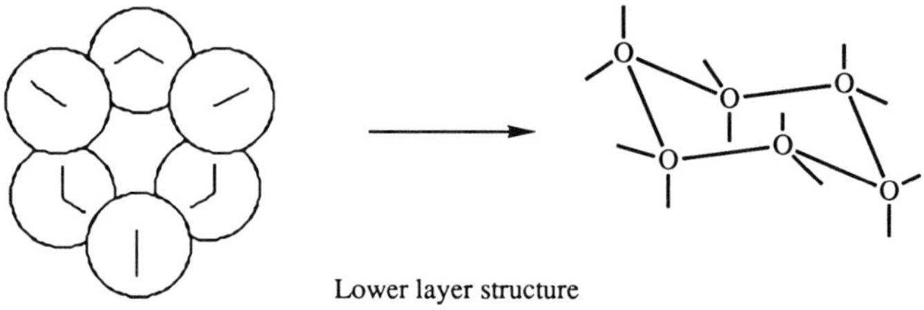

Lower layer structure

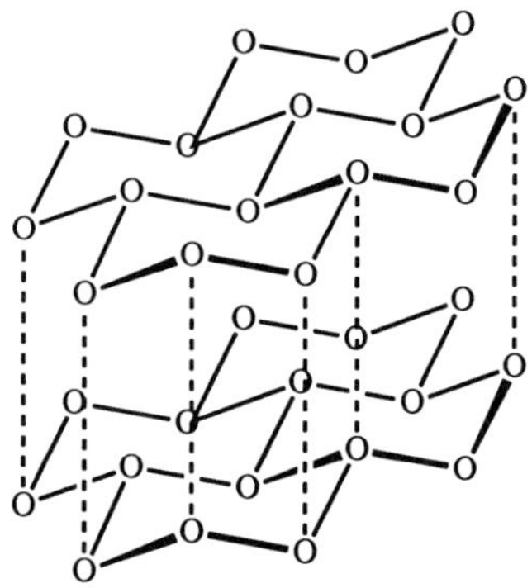

Tridymite continuum

FIG. 3.4. The tridymite structure of water.

depolymerize to form monomolecular and gaseous forms. In an analogous continuum model (Lennard-Jones & Pople, 1951) the oxygen atoms retain their four-coordination feature and there is continuous distribution of bond energies. Some possibilities for four-coordinated water are described by Minton (1970) and shown in Fig. 3.6. This continuum model is frequently employed to explain the physical

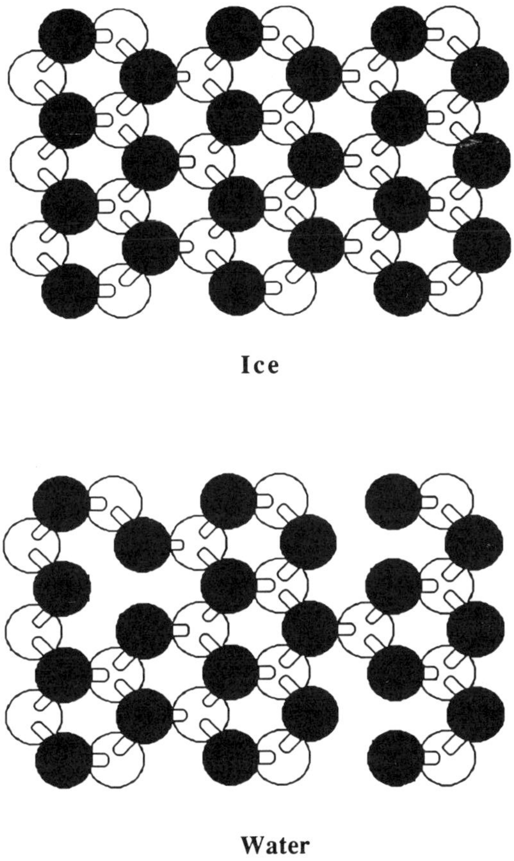

FIG. 3.5. Static model of tridymite structure of ice and water (Sienko & Plane, 1974).

properties of water, such as maximum density at 4°C. Parenthetically, the ring forms of water are similar to the conformational forms of cyclohexane, and sugar pyranose and furanose rings (Section 5.1.2.2). The existence of five- and six-membered ring forms presents the possibility of an infinitesimal number of conformational diastereoisomers where the extracyclic hydrogen atoms may assume equatorial, axial, quasi-equatorial, and quasi-axial dispositions.

There is little evidence for the static tridymite structure, and there is very little evidence to support the view that there are polygonic forms of

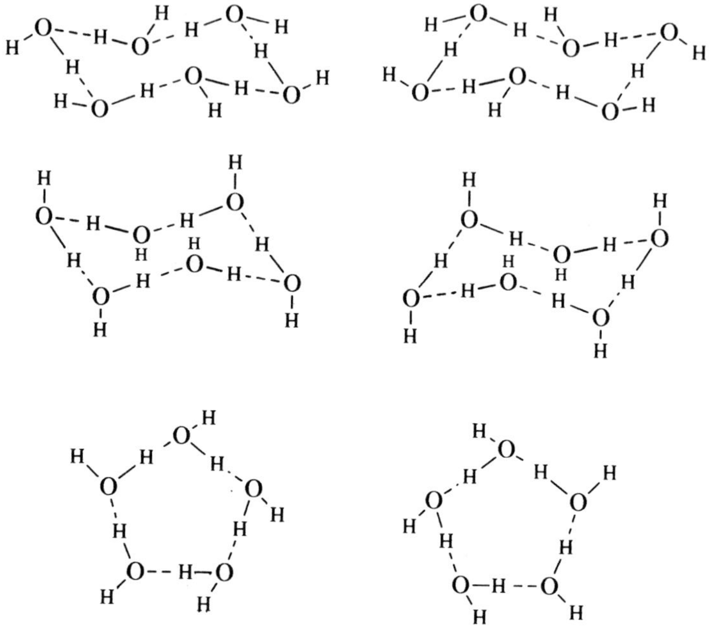

FIG. 3.6. Postulated ring forms for the structure of water.

water of various sizes, either. Nevertheless, the so-called static continuum structures have pedagogical and pneumonic significance (Chapter 6). A review of the various molecular species proposed for water up to 1968 was prepared by Stevenson (1968). It is concluded that a description of water in terms of sharply defined hydrogen bonded species is not consistent with its spectroscopic properties. Indeed, whether water, as indicated by spectroscopic studies, is a mixture of substances containing two different kinds of hydrogen-bonded groups has been argued for some time (Frank, 1970).

3.1.2.2 *Mixed and Dynamic Models*

Mixed and dynamic models for the 'structure' of water use the notion of clusters of three- or four-coordinated water molecules. They have a mean lifetime of about 1 ns (Jeffrey & Saenger, 1991). A flickering cluster model (Frank, 1958; Pauling, 1960) forms zero to four hydrogen bonds with its nearest neighbors. These are randomly arranged, with little or no periodicity. An instantaneous view from the central oxygen would reveal no sense of order beyond that of the first nearest neighbors. There is also a hydrogen-bonded network with apex-linked polygons with four, five, six, and more hydrogen bonds.

The model of Frank (1958) is composed of clusters with skewed hexagonal and pentagonal forms admixed with water molecules with four, three, two, one, or no hydrogen bonds (Fig. 3.7). In addition to the four-centered model for water, an oxygen atom with three-centered bonds has been proposed (Giguère, 1987; Giguère & Pigeon-Gosselin, 1988) for water and amorphous ice to afford better interpretation of Raman spectra. Note that some of the oxygen atoms in the clusters in Fig. 3.7 are both three- and four-coordinated (cf. Jeffrey & Saenger, 1991). In the dynamic model, the random three-dimensional networks of hydrogen bonds rapidly switch allegiance, continuously altering the network topology (Stillinger, 1980). According to Narten & Levy (1969), the bulk phase, or cluster, consists of somewhat ordered hydrogen-bonded molecules, and the other phase is more disordered and denser. The ordered cluster is short-lived (10^{-10}–10^{-11} s), and is continually exchanging molecules with the adjacent unstructured phase. To Jeffrey & Saenger (1991), when the concepts of three-center bonds, the

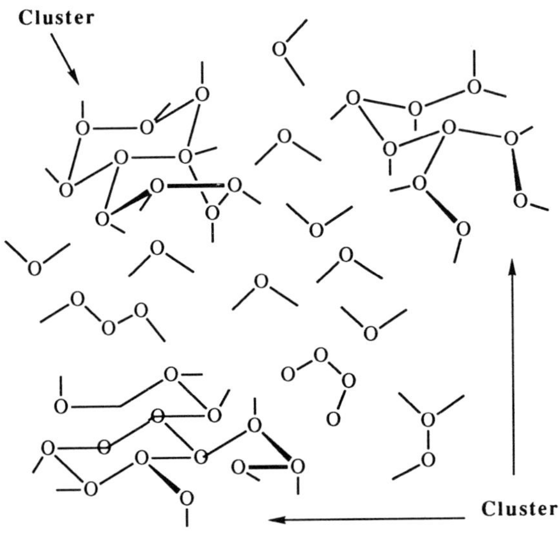

FIG. 3.7. The flickering cluster model for the 'structure' of water.

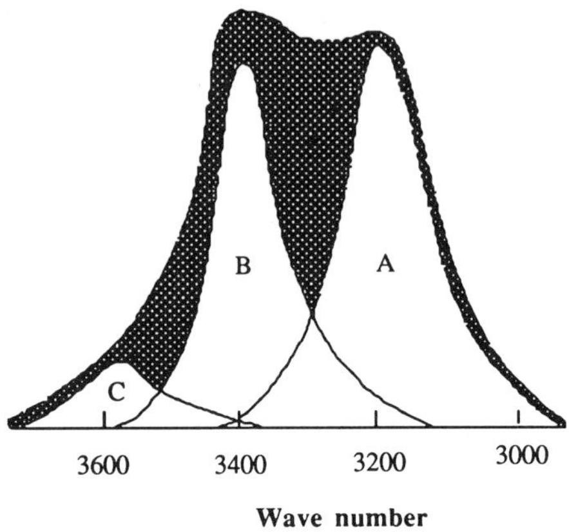

FIG. 3.8. Diagrammatic representation of mathematically resolved (deconvoluted) laser—Raman spectrum of water (Mathlouthi & Seuvre, 1988). The minor band D (1·5% of the total), occurring slightly above 3600 cm^{-1} is not shown.

cooperativities of hydrogen bonds (Section 2.7.3.4), cyclic structures, and flip-flop dynamics are added to the picture, it is not surprising that attempts to describe the structure and thermodynamic properties of water in terms of a single conceptual model have proved elusive.

Essentially, Raman and infrared spectra of water show two major absorption bands (Narten & Levy, 1969). A mathematically resolved (deconvoluted) depiction of a laser-Raman spectrum for water (Mathlouthi & Seuvre, 1988) is shown in Fig. 3.8 above. The three resolved bands shown are believed to represent different species of water, and a smaller one that is believed to be due to free water is not shown. Band assignments for the different wave numbers (ν) proposed by Mathlouthi & Seuvre (1988) are:

A. $\nu \approx 3223\,cm^{-1} \rightarrow O\cdots O \rightarrow$ Quasi-crystalline water.

B. $\nu \approx 3413\,cm^{-1} \rightarrow O\cdots O \rightarrow$ Amorphous (solid-like) water.

C. $\nu \approx 3547\,cm^{-1} \rightarrow O\cdots O \rightarrow$ Amorphous (liquid-like) water.

D. $\nu \approx 3638\,cm^{-1} \rightarrow O\cdots O \rightarrow$ Free water.

According to the interpretation of the spectrum shown in Fig. 3.8 by Luu *et al.* (1982), one major band represents crystalline ('pentamére' or 'molécule de glace') water, and the other is 'dimére'. The third is 'trimére' and the fourth is free water. Luu *et al.* (1982), believe that their assessment of the absorption bands is in accord, not only with the flickering cluster theory, but also with the uniform continuum model of Bernal & Fowler (1933).

3.1.3 Hydrogen Bonding of Water with other Functional Groups

The tetrahedral structure model of water (Fig. 3.1) has been criticized (Donohue, 1968). While it is attractive for explaining the structure of water and ice, Donohue believes it is unimportant for hydrogen bonding not only between water molecules, but also for other substances. For water of hydration in crystals, the disposition of the water oxygen atom, the two hydrogen acceptor atoms, and the one donor hydrogen-bonded atom, is approximately plane-trigonal. The tetrahedron structure of water is then quite irregular, and often skewed in different directions, as shown in Fig. 3.9. The skewing of the water molecule, in binding with other substances, tends to minimize the importance of the tenet that hydrogen bonds must be ideally linear, particularly when the interaction of taste substances might well be mediated through water structures. Linearity may be a prerequisite for stronger hydrogen-bonded interactions, but is not necessary for taste chemistry recognition phenomena. In this sense, then, the four-coordinated non-linear hydrogen bonds can be considered to be 'soft' bonds (Jeffrey & Saenger, 1991).

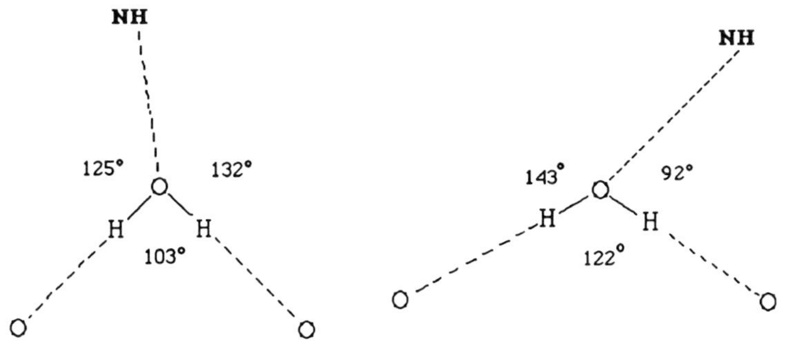

I

II

FIG. 3.9. Angles about the oxygen atom of the hydration water molecule participating in crystal hydrogen bonding in glycyltryptophan dihydrate (I) and in asparagine monohydrate (II) (Donohue, 1968).

3.1.4 Taste of Water

The tridymite form of water would seem to possess a primary and a secondary structure (adjacent hydroxyl groups in proper orientation) that might be capable of eliciting sweet taste, if it did indeed, exist. The receptor for sweetness is already hydrated, and free water, *per se*, apparently does not possess potential for interaction with the receptor for sweetness. Distilled water has a weak, insipid taste variously described as 'indefinable', 'slightly disagreeable', 'metallic', or 'bitter'. Freshly distilled water, due to the absence of carbon dioxide, has a taste similar to that of very dilute alkali solutions (Moncrieff, 1967). The weak taste of distilled water is associated with 'tastelessness', and most investigators concede that while distilled water is not completely tasteless, a taste panel can be trained to equate it with tastelessness (cf. Trout & Sharp, 1937). The taste of drinking water is another matter, and varies with the presence or absence of microorganisms, concentration of carbon dioxide, and presence or absence of ions such as sodium, calcium, iron, and chlorine.

The taste of water also varies with the taste of previously sampled substances, and the taste is a consequence of adaptation. After adaptation to sodium chloride, water tastes sour or bitter (Bartoshuk *et al.*, 1964), and after adaptation to urea, water tastes salty (McBurney, 1969). After adaptation to acid or bitter tastes, water tastes sweet (Bartoshuk, 1968; McBurney, 1969; McBurney & Bartoshuk, 1972). After substances with persistent sweet taste (dihydrochalcones, protein sweeteners) have been tasted, it is found that water also seems to taste sweet (Bartoshuk, 1987). In the latter case material bound to extraneous regions of the oral cavity is released by water and thus made available for interaction with the sweetness receptor.

3.2 ALKALI METAL AND HALOGEN SALTS

In dilute solution sodium and potassium hydroxide, and the halogen salts of sodium, potassium, and lithium taste sweet. In more concentrated solution, their tastes vary, but with exception of the hydroxides, generally taste salty. The sodium, potassium, and lithium metals belong to the Group I class of elements, known as the alkali metals. The variable taste of their salts is clearly related to the manner with which they interact with water, and those interactions are of fundamental importance to taste chemistry. The salts of the alkali metals dissociate completely in water solution, and their water solubility depends on a fine balance between

TABLE 3.1
Thermodynamic parameters for the solvation of NaCl at 298 K

Step		ΔH	ΔS	ΔG
1. NaCl(s)	$\longrightarrow Na^+(g) + Cl^-(g)$	+770	−0·277	+702
2. $Na^+(g) + Cl^-(g)$	$\longrightarrow Na^+(aq) + Cl^-(aq)$	−766	−0·184	−711
3. NaCl(s)	$\longrightarrow Na^+(aq) + Cl^-(aq)$	+4	+0·043	−9

crystalline lattice energy, hydration energy, and the entropy of the ions. When a crystalline salt (MX) is dissolved, a two-step crystalline lattice breakup process begins (Sienko & Plane, 1974). First, gaseous (g) ions are formed, or $MX(s) \rightarrow M^{n+}(g) + X^{n-}(g)$. The gaseous ions are then hydrated, and $M^{n+}(g) + X^{n-}(g) \rightarrow M^{n+}(aq) + X^{n-}(aq)$. The overall process is $MX(s) \rightarrow M^{n+}(aq) + X^{n-}(aq)$.

The course of events for the solvation of NaCl has particular significance to taste chemistry principles, and is presented herein in the format employed by Sienko & Plane (1974). The stepwise solvation process for NaCl, along with thermodynamic parameters for each step, are shown in Table 3.1. Joules are used instead of calories in this example in keeping with international chemical convention (Sienko & Plane, 1974). For reference, 1 calorie = 4·184 J and 1 J = 0·2390 cal.

The term ΔH, as kJ/mol^{-1}, is the enthalpy change; ΔS, as $kJ\,mol^{-1}\,K^{-1}$, is the entropy change; ΔG, as kJ/mol^{-1}, is the change in free energy. For NaCl the lattice energy is +770 kJ/mol and the hydration energy is −766 kJ/mol, so the solvation process for NaCl appears to be energetically unfavorable. However, the process of dissolving occurs because there is a favorable entropy change. In Table 3.1, the bottom line is the sum of the two solvation steps. The ΔH represents the increase in enthalpy of the system. It is equal in magnitude but opposite in sign for the 'heat of reaction'. Step 2 has associated with it $\Delta H = -766$ kJ/mol, and the enthalpy of the Na^+ and Cl^- system has decreased by 766 kJ in going from the gaseous to the hydrated state. The decrease in H came about because of liberation of 766 kJ to the surroundings. The heat of the reaction (heat evolved to the surroundings) is +766 kJ, but the ΔH of the system is −766 kJ. The ΔS for the second step equals $-0{\cdot}184\,kJ\,mol^{-1}\,deg^{-1}$ indicating that the ion system decreases in entropy by 0·184 kJ in going from the gaseous to the hydrated states.

When ΔH and ΔS are known, ΔG can be calculated from

TABLE 3.2
Standard enthalpies (ΔH°) change for the breakup of selected halogen salts of Group I metals ($kJ\,mol^{-1}$)

LiF, 1008	NaF, 904	KF, 803
LiCl, 828	NaCl, 770	KCl, 699
LiBr, 791	NaBr, 736	KBr, 674
LiI, 732	NaI, 690	KI, 632

$\Delta G = \Delta H - T\Delta S$. With $T = 298\,K$, substitution in the equation gives $-766 - (298)(-0{\cdot}184) = -711\,kJ/mol^{-1}$. Thus, when $Na^+(g)$ and $Cl^-(g)$ become hydrated, there is a decrease in the free energy of the system by 711 kJ. The significant point contained in Table 3.1 is that for the final overall process, although ΔH is positive, and energetically unfavorable, ΔG still turns out to be negative; therefore the overall dissolving reaction is favored. The favorable ΔG comes about because the $T\Delta S$ term, which is equal to $(298\,K)(+0{\cdot}043kJ\,mol^{-1}\,deg^{-1}) = 13\,kJ/mol^{-1}$. This subtracts from the unfavorable ΔH of $+4\,kJ\,mol^{-1}$ to give $-9\,kJ\,mol^{-1}$. The positive ΔS term arises from the fact that ions in the solution, although more restricted than in the gaseous state, have considerably more freedom than they do in solid NaCl. The overall dissolving change is one of increasing disorder and hence of increasing entropy.

The ΔH° (standard enthalpy change) for the lattice breakup is equivalent to the lattice energy. For selected substances important to the subject matter in this treatise the ΔH° for selected halogen salts with characteristic tastes is given in Table 3.2. For the crystalline alkali halides, lattice energy decreases with increasing anion and cation size. If a salt is to have appreciable water solubility, the energy needed for the breakup of the lattice needs to be offset by hydration of the ions. The standard enthalpy change (ΔH°) in $kJ\,mol^{-1}$ and the standard entropy change (ΔS°) in $kJ\,deg^{-1}\,mol^{-1}$ for selected ions is shown in Table 3.3. The negative values for ΔH° and ΔS° in Table 3.3 indicate that heat is liberated upon solvation, and the smaller the radius of the ion, the smaller the ΔH° of hydration. The ΔS° of hydration applies to the process $M^+(g) \rightarrow M^+(aq)$ and $X^-(g) \rightarrow X^-(aq)$. The fact that the values are all negative indicates that the ions in the hydrated state have a lower entropy (smaller number of configurations) than in the gaseous state, i.e. they have a higher level of free energy (ΔG°). An interesting feature with respect to the values for ΔS° given in Table 3.3 that has application to the taste of ions is that they are smaller in magnitude than

TABLE 3.3
Standard enthalpy and entropy change for selected ions

Ion	ΔH°	ΔS°
Li^+	−516	−118·8
Na^+	−397	−87·4
K^+	−314	−51·9
Rb^+	−298	−40·2
Cs^+	−255	−36·8
F^-	−506	−155·2
Cl^-	−377	−96·7
Br^-	−343	−83·3
I^-	−297	−60·7
OH^-	−502	−108·8

might be expected from calculated values. For $Na^+(g) \rightarrow Na^+(aq)$, the calculated value of ΔS° is $-120\,J\,deg^{-1}\,mol^{-1}$, but only $-87{\cdot}4\,J\,deg^{-1}\,mol^{-1}$ is observed. This seems to be a manifestation of the breakup of the structure of water due to the orientation of water molecules about the ions to generate 'melted zones'.

3.2.1 Reactivity of the Group I Metal Salts

The reactivity of the salts also has application to their taste. The Group I metals are quite reactive because they have an s^1 outer electronic configuration, and the electron is easily lost to form a +1 ion in solution. As a consequence, the ions never occur naturally in free form, but in combination with elements such as oxygen and chlorine, and always in the +1 oxidation state. All the compounds of the Group are ionic and the ionization energy and electronegativity decrease as the distance of the outer electron shells from the nucleus increases, from Li to Cs. Nearly any substance capable of being reduced in solution will be reduced by the alkali metals. In turn, the metals are oxidized through loss of electrons.

The reducing property of the alkali metals is described through reference to the electrode potentials given in Table 3.4. All of the potentials are large and negative indicating good tendency for the half reaction

$$M(s) \rightarrow M^+(aq) + e^-$$

to proceed to the right. The 'whole' reaction involves the chemistry of the anion also. Lithium has the greatest propensity to be reduced and sodium the least, although the ionization potentials shown in Table 3.4 suggest

TABLE 3.4
Alkali metal ionization properties

Metal	*Electronic configuration and (Core e^-)*	*1st ionization potential (eV)*	*Ionic radius (Å)*	*Electrode potential (eV)*	$\Delta G^{\circ\,a}$
Li	(2) $2s^1$	5·39	0·68	−3·05	+162
Na	(10) $3s^1$	5·14	0·97	−2·71	+203
K	(18) $4s^1$	4·34	1·33	−2·93	+181
Rb	(36) $5s^1$	4·18	1·47	−2·93	+182
Cs	(54) $6s^1$	3·89	1·67	−2·92	+182

[a] Change in standard free energy in kJ/mol^{-1} for the process $M(s) \rightarrow M^+(aq) + e^-$.

that it should be more difficult to extract an electron from lithium than from sodium. As with the case of NaCl solubility, this apparent anomaly is a manifestation of the difference between gas phase behavior and solution behavior. The ionization potential applies to the gaseous behavior of the atom, whereas the electrode potential applies to the solid metal and its ionic species in solution. The component parts of the aforementioned reaction can be treated as being composed of the three steps shown in Table 3.5, along with ΔG° (kJ/mol^{-1}). The total free energy (ΔG°) required for the three steps combined is shown in the last column of Table 3.4 for the alkali metals. In the first step in Table 3.5 the atoms are transformed (sublimed) to the gaseous state. There, they are independent of each other. In step 2 an electron is subtracted from the M(g) atom to give a gaseous ion $M^+(g)$. The energy required to effect this step is the ionization potential, which is largest for Li. The fact that step 2 is most difficult for Li is compensated for by step 3. The hydration energy of Li is so great (−480 kJ/mol) that the reaction $Li(s) \rightarrow Li^+ + e^-$ has the greatest tendency to occur. The energy for the reaction to proceed (the positive ΔG° values in Table 3.4 indicate that this must be so) is supplied by the energy of hydration.

TABLE 3.5
ΔG° (kJ/mol^{-1}) for the stepwise ionization of metals

	Li	*Na*	*K*
1. $M(g) \longrightarrow M(g)$	+122	+78	+61
2. $M(g) \longrightarrow M^+(g) + e^-$	+520	+496	+419
3. $M^+(g) \longrightarrow M^+(aq)$	−480	−371	−299

TABLE 3.6
Halogen properties

Halogen	*Electronic configuration and (Core e^-)*	*Ionization potential (eV)*	*Ionic radius (Å)*	*Electrode potential (eV)*	$\Delta G^{\circ a}$
F	(2) $2s^2 2p^5$	17·42	1·33	+2·87	−730
Cl	(10) $3s^2 3p^5$	13·01	1·81	+1·36	−591
Br	(28) $4s^2 4p^5$	11·84	1·96	+1·09	−563
I	(46) $3s^2 5p^5$	10·44	2·20	+0·54	−515

[a] Change in standard free energy in kJ/mol^{-1} for $1/2\,X_2(g) + e^- \rightarrow X^-(aq)$.

Because a Li^+ ion is so small, water molecules approach the center of the ion so closely as to make the hydrated ion the most stable. Sodium and potassium follow suit although sodium, among the group, as indicated by the electrode potential in Table 3.4, has the least tendency to lose an electron. While this is a property of the metal, and not the ion, it affects the properties of the ion in an indirect way that also seems to influence taste properties. Like the lithium ion, the sodium ion also has good ability to interact with water.

3.2.2 Halogens

The halogen ions are the counterpart structures of the Group I metal salts that have significant taste properties. The halogens are members of the Group VII elements, and have high electronegativity. The electrode potentials are positive (Table 3.6). Therefore the substances are good oxidizing agents. The stepwise solvation of the halogens and the corresponding free energy, ΔG° (kJ mol^{-1}) for each step are given in Table 3.7. Among the halogens, chlorine requires the most energy to break up the X_2 molecule (Step 1), and it is also easiest to add an electron to chlorine (Step 2). Totals for the entire solvation process for each halogen make up the last column of Table 3.6. Again, while these attributes apply to the element, and not the ion, they have application to

TABLE 3.7
ΔG° (kJ/mol) for the solvation of halogens

	F	*Cl*	*Br*	*I*
1. $1/2\,X_2(g) \longrightarrow X(g)$	+63	+105	+79	+59
2. $X(g) \longrightarrow X^-(g)$	−133	−348	−324	−295
3. $X^-(g) \longrightarrow X^-(aq)$	−460	−348	−318	−279

the ionic nature of the chlorine ion. As with the alkali metals, the difference among these elements is the hydration step in the solvation of them, and ΔG° is inversely related to ion size.

3.2.3 Ionic Structure of Sodium Chloride

Sodium chloride (NaCl), or common salt, is an essential human nutrient. The white crystalline material melts at 804°C. In the crystalline lattice structure (Fig. 3.10), each sodium atom is associated with six chlorine atoms, and each chlorine atom is associated with six sodium atoms. In solution, the crystalline lattice structure is broken, and the sodium and chlorine ions (Na^+ and Cl^-) are completely dissociated, and hydrated. The hydrated sodium ion and the hydrated chlorine ion, each with a coordination number of four, are shown in Fig. 3.11. The coordination number for hydration is reduced from that of the crystals because of the size of the water molecule, and the number that can be packed in mutual contact around the ions. The net effect of the hydration shells is to generate 'melted zones' in the tridymite structure of water. The oxygen atom of water is directed toward the sodium ion, and the positioning of the water molecule around the chlorine ion is reversed. Neither of these configurations match the fairly rigid, albeit transient, structure of water, and they 'disrupt' it.

The ability of sodium chloride, and other solutes such as sugar, to

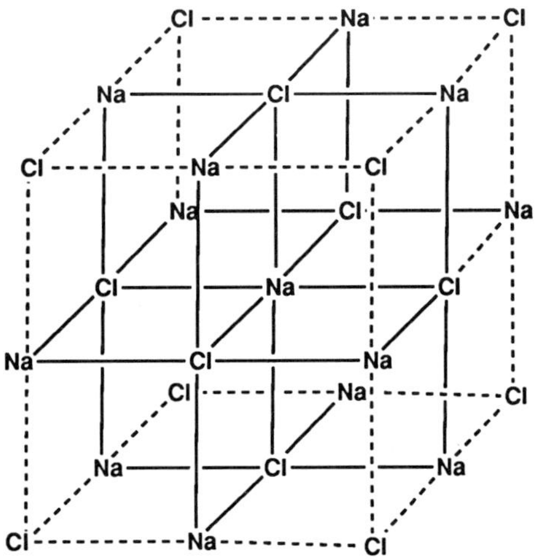

FIG. 3.10. Arrangement of sodium and chlorine ions in the sodium chloride crystal.

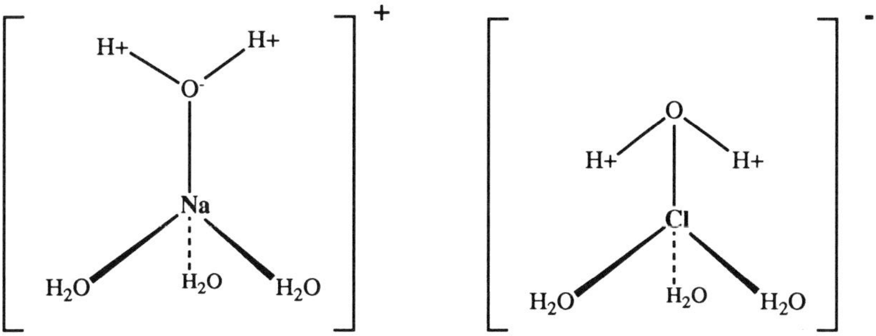

FIG. 3.11. Diagrammatic representation of fully hydrated sodium and chlorine ions.

lower the freezing point and elevate the boiling point of water is partly due to the interaction between the water molecules and the solute. The latter is stronger than the attraction among water molecules. On the whole, the phenomena of boiling point elevation and freezing point lowering are due to reduction in the concentration of water molecules, *per se*. The entropy of water in the solution is also greater than in pure water, increasing the tendency for water to remain in the condensed state. Of particular significance to the subject of this text, is the behavior of dilute solutions of halide salts.

3.2.4 Solution Properties of Group I Salts

Solutions of electrolytes show anomalous behavior with respect to conductivity, freezing point depression, and osmotic pressure. Plots of freezing point depression versus electrolyte concentration show an inflection at low concentrations. The minimum in molecular lowering of the freezing point of water for NaCl and KCl occurs between concentrations of 0·01 and 0·001M. While it was believed by Arhennius that the anomalous behavior could be explained by incomplete electrolyte dissociation, Jones and coworkers (Jones, 1914) attributed the variable properties to the formation of hydrates, which is thought to be greatest in dilute solutions.

It was also observed that the molecular conductivity of electrolytes increased as the dilution increased. The conductivity of KCl at different concentrations shown in Fig. 3.12 illustrates that while the conducting power increases with dilution, there is a marked inflection in the curve. At concentrations below 0·015M KCl, there is negligible effect on conductivity, while at concentrations greater than 0·015M, there is rapid

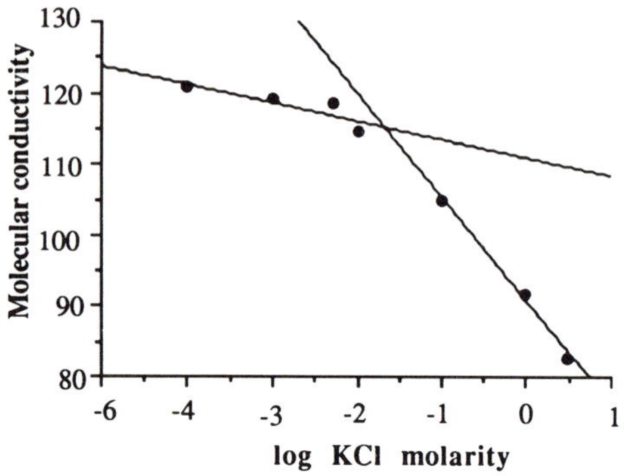

FIG. 3.12. Change in the molar conductivity of KCl with increasing molar concentrations of KCl (Jones, 1914).

decrease in conductivity. There are few exceptions to this generalization, and NaCl solutions exhibit the same biphasic effect. From conductivity data, and upon observing the effect of increased temperatures on conductivity coefficients, it is concluded that ion-hydrate associations are more complex at lower solution concentrations and at lower temperatures. While the usual coordination number for water of hydration for cations is four (Bernal & Fowler, 1933), at lower concentrations it seems to be higher. The more fully hydrated ions also seem to behave in an anomalous fashion. It would seem that there must be hydration shell species at low concentrations that have lowered conductivity potential. Hence the slow decrease in molecular conductivity with increasing concentrations of KCl (Fig. 3.12).

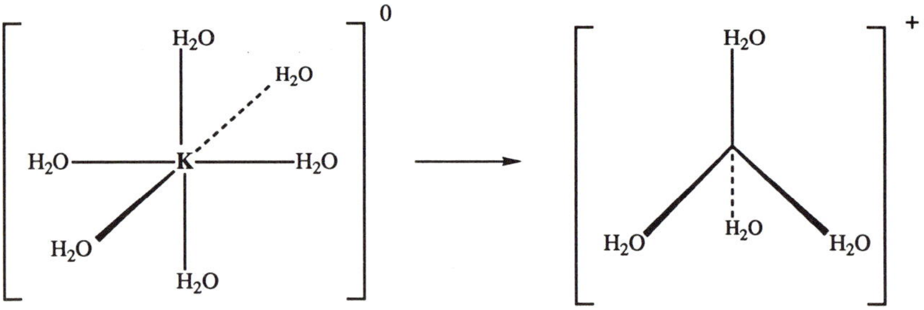

Potassium ion hydration shells

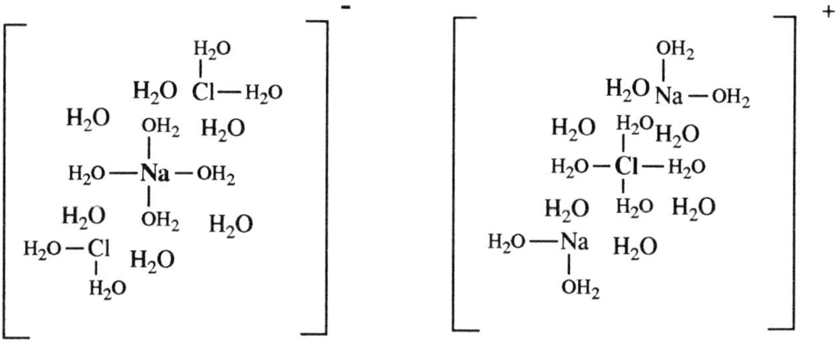

FIG. 3.13. Cluster ions of sodium chloride in water solution.

Another interpretation for the loss in molecular conductivity with increasing concentration phenomena is that it arises from interionic attractions between Na^+ and Cl^- in the solution state (cf. Sienko & Plane, 1974). As the average distance between ions is decreased by increased concentration, attractive force between positive and negative ions increases, and draws them together. Thus, in an aqueous solution there is more chance of finding more less-completely hydrated Cl^- ions near a positively charged sodium ion, and *vice versa.* The hydrated sodium ion with the attendant chlorine ions form a negatively charged cluster, and the hydrated chlorine ion with the attendant sodium ions form a positively charged cluster, and they draw together, as shown in Fig. 3.13. With each ion rich in an 'atmosphere' of its counterion, movement in an electric field would be greatly restricted. Thus, there is decrease in molecular conductivity with increased concentration. The proposal accounts for observed freezing point lowering phenomena, and anomalous behavior can be attributed to an insufficient number of water molecules available to hydrate fully all ions.

3.3 LEAD AND BERYLLIUM SALTS

The salts of two metals, one belonging to the Group II elements, and the other to Group IV, generally taste sweet although salts of beryllium are incredibly toxic, and salts of lead are cumulatively toxic.

3.3.1 Salts of Beryllium

Beryllium (glucinium, glucinum) belongs to the alkaline earth group of elements generally characterized by water insolubility, heat stability, and an alkaline reaction in water solution (calcium and magnesium oxides). Among the Group II metals, solutions of beryllium compounds are acidic, and beryllium is the 'odd-man-out' (Hashimoto *et al.*, 1987). Beryllium has two electrons in its outer shell (the core configuration is (2) $2s^2$), and the ionic radius is 0·35 Å. It has an electrode potential of −1·85 V.

Usually, Group II elements are not as good reducing agents as are those of Group I (the electrode potential of lithium is −3·04 V). Like the Group I elements, the net process

$$M(s) \rightarrow M^2 + (aq) + 2e^-$$

has great tendency to occur. The doubly charged ion interacts strongly with water to form a hydrated ion. Like lithium, beryllium has a very small ionic radius, and the double charge and small ionic radius combine to distinguish beryllium from the other Group II elements.

Because of the ability of the Be^{2+} ion to induce changes in water molecules, beryllium chloride serves as a model for studying the behavior of electrolytes (Probst *et al.*, 1986). Solid beryllium chloride is a unique substance. It is composed of covalently bonded $BeCl_2$ chains held together by weak intermolecular (Van der Waals) forces. Like LiCl, it is very deliquescent, but has a melting point to be expected of a covalent compound instead of an ionic solid such as LiCl or $CaCl_2$. Liquid $BeCl_2$ does not conduct electricity, indicating the absence of ions. While very soluble in water, solutions of $BeCl_2$ are strongly acidic (LiCl is weakly alkaline) and the proposed reaction is

$$Be^{2+} + 4H_2O \rightarrow Be(OH)_4{}^{2-} + 4H^+$$

Confirmation that the primary hydration number for the beryllium ion is four is obtained from X-ray diffraction studies and *ab initio* molecular orbital calculations (Probst *et al.*, 1986; Hashimoto *et al.*, 1987). The Be^{2+} ion is very stable when hydrated because of electron transfer from the water molecule to the Be^{2+} ion. The hydrogen atoms of the hydrated complex are very positive (naked) and the O–H bonds more polar than those for water. The Be^{2+} ion increases the negative partial charge on the oxygen atom and the positive partial charge on the hydrogen atom. The hydration enthalpy change of the Be^{2+} ion is very large (−2 470 kJ/mol), and when divided by the hydration number, the strong hydration around

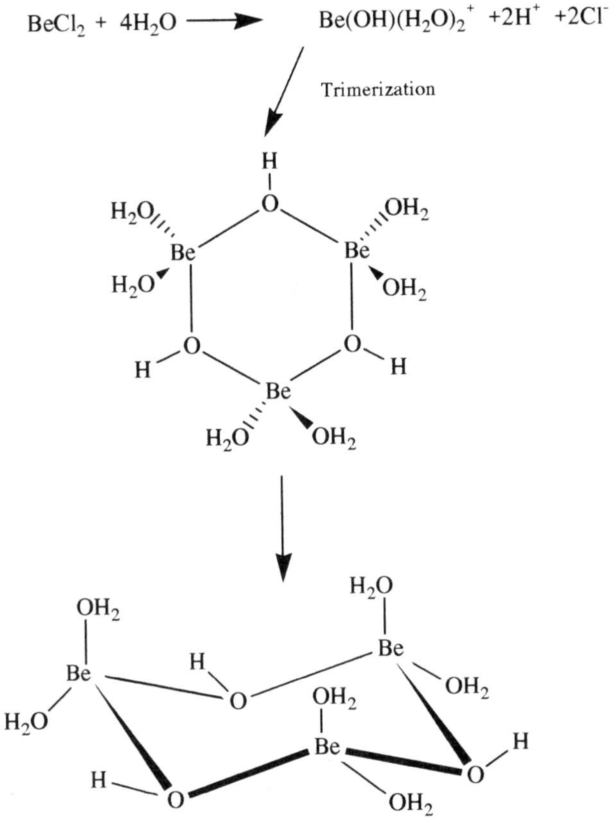

FIG. 3.14. Trimerization of beryllium ion hydrates.

the ion becomes even larger. The beryllium ion, *per se*, is an electron accepting site (Hashimoto *et al.*, 1987).

The final property of the hydrated Be^{2+} bond is that the hydrated complex polymerizes, and the most abundant form is trimeric $[Be(OH)(H_2O)_2]^+$ (Kakihana & Sillén, 1956; Bertin & Derouault, 1979). Taking the tetrahedral nature of the Be^{2+} ion, the oxygen atoms of the hydrating water molecules, and O–H stretching into account, an overall approximation of the interaction between $BeCl_2$ with water can be made. The resulting structure of the trimeric complex is shown in Fig. 3.14.

3.3.2 Lead Acetate (Sugar of Lead)

Lead belongs to the group IV elements, and most of the compounds have the +2 (plumbous) state. It too is capable of forming many complex ions. A conductivity study (Britton & Meek, 1931) of lead acetate shows an abnormally small degree of ionization, indicating that the Pb^{2+} ion probably undergoes hydration. It is believed that in aqueous solution the metallic acetate must furnish, through hydration, free acetic acid. The acetate, then rendered basic, ionizes to provide the solution with enhanced concentrations of acetate ions.

$$Pb(OAc)_2 + H_2O \rightarrow Pb(OH)_x(OAc)_{2-x} + xHOAc$$

$$Pb(OH)_x(OAc)_{2-x} \rightarrow [Pb(OH)_2(OAc)_{2-x-y}] + yOAc$$

3.4 TASTE OF SALTS

Only sodium chloride elicits a purely salty taste, and attempts to relate the salt taste quality and intensity to anions and cations have a long history. Höber & Kiesow (1898) concluded that the salt taste quality was essentially caused by halide anions (Cl^-, Br^-, I^-). Kionka & Strätz (1922) also concluded that saltiness quality is due to the anion, because all chlorides are primarily salty. They then proposed that the intensity of salt taste was determined by the cation, but with cations other than sodium, different tastes with varying intensity also occur (Höber & Kiesow, 1898; Kionka & Strätz, 1922). Skramlik (1921) attributed the traditional saltiness modality to anions, particularly Cl^-, but the bitterness of KCl was attributed to the potassium cation. On the other hand, the bitterness of sodium iodide was attributed to the iodine anion, as NaCl is not bitter. Beidler (1954) also suggested that the salt taste quality is largely due to the cations, but anion binding helped determine the magnitude of the response.

In view of the inability to assign saltiness to either anions or cations alone, several authors suggested that the fundamental quality is due to both anions and cations. Renqvist (1918) stated that the taste quality of salts at all concentrations is due to action of the salt as a whole, i.e. the anions, cations and undissociated species. The latter generalization gains some support from the observation that sodium salts of the organic acids (sodium acetate, di- and trisodium citrate, monosodium glutamate) have a saline taste and intensity related to their ability to dissociate.

Moncrieff (1967) also believed that both the cation and anion contribute to the overall taste quality of salts. Among salts, when the cation is held constant, e.g. Na^+, there is tendency for salt taste to become increasingly bitter with the increasing size of the anion. When the anion is held constant, e.g. Cl^-, increasing the size of the cation from Li^+, through Na^+, K^+, Rb^+, and Cs^+ again progressively diminishes saltiness, and increases bitterness. Attempting to ignore other tastes, and assigning a relative value of 1·0 for NaCl saltiness intensity, Skramlik (1926) recorded the saltiness intensity of a series of chloride cation compounds to be NH_4 (2·83) > K (1·36) > Ca (1·23) > Na (1·00) > Li (0·44) > Mg (0·20).

Apart from NaCl, the admixture of saltiness with other tastes (usually bitterness) increases in intensity with increasing salt concentration (Renqvist, 1918). The exception is LiCl, which is reported to increase in sourness with increasing concentration, but psychological confusion occurs often between the bitterness, sourness, and other unpleasant modalities. The increasing sourness reported by Dzendolet & Meiselman (1967) for LiCl is, probably, increasing unpleasantness, as LiCl salts have an alkaline pH, and the potential for alkaline taste. While $BeCl_2$ is sweet at low concentrations, it is distinctly sour at higher concentrations (Höber & Kiesow, 1898), which is consistent with the observation that upon hydration of the Be^{2+} ion during solvation, excess H^+ is generated.

A number of inorganic substances that completely dissociate in water solution taste sweet in dilute concentrations. Among them are sodium and potassium hydroxide and the chlorine, bromine and iodine salts of sodium, potassium, and lithium (Höber & Kiesow, 1898; Renqvist, 1918; Kionka & Strätz, 1922). Some of these substances are capable of eliciting all four basic tastes, but NaCl apparently elicits only two.

3.4.1 Sodium Chloride

Sodium chloride at different solution concentrations first elicits sweetness, and then saltiness. The changing taste of NaCl with increasing concentration (Skramlik, 1926) is as follows:

0·009M No taste
0·010M Slight sweet
0·015M Sweeter
0·020M Sweet
0·030M Strong sweet
0·040M Salty sweet
0·050M Salty
0·500M Pure salty

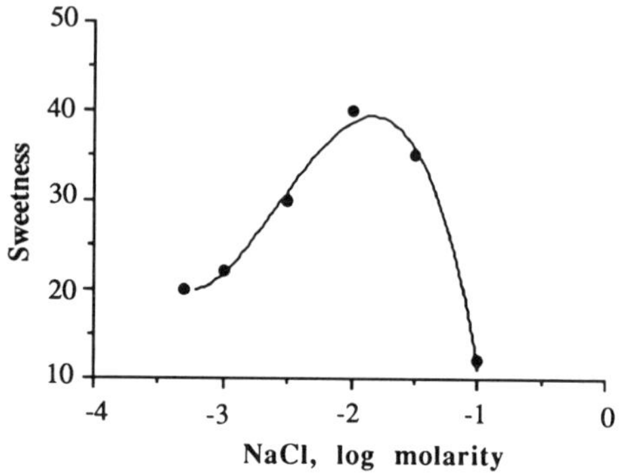

FIG. 3.15. The sweetness of sodium chloride in dilute solutions (Bartoshuk *et al.*, 1978).

Bartoshuk *et al.* (1978) also investigated the sweet taste of dilute sodium chloride solutions, and employed magnitude estimates of sweetness, bitterness, sourness, and saltiness. After the subject's sense of taste had first been adapted to distilled water, 0·032M sucrose, and an extract from *Gynmema sylvestre* were tasted. Their data for the sweetness of NaCl is shown in Fig. 3.15. As with conductivity (Section 3.2.4), there is sharp inflection in taste at about 0·015M concentrations of NaCl. Presumably, there is also a marked change in the polarity of hydration water at that concentration. Generally, the panel did not detect any significant sour or bitter taste. After adaptation to sucrose and the *Gynmema sylvestre* extract, the panel did not respond to the sweetness of NaCl, indicating

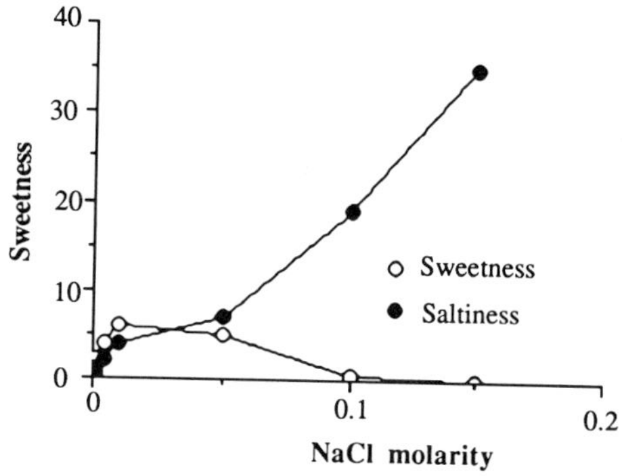

FIG. 3.16. Taste of sodium chloride at various concentrations (Bartoshuk *et al.*, 1978).

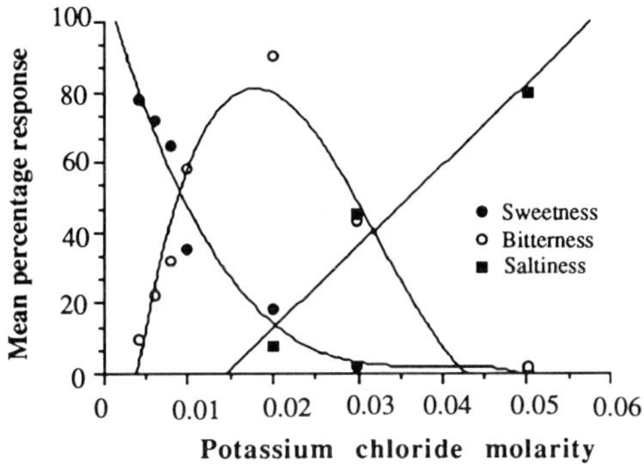

FIG. 3.17. Variation in the taste of KCl at different concentrations (Dzendolet & Meiselman, 1967).

that the response reported in Fig. 3.15 after adaptation to water, is indeed an authentic sweetness response. The response of the panel to the total taste of sodium chloride at different concentrations is shown in Fig. 3.16. At very low concentrations, the panel could not readily differentiate whether the perceived taste of NaCl was due to saltiness or sweetness. At the point of maximum NaCl sweetness, about 20% of the sweetness score might be due to the saltiness of NaCl, a taste response that was not changed by adaptation to either sucrose or the *Gynmema sylvestre* extract.

3.4.2 Potassium Chloride

Potassium chloride first exhibits a sweet taste, with slight bitterness, that then becomes bitter–sweet (Dzendolet & Meiselman, 1967). With increasing concentration bitterness decreases and saltiness increases. For all practical purposes, little sourness is associated with increasing concentrations of KCl, except perhaps at the highest concentrations. This seems to be confused with the bitterness attribute, on the part of the panel. A plot of panel scores obtained by Dzendolet & Meiselman (1967) is given in Fig. 3.17.

3.4.3 Lithium Chloride

Lithium chloride at first tastes predominately sweet, but in contrast with KCl, there is an increase in sourness with increasing concentration. Like KCl, saltiness rapidly increases with higher concentrations, but bitterness is generally absent. These data are shown in Fig. 3.18. The increasing sourness attribute of LiCl is particularly interesting as the compound,

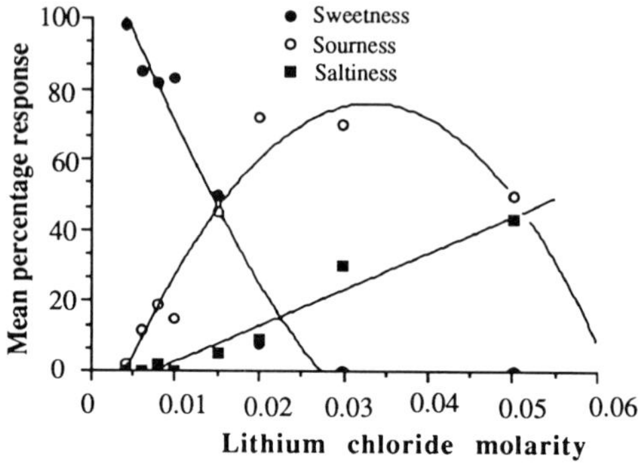

FIG. 3.18. Variation in the taste of LiCl with concentration (Dzendolet & Meiselman, 1967).

along with lithium tartrate and citrate, is slightly alkaline (pH 8). As mentioned above, this result may be due to the well-known psychophysical confusion betwen sourness and bitterness (Pfaffmann *et al.*, 1971) as sourness is primarily due to the concentration of the hydrogen ion.

3.4.4 Lithium Sulfate

Lithium sulfate is sweet at first with some bitterness, but the bitterness diminishes with increasing concentration. As with increasing concentrations of LiCl, sourness increases in opposite fashion to sweetness (Fig. 3.19). As found with LiCl, bitterness is generally lacking.

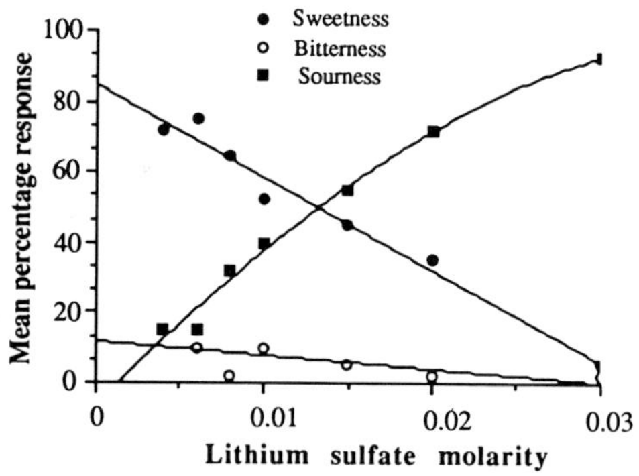

FIG. 3.19. Variation in the taste of lithium sulfate with concentration (Dzendolet & Meiselman, 1967).

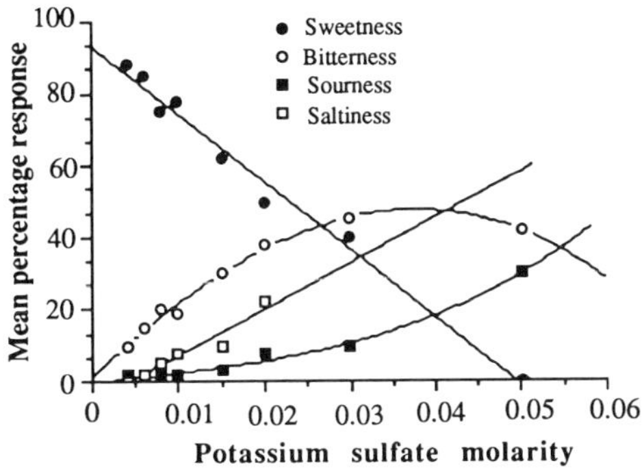

FIG. 3.20. Variation in the taste of potassium sulfate with concentration (Dzendolet & Meiselman, 1967).

3.4.5 Potassium Sulfate

Potassium sulfate taste with increasing concentration is shown in Fig. 3.20. This compound is distinguished from the others by having significant scores for all four basic taste qualities. Three increased with increasing concentration while one (sweetness) diminished. The mixed tastes appear to be a combination of the fundamental properties of individual ions, and clearly several salts are capable of eliciting all four basic tastes.

3.4.6 Relative Sweetness of Salts

Because of overlapping sweetness and saltiness attributes, Bartoshuk *et al.* (1978) believed that derivation of a relative sweetness score for dilute solutions of NaCl was not warranted, but sodium chloride and sucrose seem to be about equally sweet. As indicated by recognition thresholds, KCl is not as sweet as NaCl, but lithium chloride has about the same sweetness potential as sucrose. Little can be said about the relative sweetness of beryllium and lead salts other than that there are reports that they do taste sweet, but any attempt to confirm those reports would be extremely unwise. Interestingly, based on threshold data, beryllium chloride, with a sweetness recognition threshold of 0·0006M, seems to be about 30 times as sweet as sucrose. There is also little data available on the relative sweetness of lead salts. They seem to be at least as sweet as sucrose. Children eat paint chips containing hydrated lead hydroxycarbonate ('white lead') because they taste sweet.

3.5 INORGANIC ACIDS

It might be expected that the inorganic acids and acid salts would taste only sour or salty, respectively. This is largely true, but sweetness and bitterness attributes enter into the picture, even for the pure acids, and their sodium salts.

3.5.1 Hydrochloric Acid

Hydrochloric acid (muriatic acid, mol. wt 36·0) is a solution of hydrogen chloride gas in water. It has a pungent odor, and solutions have a sour, and faintly bitter taste (Harvey, 1920). It is completely dissociated in water, and a 1N solution has a pH of 0·10. A one-tenth normal solution has a pH of about 1·10. The ions, while hydrated, do not have any other particular structure besides the hydration structure.

3.5.2 Sulfuric Acid

Sulfuric acid (oil of vitriol, H_2SO_4, mol. wt 98·08) is a colorless, odorless, oily liquid with great affinity for water. It too is completely dissociated in water solution. The SO_4^{-2} anion also has a tetrahedral structure, in addition to the hydration structure, and solutions of H_2SO_4 can be depicted as shown below. Persons who have drawn the concentrated acid into a pipet with oral suction, and obtained the acid on the tongue, through poor technique, report that it tastes sweet (before the onset of corrosive action, of course). The sodium salt of sulfuric acid (Na_2SO_4) tastes salty and bitter.

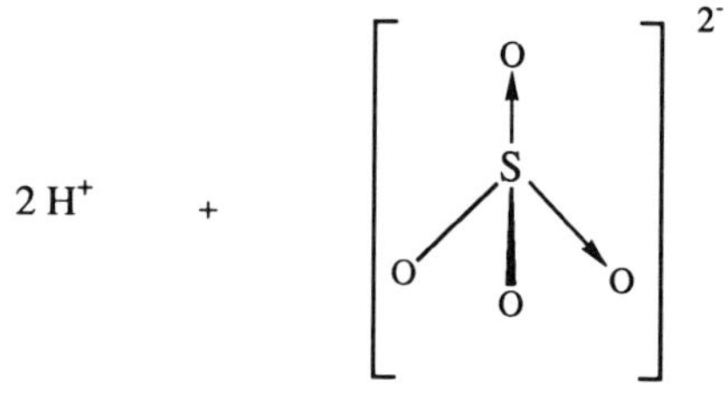

Sulfuric acid

3.5.3 Phosphoric Acid

Phosphoric acid (orthophosphoric acid, H_3PO_4) is a clear syrupy liquid said to have a pleasing (sweet?) acid taste in suitable dilution. The tribasic stepwise dissociation constants are about $K_1 = 7{\cdot}11 \times 10^{-3}$; $K_2 = 7{\cdot}99 \times 10^{-8}$; $K_3 = 4{\cdot}8 \times 10^{-13}$. Because of the tribasic nature, three different salts of phosphoric acid are known, dihydrogen sodium phosphate (H_2NaPO_4), hydrogen disodium phosphate (HNa_2PO_4), and trisodium phosphate (Na_3PO_4). These substances have varying degrees of sourness, saltiness, and perhaps sweetness. Like the sulfate ion, the phosphate ion is tetrahedral.

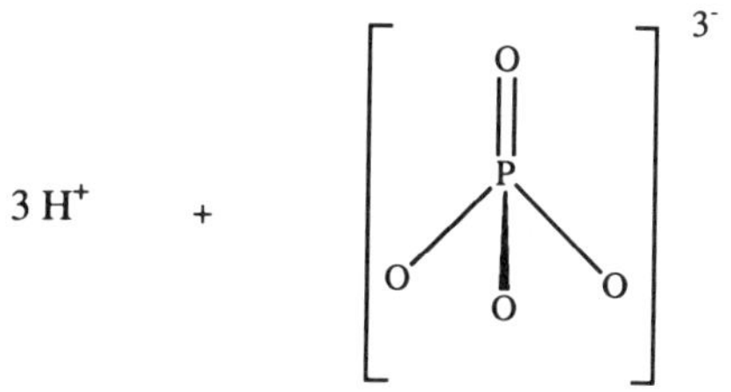

Orthophosphoric acid

3.6 SOURNESS OF INORGANIC ACIDS

The sourness of inorganic acids is directly proportionate to the normality of their solutions, and a one-tenth molar solution of H_2SO_4 is about twice as sour as a one-tenth molar solution of HCl. Neutralization of an acid solution quenches the sour taste (Richards, 1898). In addition to the hydrogen ion concentration of acid solutions, the tartness of acids is related to the total potential hydrogen ion concentration of their solutions. To illustrate this important point, the data of Harvey (1920) are shown in Fig. 3.21.

With respect to normality, the thresholds for the inorganic acids are nearly identical. Pfaffmann (1959) reports the threshold for the hydrochloric acid to be 0·0009N, for sulfuric acid to be 0·0011N, and for nitric acid to be 0·001N.

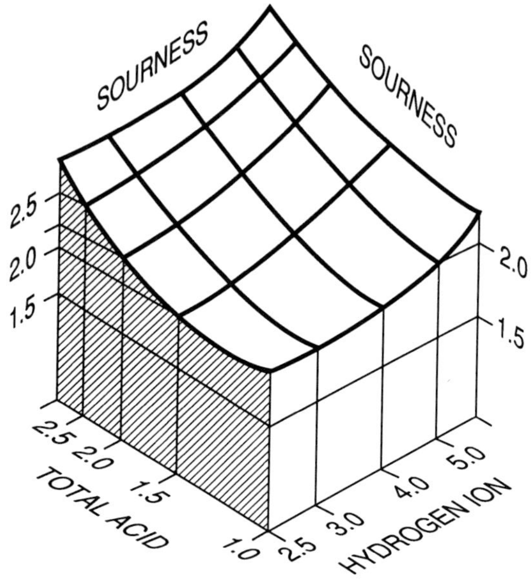

FIG. 3.21. Effect of total acid and pH on the sourness of buffered solutions compared to solutions of HCl with equal sourness (Harvey, 1920).

CHAPTER 4

Polyhydroxy Alcohols, Cyclitols, and Carbonyl Compounds

This chapter is the first of a series with the primary purpose of demonstrating the structure of compounds that possess distinct taste. Most monohydroxy alcohols have feeble tastes, but in certain cases, unsaturated monohydroxy alcohols with the general structure shown below taste sweet.

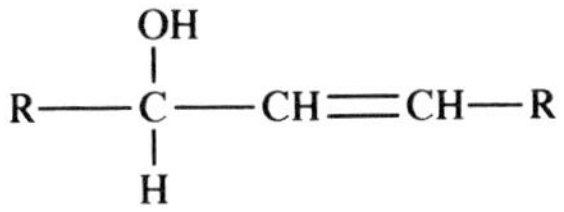

Examples (Moncrieff, 1967) are 1-octene-3-ol, 2-octene-2-ol, and 3-methyl-2-heptene-4-ol. It is not clear from the original data whether the substances are distinctly sweet, or have a sweetish taste and other common chemical sense qualities such as harshness and oilyness. On the other hand, saturated polyhydroxy alcohols taste sweet (Cohn, 1914), but the sweetness potency and common chemical sense attributes vary with the carbon atom-hydroxyl group ratio. Sweet tasting alcohols usually have a 1,2 diol group, or multiple vicinal hydroxyl groups. When polyhydroxy alcohols have hydroxyl configurations other than the 1,2 diol (α-glycol) configuration, the taste is usually designated as 'sweetish', implying additional and/or other taste and common chemical sense attributes. Resorcinol tastes 'sweetish', but phloroglucinol is distinctly

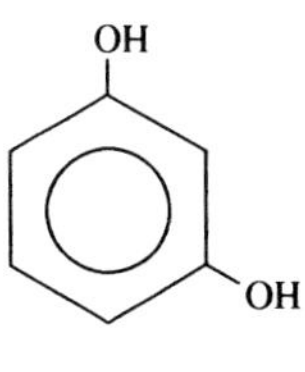

Resorcinol

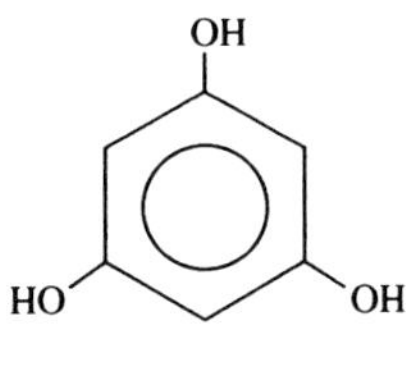

Phloroglucinol

sweet, and the suffix 'glucinol' indicates the sweetness. Among the polyhydroxy alcohols, as well as other substances, the Greek prefixes for sweet (*glyc-*, *gluc-*) are often encountered in the trivial name for the substance, i.e. the name has little structural significance. Some of them are glycol, glycerol (glycerin), glucitol (sorbitol), and glucinium (beryllium chloride).

4.1 ACYCLIC POLYHYDROXY ALCOHOLS

Oertly & Myers (1919) traced the literature on polyhydroxy alcohols back to the eighteenth century. They list glycerol to be sweet, and 1,2-propanediol as 'sweetish'. 1,2-Butanediol is also reported to be 'sweetish', although 1,3-butanediol is not. In contrast, 1,2,3-butanetriol, 1,2,4-butanetriol, and 1,2,3-pentanetriol are reported to be sweet (as opposed to 'sweetish'). Apparently, as the hydrocarbon chain length is increased, the taste of α-glycol groups becomes admixed with sensory attributes that cannot easily be related to the basic tastes. The structures of many of the aliphatic alcohols are given in Fig. 4.1, and their relative sweetness scores in Table 4.1.

4.1.1 Ethylene Glycol

Ethylene glycol (1,2 ethanediol, glycol) is prepared by hydration of ethylene oxide, and does not seem to be naturally occurring. It is a slightly viscous liquid, and although it tastes sweet (1·3×), ingestion of about 100 ml is lethal. It is also very hygroscopic, and at 100% relative humidity absorbs twice its weight of water. The dipole moment (2·20) is high, and it is also slightly soluble (1 : 200) in ether. Structurally, ethylene glycol is composed of an infinitesimal mixture of rotomer conformations, which are not symmetrical, but constitute a racemic mixture of conformations, some of which are shown below.

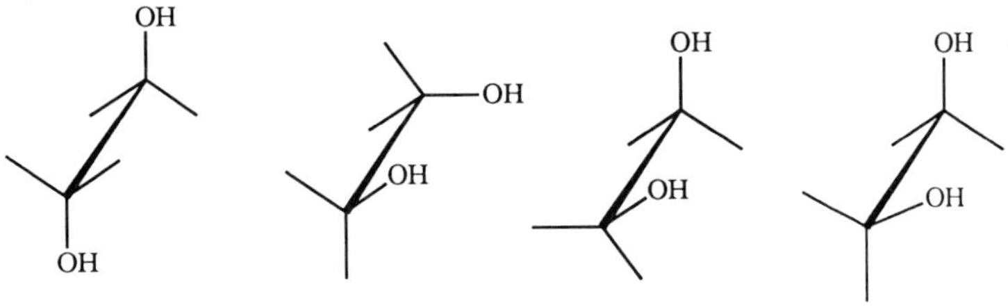

Glycol conformations

TABLE 4.1
Relative sweetness (sucrose = 100) of polyols (aliphatic alcohols)

Compound	*Sweetness*	*Investigator(s)*	*Method*
Glycol	130	Carr *et al.* (1936)	Recognition threshold.
	49	Paul (1922)	3% Sucrose equivalent.
Glycerol	108	Carr *et al.* (1936)	Recognition threshold.
	56–64	Cameron (1947)	Equal sweetness at various sucrose concentrations.
	48	Paul (1922)	3% Sucrose equivalent.
D,L-Erythritol	240	Paul (1922)	3% Sucrose equivalent.
L-Arabitol	100	Carr *et al.* (1936)	Recognition threshold.
Xylitol	96–100–117	Yamaguchi *et al.* (1970)	Comparison with increasing sucrose concentrations.
Sorbitol	54	Nieman (1958)	Best value, literature survey.
	41	Moskowitz (1970)	Magnitude estimation.
	55–58	Yamaguchi *et al.* (1970)	Comparison with increasing sucrose concentrations.
Mannitol	57	Carr *et al.* (1936)	Recognition threshold.
	69	Nieman (1958)	Best value, literature survey.
	33	Moskowitz (1970)	Magnitude estimation.
	51–62	Yamaguchi *et al.* (1970)	Comparison with increasing sucrose concentrations.
Dulcitol	41	Nieman (1958)	Best value, literature survey.
	46	Moskowitz (1970)	Magnitude estimation.
Lactitol	50	Linko (1982)	?
Maltitol	90	Hough (1979)	?
Palatinit®	(25)	Sträter (1986)	10% Sucrose equivalent

4.1.2 Glycerol

Glycerol (1,2,3-propanetriol, glycerin; trihydroxypropane) is a crystalline substance with a very low melting point (18°C). It is a constituent of oils and fats, and is obtained from them by fermentation. In that sense, glycerol is naturally occurring. It is not as sweet as ethylene glycol, but also absorbs moisture from the atmosphere.

4.1.3 Erythritol

Erythritol (1,2,3,4-butanetetrol, D,L-erythritol, erythrite). This four-

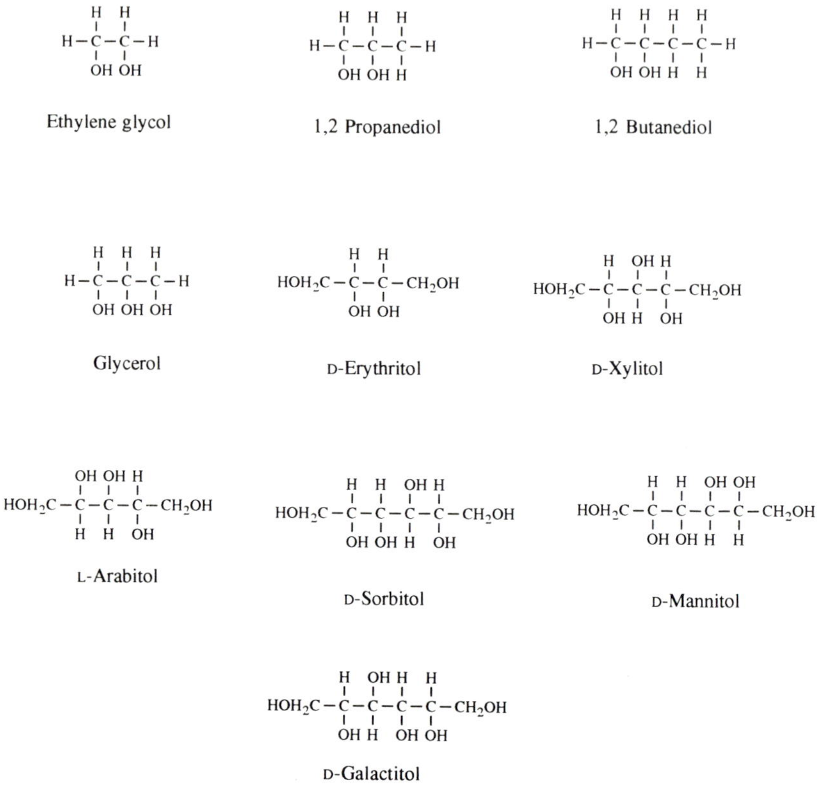

FIG. 4.1. Structure of acyclic polyhydroxy alcohols.

carbon polyol occurs in lichens, algae, and grasses. Its melting point is 121·5°C, and it is very soluble in water.

4.1.4 Arabitol

Arabitol (D-arabitol, arabinitol) is not often found in nature, but is present in some lichens and mushrooms. *Fistulina hepatica* is reported to contain 9·5% on a dry basis. Arabitol can also be obtained by reduction of either D-arabinose or D-lyxose. Its crystals melt at 102°C and the specific rotation in borax solution is +7·8°. L-Arabitol does not occur in nature, but has been synthesized. Both the D- and the L-forms taste sweet.

4.1.5 Xylitol

Xylitol is a five-carbon sugar alcohol found in fruits and vegetables, but the concentration is usually less than 1%. Its parent sugar, xylose, is a component of hemicellulose in wood and corn cobs. Crystalline xylitol (stable form) melts at 94°C, and the crystals have a negative heat of solution, leading to the common chemical sense attribute of 'coolness' when tasted (Sicard, 1982). The conformation of xylitol in solution has been determined to be a planar zig-zag structure with an oxygen–oxygen distance of 2·9–3·0 Å (Lindley *et al*., 1976*a*).

4.1.6 Galactitol (Dulcitol, Dulcite)

Galactitol is rather widely distributed in plants, and occurs in high concentration in the manna of *Melanpyrum nemorosum*. It may also be synthesized by reduction of galactose. Its crystals melt at 188–189°C. It is a *meso* compound without optical activity and, of course, without an enantiomeric form.

4.1.7 Sorbitol (D-Sorbitol, D-Glucitol)

Sorbitol is widely distributed in plants, particularly in fruits (apples, cherries, peaches, etc.) and berries. It may be synthesized by the reduction of D-glucose. The crystals melt at 93°C and the specific rotation is −1·98°. The L-form is known, but does not occur naturally. Sorbitol also has a negative heat of solution, leading to a cooling sensation when tasted.

4.1.8 Mannitol (D-Mannitol, Mannite)

The sugar alcohol, mannitol, is naturally occurring, and is found in exudates of olive and plane trees, and in seaweed. It may be prepared by the reduction of mannose. The crystals melt at 166–168°C, with a negative heat of solution. The specific rotation is positive when treated with borax. The L-form is known, but does not occur in nature.

4.1.9 Maltitol (4-*O*-α-D-glucopyranosyl-D-sorbitol)

Maltitol is prepared by hydrogenation of glucose syrups, and is a component of them (Lycasin®), or by hydrogenation of maltose. In the latter case it is amorphous, but the 6-*O*-α-D-glucopyranosyl-D-sorbitol isomer has been obtained in crystalline form (Hough, 1979).

4.1.10 Lactitol (4-*O*-β-D-galactopyranosyl-D-sorbitol)

Lactitol is prepared by hydrogenation of lactose. It is a crystalline substance, and the crystalline monohydrate is reported to be nonhygroscopic, although the anhydrous crystalline form is hygroscopic (Linko, 1982).

4.1.11 Palatinit®

Palatinit® is a mixture of two isomers. They are 6-*O*-α-D-glucopyranosyl-D-sorbitol and 6-*O*-α-D-glucopyranosyl-D-mannitol. The structure of palatinit is shown below. The product is crystalline, and nonhygroscopic (Sträter, 1986).

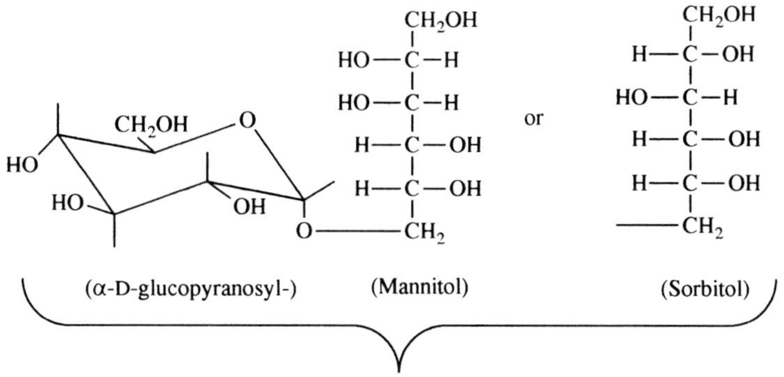

Palatinit ®

4.2 CYCLIC SUGAR ALCOHOLS

Anhydro sugars can be named as either cyclic sugar alcohols, or 1-deoxyhexoses. For example, 1,5-*anhydro*-mannitol is also 1-deoxymannose. An example of how these substances are formed is shown in Fig. 4.2. Many of the anhydro-sugar alcohols, such as 1,4-*anhydro*-ribitol, xylitol, mannitol, and glucitol are sweet, but 1,4-*anhydro*-erythritol is not. The structures and taste of these substances are shown in Fig. 4.3. These compounds subsequently serve as examples of the 1-deoxy-aldohexoses (Chapter 6). To continue the interrelations among sugars, it should also be noted that 1,5-*anhydro*-mannitol, in addition to being 1-deoxy-D-

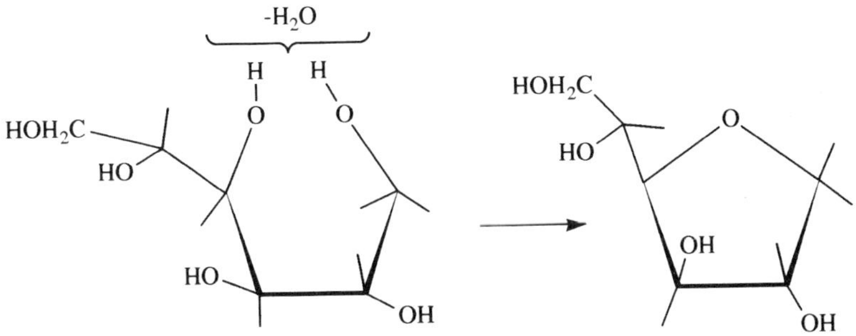

FIG. 4.2. Formation of 1,4-*anhydro*-glucitol from sorbitol.

mannose, is also 2-deoxy-D-fructopyranose (cf. Birch, 1976), and 2,5-*anhydro*-D-mannitol is also 2,6-*anhydro*-β-D-fructofuranose. The latter compound is reported to taste sweet and bitter (Shamil *et al.*, 1989).

4.3 CYCLITOLS

Many cyclic alcohols are reported to taste sweet, and Goodwin *et al.* (1981) report that *cis*- and *trans*-cyclopentane diols are about 0·07 and 0·4 times as sweet as sucrose, respectively. *cis*- and *trans*-1,3-Cyclohexane diol have about 30 percent of the sweetness of sucrose. *cis*-1,4-Cyclohexane diol is also reported to have 30% the sweetness of sucrose, and *trans*-1,4-cyclohexane diol appears to be nearly tasteless. These taste assessments have yet to be confirmed. It apppears that the taste panel used equated the common chemical sense attributes of these substances with sweetness. In an earlier study, Birch & Lindley (1973) found those substances to be devoid of sweetness. In general, the cyclohexane diols are not pleasant to taste.

Birch & Lindley (1973) also found that the symmetrical cyclohexane 1,2/4,5 tetrol was tasteless. However, the cyclohexane pentols (quercitols), and the inositols, had sweet tastes, unencumbered by any bitter note. Several of these substances are shown in Fig. 4.4. The quercitols are naturally occurring in acorns.

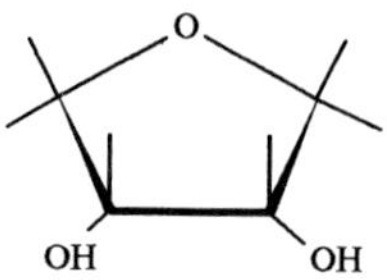

1,4-*Anhydro*-erythritol
(Not sweet)

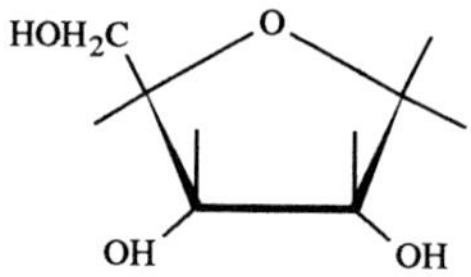

1,4-*Anhydro*-ribitol
(Sweet)

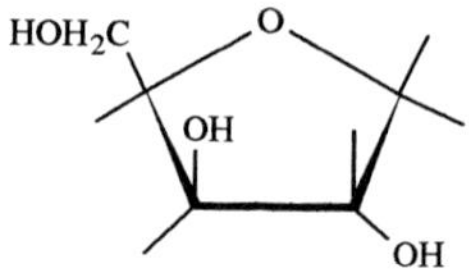

1,4-*Anhydro*-xylitol
(Sweet)

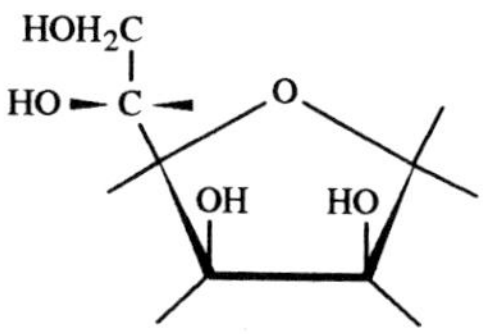

1,4-*Anhydro*-mannitol
(Sweet)

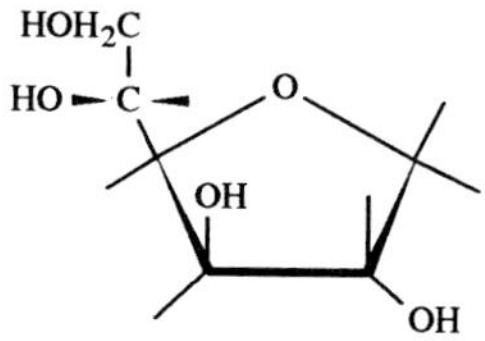

1,4-*Anhydro*-glucitol
(Sweet)

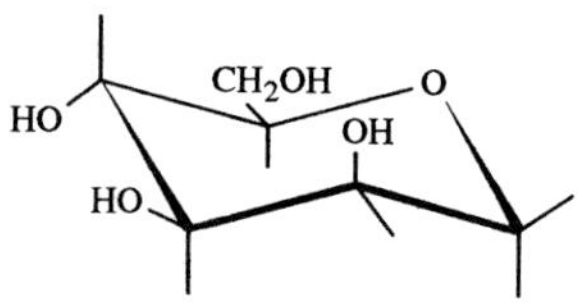

1,5-*Anhydro*-mannitol
(Sweet)

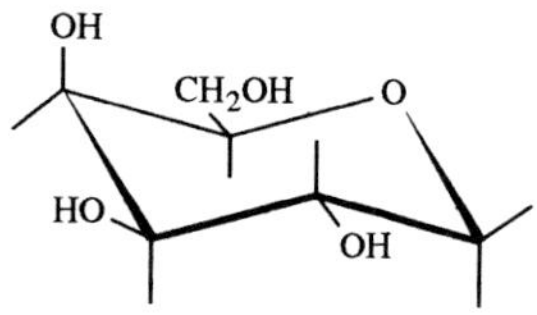

1,5-*Anhydro*-galactitol
(Sweet)

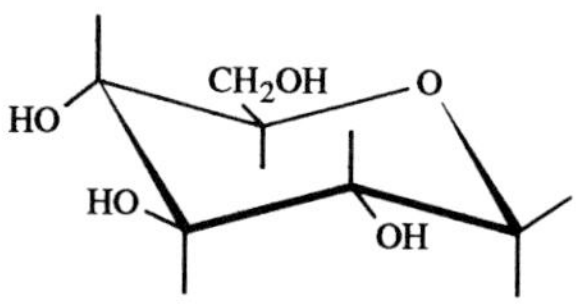

1,5-*Anhydro*-glucitol
(Sweet)

FIG. 4.3. Structure and taste of *anhydro*-pentitols and hexitols (1-deoxy sugars) (Birch & Lee, 1971).

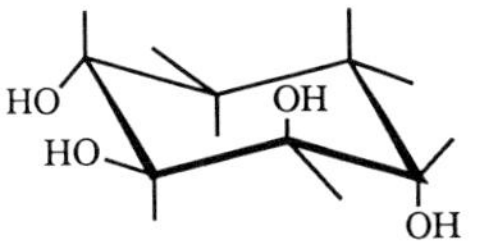

(+) Protoquercitol (Trace)

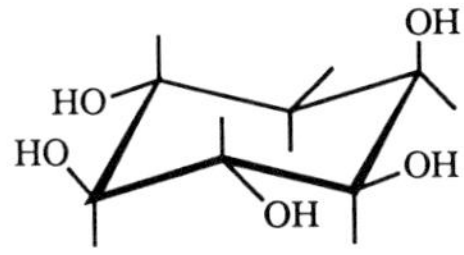

(-)-Viboquercitol (Trace?)

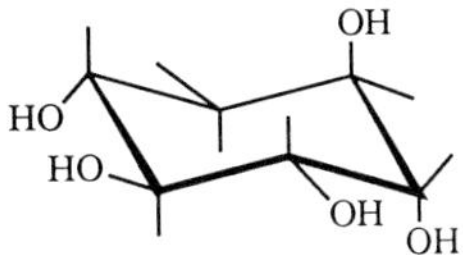

(+)-Viboquercitol (Sweet)

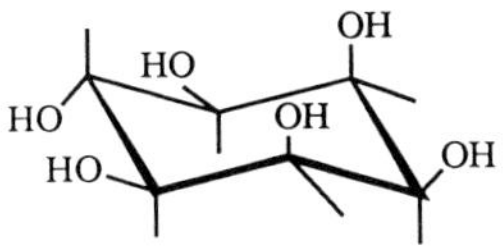

Epi-inositol (Trace)

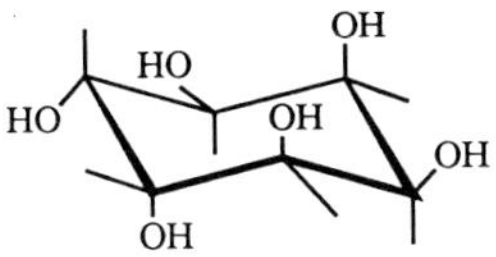

Allo-inositol (Sweet)

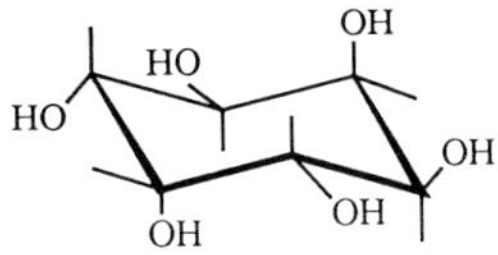

Neo-inositol (Trace)

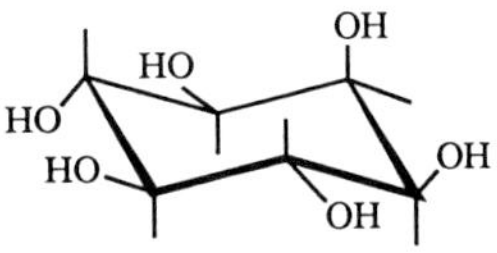

Myo-inositol (Trace)

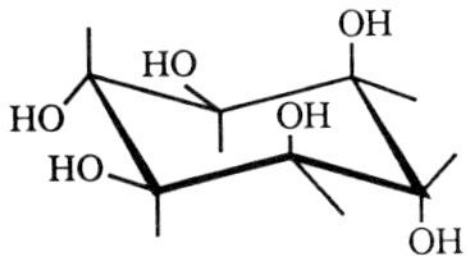

Muco-inositol (Trace)

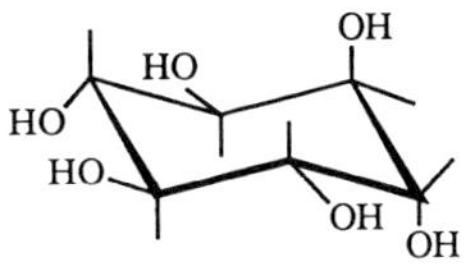

(+) Chiro-inositol (Sweet)

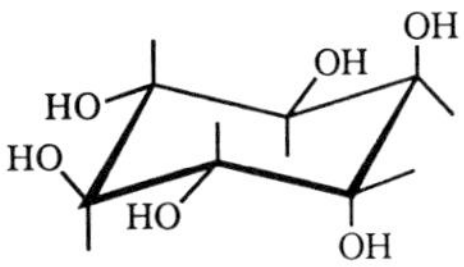

(-)-Chiro-inositol (Sweet)

FIG. 4.4. Structure and taste of cyclitols.

4.4 HYDROXY CARBONYL COMPOUNDS

The simplest carbonyl compound is formaldehyde ($H_2C{=}O$), which is a pungent and toxic gas at ordinary temperature and pressure. A few of the lower molecular weight hydroxy carbonyl compounds taste sweet. Higher molecular weight hydroxy carbonyl compounds are sugars, and are presented in the following chapter. The lower molecular weight hydroxy carbonyl compounds are prone to form dimeric hemiacetal ring structures. A hemiacetal is the product produced by the reaction of a carbonyl group with an alcohol group. When the alcohol and carbonyl functions occur in one compound, the reaction can be bimolecular, with the formation of dimers. The reaction is intramolecular whenever the carbon chain for the compound is sufficiently long to allow the two groups to approach each other through rotation about single bonds.

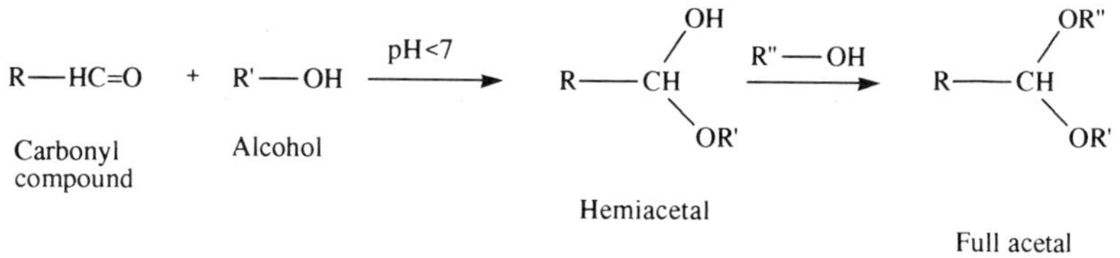

Hemiacetals are moderately stable, but can be reacted with an alcohol group of another molecule to form a much more stable acetal. Sugar carbonyl straight-chain structures react intramolecularly to form a moderately stable hemiacetal. The sugar hemiacetal can also react with a second molecule of alcohol, e.g. methanol, to form stable full acetals. They are known as glycosides. The structure of several low molecular weight dimers, obtained through hemiacetal formation, are shown in Fig. 4.5.

4.4.1 Glycolaldehyde

This simplest hydroxyaldehyde, glycolaldehyde (glycollic aldehyde), is reported to taste distinctly sweet by Abderhalden (cf. Oertly & Myers, 1919) and by Fenton & Jackson (1899). The crystalline material, obtained with difficulty, melts at 95–97°C and sublimes at 100–130°C. Its crystalline structure is a dimer (Summerbell & Rochen, 1941), and crystalline dimeric glycoaldehyde is also known as 2,5-dihydroxydioxane. In solution, the dimeric structure decomposes to the monomeric form. It does not seem to naturally occur.

4.4.2 Glyceraldehyde

Glyceraldehyde (glyceric aldehyde, aldotriose, glycerose, 2,3-dihydroxy-propionaldehyde) is the reference standard for sugar configurational nomenclature. It is also known in the D-, L-, and racemic crystalline forms, all of which are apparently dimers. In solution, the dimeric structure is a minor component of the monomer-dimer equilibrium.

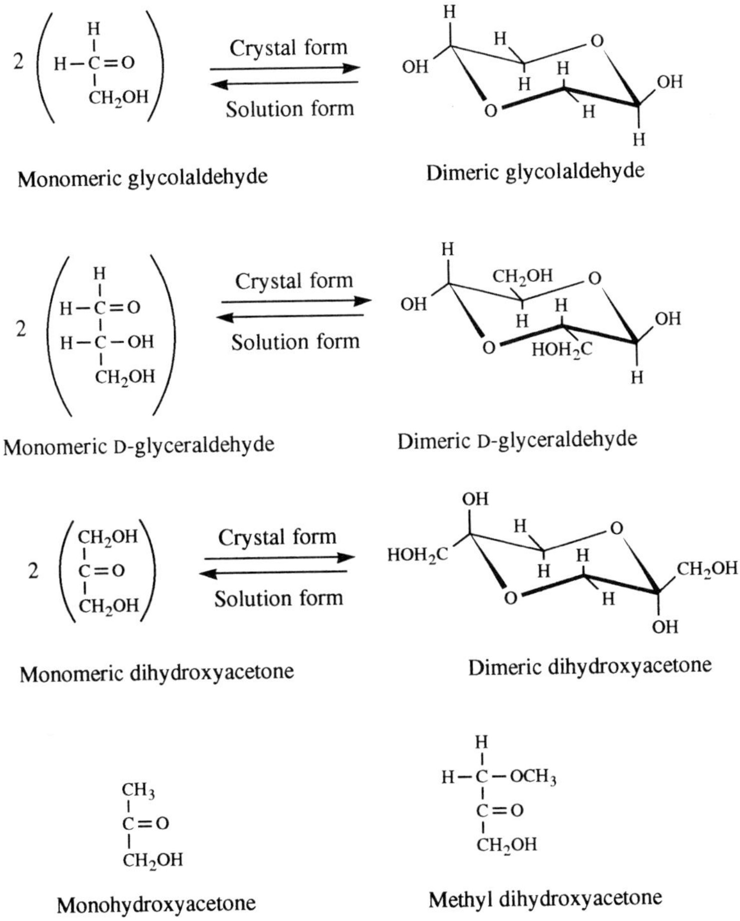

FIG. 4.5. Sweet-tasting low molecular weight aldehydes.

According to Wohl (cf. Oertly & Myers, 1919) monomolecular glyceraldehyde is sweet and bitter, but dimeric (bimolecular) glyceraldehyde is slightly sweet. The crystalline racemic form melts at 145°C and the crystalline D-form establishes an equilibrium between the monomeric and dimeric forms (Feeley *et al.*, 1968). It also exhibits thermal mutarotation (Baer & Fischer, 1939). Free glyceraldehyde does not occur naturally, but can be prepared by mild oxidation of glycerol (Witzemann, 1914). In this case an equilibrium mixture of glyceraldehyde and dihydroxyacetone is obtained, and the mixture is also called *glycerose*.

4.4.3 Dihydroxyacetone

Dihydroxyacetone (1,3-dihydroxymethyl ketone; 1,3-dihydroxy-2-propanone) is a white crystalline powder, m.p. 75–80°C. In crystalline form, which is very hygroscopic, it is a dimer that reverts to the monomer upon solvation with water. The crystals taste sweet and cool, and it is reported (Birch, 1976) that in concentrated solution some sweetness is detectable, presumably because some dimer is present. Dihydroxyacetone serves as a phosphorylated intermediate in the alcoholic fermentation of sugars. Other derivatives of acetone that taste sweet are monohydroxyacetone and methyl dihydroxyacetone.

4.4.4 Importance to Structure-Activity Relationships

The relation between structure, chemistry, and the sweetness of the low molecular weight hydroxy carbonyl substances is one of the first instances that point out the subtle ease with which weak chemical interactions affect taste. The sweetness of the low molecular weight hydroxycarbonyl substances is undoubtedly derived from the structure of the dimeric forms. These forms are generated, in turn, by mild intermolecular interactions that are only about $-1\,\text{kcal}\,\text{mol}^{-1}$, but nevertheless have some covalent character. Furthermore, the intermolecular hemiacetal interactions have a counterpart in the mutarotation of the simple sugars (Section 5.3.3), but in this case the weak hemiacetal linkage with covalent-like character is intramolecular, but again, rather weak ($-1{\cdot}7\,\text{kcal}\,\text{mol}^{-1}$).

CHAPTER 5

Structure, Reactions and Properties of Sugars

From the nutritional point of view, sugars are the primary source of energy for life processes. The sugars also taste purely sweet, and their structures serve as models for the derivation of stereochemical principles. When it was found that the molecular formula for most monosaccharides was $C_6H_{12}O_6$, it was also recognized that the formula was sterically inadequate. Some of those substances were sweet, others were very sweet, and still others were devoid of sweetness. The monosaccharides, composed of the hexose sugars alone, are made up of 16 configurational isomers that may have either 5-membered or 6-membered ring forms. The 16 configurational isomers also belong to one enantiomeric class, and 16 others belong to another enantiomeric class. The monosaccharides form a complete set of configurational isomers, ring isomers of different size, antipodal forms, and variable conformations. As such, they form an ideal set of substances to explore the role of structure and stereoisomerism in sweetness (Chapter 9).

5.1 SIMPLE SUGARS (MONOSACCHARIDES)

The sugars are the simplest forms of carbohydrates, and their empirical formula is $C_n(H_2O)_n$ ('hydrates of carbon' or 'carbon hydrates'). They are also the building blocks for plant structural substances (cellulose, pectins) and inert reserve energy sources (starches). The major monosaccharides are pentoses ($C_5H_{10}O_5$), and hexoses ($C_6H_{12}O_6$). Compound sugars (oligosaccharides) contain several monomeric units. The monosaccharide glucose is also the repeating monomeric unit in the plant substances starch and cellulose, which are distinguished from each other by the nature of the glucose form.

5.1.1 Acyclic Structure of the Sugars

The various configurational forms of the monosaccharides are most clearly depicted by the acyclic structure. These 'straight chain' structures also emphasize that the acyclic forms of sugars are aldehydes (RHC=O) and/or ketones (R–(C=O)–R).

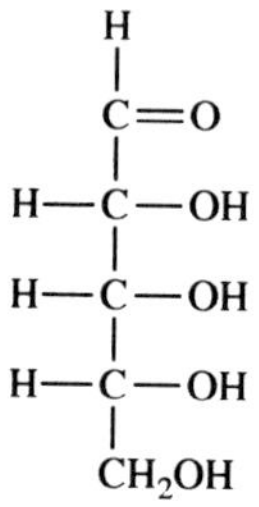

Aldehydo-form

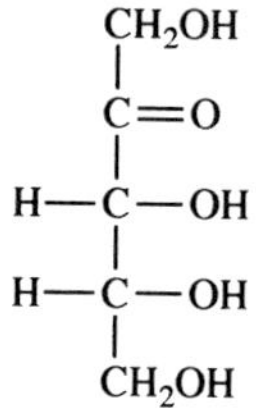

Keto-form

5.1.1.1 *Fischer Projection Structures*

All the acyclic *aldehydo*- and *keto*-pentose forms possible for the D-series of these sugars are shown in Fig. 5.1. The acyclic configurational structure of the hexose sugars is shown in Fig. 5.2. The structures are known as Fischer projection structures, and signify that all horizontal bonds are directed toward the observer (see Section 2.11.3.3), even though that feature may not be emphasized.

When the theory of the asymmetric carbon atom (Section 2.11.3.3) is applied to the sugars, it is found that there are tetrahedrally substituted carbon atoms with four groups of different size. An example of how carbon atom number 2 is rendered asymmetric by having four substituents of different size ($a > b > c > d$) appended to it is shown for an aldopentose in Fig. 5.3.

Because they have three asymmetric carbon atoms, the 2,3,4,5-tetrahydroxypentanals have 2^3 or 8 possible configurational forms. The *keto*-structural analogs have 2^2 or 4 configurational forms. Four of the *aldehydo*- and two of the *keto*-forms belong to one chiral family of the sugars (D- or L-) and the remainder to the other. The tetrahydroxy-pentanals are D- and L-arabinose, xylose, lyxnose, and ribose. The *keto*-analogs are D- and L-*erythro*-pentulose and D- and L-*threo*-pentulose with the suffix *ulose* used to indicate the *keto*-structure.

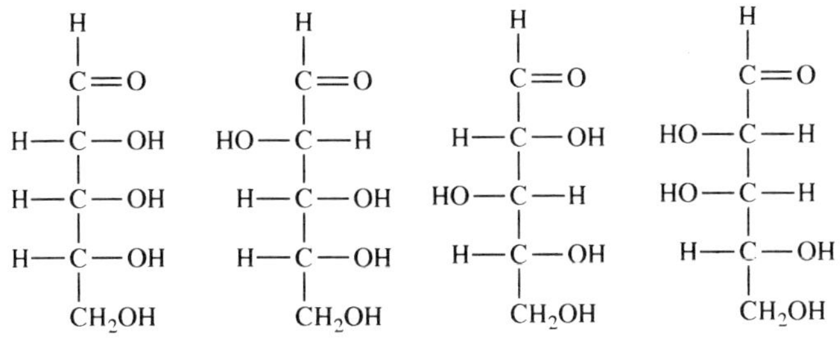

Ribose Arabinose Xylose Lyxose

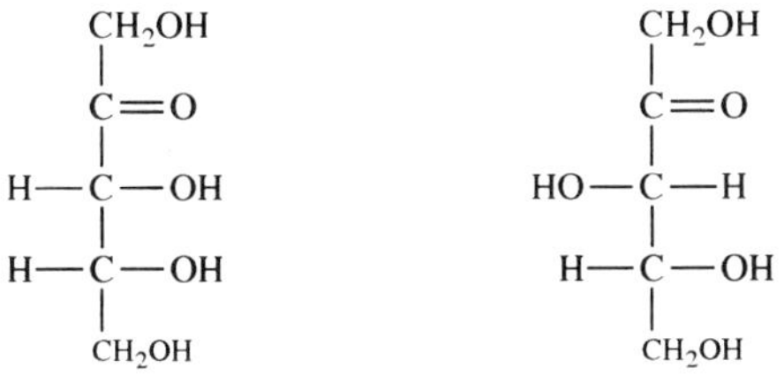

erythro-Pentulose *threo*-Pentulose

FIG. 5.1. Acyclic configurational structure of the D-series of *aldo*- and *keto*-pentoses.

The 2^4 (sixteen) 2,3,4,5,6-pentahydroxyhexanals are D- and L-glucose, galactose, mannose, gulose, allose, altrose, idose, and talose (Fig. 5.2). The eight (2^3) *keto*-analogs are D- and L-allulose, fructose, sorbose and tagatose (Fig. 5.2). All the substances in a given D- or L-series are configurational *diastereoisomers*, or *allosteric* forms of one another. The latter term, popular in biochemical nomenclature, is a synonym (allostereoisomer) for diastereoisomer, or isomers that are not enantiomers (Eliel, 1971; Shallenberger & Wienen, 1989).

Configurational diastereoisomers differ in that the position of two substituents about one or more asymmetric carbon atoms is transposed. These configurational isomers are known as *epimers*. In Fig. 5.2 mannose is the 2-epimer of glucose, and galactose is the 4-epimer of glucose. To convert the type of diastereoisomeric form for a given asymmetric carbon atom to that of its mirror image form, only two groups need to be

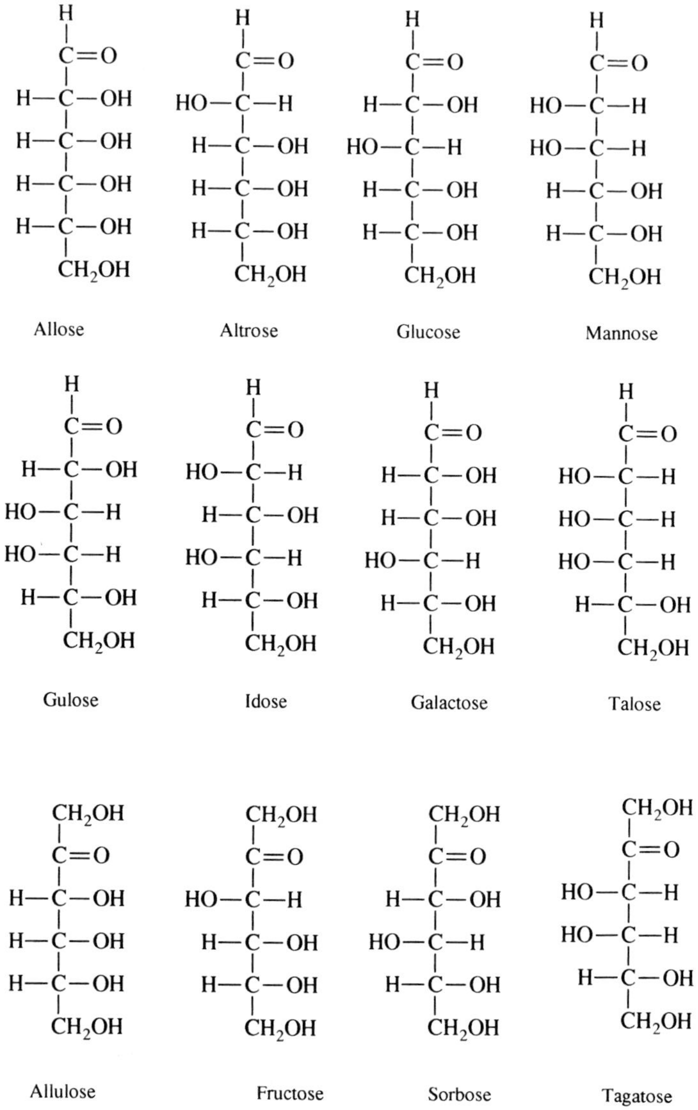

FIG 5.2. Acyclic configurational structure of the D-series of *aldo*- and *keto*-hexose sugars.

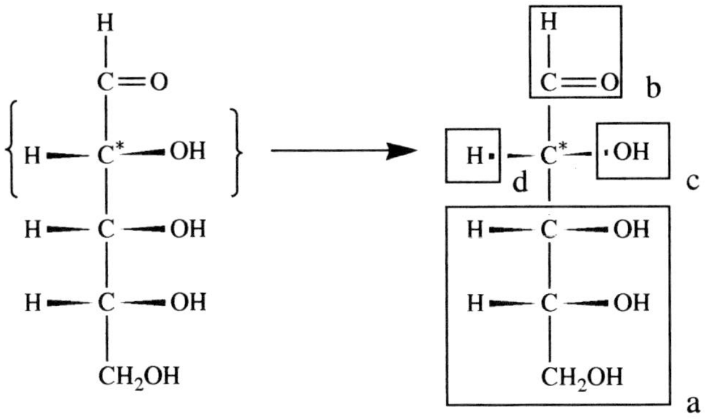

FIG 5.3. The four groups of different size appended to carbon atom No. 2 of an *aldo*-pentose sugar.

transposed. When the configuration about all the asymmetric carbons is transposed, then the compounds are enantiomers, or complete nonsuperposable mirror image structures. One will be the D-form, and the other the L-form.

5.1.1.2 *Specification of the Chiral Family* (D- *or* L-)

The reference structures for specification of a sugar structure's chiral family are D- and L-glyceraldehyde. The convention that is applied to the sugars is due to Rosanoff (1906). According to that convention, the *aldehydo*-carbon atom is assigned the number one (i.e., top priority). Then, the acyclic structure for the sugar is oriented in Fischer projection. When the *highest numbered* asymmetric carbon has the hydroxyl group written to the right in Fischer projection, the sugar belongs to the D-(pronounced 'dee') series of sugars. When it is written to the left it belongs to the L- (pronounced 'ell') series. The structures shown in Fig. 5.2 belong to the D-series.

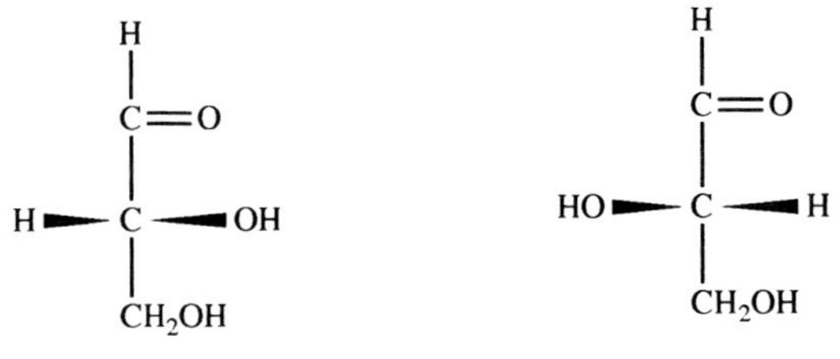

D-Glyceraldehyde L-Glyceraldehyde

Fischer (1891) arbitrarily wrote the hydroxyl group of the reference asymmetric carbon atom to the right, and used the prefixes *d*- and *l*-, which are optical rotatory (*dextro*-, *levo*-) criteria, to distinguish between the chiral families. The original specification for the usual form of glucose was *d*-. Because of the contradictions that subsequently arose, Rosanoff introduced his convention, and it gradually became apparent that optical rotatory power and the structural nature of a chiral family are not necessarily related. What is perhaps rather intriguing is that Fischer's arbitrary choice is now known to be absolute (Bijvoet *et al.*, 1951). As a result, the known configuration of the amino acids is also absolute (Section 7.1.1).

The problem of specifying whether a compound is dextrorotatory (+) or levorotatory (−) is solved by incorporating the sign for its optical rotatory power in the name. The glucose structure shown in Fig. 5.2 is D-(+)-glucose to indicate that it is dextrorotatory and belongs to the D-family of sugars. On the other hand, the fructose structure shown in Fig. 5.2 is D-(−)-fructose to indicate that while it belongs to the D-family of sugars, it is levorotatory. Fructose is also the ketone analogue of D-glucose. The Fischer projection formulas for the sugars are ideal for depicting the configuration about each asymmetric carbon atom. Using (+) signs and (−) signs to specify configuration about each asymmetric (chiral) center, D-glucose is 2+, 3−, 4+, 5+ pentahydroxyhexanal. Using R and S notation, D-glucose is 2R, 3S, 4R, 5R pentahydroxyhexanal. The latter specifications are the result of yet another convention.

5.1.1.3 *Specification of Sugar Multiple Configuration*

The acyclic structures of the sugars have an infinitesimal number of forms due to the free rotation possible for contiguous carbon atoms. Structures shown in Figs 5.1 and 5.2 are not truly straight chain structures for the sugars. Instead, they are a linear projection of a *pseudo*-cyclic form onto the plane of a sheet of paper. In a true extended linear depiction of the D-glucose molecule, individual carbon atoms alternately project back through the plane of the paper away from the observer, and above the plane of the paper toward the observer. This is shown in a linear extension of D-glucose in Fig. 5.4A; where it superficially resembles the structure of D-talose.

From molecular models, it can be determined that a *pseudo*-cyclic form can be obtained from the structure represented by Fig. 5.4A by bringing the carbonyl function at C-1 and the primary alcohol group at C-6 together. Each asymmetric carbon atom is then viewed separately and

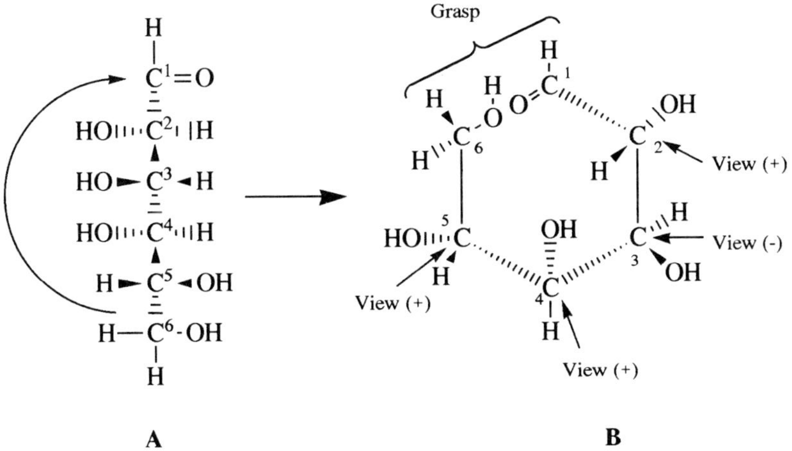

FIG. 5.4. Linear extension of the D-glucose molecule (A), and formation of a *pseudo*-cyclic ring (B) by bringing the hydroxymethylene group and the carbonyl group together.

directly. This is brought about by rotating the model, as shown in Fig. 5.4B. When rotating the molecule, if the OH group at a specific carbon atom appears on the viewer's right, it is written to the right. If it is on the viewer's left, it is written on the left of the molecule. In fact then, a (+) designation indicates a configuration wherein the OH group is on the left side of the molecule. Another way to comprehend this procedure is to visualize the *pseudo*-cyclic model as a cylindrical rubber stamp. When inked, and rolled over a piece of paper, the structure for D-glucose shown

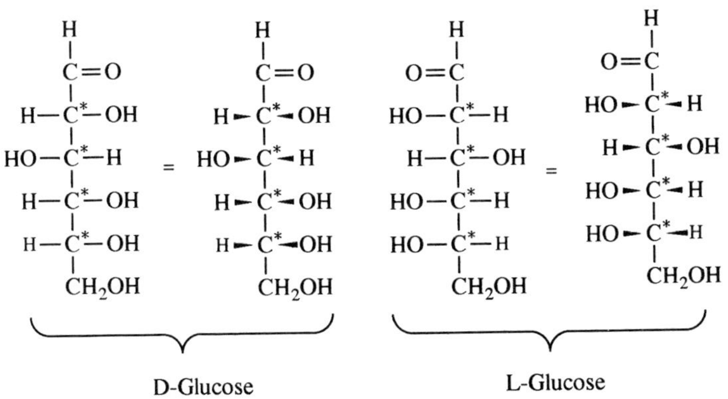

FIG. 5.5. Fischer projection structures of D- and L-glucose.

in Fig. 5.2 appears. It can only result if all horizontal carbon atom substituents project toward the observer when the *pseudo*-cyclic form is rotated. The distinction between D- and L-glucose in Fischer projection is shown in Fig. 5.5. The drawings of D- and L-glucose in Fig. 5.5 illustrate why Fischer projection structures of the sugars, as depicted in Fig. 5.2, cannot be 'lifted from the plane of the paper and turned over.' The forward direction of the horizontal single covalent bonds are implied only. Such an operation would seem to generate L-glucose, because the significance of a solid horizontal bond is violated. Turning glucose over would lead to projecting the bonds back through the plane of the paper. If this were to be specified, by use of dashed single covalent bonds, it would be acceptable, and the sugar would still be D-glucose.

A final convention used in sugar chemistry is that the structures can be rotated 180° in the plane of the paper. Now, however, it is the vertical bonds that are to be construed as projected towards the observer, and the configurational identity of the original structure remains.

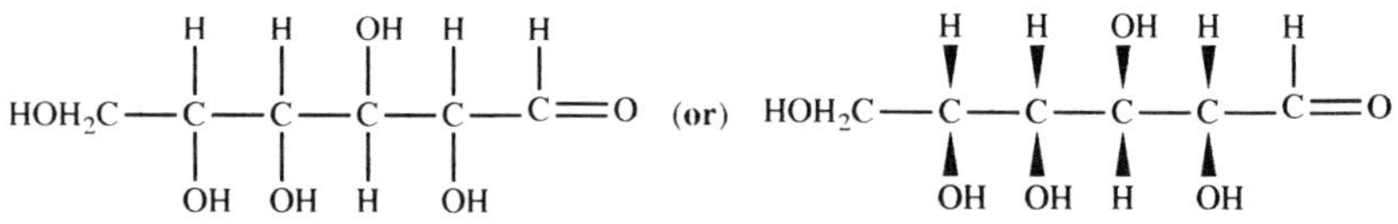

Horizontal depiction of the D-glucose structure

The structures for the entire series (family tree) for the *aldehydo*- and *keto*-sugars are not given here, but may be obtained from any standard reference text on the carbohydrates, such as Pigman & Horton (1972) or Shallenberger (1983*a*).

5.1.2 Ring Forms of the Sugars

The acyclic forms of the sugars are minor components of a sugar solution. Ring forms predominate, because flexibility of the acyclic forms of the sugars brings hydroxyl groups four or five carbon atoms removed into proximity with the carbonyl function. When this happens, sugar carbonyl and hydroxyl groups react intramolecularly to form a hemiacetal (Section 4.4) ring structure. Sugar hemiacetal ring structures are moderately stable. However, they can be reacted with another molecule of alcohol, e.g. methanol, to form a stable *full* acetal. The full acetal is known as a *glycoside*.

5.1.2.1 *Haworth Structures*

For the free sugars, the ability of OH groups at carbon atom No. 4 or 5 to form a five- or six-membered ring compound is shown in Fig. 5.6. The structures shown for glucose in Fig. 5.6 are the Haworth (1929) method of depicting the five-membered (furanose) and six-membered (pyranose) ring structures. As with the Fischer acyclic projection structures, Haworth ring structures contain several conventions. The first is that ring structures are not construed to be written on the plane of the paper on which they are shown. For the six-membered form, only carbon atom No. 5 and the single bond to the ring oxygen atom, and the ring oxygen atom, lies in the plane of the paper. The front edge of the Haworth ring is usually drawn heavily, as in Fig. 5.6, to indicate this convention. The three-dimensional significance of the bold lines is shown for a hexagonal drawing of the Haworth formula for α-D-glucopyranose in Fig. 5.7. Through usage, the bold front edge of Haworth formulas is often omitted, but the convention must be remembered. Otherwise, D- and L-sugars in the Haworth formulas are indistinguishable.

There is a straightforward relation between the configuration of the asymmetric carbon as depicted in Fischer projection and in Haworth perspective. Namely, those groups written to the right in Fischer projection are written toward the top of the paper in Haworth

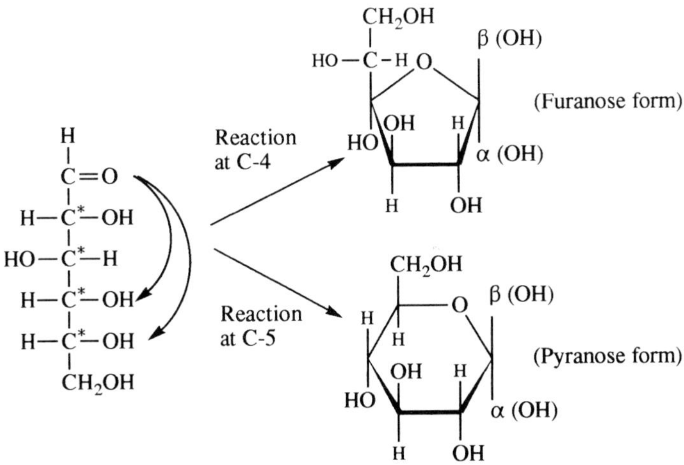

FIG. 5.6. Reaction of the D-glucose aldehyde group with an alcohol group at C-4 or C-5 to form a five- or six-membered hemiacetal ring structure.

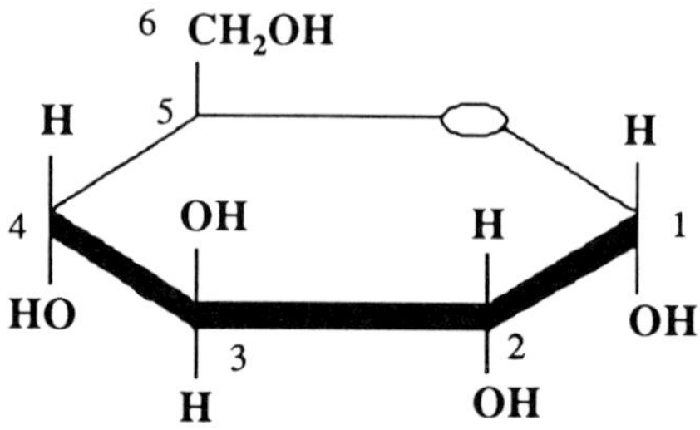

FIG. 5.7. Haworth formula for α-D-glucopyranose shown in perspective.

perspective, and are on the top of the ring. Those written to the left in Fischer projection are on the bottom of the ring.

Due to ring closure, carbon atom No. 1 of the *aldehydo*-sugars becomes asymmetric. Two configurational forms for the disposition of those H, OH groups are then possible. The compounds are known as *anomers*, or diastereoisomers that are configurational isomers at the lowest numbered asymmetric carbon atom (the *anomeric carbon atom*). As carbon atom No. 1 is made asymmetric by ring formation, the configuration of the new chiral center is specified by analogy to the configuration of the chiral family reference carbon atom. When the anomeric carbon atom has the same configuration as that of the reference carbon for chiral family specification, the sugar is designated as *alpha*-(α-). When it has opposite configuration, it is specified as *beta*-(β-). As the reference carbon atom is involved in ring closure, it is not readily apparent how the above convention applies to the Haworth structures. Nevertheless, in keeping with the relation between the Haworth and Fischer structures, in Haworth perspective the OH group written toward

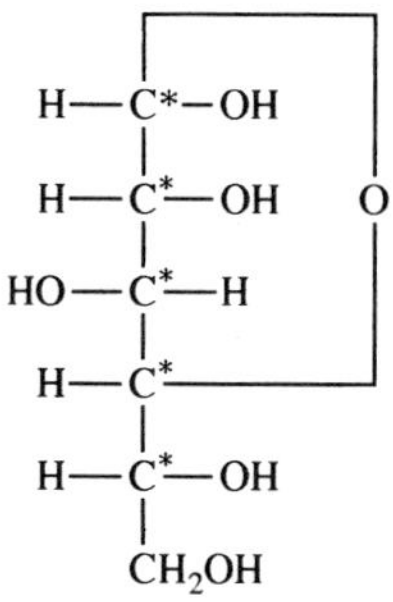

α-D-Glucofuranose

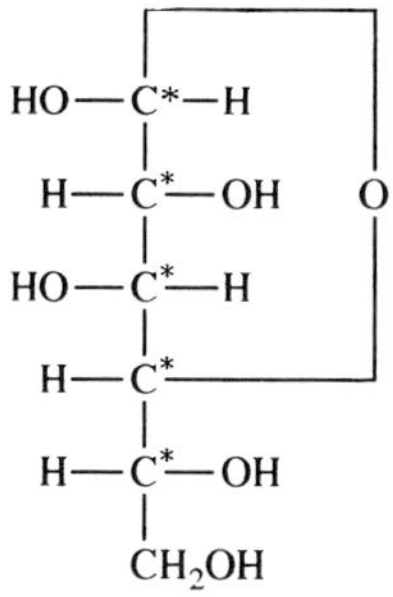

β-D-Glucofuranose

the top of the paper is designated β. The one written toward the bottom of the paper is designated α. The indication is that they assume positions on the top and bottom, respectively, of the perspective drawing.

As two ring forms are possible, the total number of diastereoisomeric (allosteric) forms now possible for the D-series of sugars is 32. There are, of course, 32 enantiomorphic L-forms also possible.

In a water solution of the monosaccharides, all five forms are present to a greater or lesser extent, and are in equilibrium, as shown for D-glucose in Fig. 5.8. The five forms are known as tautomers, or easily interconverted

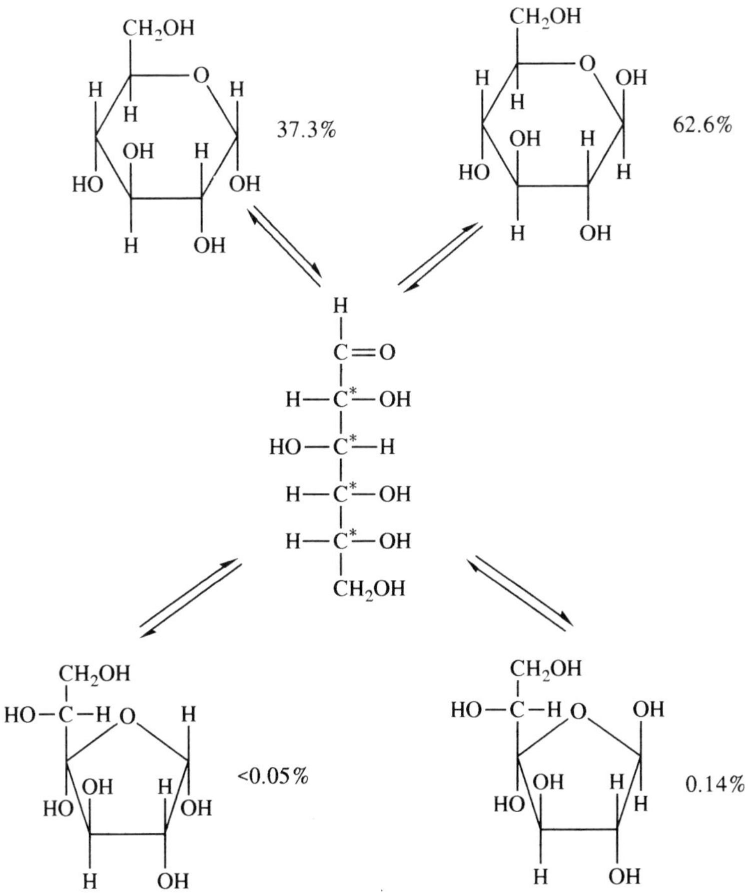

FIG. 5.8. The five tautomeric forms of D-glucose possible in water solution. The percent distribution of tautomers assigned is from Williams & Allerhand (1977).

isomers. Viewing the three-dimensional furanose and pyranose rings in perspective, the two Haworth ring structures for glucose become indicated as shown in Fig. 5.8, i.e. the bold-line convention is often omitted. For the furanose structure, only the ring oxygen atom is to be construed as written on the plane of the paper. All others are directed toward the observer to create the illusion that the planar structure for both the furanose and pyranose ring are perpendicular to the plane of the paper.

5.1.2.2 *Conformational Structures*

Ultimately, five- and six-membered ring structures are not planar. Contiguous carbon–carbon atom bond angles have a projected angle of 109°28′, and the carbon to ring oxygen atom bond angle is about the same (104°). Thus, the ring is quite puckered, a fact that Haworth (1929) recognized, but left the task of establishing the significance of the puckering to future generations of stereochemists. The sugars now are known to have configurational diastereoisomeric and enantiomeric forms, ring-size diastereomeric forms, and conformational diastereoisomeric and enantiomeric forms. For a given six-membered sugar ring structure, such as β-D-glucopyranose, there are two basic *chair* forms and six basic *boat* forms (Fig. 5.9). Two basic chair forms are possible because the combined operations of numbering and puckering of an otherwise symmetrical ring make it chiral (Section 2.11.3.5).

β-D-Glucopyranose in the $^4C_1$ conformation (Fig. 5.10A) now can have a ring-only mirror image diastereomeric $^1C_4$ form (Fig. 5.10B). In this case the $^4C_1$ ring has merely been turned inside out (everted) by rotation about single covalent ring bonds. No configurational transformations have taken place as no single covalent bonds have been severed. The sugar is nevertheless β-D-glucopyranose, but in a highly unfavored, bulky form. The enantiomer of β-D-glucopyranose is the complete configurational and conformational mirror image structure (Fig. 5.10C). This structure is formed by transposing the hydrogen and hydroxyl substituents about each asymmetric carbon atom, and then everting the ring conformation. Configurational and conformational chiral operations are necessary to form C from A. Alternately, reflecting A by placing it upon a mirror table will generate image (C) by reflection. This operation is a total chiral transposition. It bears the same relation to A as the reflection of the right hand in a mirror, and the reflected image is the structure of the left hand.

The chair structures for the $^4C_1$ conformational forms of the *aldehydo*-sugars are given in Fig. 5.11. The conformational structures of the sugars,

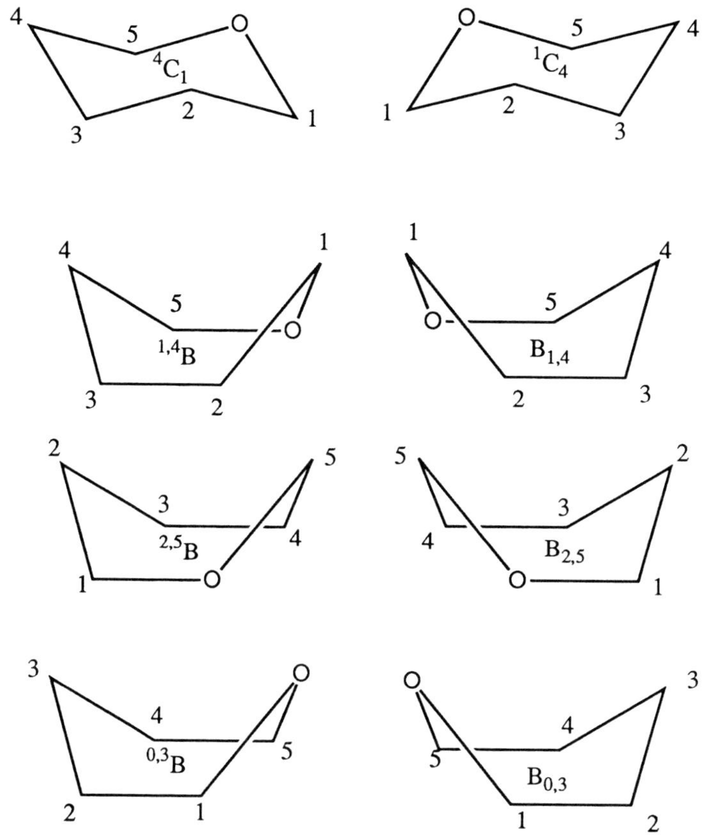

FIG. 5.9. Reeves (1949) basic chair and boat forms of the pyranoid ring.

like Haworth structures, are three-dimensional representations drawn in perspective. The front edge of the ring may or may not be shaded for emphasis. The relation of these structures to the 5- and 6-carbon sugars in Fischer projection, or in the Haworth pyranose structure, is straightforward. Any substituent written to the right in Fischer projection appears on the top of the Haworth ring, and any substituent written to the left appears on the bottom. Conformational structures contain additional information with respect to the disposition of ring carbon atom substituents. In conformational structures, there are two carbon atom substituent positions corresponding to the top or bottom designation. The first is *equatorial*, or in about the same average plane as that for the ring, and

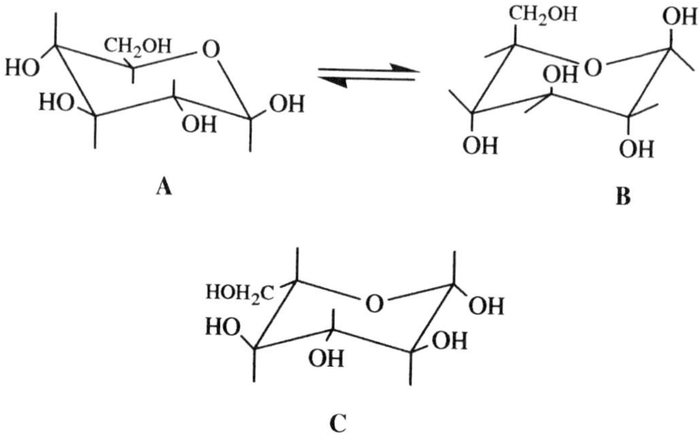

FIG. 5.10. Enantiomeric and chair conformational forms of β-D-glucopyranose.

the second is *polar*, or *axial*, or perpendicular to the average plane for the ring. In Fig. 5.10, the β-D-glucopyranose anomeric –OH group at C-1 is above the average plane for the ring in both chair conformations, but equatorially disposed in one and axially disposed in the other. It can also be seen in Fig. 5.10 that transposing from one chair conformation to the other by eversion interchanges axially and equatorially disposed substituents.

For reference, the favored chair conformation for the *ketohexa*-pyranose sugars is given in Fig. 5.12. The aforementioned relationship among the Fischer, Haworth, and conformational structures applies to ketose sugars also. In addition to the chair forms for the sugars, there are a great number of boat forms that are energetically unfavorable because of relative flexibility, and will not be considered in detail here. It is possible, however, that certain sugar boat conformations serve as intermediates in some sugar reactions.

The furanose rings of the sugars have a great number of possible conformations that are variations on standard envelope (E) and twist (T) forms (Hall, 1963). Examples are shown in Fig. 5.13. The spatial volume that the puckered modes of these forms pass through upon changing conformation is only about one-third of that for the chair forms (i.e., furanoid conformations are relatively planar). The puckering modes for the drawings in Fig. 5.13 are exaggerated. Like the boat and skewed boat forms, however, furanoid conformations serve as intermediates in certain chemical reactions. Conformational structures have application to the

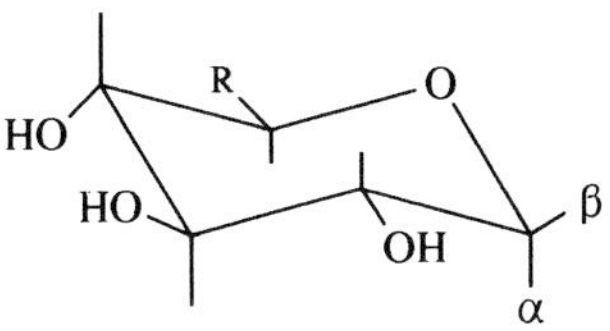

R=H, D-Xylose
R=CH_2OH, D-Glucose

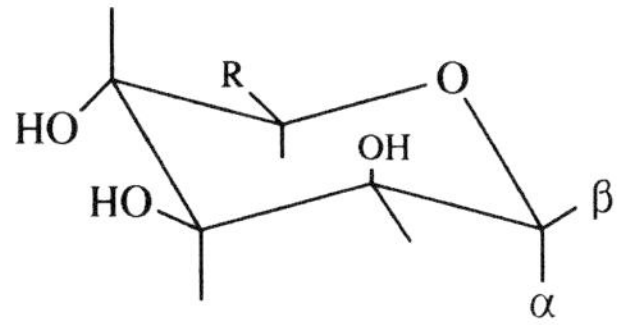

R=H, D-Lyxose
R=CH_2OH, D-Mannose

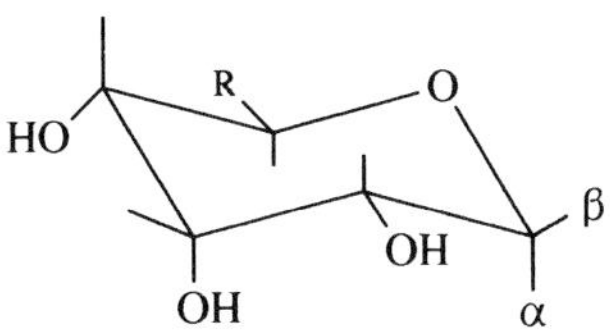

R=H, D-Ribose
R=CH_2OH, D-Allose

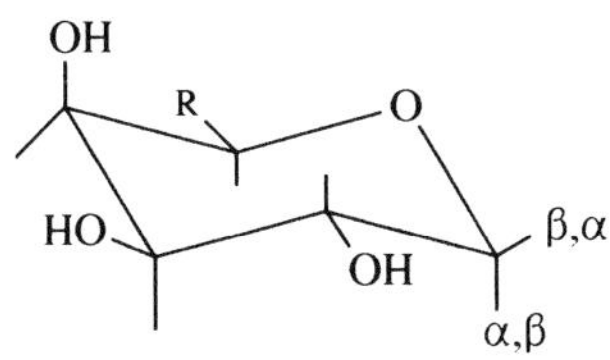

R=H, L-Arabinose
R=CH_2OH, D-Galactose

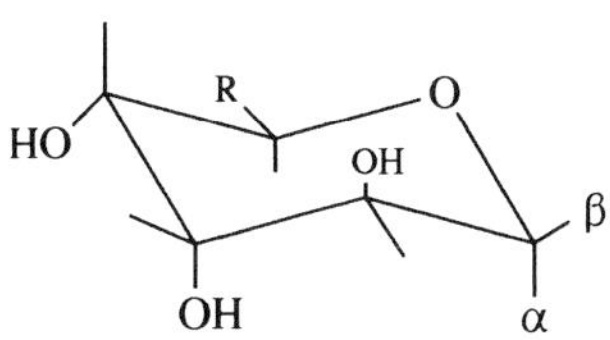

R=H, D-Arabinose
R=CH_2OH, D-Altrose

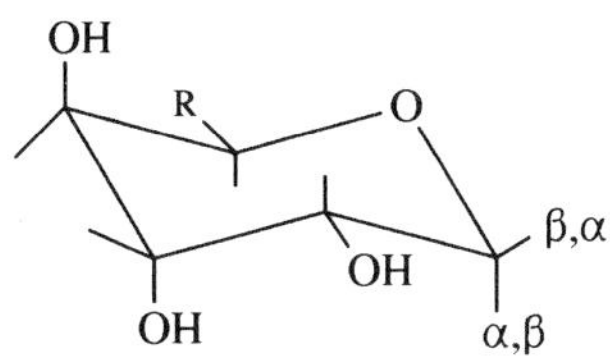

R=H, L-Lyxose
R=CH_2OH, D-Gulose

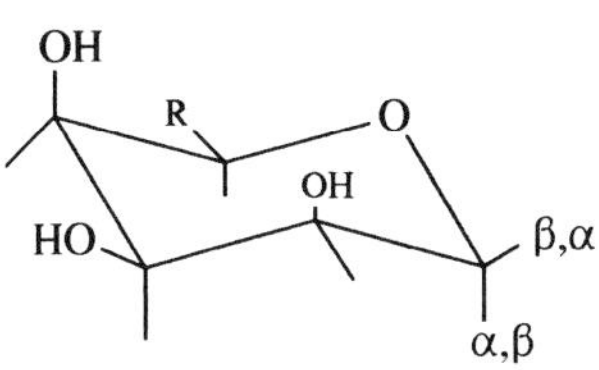

R=H, L-Ribose
R=CH_2OH, D-Talose

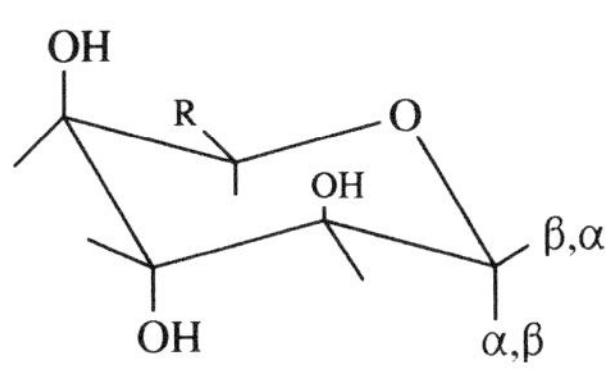

R=H, L-Xylose
R=CH_2OH, D-Idose

FIG. 5.11. The $^4C_1$ conformation for *aldohexa-* and *aldopenta-*pyranose sugars. The anomeric specification for the L-sugars (α- or β-) follows the comma at the anomeric center.

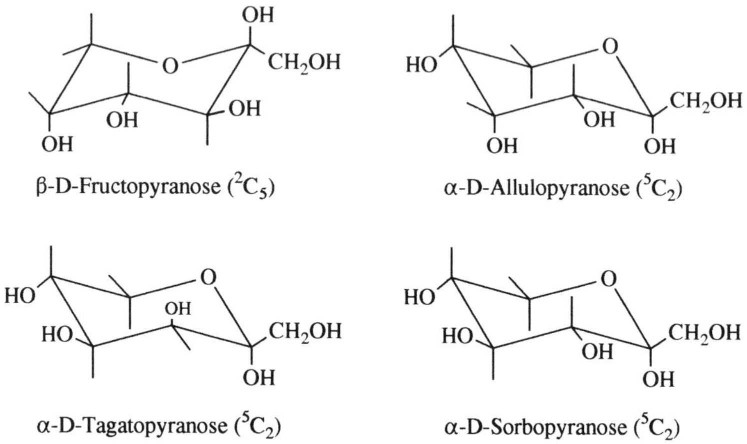

FIG. 5.12. Favored conformations for *ketohexa*-pyranose sugars.

sweetness of the sugars, and an estimate of their distribution in water solution is given in Table 5.1.

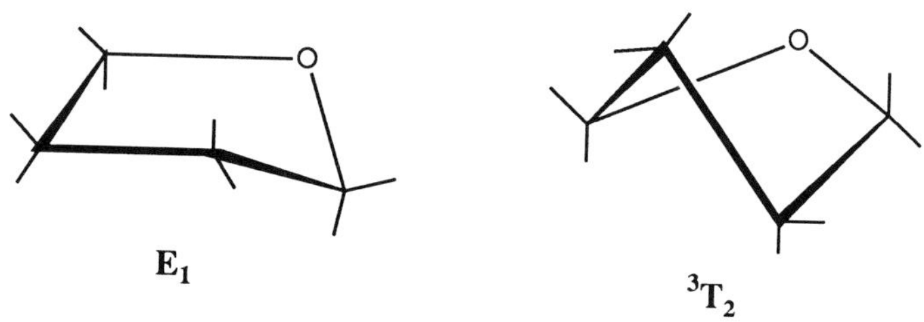

FIG. 5.13. An envelope (E) and a twist (T) conformation for furanoid sugars.

5.1.3 Naturally Occurring Monosaccharides

5.1.3.1 *Arabinose*

By application of carbohydrate nomenclature rules, L-arabinose is the naturally occurring form of this pentose sugar, and it is structurally related to D-galactose. Like D-galactose, it is widely distributed in glycosidic union in gums, hemicelluloses, and pectic substances. The β-L-pyranose anomer, which has the same ring configuration as α-D-galactopyranose, melts at 160°C and has a specific rotation of +190·6° initially, which comes to mutarotational equilibrium at +104·5°. D-Arabinose is extremely rare although it has the same configuration as D-glucose at the lower five carbon atoms.

TABLE 5.1
Calculated conformational energies (E as kcal mol^{-1}) and predominate conformations of sugars in water solution as indicated by E or found by n.m.r. spectroscopy (Angyal, 1968; Angyal & Bethell, 1976)

Sugar	$^4C_1$ *or* $^5C_2$	$^1C_4$ *or* $^2C_5$	*Calculated*	*Found*
α-D-Allose	3·9	5·35	$^4C_1$	$^4C_1$
β-D-Allose	2·95	6·05	$^4C_1$	$^4C_1$
α-D-Altrose	3·65	3·85	$^4C_1$,$^1C_4$	$^4C_1$,$^1C_4$
β-D-Altrose	3·35	5·35	$^4C_1$	$^4C_1$
α-D-Galactose	2·85	6·3	$^4C_1$	$^4C_1$
β-D-Galactose	2·5	7·75	$^4C_1$	$^4C_1$
α-D-Glucose	2·4	6·65	$^4C_1$	$^4C_1$
β-D-Gluocse	2·05	8·0	$^4C_1$	$^4C_1$
α-D-Gulose	4·0	4·75	$^4C_1$	—
β-D-Gulose	3·05	5·45	$^4C_1$	$^4C_1$
α-D-Idose	4·35	3·85	$^4C_1$,$^1C_4$	$^4C_1$,$^1C_4$
β-D-Idose	4·05	5·35	$^4C_1$	—
α-D-Mannose	2·5	5·55	$^4C_1$	$^4C_1$
β-D-Mannose	2·95	7·65	$^4C_1$	$^4C_1$
α-D-Talose	3·55	5·9	$^4C_1$	$^4C_1$
β-D-Talose	4·0	8·0	$^4C_1$	—
α-D-Arabinose	3·2	2·05	$^1C_4$	$^1C_4$
β-D-Arabinose	2·9	2·4	$^4C_1$,$^1C_4$	—
α-D-Lyxose	2·05	2·6	$^4C_1$,$^1C_4$	$^4C_1$,$^1C_4$
β-DLyxose	2·5	3·55	$^4C_1$	$^4C_1$
α-D-Ribose	3·45	3·55	$^4C_1$,$^1C_4$	$^4C_1$,$^1C_4$
β-D-Ribose	2·5	3·1	$^4C_1$,$^1C_4$	$^4C_1$,$^1C_4$
α-D-Xylose	1·95	3·6	$^4C_1$	$^4C_1$
β-D-Xylose	1·6	3·9	$^4C_1$	—
α-D-Fructopyranose	3·65	4·3	$^2C_5$,$^5C_2$	—
β-D-Fructopyranose	5·85	2·95	$^2C_5$	—
α-D-Sorbopyranose	2·4	6·55	$^5C_2$	—
β-D-Sorbopyranose	3·85	4·35	$^5C_2$	—
α-D-Tagatopyranose	2·5	6·0	$^5C_2$	—
β-D-Tagatopyranose	4·3	4·0	$^2C_5$,$^5C_2$	—
α-D-Psicopyranose	3·9	5·35	$^5C_2$	—
β-D-Psicopyranose	5·9	3·55	$^2C_5$	—

5.1.3.2 *Ribose*

D-Ribose, and its 2-deoxy analogs are components of nucleic acids, but occurrence in the free state is rare. The crystals melt at 87°C and the specific rotation is about −23·4°, with minimal mutarotation. The L-form is known.

5.1.3.3 *Xylose* (*Wood Sugar*)

D-Xylose does not occur as the free sugar in nature in significant concentration, but is widely distributed as a component of plant hemicellulose. The principal source is from corn cobs. D-Xylose is configurationally related to D-glucose. The L-form is also known.

5.1.3.4 *Glucose*

D-Glucose (grape or *trauben* sugar, starch sugar, urine sugar, blood sugar, right-handed sugar, dextrose) is the ubiquitous monosaccharide, occurring free in the sap of plants and blood of animals. It is the major component of starch, and the repeating monosaccharide unit is α-D-glucopyranose. D-Glucose is also the major component of cellulose, but here the repeating monosaccharide unit is β-D-glucopyranose. The monosaccharide crystallizes from water at concentrations of 30–70% at temperatures from 0° to 50°C as α-D-glucopyranose monohydrate. The melting point is 83°C and the specific rotation is initially +102° changing with time to +47°. Anhydrous α-D-glucopyranose is obtained by evaporating solutions of D-glucose at temperatures above 50°C. The initial specific rotation is +112°, and the equilibrium specific rotation is +52°. The crystals melt at 146°C. When D-glucose solutions are evaporated at a temperature of 100°C, the crystalline form obtained is anhydrous β-D-glucopyranose. This tautomeric form is more easily soluble than the anhydrous α-anomer, which in turn is more soluble than the monohydrate of the α-anomer. β-D-Glucopyranose melts at 148–150°C and the specific rotation is initially +19°, changing over time (mutarotation) to + 52°.

Anhydrous α-D-glucopyranose is available in crystalline form for use as a sweetening agent. D-Glucose is also a component of the so-called glucose syrups, along with the maltose series of oligosaccharides.

5.1.3.5 *Galactose*

D-Galactose does not occur as the free sugar in either plants or animals to any great extent, but is found in trace amounts in plants of the legume and other families. Many plant gums and mucilages, such as agar and gum arabic, yield D-galactose upon hydrolysis. Upon hydrolysis, other polysaccharides, such as flaxseed mucilage, and snail galactogen yield D- and L-galactose. The β-D-anomer is a component of lactose, the disaccharide of mammalian milk. In glycosidic linkage either α-D-galactose or α-D-galacturonic acid (or its methyl ester) are the repeating monomeric units of the plant pectic substances. These are known for their gel-forming properties. α-D-Galactopyranose and β-D-galactopyranose are known as crystalline substances. Anhydrous crystals of α-D-galactopyranose are obtained by crystallization from water at ambient temperatures. They melt at 167°C, and the initial specific rotation is +150·7° changing to +80·2°. β-D-Galactopyranose crystallizes from water solution near 0°C to form anhydrous crystals that melt at 143–145°C. The specific rotation is +52·8° changing to +80·2°.

5.1.3.6 *Mannose*

α-D-Mannopyranose and β-D-mannopyranose are known in crystalline form. The occurrence of mannose as the free sugar, even though it is the 2-epimer of glucose, is rare. The former is crystallized from vegetable ivory nut shavings (a waste product of the button industry), after hydrolysis by sulfuric acid. Another source is spruce hemicellulose. α-D-Mannopyranose melts at 133°C and the specific rotation is +29·3° with mutarotation to +14·5°. It is the favored crystalline form. However, the first anomer of mannose to be crystallized was β-D-mannopyranose which, because of an unfavorable structure, is not now generally available commercially. It melts at 132°C, and the specific rotation is −16·3° with mutarotation to +14·5°. The L-form of the sugar has been synthesized.

5.1.3.7 *Fructose*

(β-D-*arabino*-hexulose, levulose, fruit sugar, the uncrystallizable sugar, *sucre dé liquidé* and 'left-handed glucose'). This ketohexose is widely distributed in plants as a free sugar, but only one anhydrous crystalline form (β-D-fructopyranose) is known. Crystals of β-D-fructopyranose have a melting point of 102–104°C and a specific rotation of −133·5° with rapid mutarotation to −92°. In glycosidic linkage the form of fructose found in crystalline disaccharides such as sucrose, and higher molecular weight oligosaccharides and polysaccharides, is invariably that of the β-D-fructofuranose tautomer. Fructose, like sucrose, but unlike glucose, is widely distributed only in the plant kingdom. It is the major sugar of apples and pears. It is also the major component of inulin, the carbohydrate reserve in plants of the Compositae family, such as the Jerusalem artichoke, dandelion, and dahlia.

5.2 OLIGOSACCHARIDES

Compound sugars (oligosaccharides) are made up of from two to about ten monosaccharides. They are linked through the anomeric hydroxyl group, and either another anomeric hydroxyl group, or an OH group elsewhere in the second sugar molecule. The sugar donating the anomeric OH group is designated as an R–C–O^- (*glycosyl*) radical. A molecule of water is eliminated between the two OH groups, R–C–O–(H, H′–O′)–C′–R′=R–C–O–C′–R′ + H–O′–H. The product of the reaction is a *glycosidic* (*ether*) *bond*, R–C–O–C′–R′ between the glycosyl radical and the OH group of the second molecule, plus water.

5.2.1 Formation of Oligosaccharides

If the reaction between the two monosaccharides is between two anomeric hydroxyl groups, the disaccharide formed is called a *glycosyl glycoside*, and is *nonreducing*. If it occurs between the anomeric OH group of one sugar and an OH group of another that is not the anomeric OH group, the compound is a *glycosyl glycose*, and is a *reducing* oligosaccharide. The difference between the two is that the first does not possess a potentially free carbonyl group to partake in reducing reactions, such as the reduction (Section 5.3.1) of alkaline cupric hydroxide to cuprous oxide, while the latter does. The structure of several significant oligosaccharides is shown in Fig. 5.14.

5.2.2 Oligosaccharide Nomenclature

The names given to the oligosaccharides in Fig. 5.14 are trivial names established by usage. Generic names that completely identify the compounds use configurational and linkage specification designations that become progressively more cumbersome to write or to pronounce. Oligosaccharides, as mentioned in Section 5.2.1, are generically specified as being either glycosyl glycoses (reducing oligosaccharides) or glucosyl glycosides (nonreducing oligosaccharides) using several methods to specify positions for the glycosidic linkage(s). Lactose is a reducing sugar, β-D-galactopyranosyl (1 → 4) D-glucose, or 4-*O*-β-D-galacto-pyranosyl-D-glucose, and has α-D- and β-D-anomeric forms. Sucrose is α-D-glucopyranosyl β-D-fructofuranoside, and is nonreducing. Alternatively, sucrose can be named β-D-fructofuranosyl α-D-glycopyranoside and/or α-D-glucopyranosyl (1 → 2) β-D-fructopyranoside.

5.2.3 Naturally Occurring Oligosaccharides

5.2.3.1 *Sucrose*

Sucrose (α-D-glucopyranosyl β-D-fructofuranoside, cane sugar, beet sugar, table sugar, saccharose) is the familiar disaccharide of commerce. It is

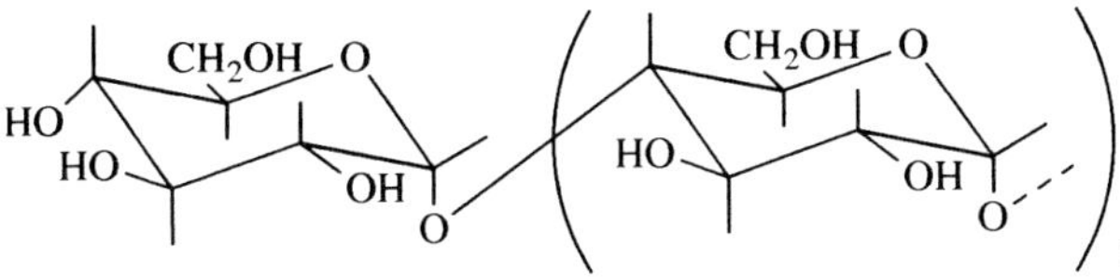

n = 1. Maltose.
n = 2, Maltotriose
n = 3, Maltotetraose
etc.

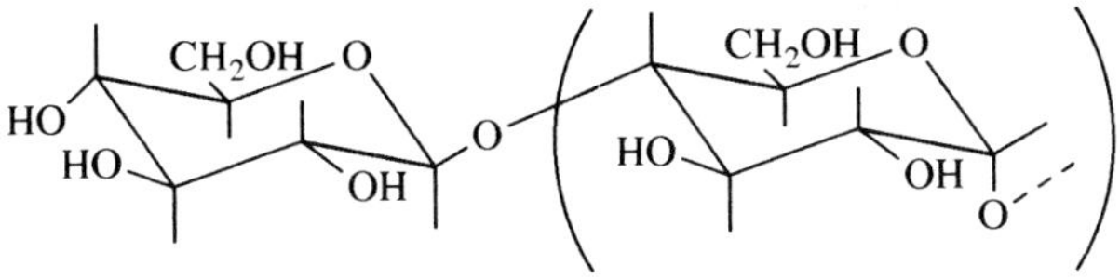

n = 1. Cellobiose.
n = 2, Cellotriose
n = 3, Cellotetraose
etc.

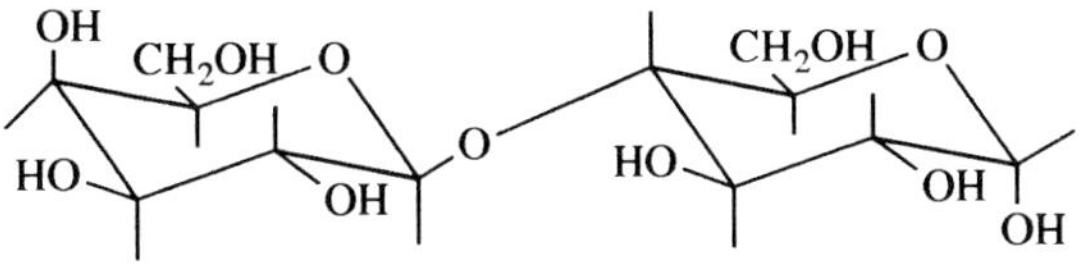

α-D-Lactose

FIG. 5.14. Structure of selected oligosaccharides.

composed of one molecule of α-D-glucopyranose and one molecule of β-D-fructofuranose joined through a glycosidic linkage that involves both anomeric carbon atoms. The Haworth formulation is shown in Fig. 5.15.

In the crystal structure (cf. Lichtenthaler *et al.*, 1991), the glucopyranose moiety has the $^{4}C_{1}$ conformation, and the fructofuranose moiety the $^{4}T_{3}$ twist conformation. The hexose moieties are approximately at right angles to each other. The structure is stabilized by two strong intramolecular hydrogen bonds. One is between the glucose ring oxygen atom and the fructose 6′-OH. The other is between the glucose O-2 and the fructose 1′-OH. The conformation of sucrose in solution is essentially the same as for the crystal, but in equilibrium with two forms.

The two favored conformational structures (Christofides & Davies, 1985; Davies & Christofides, 1987) are also shown in Fig. 5.15. The conformational structures are characterized by a cooperative hydrogen bonding network that is shown, however, only for the favored structure. Mathlouthi (1981) showed that there are no intramolecular hydrogen bonds in dilute solutions of sucrose, and Hervé du Penhoat *et al.* (1991) believe that the hydrogen bonding is not particularly important in determining sucrose conformations in aqueous solution. Molecular dynamic simulations (Tran & Brady, 1990) suggest that the sucrose molecule is relatively rigid. However, Pérez *et al.* (1993), believe that the description of sucrose as a rigid molecule under all conditions has to be mistrusted.

The melting point of crystalline sucrose is 188°C, and the specific rotation is +66·5°. Sucrose, along with its component monosaccharides, are the principal sugars in fruits and vegetables. Each monosaccharide is found in about the same proportion as found in sucrose (1 : 1). Several compilations of the distribution of free sugars in fruits and vegetables are available (Lee *et al.*, 1970; Wrolstad & Shallenberger, 1981). Sucrose is also the reference standard for the relative sweetness of substances. It serves admirably for this purpose because of the high degree of purity (> 99·93%) with which it can be obtained, and because it is nonreducing. The latter property mandates that it prevails in solution as a single configurational entity. The L-form is also known (Szarek & Jones, 1980).

5.2.3.2 *Lactose*

Lactose (milk sugar; 4-*O*-β-D-galactopyranosyl-D-glucose) occurs in mammalian milk to the extent of about 5%. It is a reducing disaccharide that crystallizes in three forms. The most common form is α-D-lactose monohydrate (4-*O*-β-D-galactopyranosyl-α-D-glucose · H_2O). After first transforming to the anhydrous form at 120°C, it melts at 201–202°C. The specific rotation is +85·0° initially, which slowly (22 h) comes to an equilibrium at +52·6°. The specific rotation of the anhydrous form of the α-D-tautomer is +92·6° changing slowly with time to +52·3°. The anhydrous β-D-tautomer crystallizes from solutions at temperatures above 93·5°C and has an initial specific rotation of +34° changing over 22 h to +53·6°. It is reported to be sweeter than the α-forms. The L-form is not known.

5.2.3.3 *Maltose*

Maltose (malt sugar; 4-*O*-α-D-glucopyranosyl-D-glucose), the only known crystalline form of this reducing disaccharide, is the monohydrate of the β-D-anomer. The crystals melt at 102–103°C, and the initial specific rotation is +111·7° with mutarotation to +130·4°. It is

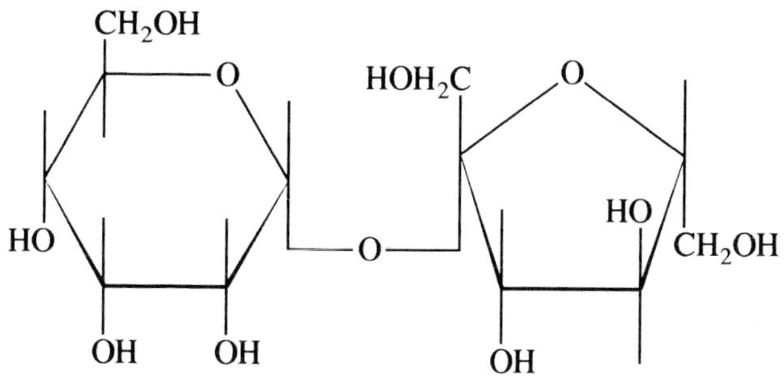

Haworth structure

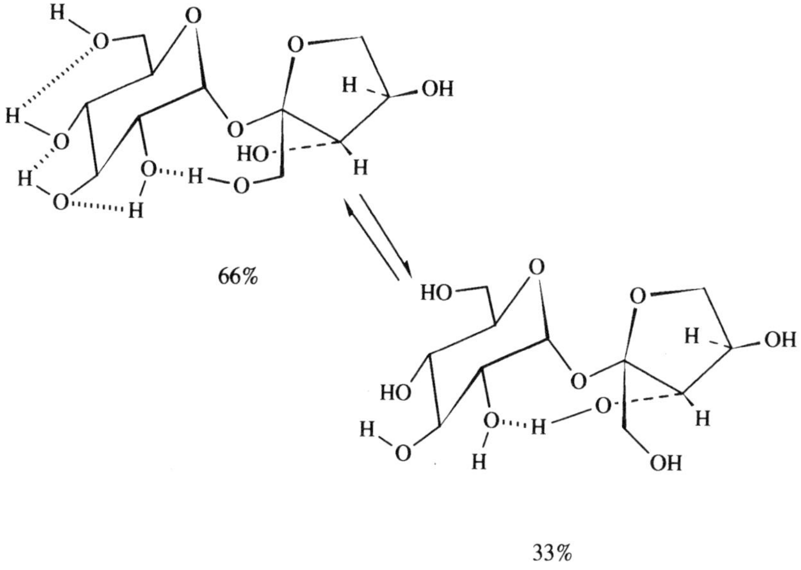

Favored conformations for sucrose

FIG. 5.15. Haworth and conformational structures of sucrose (α-D-glucopyranosyl β-D-fructofuranoside).

found free in nature on occasion, but is a major component, along with starch, of germinated and dried barley seeds (malt). The L-form is not known. Maltotriose, maltotetraose, and maltopentaose, etc., are components of the so-called 'glucose syrups' prepared by acid and/or enzyme hydrolysis of starch.

5.2.3.4 *Raffinose*

Raffinose (6-*O*-α-D-galactopyranosyl-6-*O*-α-D-glucopyranosyl β-D-fructofuranoside; α-galactosyl sucrose) is a nonreducing trisaccharide found in the free state in plants of the *Leguminosae* family (peas, beans, etc.), along with the tetrasaccharide stachyose (Fig. 5.16). Crystalline raffinose is a pentahydrate that melts at 80°C, and has a specific rotation of +105°. It is hydrolyzed by yeast invertase (β-D-fructosido-invertase) to melibiose and fructose. As the human digestive tract does not possess a fructosido-invertase, nor an α-D-galactosidase, neither raffinose nor melibiose can be metabolized. The L-form of raffinose is not known.

5.2.3.5 *Stachyose*

Stachyose (6-*O*-α-D-galactopyranosyl-6-*O*-α-D-galactopyranosyl-6-*O*-α-D-glucopyranosyl β-D-fructofuranoside) (Fig. 5.16) is the classic tetrasaccharide. It occurs as the free sugar in many plants, but especially those of the legume family, accompanied by sucrose and raffinose. It crystallizes as a tetrahydrate with a melting point of 167–170°C, and has a specific rotation of +148°. It is hydrolyzed by yeast invertase to manninotriose and fructose, but neither stachyose nor the manninotriose unit, upon ingestion, is metabolized. The L-form is not known.

5.2.3.6 α,α-*Trehalose*

Trehalose (1-*O*-α-D-glucopyranosyl α-D-glucopyranoside; mycose, mushroom sugar) is a nonreducing disaccharide made up of two molecules of α-D-glucose linked through their anomeric centers. Although it has a twofold axis of symmetry, and has been called the 'symmetrical disaccharide,' it is chiral, and is a structural palindrome. It occurs in fungi including immature mushrooms, but is converted to mannitol in mature mushrooms. The structure is shown below. α,α-Trehalose has a melting point of 97°C and a substantial specific rotation (+178·3°). The isomers β,β- and α,β-trehalose have been synthesized.

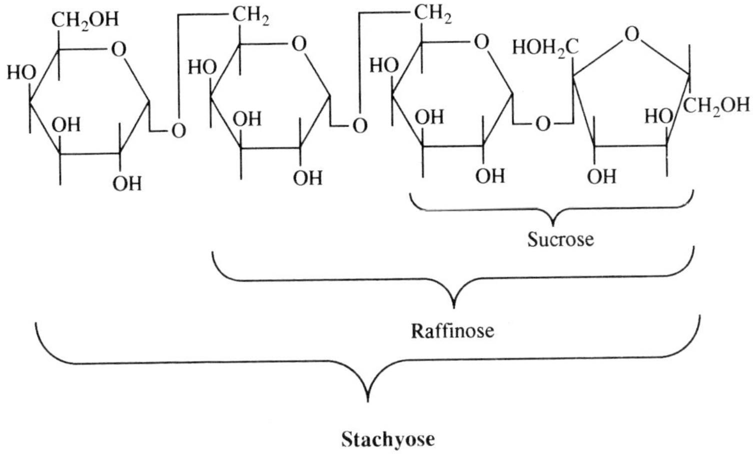

FIG. 5.16. Haworth structure for stachyose, including component nonreducing sugars.

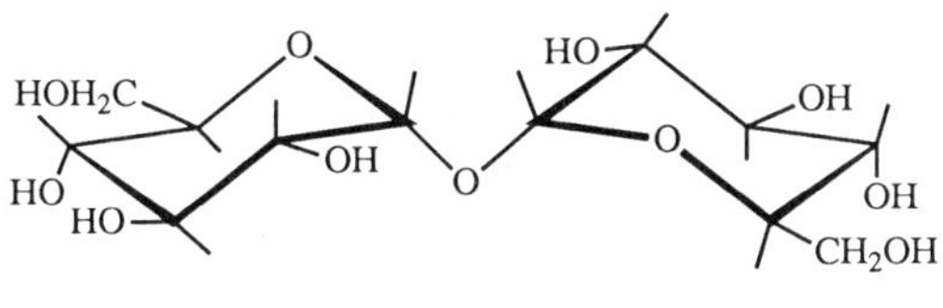

α,α -Trehalose

5.3 REACTIONS OF SUGARS

5.3.1 Oxidation, Reduction

The free monosaccharides are reducing sugars. The term *reducing* applies to the chemical definition of oxidation (gain in positive valence through loss of electrons) and reduction (loss of positive valence through gain of electrons), and is strictly a chemical definition. In the presence of alkaline cupric hydroxide (Fehling's solution), and upon application of heat, the carbonyl function of a reducing sugar will reduce cupric hydroxide to cuprous oxide. The sugar is oxidized to an acid. The reaction serves as a qualitative and quantitative method for the reducing and nonreducing sugar content of materials. The oxidation-reduction ability of sugar carbonyl functions is induced by high temperature at alkaline pH.

Oligosaccharides may be either reducing or nonreducing substances (Section 5.2.1), and lactose is an example of the former, while sucrose is the classic nonreducing sugar.

$$2\,Cu^{2+}(OH)_2 + RHC{=}O \longrightarrow \underset{\text{(Precipitation)}}{\underset{\downarrow}{Cu_2^{+}O}} + 2H_2O + RCOOH$$

5.3.2 Reactions of Hydroxyl Groups

The hydroxyl groups of the sugars vary in reactivity depending upon whether they are primary or secondary alcoholic groups. Reactivity also depends on their position in the molecule. Although the acidity of the different hydroxyl groups varies, solutions of sugars are neutral. The most acidic hydroxyl group is the one at the anomeric center (OH-1 for aldoses; OH-2 for ketoses). In order of reactivity, the general listing for an aldose is $1 > 6 > 4 > 3 > 2$. The order has bearing on the ability of the different groups to participate in hydrogen bonding.

Intrinsic reactions and properties of hydroxyl groups govern their use in foods as sweetening agents, and also govern the sweetness of the sugar. The use of sugars in foods is related to the ability of hydroxyl groups, through hydrogen bonding, to alter the properties of water (viscosity, osmotic pressure, lowering of the freezing point, etc.), and is covered in the next chapter. One characteristic reaction of a hydroxyl group has already been mentioned, i.e. the ability to react with another hydroxyl group to form an acetal or hemiacetal, and the reaction may be either intra- or intermolecular (Section 4.4). The reaction proceeds so easily with the sugars that the carbonyl form is nearly undetectable in water solution, and the equilibrium strongly favors the cyclic forms. On the other hand, the hemiacetal sugar ring bond is so weak ($-1{\cdot}7\,\text{kcal mol}^{-1}$), it is cleaved merely by dissolving sugar crystals in water, leading to an equilibrium between the various hemiacetal forms. The phenomenon is accompanied with a change in the optical rotation, and is known as the *mutarotation* reaction. Substitution for, or on, individual hydroxyl groups alters sugar sweetness.

5.3.3 Mutarotation

Upon solvation in an ampholytic solvent, such as water, reducing sugars immediately begin to mutarotate. There is progressive change in the optical rotatory characteristics, and eventual establishment of an equilibrium. The change in optical rotation is due to transposition of various tautomeric sugar forms. The phenomenon is not unique to reducing sugars, but is characteristic of them. Mutarotation may also be observed with an increase in the concentration of an optically active solute, or change in the temperature (*thermal mutarotation*). The change in optical rotation upon solvation of a crystalline sugar is a manifestation of the fact that the simple sugars, in solution, are not a single structural entity. They assume several different forms that are, ultimately, different sugars (tautomers). The mutarotation observed with changing concentration or temperature may be either a change in tautomer distribution, change in conformation (e.g., sucrose), or both.

Sucrose has a relatively constant and stable specific rotation of +66·5°. The specific rotation of reducing sugars has an initial value, obtained by extrapolation, and an equilibrium value. The optical rotatory power of a substance is the observed rotation (α), at the sodium D line (589 nm), in angular degrees, which is either to the right or left as the observer faces the light source of the polarimeter. The magnitude is a function of concentration, length of the tube containing the solution, and temperature (t). Specific rotation is a calculation that standardizes the measurement, and is obtained from the relation (α)/g per ml × decimeter length of the polarimeter tube. It is then expressed as $[\alpha]_D^t$.

Starting with α-D-glucopyranose, mutarotation is downward from an initial value of about +112° to an equilibrium value, in about 4 h, of approximately +52°. Starting with β-D-glucopyranose, mutarotation is upward from an initial value of +19° to the same equilibrium value. The kinetics of the reaction are first-order (Section 2.5.2), indicating that a two-component system prevails at equilibrium, and a simple calculation, from either rate or equilibrium constants (Sections 2.5.3 and 2.5.4), reveals that a solution of D-glucose contains about 64% β-D-glucopyranose and 36% α-D-glycopyranose. The two possible furanose forms, and the acyclic forms, are present only in trace amounts (Fig. 5.8). Neither change in temperature nor change in concentration has significant effect on the position of the D-glucose mutarotation equilibrium.

Crystalline β-D-fructopyranose mutarotates rapidly, obtaining equilibrium in about 30 min. By comparison, D-lactose requires about 24 h to come to equilibrium, and D-galactose requires about 3 h. Because of the importance of the position of the mutarotation reaction in assessing the sweetness of sugars (cf. Pangborn & Gee, 1961), the approximate

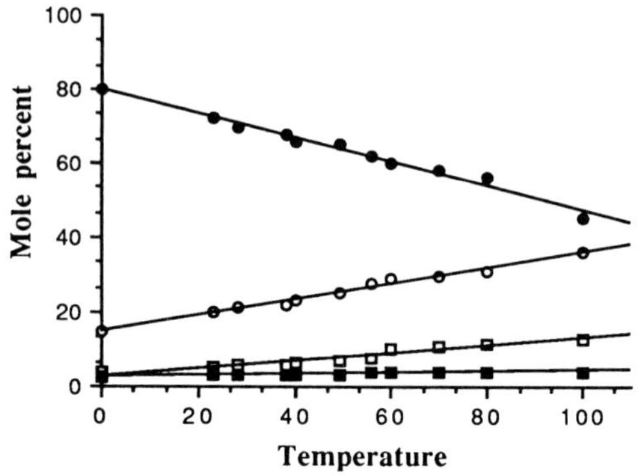

FIG. 5.17. Change in distribution of fructose tautomers with change in temperature. Solid circles, β-D-fructopyranose. Open circles, β-D-fructofuranose. Open squares, α-D-fructofuranose. Solid squares, α-D-fructopyranose (Schneider *et al.*, 1985).

position of the equilibrium for selected sugars is given in Table 5.2. The values were obtained by different methods, which are mentioned in the footnotes to Table 5.2. Unlike glucose, the composition of a solution of fructose changes considerably with temperature. A plot of the shift that occurs among the cyclic fructose tautomers, with change in temperature, is shown in Fig. 5.17.

The position of the fructose mutarotation equilibrium changes somewhat with concentration also. With change in temperature, the specific rotation of fructose becomes more negative according to the expression $[\alpha]_D^{20} = (-91{\cdot}5 + 0{\cdot}133x)$, where x is the weight per volume percentage of D-fructose (Browne & Zerban, 1941). The concentration of the β-D-fructopyranose tautomer increases by about 5% over the range 5–80% fructose (Hyvönen *et al.*, 1977*a*).

5.3.4 Sucrose Inversion

The sucrose molecule's glycosidic linkage between the two monosaccharides is cleaved through the addition of water (hydrolysis), by the action of either acids or enzymes (invertases), to give *invert sugar*. Invert sugar is a mixture containing near-equal parts of glucose and fructose. The reaction is important to the function of sucrose as a sweetener, and is first-order (Section 2.5.2) with respect to sucrose. The inversion is independent of the type of acid, and the general equation for inversion rate (Wienen & Shallenberger, 1988) is: $\log_{10}$ (hours to 99·99% inversion) $= (\text{pH} - 0{\cdot}5) + (39{\cdot}0)(\text{temperature})^{-1}$. A solution of sucrose at pH 1·5 at

TABLE 5.2
Percent tautomer distribution of reducing sugars at equilibrium in water

Sugar	*Temp.*	*α-Pyranose*	*β-Pyranose*	*α-Furanose*	*β-Furanose*	*Acyclic*
Glucose	0[a]	36·2	63·8	—	—	—
	31[b]	38	62	<0·05	0·14	0·002[c]
Gulose	22[c]	16	81	—	3	—
Galactose	0[a]	29·6	70·4	—	—	—
	31[c]	30	64	2·5	3·5	0·020[c]
	25[d]	32·0	63·9	1·0	3·1	—
Mannose	0[a]	68·8	31·2	—	—	—
	44[c]	65·5	34·5	0·6	0·3	0·005[c]
Arabinose	0[a]	67·6	26·5	—	—	—
	31[c]	60	35·5	2·5	2	0·030[c]
	20[e]	55	32	7	6	—
Lyxose	0[a]	76·0	24·0	—	—	—
	31[c]	70	28	1·5	0·5	0·03[c]
Ribose	31[c]	21·5	58·5	6·5	13·5	0·05[c]
	31[f]	19	64	6	11	—
Xylose	0[a]	34·8	65·2	—	—	—
Lactose	0[a]	36·8	63·2	—	—	—
Maltose	20[a]	36·0	64·0	—	—	—
Fructose	20[g]	2	73	5	20	—
	20[h]	0	75	5	20	—
	31[c]	2·5	65	6·5	25	0·8[c]
Allulose	27[i]	22	24	39	15	0·3
Sorbose	27[i]	98	—	2	—	0·2
Tagatose	27[i]	79	16	1	4	0·6

[a]Bates *et al.* (1942), optical rotation; [b]Williams & Allerhand (1977), nuclear magnetic resonance (n.m.r.); [c]Angyal (1968), n.m.r.; [d]Acree *et al.* (1969), gas-liquid chromatography; [e]Liptak *et al.* (1982), n.m.r.; [f]Horton & Walaszek (1982), n.m.r.; [g]Schneider *et al.* (1985), n.m.r.; [h]Shallenberger *et al.* (1977), g.l.c.; [i]Angyal & Bethell (1976), n.m.r.

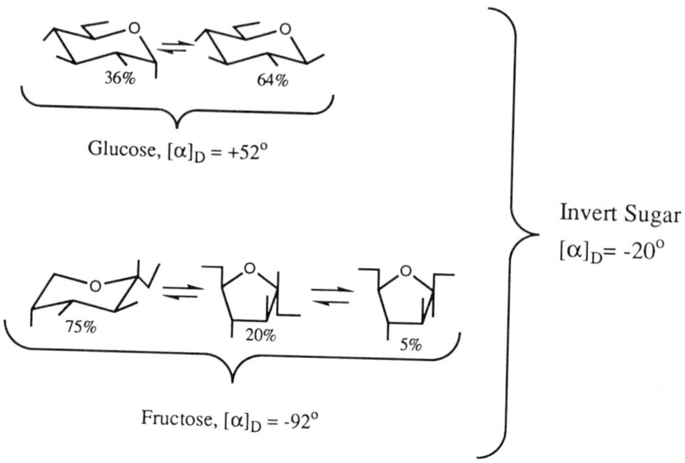

FIG. 5.18. Composition of invert sugar.

30° requires about 200 h for complete inversion. The general tautomeric composition of invert sugar is given in Fig. 5.18. As sucrose has a molecular weight of 342, and the free monosaccharides have a combined molecular weight of 360, the inversion reaction increases the percentage sugars, by weight, about 5%. The water activity of the system is also altered. The reversal of the sign for the optical rotation is the reason for the term 'invert sugar'.

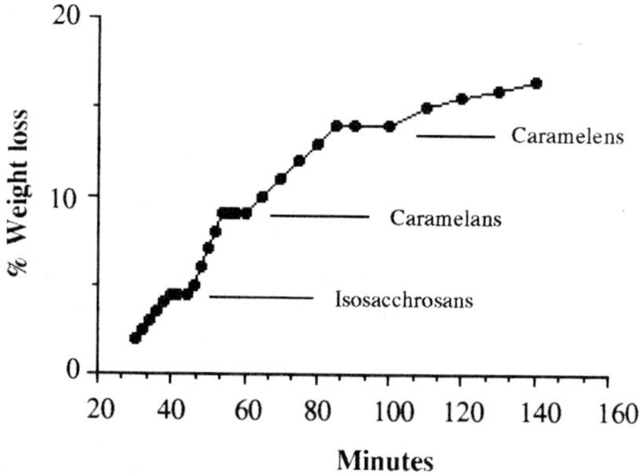

FIG. 5.19. Formation of caramel substances from sucrose at 200°C (Shumaker & Buchanan, 1932).

5.3.5 Caramelization

When crystalline sucrose (or crystalline glucose and fructose) is held at a temperature of 200°C, a series of endothermic events occurs as shown in Fig. 5.19. When the sucrose has melted, the continued application of heat leads to sugar dehydration and foaming. After a weight loss of 4·5%, which corresponds to loss of one molecule of water per molecule of sucrose, the foaming temporarily subsides. The average molecular weights of the materials present, known as *isosacchrosans*, corresponds to $C_{12}H_{22}O_{12}-H_2O$, or $C_{12}H_{20}O_{11}$. Soon after, foaming begins again, and subsides when an additional molecule of water has been lost. The dehydration products have an average molecular weight corresponding to two molecules of sucrose minus four molecules of water, or $C_{24}H_{36}O_{18}$. These materials are known as *caramelans*. A third stage of foaming corresponds to $3\,C_{12}H_{22}O_{11}$ minus $8\,H_2O{=}C_{36}H_{50}O_{25}$. These materials are known as *caramelens*. As the reaction progresses further, the percentage carbon in the materials increases, and are known as *humins* or *caramelins*. With time, the product increases in color intensity from pale yellow to brown, and the products at the various stages of the dehydration phenomenon are *caramels*. At any stage of the reaction, varying amounts of the aforementioned dehydration products are present.

The caramel pigments are polymers with pyran and furan nuclei, and the color is due to unconjugate unsaturation. They are colloidal substances, and have isoelectric points that vary with the method of preparation. The caramel pigments possess carbonyl, carboxyl, and phenolic hydroxyl groups. The occurrence of the latter seems to explain why caramels, used as colorants, complex with iron to give deeper colors. The caramel pigments, *per se*, are bitter tasting, and herein lies an important function in foods. Whether added, or formed *in situ*, the caramel pigments, plus a sweetening agent (sugar or otherwise), impart the unique flavor attribute associated with the taste of a proper sugar–bitterness ratio (e.g., chocolate candy). Many of the caramels in the food industry are prepared from glucose syrups employing ammonia as a catalyst, and are *melanoidins*, as they contain nitrogen. The volatile compounds produced during caramelization have special flavors, depending upon the temperature employed and time of heating. Caramels produced from either sucrose, fructose, glucose, or maltose are similar (Monte & Maga, 1982).

5.3.6 Anhydride formation

On standing in acid solution, monosaccharides undergo a series of dehydration and polymerization reactions that affect their taste. For the aldose sugars, these reactions are characterized by the elimination of a molecule of water between the β-D-anomeric hydroxyl group and the primary alcohol group at C-6 to form bicyclic compounds. The substance formed from β-D-glucopyranose is known as *levoglucosan* as it is levorotatory (−67°). The transformation is an example of a conformationally-directed reaction. The amount of 1,6-*anhydro* form present in solution at 100°C is directly proportionate to the conformational stability of the parent sugar. While only about 1 mol percentage levoglucosan is found for glucose, about 80% is found for 1,6-*anhydro*-β-D-idopyranose (Angyal & Dawes, 1968). In crystalline form, levoglucosan possesses little sweetness, and the crystals yield a cooling sensation. In solution, it is reported that they taste sweet (Lee, 1987).

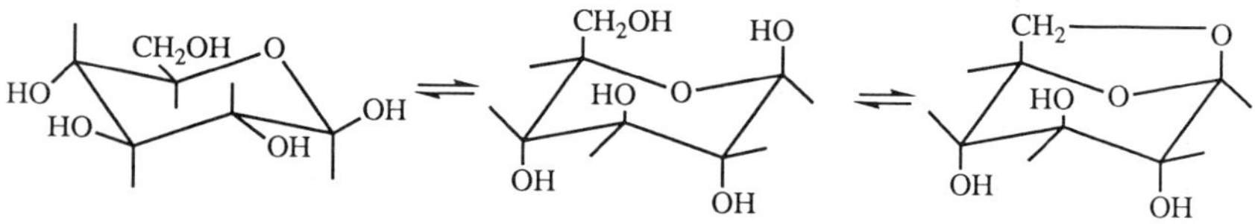

β-D-Glucopyranose, $^4C_1$ β-D-Glucopyranose, $^1C_4$ 1,6 anhydro-β-D-Glucopyranose (Levoglucosan)

With the *anhydro* aldose sugars, dehydration leads to the formation of *anhydro*-glucose polymers (*glucosans*) with variable properties and molecular weights. The process is known as *reversion*. The reversion products of glucose have applications for replacing some of the bulk and viscometric properties of sugars in foods when some or all the sugar has been replaced by a high-potency sweetening agent.

When fructose is held in acid solution, dehydration occurs with the formation of a series of dimers known as the *di-D-fructose dianhydrides*, and/or *diheterolevulosans* (Binkley *et al.*, 1973). All combinations of the anomeric pyranose and furanose forms are possible, and most have been characterized. The structure of diheterolevulosan II (1′,2-*anhydro*-1-*O*-D-fructopyranosyl-D-fructofuranose) is shown below. As the fructose dianhydrides and diheterolevulosans have a central dioxane ring, formed by removal of a molecule of water from the anomeric carbon atom of each sugar, they are nonreducing. Di-D-fructose anhydride I (1′,2-*anhydro*-[1-(α-D-fructofuranosyl]-β-D-fructofuranose), prepared by Lemieux &

Nagarajan (1964), is tasteless. Polymerization in acid solution then halts after dimer formation, but it may proceed again with continued application of heat, to form caramels. At this stage there is little, if any, distinction between caramels formed from aldoses, ketoses, or disaccharides.

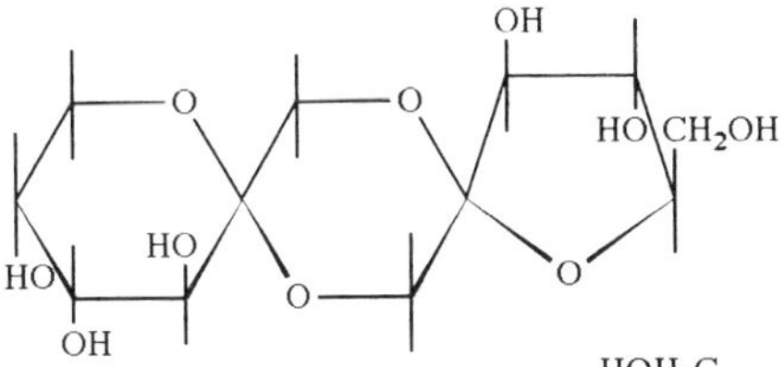

Diheterolevulosan II

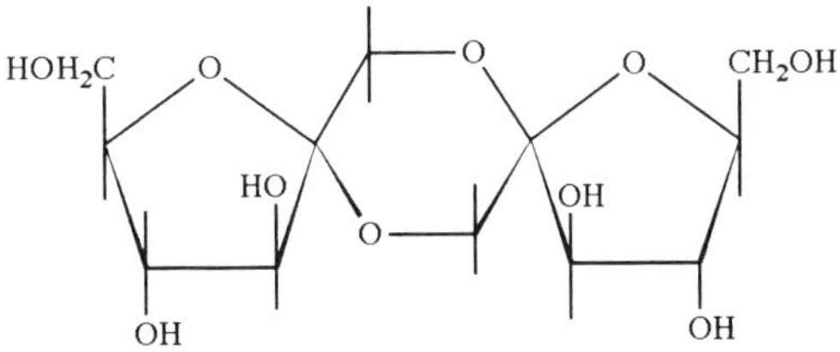

Di-D-Fructose anhydride I

5.4 PROPERTIES OF SUGARS

The physical properties of the sugars have application to taste chemistry theory, the mechanism of sweet taste, their subsidiary attributes as sweetening agents, practical use as sweetening agents, and components of food.

5.4.1 Water Solubility

The sugars are so soluble in water that thinking of their solubility in the conventional sense of the grams of sugar dissolved in water to give a final volume of 100 ml solution (w/v) is not practical. As can be seen from Table 5.3, the ability of the sugars to dissolve water is greater than the ability of water to dissolve sugar. For example, 1 g sucrose will dissolve in 0·5 ml (0·5g) water. It is for this reason that the percentage sugar in commercial sugar/water preparations is expressed on a weight/weight basis, and not weight/volume. Calculations, based on the data from the solubility data in Table 5.3 show that one molecule of sucrose dissolves completely and stoichiometrically in about 10 molecules of water. From the data in Table 5.3, 0·0020 mol sucrose (0·67 g/342) divided by

TABLE 5.3
Solubility of the sugars

Sugar	*ml H_2O to dissolve one g sugar*	*Grams sugar in one g final syrup*	*Moles water/mole sugar*
Sucrose	0·50	0·67	9·15
α-D-Glucopyranose-H_2O	1·15	0·47	—
α-D-Glucopyranose	0·50	0·67	4·96
β-D-Glucopyranose[a]	(0·50)	(0·67)	(4·96)
Fructose	0·25	0·79	2·66

[a]The initial solubility is higher, and rate of solvation more rapid than that of the α-anomer monohydrate. Final solubility is the same as for the anhydrous α-anomer.

0·0183 mol water (0·33 g/18) = 9·15. One molecule of crystalline glucose is dissolved by, and presumably binds about 5 molecules of water. For fructose, the calculation shows that less than three molecules of water are needed to dissolve one molecule of fructose.

5.4.2 Hygroscopicity

A crystalline sugar will absorb water, and if the relative humidity is high enough, it will begin to liquify when it has absorbed about 15% (w/w) atmospheric moisture (Sokolovsky, 1937). The absorption of moisture is a mass action phenomenon, and the rate varies with the identity of the sugar. The rate at which sugars take up moisture is the same order (fructose > sucrose > glucose) as their solubility, and show first-order

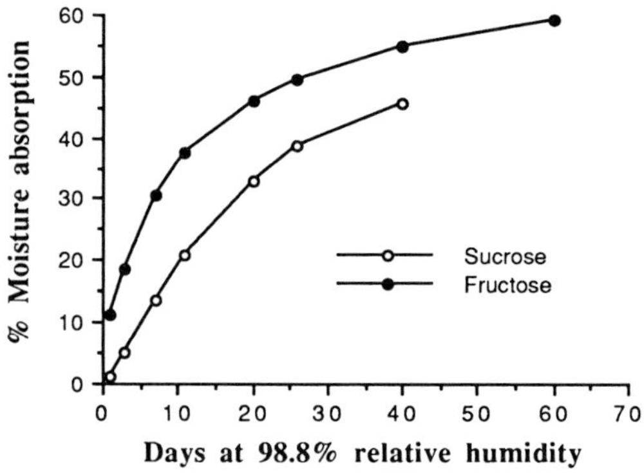

FIG. 5.20. Percent moisture absorbed by sucrose and fructose crystals at 98·8% relative humidity (Sokolovsky, 1937).

kinetics (Browne, 1922). A plot of the absorption of atmospheric moisture by sucrose and fructose is shown in Fig. 5.20.

5.4.3 Water Activity

Water activity alteration is the ability of substances to alter the vapor pressure of water, and is defined as the E.R.H. (equilibrium relative humidity of the system divided by 100). The sugars, with their ability to interact with water, significantly lower the very activity of water. The activity of pure water is unity, and 1 g molecular weight of sucrose (342 g) dissolved in 1 litre of water (55·5 g moles) yields a water activity of 55·5/56·5 = 0·9823. The vapor pressure of a solution then is equal to the vapor pressure of the pure solvent times the mole fraction of the solvent in solution. The vapor pressure lowering of water is about 1·77% (100–98·23) for a 1 molal ideal solution. However, a 1 molal solution of sucrose has a vapor pressure that is lowered about 10% more.

5.4.4 Partition Coefficients

An important physical parameter in taste chemistry is the solubility of the sugars in solvents other than water, and/or the ability of sugars to partition between water and a solvent with less polarity. Only meager data is available on the subject. Collander (1950) determined the solubility of a series of organic compounds in isobutanol. The data

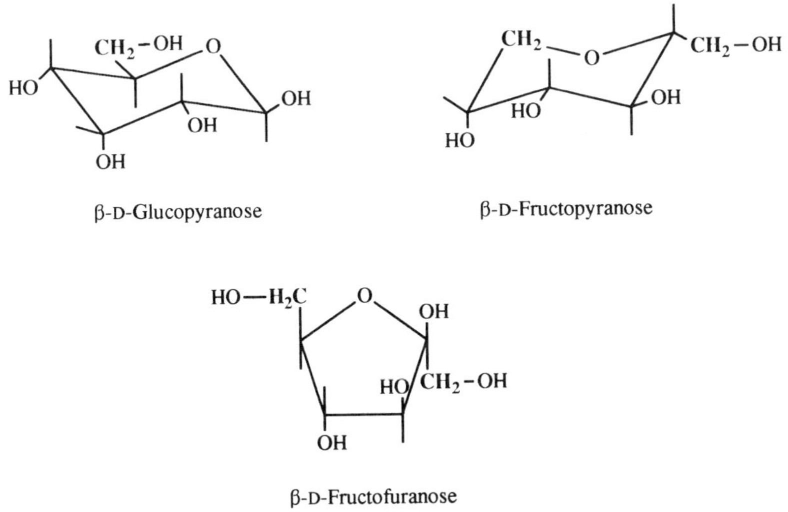

FIG. 5.21. The hydrophobic/lipophilic methylene carbon atoms in selected hexose structures.

TABLE 5.4

Partition coefficient at 20° between water and isobutanol for selected alcohols and sugars (Collander, 1950). Concentrations are in millimoles/liter

Compound	*Water conc.*	*Isobutanol conc.*	*Partition coeff.* $\times 10^{-5}$
Glycerol	2 170	0·100	4·61
Erythritol	2 300	0·037	1·61
Mannitol	962	0·014	1·46
Arabinose	2 000	0·019	0·95
Rhamnose	1 000	0·057	0·57
Glucose	1 911	0·011	0·57
Fructose	1960	0·017	0·87
Sucrose	1 300	0·005 6	0·43
Maltose	853	0·004 0	0·47

obtained for selected compounds, which varied by ±10–20%, is given in Table 5.4, along with the calculated partition coefficients. The latter were derived for the special purposes of this treatise by dividing the millimolar solubility in isobutanol by the initial millimolar concentration of the compound in water so that the coefficients for the different compounds might be compared. The partition coefficients of the sugars, as small as they are, appear to have application to their sweetness attributes, particularly impact or reaction time. Obviously there are elements of hydrophobicity/lipophilicity in the structure of the sugars that have important biological property significance. In spite of the number of polar hydroxyl groups, the hydrophobic/lipophilic centers are the methylene carbon atoms of the extracyclic methylene carbon atoms, and the methylene carbon atoms contained in the ring(s), as shown in bold type in Fig. 5.21. β-D-Fructopyranose and furanose has two such centers and glucose but one. The difference shown for their partition coefficients is probably not accurate, but nevertheless, real.

CHAPTER 6

Sweetness and Other Taste Attributes of the Sugars

Sweetness is a pleasant sensation, but the rates with which sweetness and pleasantness increase with increasing sugar concentration are not the same. Working with pure solutions, Moskowitz (1971*a*) found a linear log-log relation to suggest that pleasantness increases at only about one-quarter to one-third the rate of perceived sweetness. There is also a limit to just how much sweetness intensity is perceived as pleasant. When considering the sweetness attributes of substances alone, distinction needs to be made between perceived (*absolute*, intrinsic) sweetness, and *relative* sweetness. These two are not the same, either. The former applies to the intrinsic taste of a substance, as opposed to tastelessness and the salt, sour, and bitter attributes. The latter is a comparison that is made on a concentration basis.

6.1 INTRINSIC SWEETNESS OF SUGARS

The intrinsic sweetness attributes of a sugar include recognition threshold, reaction time, and intensity at a given concentration. Intrinsic sweetness also includes duration of the sweet sensation, 'body' (presumably due to an ability of a substance to alter the viscosity of water), and presence or absence of other tastes. Therefore intrinsic sweetness is a multidimensional sensory attribute, and single value scores for the sweetness of a compound are an integrated, dimensionless parameter, for the aforementioned components of taste. No two compounds can be expected to yield identical sigmoidal (Section 1.3.1) sweetness intensity-concentration plots.

6.1.1 Effect of Concentration

The intrinsic sweetness of sucrose, plotted against a small number, but wide range, of increasing sucrose concentrations is shown in Fig. 6.1.

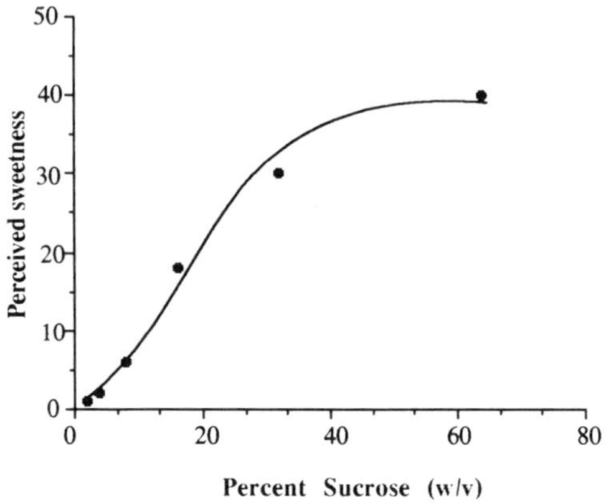

FIG. 6.1. Effect of increasing concentration on the perceived sweetness of sucrose (Moskowitz, 1970).

Elements of a sigmoidal nature for the plot are apparent in the figure. The data are average values determined from median magnitude estimates of sucrose sweetness obtained by Moskowitz (1970), but shown here on a weight per volume basis. The plot describes a 3rd order polynomial whereby the first two points indicate the weak induction phase. All points can be fitted to the usual (Stevens, 1960) sugar sweetness power function $S = kC^n$, where $n = 1{\cdot}3$ (Section 1.3.1). The perceived sweetness of sucrose diminishes rapidly when the concentration on a weight/volume basis exceeds 16%. Until a concentration of 16% sucrose is attained, the plot of sensory response versus concentration is quite linear (DuBois *et al.*, 1991). That plot is shown in Fig. 6.2. In spite of the variation observed in panel scores, the induction phase is still apparent, but insufficient data are included for the low concentrations to place the linear intercept at the negative value required of the sigmoid relation. The linear regression for the sucrose plot shown in Fig. 6.2 is Sweetness response $= 0{\cdot}80 + 0{\cdot}94x$. Linear regression equations for the data for glucose and fructose are Sweetness response $= -0{\cdot}2 + 0{\cdot}6x$, and $0{\cdot}04 + 1{\cdot}27x$, respectively. At increased concentrations, glucose may have an additional intrinsic taste that has been described as 'burning' and bitter (Cameron, 1947). It is probably due to residual acid used in the conversion of starch to glucose as acid hydrolysis is a step in the preparation of crystalline D-glucose.

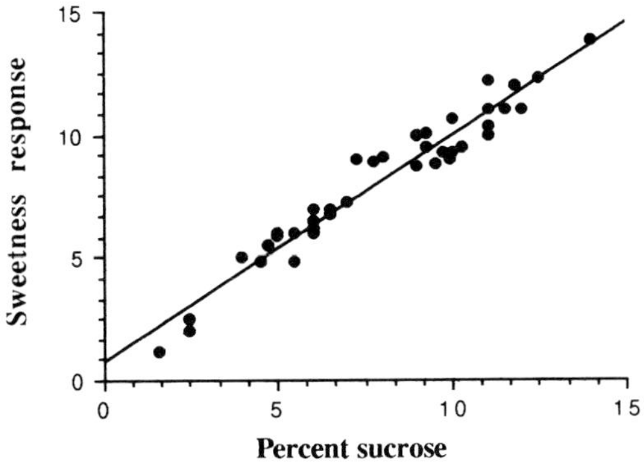

FIG. 6.2. Concentration-response plot for the increase in perceived sweetness with increasing sucrose concentration (DuBois *et al.*, 1991).

6.1.2 Effect of Temperature

At increased temperatures, the intrinsic sweetness of sucrose increases slightly. By using the magnitude scoring technique (Section 1.2.3) for the sweetness of increasing sucrose concentrations at different temperatures, Bartoshuk *et al.* (1982) found that increasing temperatures caused a slight increase in sucrose sweetness at low sucrose concentrations (0·03 M, 1% w/v), but it was not apparent at 0·5 M (17%) sucrose concentrations.

By matching a range of sucrose concentrations at various temperatures with the same range of solutions held at ambient temperature, Hyvönen *et al.* (1977*b*) found that perceived sucrose sweetness for all solutions held at 5°C is lowered by 10%. At 37°C, all sucrose solutions were the same as at ambient temperature. At 50°C all sucrose solutions were about 8% sweeter than those held at ambient temperature.

For glucose, the effect of increasing temperature on perceived sweetness is minimal, but increasing the temperature of fructose solutions results in a marked decrease in perceived sweetness. The unique effect of temperature on the intrinsic sweetness of fructose solutions also affects the relative sweetness of fructose to sucrose at different temperatures.

6.2 RELATIVE SWEETNESS OF SUGARS

Literature values for the relative sweetness of sugars using sucrose as the reference standard (unity × 100), and the method employed, are tabulated in Table 6.1. In many cases, the short description of the

TABLE 6.1
Relative sweetness of sugars, with sucrose taken as unity times 100. All scores reported are on a weight/volume basis

Sugar	*Relative sweetness*	*Method*	*Investigator(s)*
D-Glucose[a]	52	Supraliminal comparison.	Paul (1922)
(Anhydrous)	50	Threshold.	Sale & Skinner (1922)
	74	Recognition threshold.	Biester *et al.* (1925)
	62·5, 70·8[b]	Sucrose at 5, 10%	Dahlberg & Penczek (1941)
	80	Recognition threshold.	Fabian & Blum (1943)
	60, 68	Sucrose at 5, 10%	Cameron (1947)
	58	Paired comparison.	Tsuzuki & Yamazaki (1953)
	60, 67	Sucrose at 5, 10%	Yamaguchi *et al.* (1970)
	50	Magnitude scaling.	Moskowitz (1970)
	70	Paired comparison	Hyvönen *et al.* (1977*a*)
	61	Relative to standard scale.	Tunaley *et al.* (1987)
α-D-, β-D-Glucose	64, 61	Sucrose at 9·12%.	Schutz & Pilgrim (1957*a*)
α-D-, β-D-Glucose	74, 82	Magnitude scaling.[c]	Shallenberger & Acree (1971)
D-Glucose average	**64**		
L-Glucose	64[d]	Magnitude estimation.	Shallenberger & Lindley (1977)
	64[d]	Simply tasted.	Fischer (1890)
D-Fructose	103	Supraliminal comparison.	Paul (1922)
	150	Threshold.	Sale & Skinner (1922)
	175	Recognition threshold.	Biester *et al.* (1925)
	111, 115	Sucrose at 5, 10%	Dahlberg & Penczek (1941)
	110	Sucrose at 5, 10%	Cameron (1947)
	138	Recognition threshold.	Fabian & Blum (1943)
	127	Paired comparison.	Tsuzuki & Yamazaki (1953)
	119	Paired comparison.	Hyvönen *et al.* (1977*a*)
	100	Magnitude scaling.	Moskowitz (1970)

	132, 133	Sucrose at 3·3, 12·3%.	Yamaguchi *et al.* (1970)
	115	Sucrose at 9·12%.	Schutz & Pilgrim (1957*a*)
	180	Magnitude scaling.[c]	Shallenberger & Acree (1971)
	120	Relative to standard scale.	Tunaley *et al.* (1987)
D-Fructose average	**120·5**[e]		
L-Fructose	(120·5)[f]	Simply tasted.	Wolfrom & Thompson (1946)
	< 120·5[f]	Paired comparison.	Ogawa *et al.* (1987)
Invert sugar	85	Rank order.	Sale & Skinner (1922)
	> 100	Paired comparison.	Humphriss (1940)
	80–111[g]	Calculated 'in terms of glucose.'	Cameron (1947)
	97	Calculated from average scores in this table.	
D-Galactose	32	Recognition threshold.	Biester *et al.* (1925)
	67	Sucrose at 5, 10%	Cameron (1947)
	48	Paired comparison.	Tsuzuki & Yamazaki (1953)
	59	Sucrose at 9·12%.	Schutz & Pilgrim (1957*a*)
	32[h]	Magnitude scaling.[c]	Shallenberger & Acree (1971)
D-Galactose average	**50**		
L-Galactose	(50)		
D-Mannose	32	Recognition threshold.	Biester *et al.* (1925)
	59	Sucrose at 9·12%	Schutz & Pilgrim (1957*a*)
α-D-Mannose	32	Magnitude scaling.[c]	Shallenberger & Acree (1971)
β-D-Mannose	(Bitter)	Magnitude scaling.[c]	Shallenberger & Acree (1971)
L-Mannose	(50)		
D-,L-Tagatose	(120)	Paired comparison.	Zehner (1988)
D-Maltose	60	Rank order.	Sale & Skinner (1922)

TABLE 6.1. Continued

Sugar	*Relative sweetness*	*Method*	*Investigator(s)*
	32	Recognition threshold.	Biester *et al.* (1925)
	37, 45	Sucrose at 5, 10%.	Dahlberg & Penczek (1941)
	45	Recognition threshold.	Fabian & Blum (1943)
	34	Paired comparison.	Tsuzuki & Yamazaki (1953)
	46	Sucrose at 9·12%.	Schutz & Pilgrim (1957*a*)
D-Maltose average	**43**		
D-Lactose	28	Supraliminal comparison.	Paul (1922)
	16	Recognition threshold.	Biester *et al.* (1925)
	37, 45	Sucrose at 5, 10%.	Dahlberg & Penczek (1941)
	40, 38	Sucrose at 5, 10%.	Cameron (1947)
	30	Sucrose at 9·12%.	Schutz & Pilgrim (1957*a*)
D-Lactose average	**33**		
Raffinose	22	Recognition threshold	Biester *et al.* (1925)
	1	Magnitude scaling.[c]	Shallenberger & Acree (1971)
Stachyose	10	Magnitude scaling.[c]	Shallenberger & Acree (1971)

[a] Generally at mutarotational equilibrium.
[b] Adjusted to an anhydrous glucose basis.
[c] Crystalline material. After averaging 10–15 ratings, scores were squared, and sucrose then taken as unity × 100.
[d] These are the same as the average D-score showing that panels can not distinguish between the sweetness potency of enantiomers.
[e] Obtained by eliminating observations where fructose is obviously not at mutarotational equilibrium.
[f] Based on observations employing D- and L-*pseudo*-fructose (fructose with a CH_2 group in place of the ring oxygen atom).
[g] Inversion of 1–20% sucrose.
[h] Calculated by reference to sweetness of glucose.

method employed by investigators does not do justice to the rigor applied in an attempt to derive significant values (e.g., Tsuzuki & Yamazaki, 1953). On the other hand, the relative score is a judgment by several persons at most, or in a few cases, merely a subjective estimate in place of an actual sweetness comparison.

6.2.1 Glucose

Relative sweetness values for glucose in solution given in Table 6.1 are for preparations that are usually at an unknown position for the anomeric α- and β-equilibrium. The scores were also obtained with glucose solutions at various ambient temperatures. The average sweetness value, usually obtained by employing 5–10% solutions of D-glucose, indicates that glucose is about two-thirds (64%) as sweet as sucrose. From the slope of the linear plots for the intrinsic sweetness of sugars (DuBois *et al.*, 1991), or 0·60/0·94, glucose also appears to be 64% as sweet as sucrose.

When the relative sweetness of freshly prepared glucose solutions are compared with solutions at mutarotational equilibrium, it is found that the α-D-glucopyranose tautomer is sweeter. This is true whether starting with freshly dissolved crystals of anhydrous α-D-glucopyranose (Schutz & Pilgrim, 1957*a*) or β-D-glucopyranose (Pangborn & Gee, 1961). The sweetness difference between solvated glucose anomers is relatively unimportant, however, as this difference is not great (< 5%). There also seems to be little difference in the sweetness potency of methyl α-D- and β-D-glucopyranoside (Birch, 1976). On the other hand, when the sweetness of crystals of the unsubstituted sugars is compared (Shallenberger & Acree, 1971), it is found that the relative sweetness scores reverse. That finding is probably related to the observation that crystals of the β-anomer dissolve more rapidly than those of the α-form. In addition, the initial solubility of the β-D-anomer is greater.

6.2.2 Fructose

As with glucose, many of the sweetness scores reported for fructose correspond to an unknown stage of the mutarotation reaction. This is quite important, as only one tautomeric form of fructose (β-D-fructopyranose) appears to taste sweet. This is the form present in crystalline fructose. The average sweetness score shown in Table 6.1 was derived by eliminating those fructose sweetness scores that were made at an unknown stage of fructose mutarotation. On a weight/volume basis, the sweetest sugar is fructose. On a molar basis, however, the sweetest sugar is sucrose.

6.2.3 Fructose syrups

In the conventional corn syrups, a portion of the glucose prepared from starch may be isomerized to fructose. Like honey, they are a form of invert sugar. The products are known as 'high fructose corn syrups' (HFCS), 'high fructose glucose syrups' (HFGS), and by various registered trade names. One such preparation, containing about 55–60% fructose and 36–41% glucose cannot readily be distinguished, organoleptically and functionally, from the invert sugar formed from sucrose by the action of acid and/or enzymes. The relative sweetness of invert sugar is presented in Section 6.4.3.

6.2.4 Galactose

Galactose is not as sweet as glucose, although the two substances are 4-epimers. It would appear, in contrast to glucose, that the α-anomer is sweeter than the β-anomer, not only in crystalline form (Shallenberger & Acree, 1971), but also in solution (Pangborn & Gee, 1961). When glucose is used as the reference standard, galactose appears to be about one-half as sweet as glucose. Galactose and glucose are components of lactose, but both galactose and glucose are sweeter than lactose.

6.2.5 Mannose

The α-D-anomer of mannose is, like galactose, only about one-half as sweet as glucose, and devoid of bitterness. On the other hand, the β-D-anomer is devoid of sweetness (Stewart *et al.*, 1971), and tastes bitter instead. At mutarotational equilibrium, the α-anomer is present in abundance, and solutions of mannose taste sweet and bitter.

6.2.6 Maltose and Maltose/Glucose Syrups

Maltose mutarotates very slowly, and the scores assigned in Table 6.1 probably refer to some unknown position of tautomer distribution. Maltose is a relatively rare crystalline sugar, and only one crystalline form is known (β-D-maltose). At equilibrium in solution, it has the same anomeric distribution as glucose. Whether this can be taken to imply that the α-anomer is sweeter than the β-anomer, either in crystalline form or in solution, is a moot point.

Maltose and the maltooligosaccharides are available commercially as glucose syrups with different 'dextrose equivalents' (DE), and with dry matter content ranging from 71 to 85%. The syrups are prepared by either acid and/or enzyme hydrolysis of starch, using enzymes known as amylases and glucoamylases. The end product of enzyme action by the

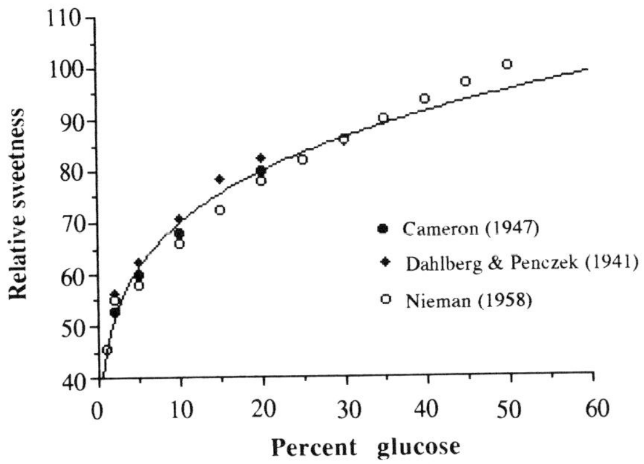

FIG. 6.3. Effect of increasing concentration on the relative sweetness of glucose.

former is maltose, and the end product by action of the latter is glucose. The glucose syrups are available with a large range of composition, and contain varying proportions of mono-, di-, tri-, and higher saccharides.

Two syrups can have the same dextrose equivalent rating, e.g. 42 (analytically *equivalent* to 42% glucose, by weight), but one can have 19% dextrose, 14% maltose, 11% maltotriose, and 56% 'higher saccharides.' The other might have 3% dextrose, 50% maltose, 10% maltotriose, and 37% 'higher saccharides.' Still others contain 92% glucose. Glucose syrups are described as having a clean, pleasant, sweet taste. Sweetness potency depends on concentration and distribution of individual saccharides.

6.2.7 Lactose

Crystals of α-D-lactose monohydrate are only faintly sweet, and the anhydrous β-D-anomer is somewhat sweeter. In solution, however (Pangborn & Gee, 1961), α-D-lactose is reported to be sweeter than an equilibrium solution at low concentrations (0·3%). The relation is reversed at higher initial concentrations of the freshly dissolved α-form (7%). Low concentrations (0·3%) of freshly dissolved β-D-lactose are not as sweet as the equilibrium mixture, but higher (7%) concentrations are sweeter. The distribution of the two anomeric forms at mutarotational equilibrium is about the same as for glucose, i.e. 64% β-D- and 34% α-D-forms. Conflicting reports for the solution sweetness of lactose anomers result when solutions are tasted within 3 h after solvation. It takes at least 22 h for lactose to attain mutarotational equilibrium.

6.3 EFFECT OF CONCENTRATION ON RELATIVE SWEETNESS

With increasing concentration over the range of 1–50%, the relative sweetness of D-glucose approximately doubles, as shown in Fig. 6.3. As the rate of relative sweetness increase is greater than the rate of concentration increase, the effect of increasing concentration appears to be synergistic. The phenomenon is not unique to glucose, and occurs with reducing oligosaccharides that have a glucose unit as the terminal reducing function, such as maltose and lactose (Dahlberg & Penczek, 1942). It also occurs with the glucose syrups. Increasing the dextrose equivalents also increases the relative sweetness of the glucose syrups (Dahlberg & Penczek, 1947; Nieman, 1958).

The increase in glucose relative sweetness with concentration is sometimes described in psychological parlance as *self-synergism*. Fructose solutions also exhibit the self-synergism with increasing concentration, but the effect is not as marked. Over the range of 5–20% fructose, Dahlberg & Penczek (1941) and Stone *et al.* (1969) report that the relative sweetness of fructose increases by 8%. The reason for glucose self-synergism is not due to tautomer shift, as increasing concentrations do not significantly change the position of the glucose mutarotational equilibrium. This is also true for fructose.

6.4 REACTIONS AFFECTING RELATIVE SWEETNESS

6.4.1 Conventional Mutarotation

When dissolved in water, solutions of glucose mutarotate, but the relative sweetness of the two tautomeric forms of glucose is about equal, and relative sweetness is about constant. When crystalline fructose (100% β-D-fructopyranose tautomer) is dissolved in water, the sugar rapidly mutarotates to establish an equilibrium with other tautomeric forms, largely β-D-fructofuranose. As the relative sweetness of fructose diminishes rapidly upon dissolving in water, it appears that the five-membered ring forms (the major tautomeric components formed) are devoid of sweetness.

This assessment of fructose sweetness variation, i.e. the variance in sweetness with shift in the distribution of tautomeric forms, is most clearly demonstrated by the effect of temperature on the sweetness of fructose solutions. The effect is known as thermal mutarotation, and is due to the shift in the position of the sugar equilibrium with change in temperature.

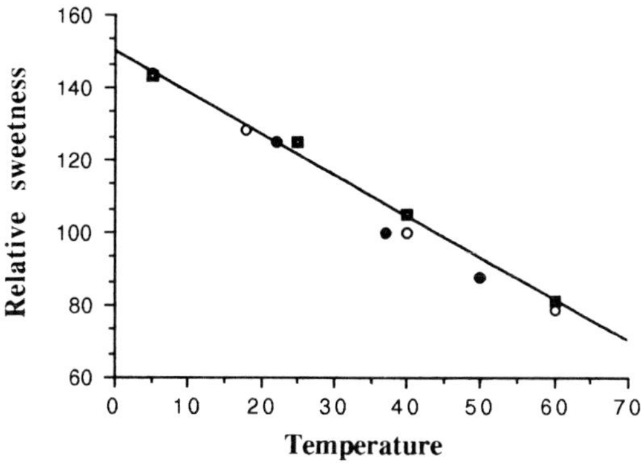

FIG. 6.4. Effect of temperature on the relative sweetness of fructose. Open circles, Tsuzuki & Yamazaki (1953). Solid squares, Fricker *et al.* (1973). Solid circles, Hyvönen *et al.* (1977*b*).

6.4.2 Thermal Mutarotation

While the effect of changing temperature on the position of the glucose mutarotational equilibrium is minimal, the effect on fructose is pronounced (Section 5.3.3). As a result, the relative sweetness of glucose is unaffected by increasing temperature, but the relative sweetness of fructose diminishes. A plot of the fructose relative sweetness score (Sucrose = 100) versus temperature is shown in Fig. 6.4. The shift in the position of the β-D-fructose tautomer with temperature demonstrated by Schneider *et al.* (1985) in Fig. 5.17, and the relative sweetness of fructose at different temperatures is shown in Fig. 6.5. Examination of the mole percentage β-D-fructopyranose versus relative sweetness reveals that crystalline fructose, or pure β-D-fructopyranose, is about 2·23× as sweet as sucrose.

Not all the decrease or increase in the sweetness of all the substances in the sugar group can be attributed to a shift in the distribution of configurational or ring tautomers. Between 23°C and 50°C the relative sweetness of xylitol drops by 20% (Hyvönen *et al.*, 1977*b*). Since sugar alcohols do not exhibit conventional mutarotation, the explanation for this phenomenon must lie elsewhere. One plausible suggestion is that it is due to a shift in conformation from the pseudo-ring form, stabilized at lower temperature by water of hydration, to acyclic forms with fewer appropriate glycol conformations (Fig. 9.10) at the higher temperatures.

6.4.3 Inversion of Sucrose

Early reports on the relative sweetness of sucrose versus invert sugar

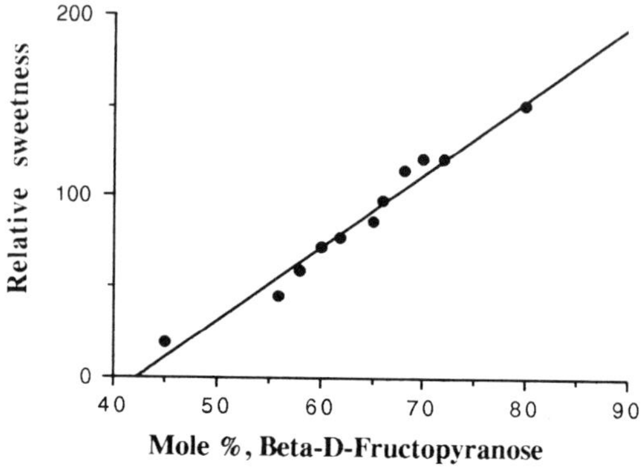

FIG. 6.5. Relation between the relative sweetness of fructose plotted against the different mol % β-D-fructopyranose present at different mutarotational equilibrium temperatures ($y = -172 \times 4{\cdot}07x$).

prepared from sucrose are variable. Cameron (1947) summarized those reports, and pointed out that the variance is due in part to measurement of relative sweetness at different concentrations. The data of Dahlberg & Penczek (1941) show that for concentrations below 10%, inversion slightly lessens the sweetness of sucrose. For concentrations above 10%, presumably because of the self-synergism demonstrated by glucose, inversion slightly increases the sweetness. Part of the problem is also because an admixture of glucose and fructose has a sweetness profile different from sucrose. This led the panel of Humphriss (1940) to conclude that while invert sugar at 20% was slightly sweeter than sucrose, the invert sugar sweetness was somehow 'different.'

When sucrose is inverted (Section 5.3.4), the concentration of solids increases by 5%. As the mere process of inversion causes a 5% increase in solids, it would seem logical to assume that invert sugar sweetness should be increased by 5%, but this is not the case. From the average data shown in Table 6.1, it can be calculated that the sweetness of invert sugar is $64 \div 2 + 121 \div 2 + 0{\cdot}05(64 \div 2 + 121 \div 2) = 97{\cdot}1$. However, an element of variability enters into the sweetness attributes of invert sugar that sucrose solutions do not possess. At a moderate temperature, e.g. 10°C, the relative sweetness score for fructose is 140. The sweetness of invert sugar due to free glucose and fructose calculates to be $(64/2) + (140/2) + 0{\cdot}05[(64/2) + (140/2)] = 107$, and invert sugar is sweeter than sucrose. However, at the temperature of a cup of coffee, fructose has a relative sweetness score of 80–90, and the relative sweetness of invert sugar is

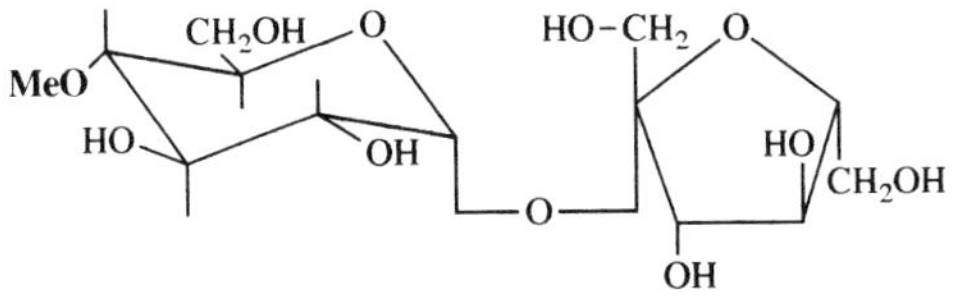

4-*O*-Methyl sucrose (less sweet)

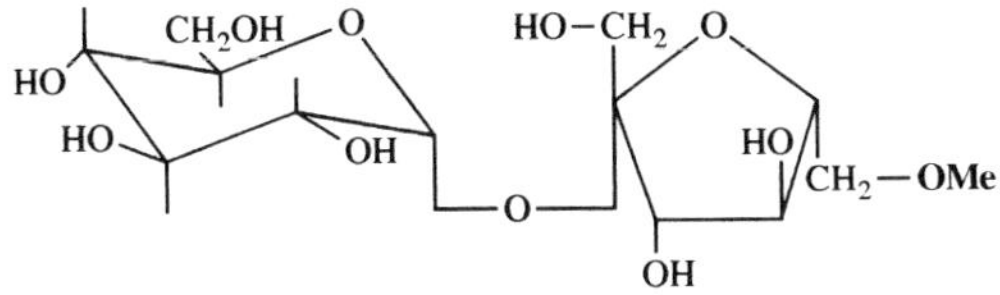

6'-*O*-Methyl sucrose (1x)

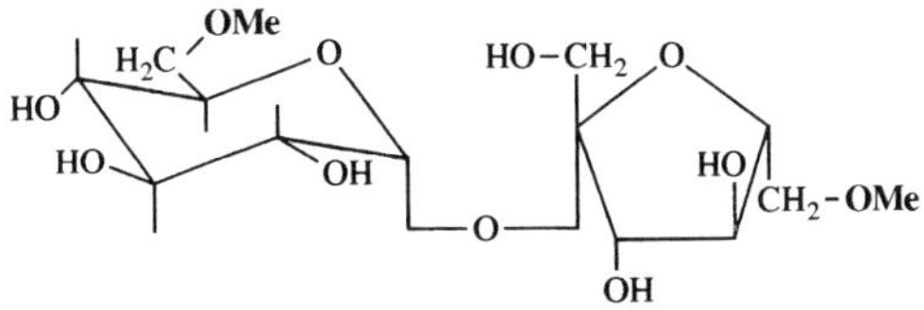

6,6'-di-*O*-Methyl sucrose (1x)

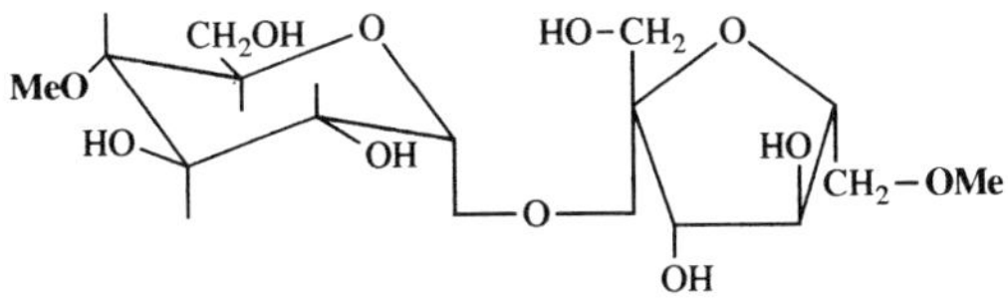

4,6'-di-*O*-Methyl sucrose (less sweet)

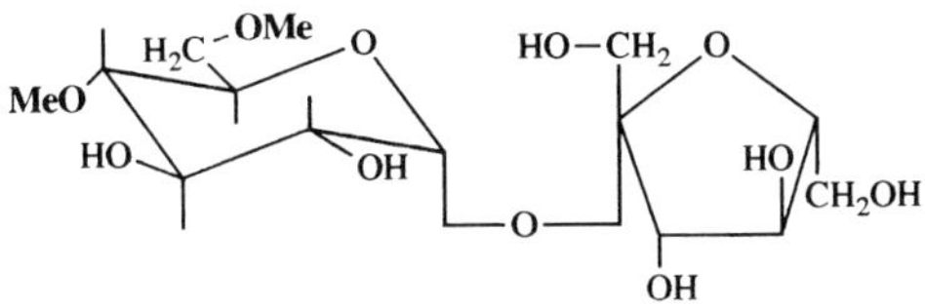

4,6-di-*O*-Methyl sucrose (less sweet)

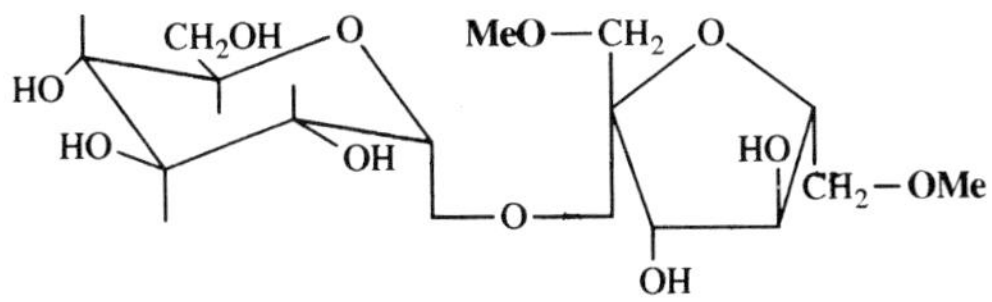

1',6'-di-*O*-Methyl sucrose (less sweet)

FIG. 6.6. Structure of methylated sucrose derivatives and preliminary assessment of their sweetness relative to that of sucrose (Lindley *et al.*, 1975).

only about 80. The temperature then also has marked influence on the sweetness of invert sugar.

6.4.4 Reversion

When sugars condense, under acidic conditions, with formation of

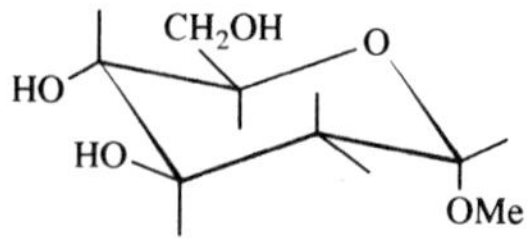

Methyl 2-deoxy-α-D-*gluco*-pyranoside (sweet)

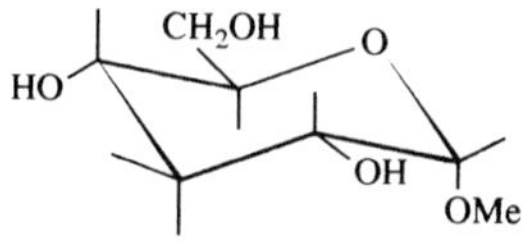

Methyl 3-deoxy-α-D-*gluco*-pyranoside (sweet)

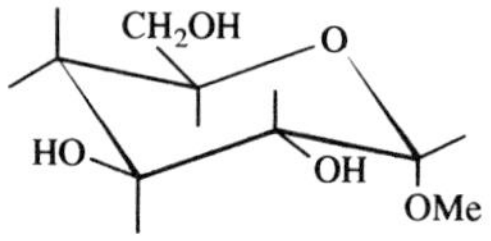

Methyl 4-deoxy-α-D-*gluco*-pyranoside (sweet)

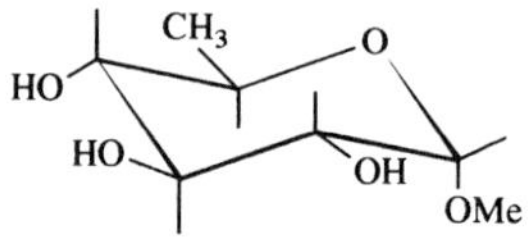

Methyl 6-deoxy-α-D-*gluco*-pyranoside (sweet)

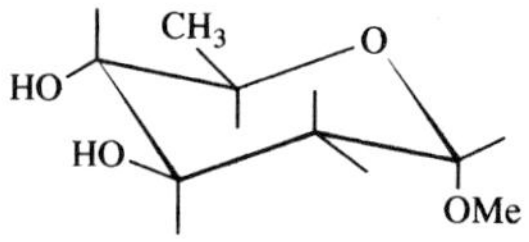

Methyl-α-D-2,6-dideoxy-*gluco*-pyranoside (bitter)

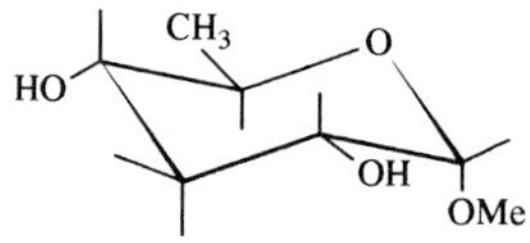

Methyl-α-D-3,6-dideoxy-*gluco*-pyranoside (bitter)

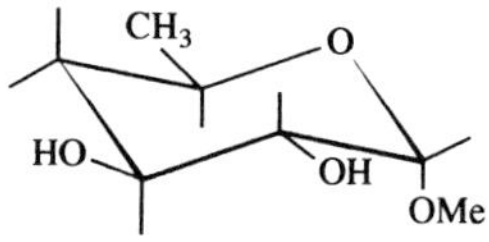

Methyl-α-D-4,6-dideoxy-*gluco*-pyranoside (bitter)

FIG. 6.7. Taste and structure of deoxy-hexose sugars derived from glucose (Birch, 1976).

anhydro forms (Section 5.3.6), sweet taste diminishes. This is true, not only for levoglucosan and polyanhydro-glucose substances, but is true for fructose reversion products also. Di-D-fructose anhydride I, or 1′2-*anhydro*-[1-(α-D-fructofuranosyl)]-β-D-fructofuranose, prepared by Lemieux & Nagarajan (1964), is tasteless (Section 5.3.6).

6.4.5 Effect of Substitution on Sweetness

6.4.5.1 *Methylation*

Methylation of the sugars significantly affects their taste. Glucose is about two-thirds as sweet as sucrose, but methyl-α-D-glucopyranoside

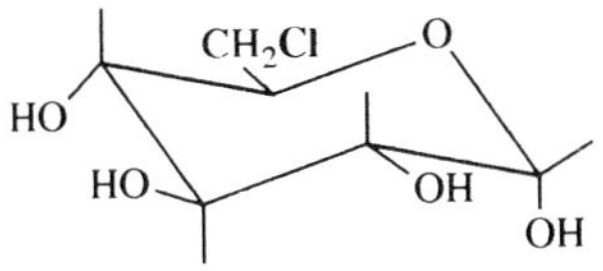

6-Chloro-6-deoxy-D-glucose (trace)

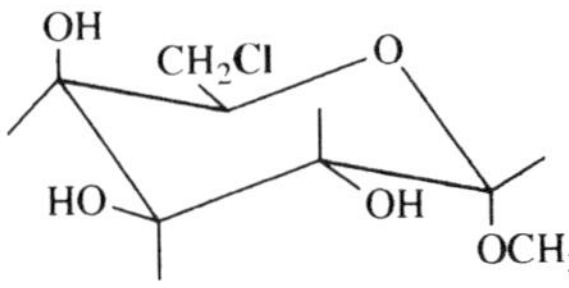

6-Chloro-6-deoxy-D-galactose (tasteless)

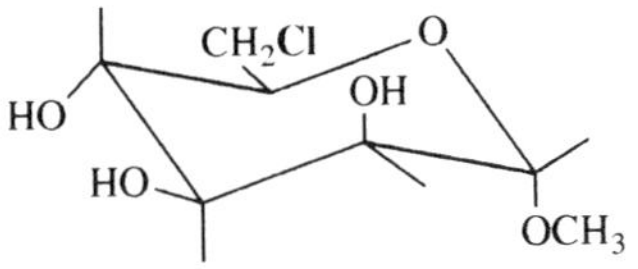

6-Chloro-6-deoxy-D-mannose (trace)

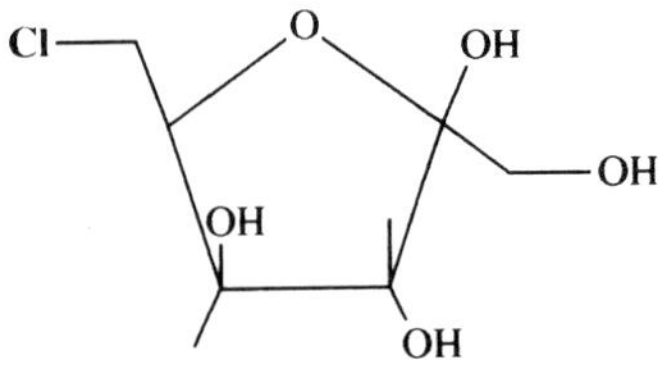

6-Chloro-6-deoxy-D-fructofuranose (<1x)

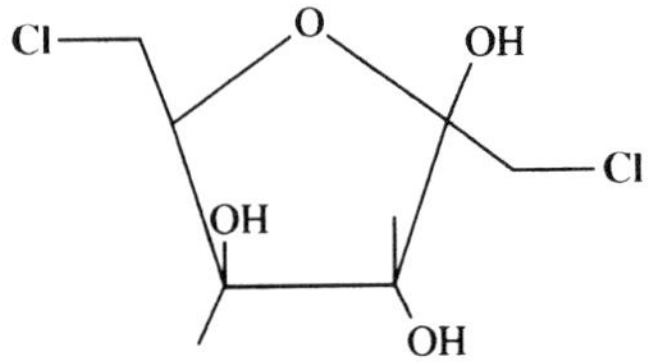

1,6-dichloro-1,6-dideoxy-D-fructofuranose (1x)

FIG. 6.8. Taste and structure of chloro-hexoses (Hough, 1990).

is only about one-quarter as sweet, and slightly bitter. Methyl-β-D-glucopyranoside is only 0·16 times as sweet as sucrose, and distinctly bitter (Birch, 1976). Presumably, it is because of the bitterness attribute that the β-anomer receives the low relative sweetness score.

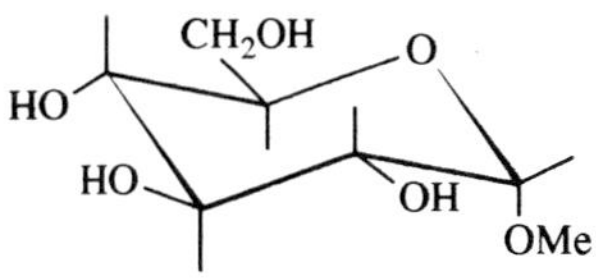

Methyl α-D-glucopyranoside (0.25x)

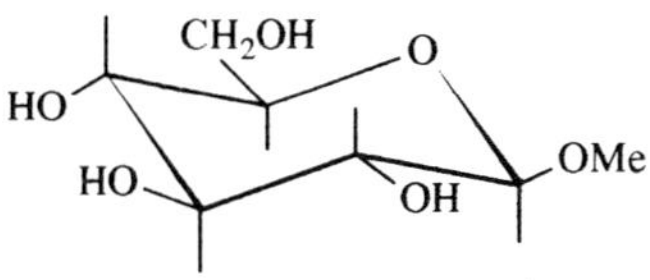

Methyl β-D-glucopyranoside (0.16x)

Mono-methylation of the remaining methyl-α-D-glucopyranoside hydroxyl groups yield di-methyl compounds. The di-*O*-methyl derivatives of methyl-α-D-glucopyranoside are devoid of sweetness, and invariably bitter. Similarly, $2,2'$-di-*O*-methyl-α,α-trehalose through $6,6'$-di-*O*-methyl-α,α-trehalose derivatives retain sweetness, but further methylation results in compounds without any sweetness, and are bitter.

Results obtained from methylation of various hydroxyl groups of sucrose (Lindley *et al.*, 1975) are shown in Fig. 6.6.

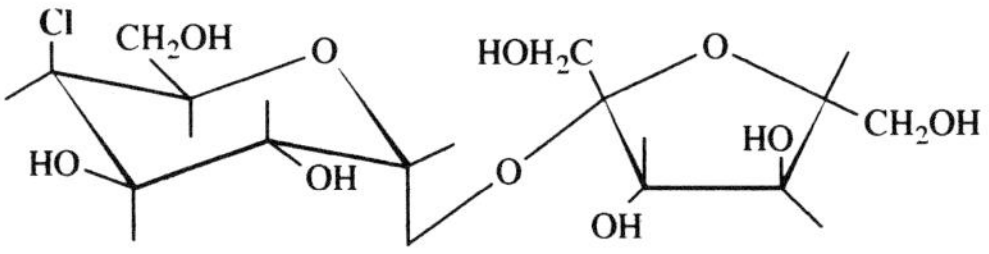

4-Chloro-*galacto*sucrose (5x)

6'-Chloro-sucrose (20x)

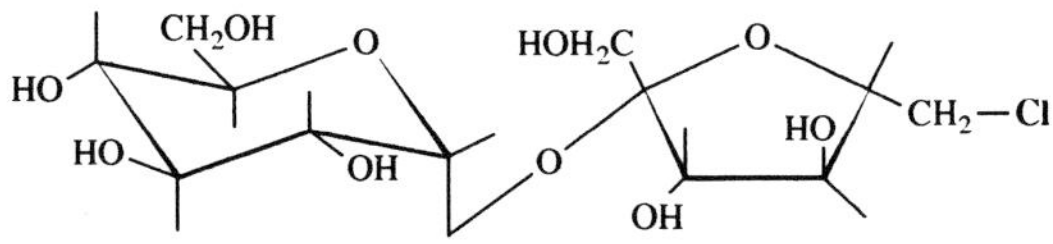

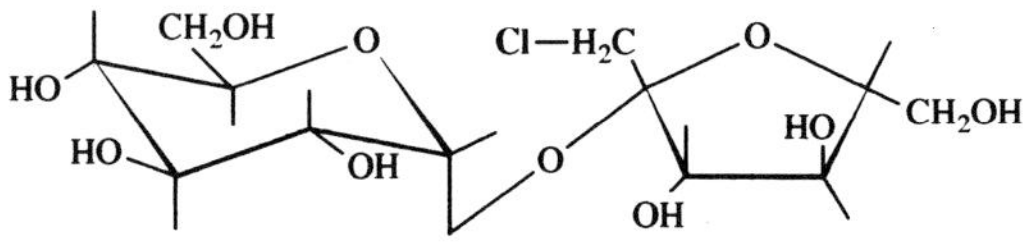

1'-Chloro-sucrose (20x)

1',6'-Dichloro-sucrose (70x)

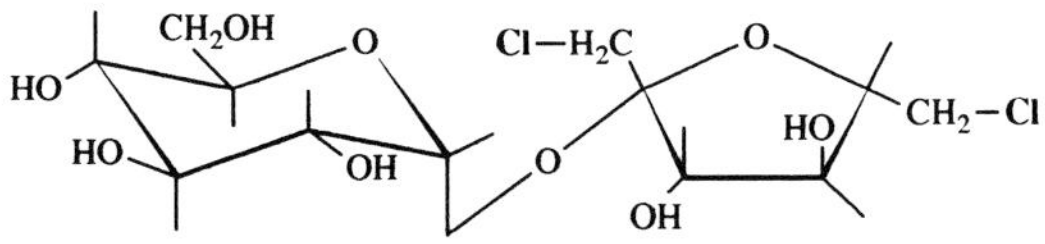

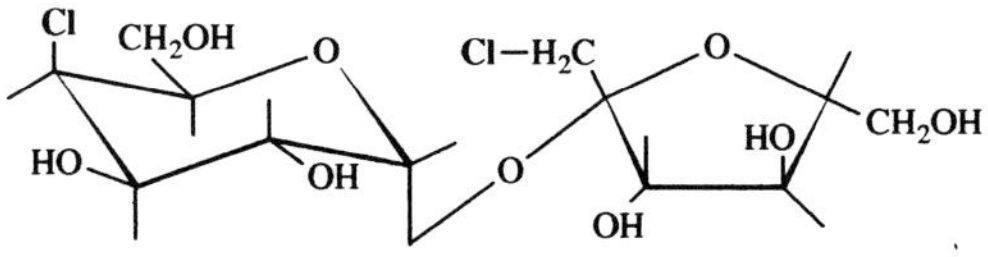

1',4-Dichloro-*galacto*sucrose (120x)

4,1',6'-Trichloro-*galacto*sucrose (650x)

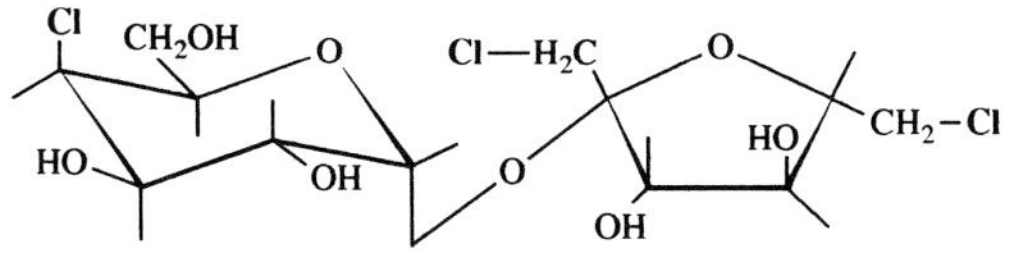

FIG. 6.9. Sweetness, relative to sucrose, of the chloro-substituted sucroses (Dziedzic & Birch, 1981).

6.4.5.2 *Deoxy-Sugars*
The taste and structure of a series of 1-deoxy-glucose sugars (Birch, 1976) is shown in Fig. 6.7. 2-Deoxy-α-D-glucose, which has about the same sweetness potency as the parent sugar, has a more rapid onset time. Removal of one hydroxyl group from the structure of the sugars has little effect on sweetness, but removal of two invariably leads to bitterness.

6.4.5.3 *Chloro- and Thio-Sugars*
A series of studies on the effect of substituting various sugar hydroxyl groups with the chlorine atom, and the resultant taste, have been reported. A summary is given by Lee (1987) and Hough (1990), and the structure and taste of chloro-hexoses is shown in Fig. 6.8. Chlorination at position six in aldohexose sugars (Dziedzic & Birch, 1981) drastically reduces sweetness, but the seemingly tasteless β-D-fructofuranose tautomer of fructose is rendered nearly as sweet as sucrose.

The structure and taste of pertinent examples of chlorinated sucrose compounds is given in Fig. 6.9. Unlike the hexoses, chlorination of sucrose leads to moderate to high-potency sweetness.

The effect of substituting sugar hydroxymethylene groups with an *S*-methyl group was studied by Gan *et al.* (1990). The authors prepared 1,6-di-*S*-methyl-1,6-dithio-D-fructofuranose with $[\alpha]_D^{25} = +35{\cdot}5°$, and 1,6-di-*S*-methyl-1,6-dithio-D-sucrose. While the sweetness of sucrose was hardly changed, the sweetness of the fructofuranose derivative was found to be 15–20 times.

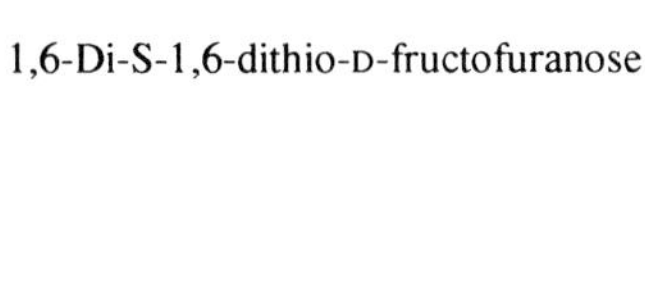

1,6-Di-S-1,6-dithio-D-fructofuranose

6.4.5.4 *Replacement of the Ring Oxygen Atom*

Replacement of the ring oxygen atom with a sulfur atom increases the sweetness of α-D-glucopyranose and β-D-fructopyranose, but only by 5% (Lindley *et al.*, 1976*b*). When the ring oxygen atom is replaced by sulfur in β-D-fructofuranose, which is presumably tasteless (Section 6.4.1), it is found that the thio-fructose analogue is also tasteless.

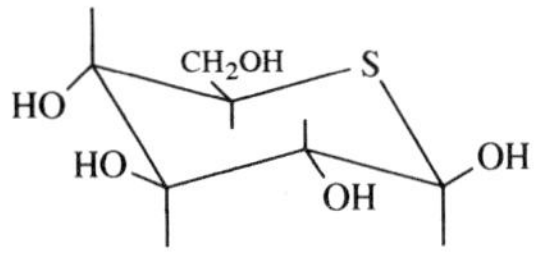

5-Thio-β-D-glucopyranose

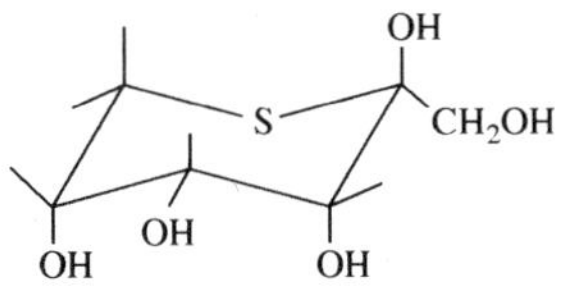

6-Thio-β-D-fructopyranose

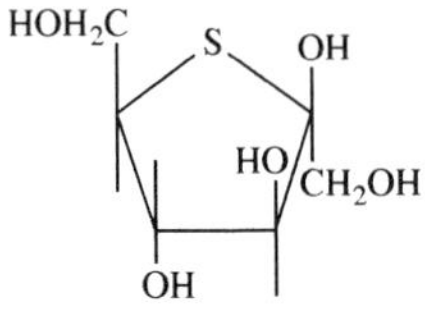

5-Thio-β-D-fructofuranose

Replacement of the ring oxygen atom of sugars with a carbon atom leads to formation of the *pseudo*-sugars. Suami *et al.* (1983, 1985) found that *pseudo*-β-D,L-fructopyranose, *pseudo*-α-D,L-glucopyranose, and *pseudo*-α-D,L-galactopyranose were as sweet as the D-compounds. It would seem that, as components of sugar rings, the carbon, sulfur, and oxygen atoms have isosteric saporous functions (Section 2.9).

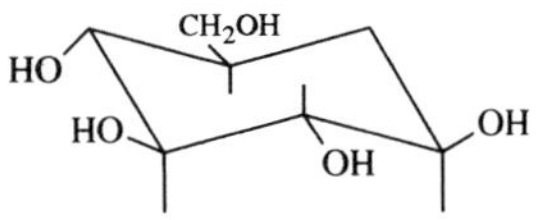

Pseudo-β-D-glucopyranose

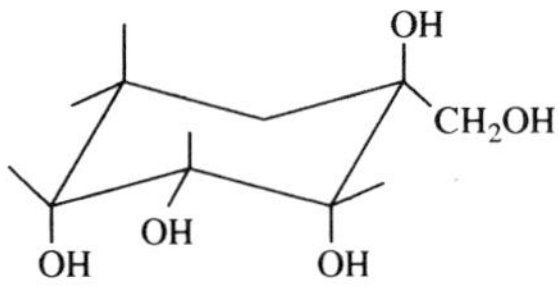

Pseudo-β-D-fructopyranose

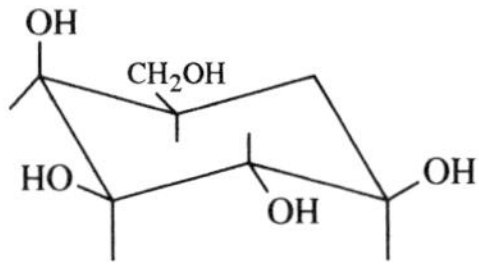

Pseudo-β-D-galactopyranose

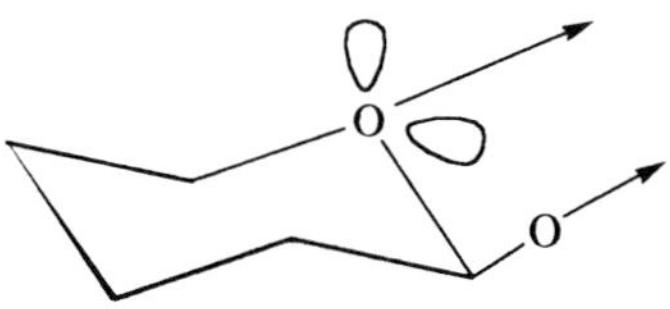

Unfavorable β-D- dipole interaction

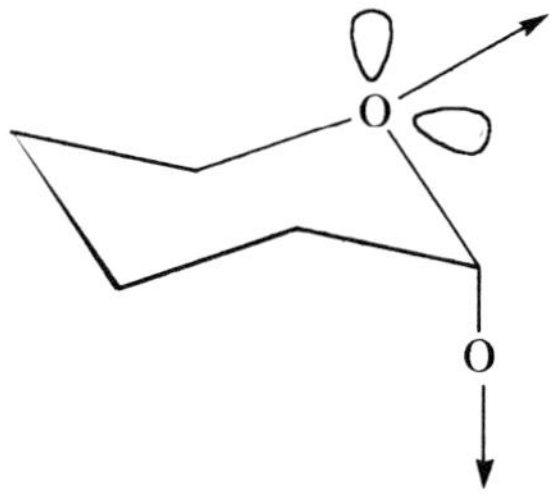

Favorable α-D- dipole interaction

FIG. 6.10. The unfavorable dipole–dipole interaction between the anomeric hydroxyl group and a ring oxygen atom (anomeric effect) in the β-D-configuration of the sugars (Edward, 1955).

6.5 RELATION BETWEEN STRUCTURE AND TASTE IN THE SUGAR SERIES

The sugars present an opportunity to examine taste from the viewpoint of multiple configuration, conformation, and ring size. In addition, the sugars have varying molecular weights, and interesting relations between structure and taste are apparent. Foremost among them, minor configurational transpositions markedly alter taste. Galactose and mannose are only the 4- and 2-epimers of glucose, respectively, and yet they are about one-half as sweet as glucose. α-D-Mannose and β-D-mannose are only configurational isomers at carbon atom number one, and yet one is sweet and the other is bitter.

The sugar fructose is the ketone analog of the aldose D-glucose, and has the same configuration about its asymmetric carbon atoms as glucose, but fructose is about 2 to 3 times as sweet. When fructose (β-D-fructopyranose) is dissolved in water, five-membered (furanose) rings are formed. As the sweetness of the solution markedly diminishes with time, the five-membered ring forms seem to be devoid of sweetness (cf. Section 6.4.1). Glucose forms an equilibrium between substances that are isomeric at the first carbon atom, but the anomers have nearly equal sweetness.

Other observations with respect to configuration are possible. Whenever the β-D-configuration appears among the structures of sugars it may be associated with bitterness. β-D-Mannose is bitter, as is gentiobiose (6-*O*-β-D-glucopyranosyl D-glucose). Methyl-β-D-glucopyranoside is more bitter than methyl-α-D-glucopyranoside. Finally, levoglucosan (1, 6-*anhydro*-β-D-glucopyranose) is sweet and bitter. The significant feature of the β-D-configuration is that the equatorial anomeric hydroxyl group imparts an unfavorable interaction with the ring oxygen atom (Edward, 1955), as shown in Fig. 6.10. The phenomenon is known as the *anomeric effect*.

The shape (conformation) of sugar rings is also important to sweetness. For example, the observation that levoglucosan is not as sweet as D-glucose suggests that the sweet-tasting conformation for the latter is $^4C_1$. Relative sweetness on a weight basis generally decreases in the order mono- > di- > tri- > tetra-saccharides, etc. However, on a molar basis, the disaccharide α,α-trehalose and methyl-α-D-glucopyranoside are equally sweet. The latter observation, along with the decrease in sweetness with increasing molecular weight, suggests that there is a stoichiometric attribute to the nature of the sweetness recognition interaction.

The effect of derivatization may be multifold. In addition to eliminating an active saporous unit, substitution may create alternate saporous units, or even induce and enhance the ability of interactive sites to elicit a given taste sensation. The latter seems to be the case for the effect of chlorination of fructose and sucrose. As judged by the progressive methylation of glucose, all mono-methylated derivatives retain an element of sweetness, but di-methylated derivatives are tasteless or bitter.

Finally, taste panel assessments of the sweetness of substituted compounds may be complicated by the presence of other tastes, or by an ability of a compound to elicit the common chemical sense. In several instances, taste panel results are affected by minute traces of solvent, such as acetone, acetic acid, ethanol, and ethyl acetate. These solvents are commonly used in purification and recrystallization procedures. They also possess significant sensory attributes.

6.6 SUPPLEMENTAL FOOD ATTRIBUTES OF SUGARS

In some respects, other food properties of the sugars are as important as, or even more significant than, sweet taste. These are described here as the supplemental (complementary, subsidiary) attributes and are associated with the ability of sugars to interact with, and alter the properties of water (Section 5.4.3). They include viscosity, raising of the boiling and lowering of the freezing point, and the osmotic pressure of sugar solutions.

For pnemonic purposes, the association of sugars with water is based on the notion of the tridymite structure of water, and the caveats mentioned in Section 3.1.2.1 for interpretation of water structure apply. Nevertheless, each oxygen atom of a water molecule exhibits a coordination number of four, and this can be related to hydrogen bonding cooperativity (Jeffrey & Saenger, 1991) between sugar hydroxyl groups and water molecules.

6.6.1 Stoichiometric Association of Sugars with Water

Calculations based on sugar solubility data in Table 5.3 show that at saturation (67%, w/w), one molecule of sucrose associates completely and stoichiometrically with nine molecules of water. A stoichiometric model of a hypothetical association of nine molecules of water with a single sucrose molecule is shown in Fig. 6.11. In addition to the hydrogen bonding cooperativity of sugar hydroxyl groups (Fig. 5.15), water

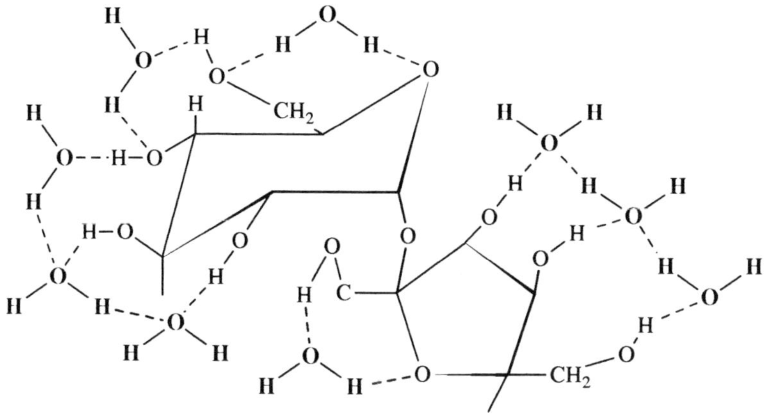

FIG. 6.11. Hypothetical model for the binding of nine molecules of water by one molecule of sucrose.

associated with each group is also cooperatively hydrogen-bonded. The combined cooperativity leads to a six-membered water ring structure and a five-membered water ring structure superimposed over/on the sucrose molecule. The cooperativity is based on the ability of carbohydrates to fit into the three-dimensional hydrogen-bonded structure of water, and the better the so-called fit, the more the sugar hydroxyl groups are spectroscopically camouflaged (cf. Galema *et al.*, 1990). The relation is on the order of an 'open-faced' sandwich.

The solubility of glucose is also 67% w/w, and one molecule of glucose is dissolved by, and presumably interacts with five molecules of water (Table 5.3). The conceptual pnemonic is again an 'open face' sandwich. The diagram shown for glucose is in accord with the observation, based on intrinsic viscosity, that the van der Waals' volume of glucose is increased by a factor of about two by hydration (cf. Kabayama & Patterson, 1958). The superficiality of static pnemonic models cannot be over-emphasized, but it must also be acknowledged that hydroxyl groups in sugars are well hydrogen-bonded to the solvent. However, the strong localization of water molecules around hydroxyl groups does not mean that solvent molecules are actually *bound*. They are known to rapidly exchange with the solvent (Brady, 1990). Typical residence times, as indicated by molecular dynamic simulations, are only a few picoseconds (less than 10 ps). Moreover, the hydroxyl groups of α-D-glucopyranose that are hydrogen-bonded to water undergo rotations that are only weakly correlated to one another (Ha *et al.*, 1988). In the light of these

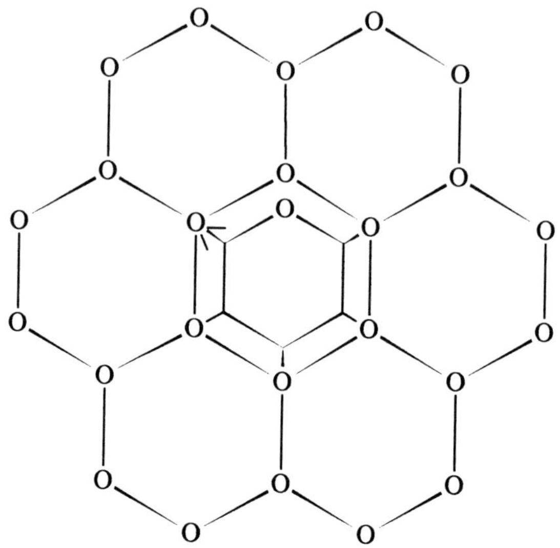

FIG. 6.12. Replacement of one molecule of tridymite water with a molecule of glucose (Kabayama & Patterson, 1958).

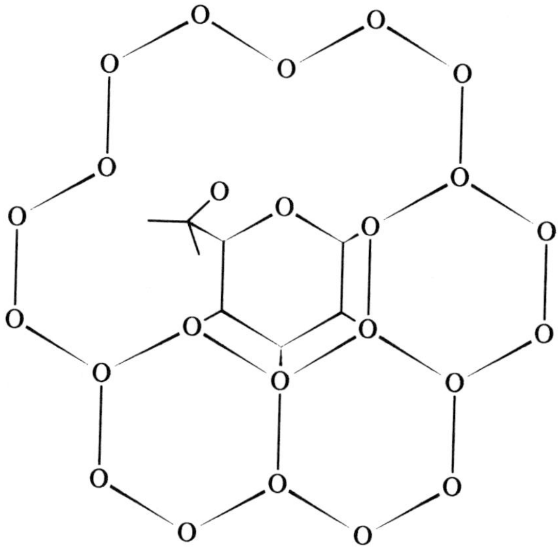

FIG. 6.13. Proposed breaking of the tridymite structure of water through repulsive effects afforded by the electronic nature of a sugar's anomeric center.

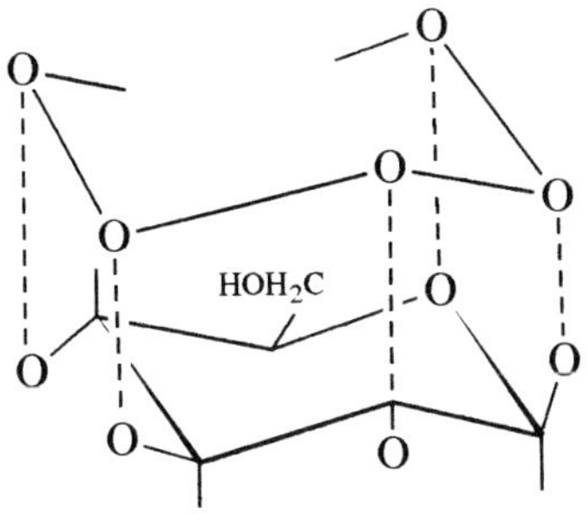

Hydrated β-D-glucose

interpretations of the interaction between sugars and water, it is clear that the interactions bear close resemblance to the notion of recognition chemistry (Section 2.12). The notion becomes all the more intriguing with the observation that the solvation chemistry of saporous substances (tastants) is also analogous to taste recognition chemistry.

For fructose, calculations show that only about three molecules of water are needed to dissolve one molecule of fructose. This is equivalent to two molecules of fructose 'binding' to six molecules of water. Pnemonically, the structure is a conventional sandwich with a water layer between two molecules of fructose.

An alternative to the above sandwich pnemonics for the association of sugar with water is that in *dilute solution*, a sugar molecule may replace a six-membered water form in the tridymite structure of water. This thought (Kabayama & Patterson, 1958), is shown in Fig. 6.12. The latter static model has application toward explaining some of the taste properties of sugars. Birch & Shamil (1988) proposed that one end of the glucose molecule is more hydrated than the other, as the anomeric center of a sugar is essentially hydrophobic, and plays no role in the sweetness response. The ring oxygen atom, through repulsion of a molecule of water in a tridymite ring, presumably breaks the water structure near the ring oxygen atom. Molecular dynamics simulations also indicate that O-6 serves to some extent as a water structure breaker (Brady, 1990). The idea is demonstrated in Fig. 6.13 for the breakage of the water structure in the same plane as that of the glucose molecule. It is possible that the structure is broken in the z plane as well. Due to the association of sugars with water, boiling point, freezing point, viscosity, osmotic pressure, density, and water activity of the solutions is altered.

When these phenomena are plotted against percentage sugar, all exhibit the same general type of decelerating exponential curve.

6.6.2 Osmotic Pressure

The presence of sugar in water leads to an osmotic pressure gradient. Osmotic pressure is the pressure that must be applied to a solution confined by a semipermeable membrane to prevent the inward diffusion of water. No pressure develops until water begins to diffuse into the sugar solution. Nevertheless, sugars at a given concentration are said to have specific osmotic pressure, or more accurately, an osmotic pressure potential.

At low sugar concentrations, traditional gas laws apply to osmotic pressure (i.e., $PV = nRT$), and a 4% sugar solution at 15°C has a calculated potential pressure equal to 199·1 cmHg. Due to sugar association with water, the experiment reveals a value of 197·4 cm. Water in an osmotic system moves toward the solution with the higher sugar concentration. Sugar solutions impede the growth of bacteria and molds either by dehydrating them, or rendering the water unavailable for growth. Deviation from the ideal gas laws at higher sugar concentrations is marked at concentrations greater than 10%. As osmotic pressure is proportionate to the number of sugar molecules present in solution at a given weight percentage sugar, invert sugar, which has about twice the number of molecules as sucrose, has twice the osmotic pressure.

6.6.3 Viscosity

The viscosity of sugar solutions varies directly with concentration and inversely with temperature, and the behavior of pure solutions is Newtonian. Viscosity plays an important role in the texture of foods, particularly with respect to mouthfeel. As temperature is increased, viscosity decreases because the hydrogen bonds forming sugar–water and water–water bonds are severed to yield complexes of lower average molecular weight. When sugar concentration is increased, the viscosity of sugar solutions increases exponentially.

6.6.4 Boiling Point Elevation, Freezing Point Depression

The ability of sugars to elevate the boiling point of water is also exponential. This attribute is important in the confectionary industry. Similarly, sugar solutions depress the freezing point of water. This has importance in the ice-cream, frozen food, and freeze-drying industry.

CHAPTER 7

Amino Acids, Peptides, and Proteins

7.1 STRUCTURE, CLASSIFICATION, AND PROPERTIES OF AMINO ACIDS

The amino acids are the building blocks of peptides and proteins. They occur both free and combined in structural proteins (keratin, etc.), and enzymes, the catalysts associated with life's processes. About ten of them are essential human nutrients. Receptors for taste and olfaction are also believed to be proteins, and amino acids, peptides, and proteins have varied tastes.

7.1.1 Structure

Approximately 20 different amino acids are commonly found in biological materials. With the exception of proline and hydroxy proline, they have the general form, in Fischer projection, shown below. The prefix α- is used in naming the amino acids to indicate that the NH_2 group is on the carbon atom adjacent to the carboxyl (COOH) group. The prefix β indicates that it is on the second carbon atom (e.g., β-alanine), and γ indicates that it is on the third carbon atom (e.g., γ-aminobutyric acid). Because the reference carbon atom for specifying amino acid structure is asymmetric, two configurational forms are possible that place the amino acids in either one or the other of two chiral families. Unlike the sugars, the reference atom for specifying the amino acid chiral family is the lowest numbered asymmetric carbon atom, and the carboxyl group takes on the priority (C-1) assigned to the sugar carbonyl function. Specification of the configuration at that carbon atom is also directly related to the structure of D- and L-glyceraldehyde (Section 5.1.1.2). As the configuration of the sugars is known to be absolute, the configurational structure of the amino acids is also absolute.

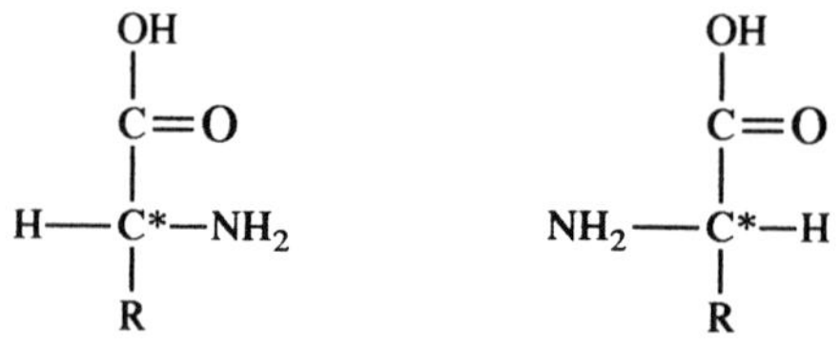

D-Amino acid L-Amino acid

The D-amino acid series does not generally occur naturally, and the NH_2 group has the same steric disposition as the highest numbered OH

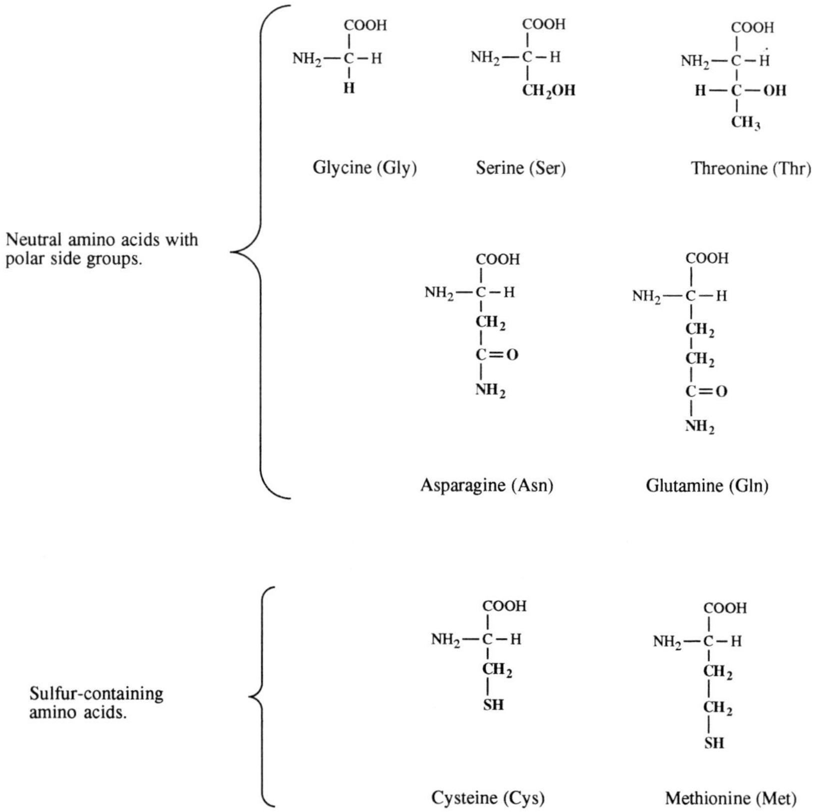

FIG. 7.1a. Neutral amino acids with uncharged side chains, and sulfur-containing amino acids. Side chains are in bold type.

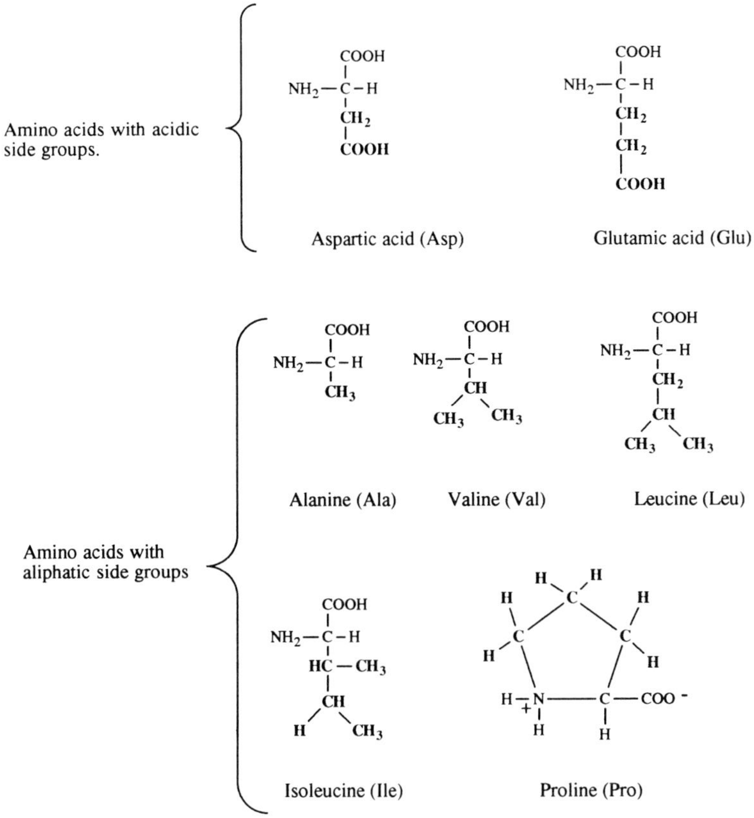

FIG. 7.1b. Amino acids with acidic and aliphatic side groups. Side chains are in bold type.

group in the sugars. Therefore it is written to the right in the Fischer projection formula. The L-amino acids, which do occur naturally, have the opposite configuration.

7.1.2 Classification

Amino acids can be divided into three groups (Dickerson *et al.*, 1974). The *neutral* amino acids, with uncharged side chains, along with sulfur-containing amino acids are shown in Fig. 7.1a. *Acidic* amino acids, i.e. those with acidic side groups are shown in Fig. 7.1b, along with amino acids with aliphatic side groups. Basic amino acids, i.e. those with basic side chains are shown in Fig. 7.1c, along with those with aromatic side groups.

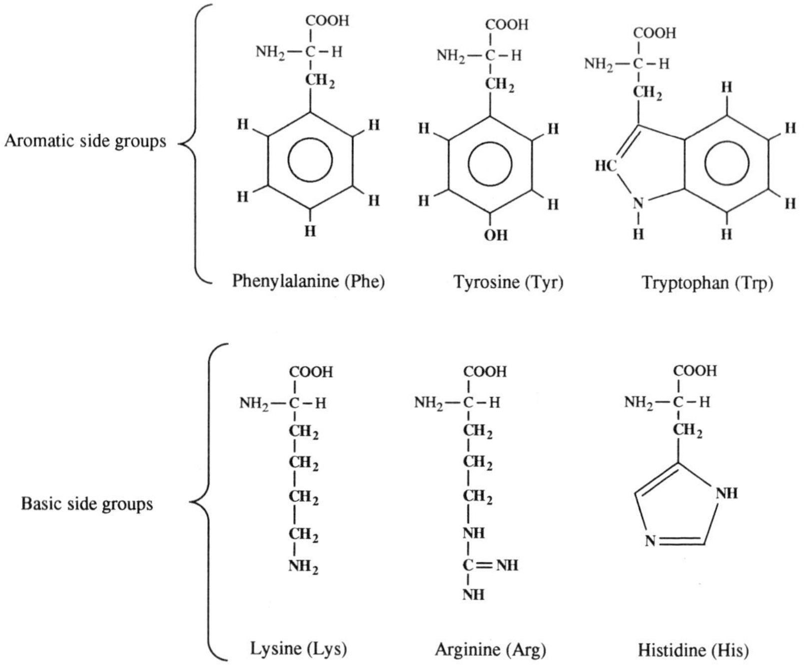

FIG. 7.1c. Amino acids with aromatic and basic side chains. Side chains are in bold type.

A second classification scheme is based on the ability of the side chain to partake in intra- and intermolecular reactions (Belitz & Grosch, 1987). It is: (1) Amino acids having nonpolar, uncharged side chains (glycine, alanine, valine, leucine, isoleucine, proline, phenylalanine, tryptophan, and methionine); (2) amino acids with uncharged, but polar side chains (serine, threonine, cysteine, tyrosine, asparagine, and glutamine); and (3) amino acids with charged side chains (aspartic and glutamic acids, histidine, lysine, and arginine). Both classification schemes have application to the subject matter of this treatise.

7.1.3 Reactions and Properties of Amino Acids

With the exception of glycine, alanine, proline, lysine, and arginine, and the two amides glutamine and asparagine, the amino acids are only sparingly soluble in water. The insolubility is due to strong intermolecular forces in the crystal. When the amino acids are dissolved in water, and with the exception of configuration, their structures are no longer fixed. Somewhat analogous to the sugars, they assume different forms that are governed by the polarity and pH of the solvent.

7.1.3.1 *Zwitterionic Forms*

In water solution, amino acids are ampholytic, and assume *zwitterionic* (dipolar) forms. In acidic solution the amine group is completely protonated and is present as the conjugate acid. In basic solution, the amine group is deprotonated, and the compound prevails as the conjugate base. In neutral solution, or at the isoelectric point of the amino acid, the dipolar zwitter ion prevails.

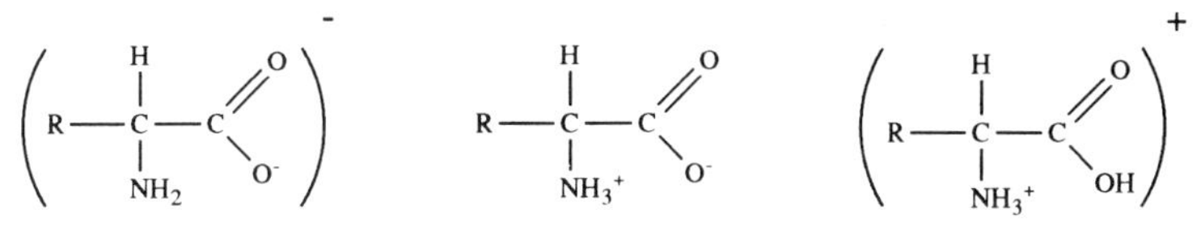

Base form in basic solution — Dipolar uncharged ion — Acid form in acid solution

When the hydrochloric acid salt of glycine is titrated with a sodium hydroxide solution (Fig. 7.2), a plot of the milliequivalents NaOH versus pH shows inflection points at pH 2·35, 6·07, and 9·78. The first inflection point (pH 2·35) occurs at half-neutralization, and is equal to the dissociation constant pK_1 for the ionization of the carboxyl group. Interestingly, glycine (sweet) is more acidic than sour-tasting acetic acid (pK_a = 4·76). The acidity is due to the large inductive effect of the NH_3^+ group. When one equivalent of base is added (pH 6·07), the major species in solution is the dipolar zwitterionic form, and this pH is the pH for a solution of the pure amino acid in water. It is also the isoelectric point of the amino acid, and the pH at which water solubility is at a minimum. Upon addition of a further half-equivalent of base, the inflection point obtained at pH 9·78 is equal to pK_2, and is the dissociation constant for the protonated amino group.

7.1.3.2 *Hydration of Zwitterions*

In solution a zwitterionic compound, with the charges sufficiently separated, both ionic centers are solvated in normal fashion, as in (a) and (b) in Fig. 7.3. When the charges are close together, solvation is less efficient. Thus, solvent dipoles (c) and (d) provide a stabilizing Coulombic

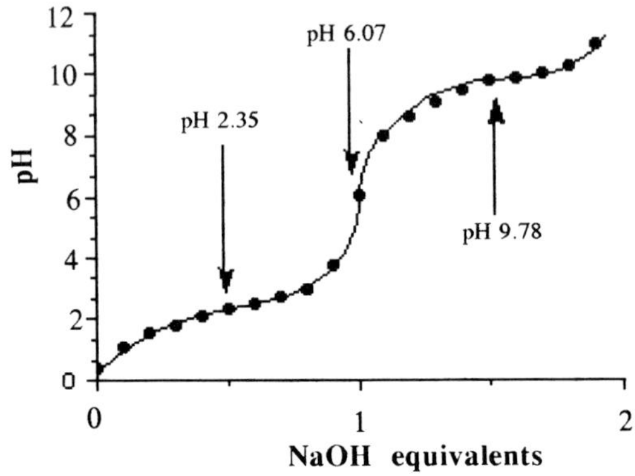

FIG. 7.2. Titration curve for glycine · HCl (Strietwieser & Heathcock, 1976).

attraction to one charge, but repulsion to the other. A dipolar system of this type is less stabilized by solvation, and is less favored in an equilibrium (Streitwieser & Heathcock, 1976). It is for this reason that the amino acids are only sparingly soluble in water.

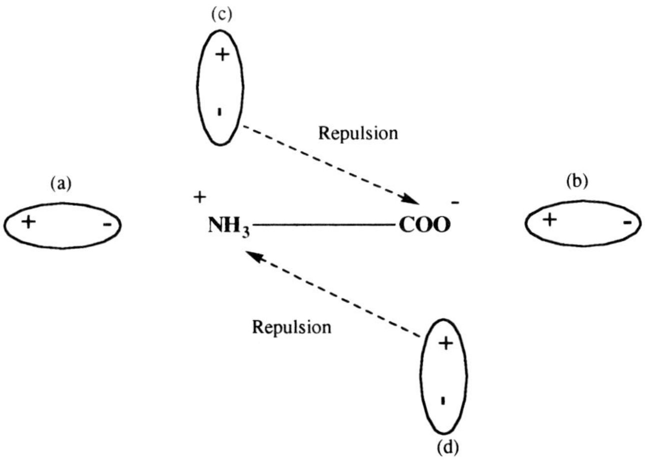

FIG. 7.3. Dipolar solvation effects on a zwitterion.

7.1.3.3 *Schiff Base Formation*

Among the facile reactions for the amino acids is the propensity of the primary amine group to react reversibly with the pair of unshared electrons of an aldehyde to form a Schiff's base. In a sense, Schiff base formation is analogous to the interaction between an alcohol, or sugar hydroxyl group, and an aldehyde to form a hemiacetal and/or acetal. The substituted amines, or amino acids are unstable, and easily undergo hydrolysis to yield the initial substances.

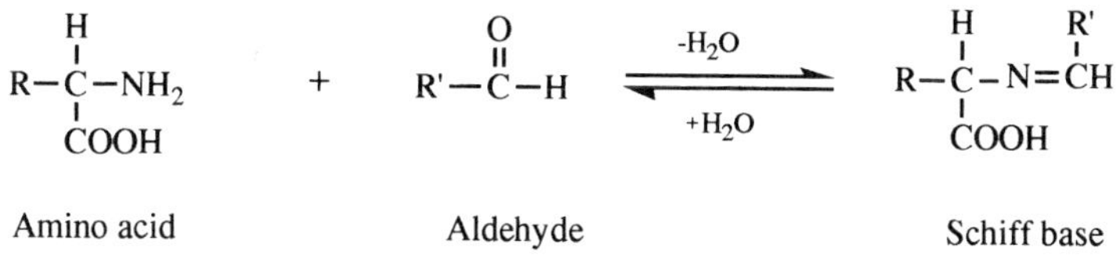

7.1.3.4 *Side Chain Free Energy (Hydrophobicity)*
With the object of assessing the varying hydrophobic nature of amino acid side chains, Tanford (1962) made use of the solubility of amino acids in ethanol and water as a measure of the side-chain free energy. The free energy change (ΔF_t) was calculated from the solubility of 1 mol of amino acid in ethanol and in water at pH 2·35, 6·07, and 9·78.

$$\Delta F_t = RT \ln\left(\frac{N_{EtOH}}{N_{HOH}}\right)$$

It was reasoned that ΔF_t is a measure of that part of the chemical potential of amino acid molecules that derives from interaction with the solvent. As ΔF_t for the various amino acids is the sum of constant contributions from the various parts of the amino acid, then

$$\Delta F_t = \Sigma\, \Delta f_t$$

As a CH_2 group increases ΔF_t by 700 cal., the rearrangement of H_2N–CO–CHR–OH to ^+H_3N–CHR–COO^-, where R is the side chain, always decreases ΔF_t by 3700 cal. To calculate the contribution of amino acid side chains, particularly the nonpolar ones, to ΔF_t, each contribution is simply the difference between ΔF_t for the whole amino acid molecule, and the value for glycine. In the latter case the contribution of Δf_t for the side chain is zero. The calculations made for different amino acids are given in Table 7.1.

The values given in Table 7.1 are about the same as those obtained by using nonpolar solvents other than ethanol (e.g., butanol, acetone). More recently, the side-chain values were modified by Bigelow & Channon (1976), and employed by Adler-Nissen (1986) for correlation with peptide structure and bitter taste. These somewhat more complete side-chain

TABLE 7.1
Free energy changes in calories mol^{-1} for transfer of an amino acid from ethanol to water at 25°C (Tanford, 1962)

Amino acid	ΔF_t (*Molecule*)	Δf_t (*Side chain*)
Non-polar side chain amino acids		
Glycine	−4630	0
Alanine	−3900	+730
Valine	−2940	+1690
Leucine	−2210	+2420
Isoleucine	−1690	+2970
Phenylalanine	−1980	+2600
Amino acids with other side chains		
Methionine	−3330	+1300
Tyrosine	−930	+2870
Threonine	−4190	+440
Serine	−4590	+40
Asparagine	−4640	−10
Glutamine	−4730	−100
Aspartic acid	−4090	+540
Glutamic acid	−4080	+550
(Tryptophan)	—	(+3000)

values, in kcal mol^{-1}, are: serine −0·3; glycine, 0·0; aspartic acid, 0·0; asparagine, 0·0; glutamic acid, 0·0; glutamine, 0·0; threonine, 0·4; histidine, 0·5; alanine, 0·5; arginine, 0·75; cysteine/cystine, 1·0; methionine, 1·3; lysine, 1·5; valine, 1·5; leucine, 1·8; hydroxyproline, 1·8; tyrosine, 2·3; phenylalanine, 2·5; proline, 2·6; isoleucine, 2·95; tryptophan, 3·4.

7.2 PEPTIDES AND PROTEINS

7.2.1 The Peptide Bond

Peptides are polymers of amino acids. Peptide bonds are formed by elimination of a molecule of water between the N-terminal end of one amino acid and the C-terminal (carboxyl) end of a second amino acid (Fig. 7.4). Because the electrons of the carbonyl double bond are delocalized by formation of the peptide bond, the carbonyl group carbon atom and the peptide bond nitrogen atom take on a partial double-bond character. The process reduces the basicity of the nitrogen atom, restricts rotation about the C–N linkage, and confines the N, C, O and H atoms

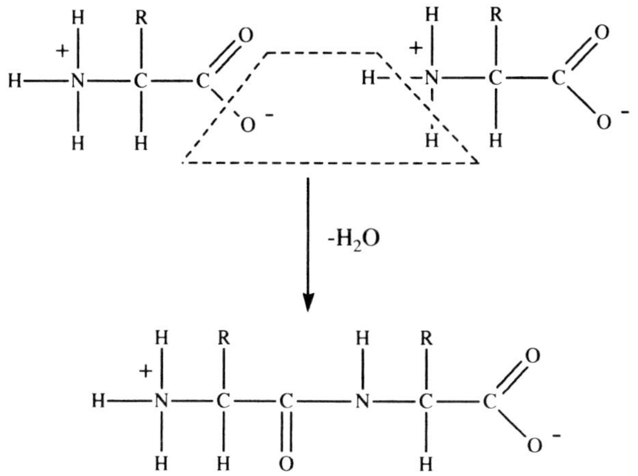

FIG. 7.4. Formation of a peptide bond.

of the peptide bond grouping to a planar configuration. The planar peptide bond plays a major role in the three-dimensional structure of peptides and proteins. A second type of covalent bond between amino acids in proteins and peptides is the disulfide link between two cysteine molecules. These too have significance in governing the three-dimensional form of peptides and proteins.

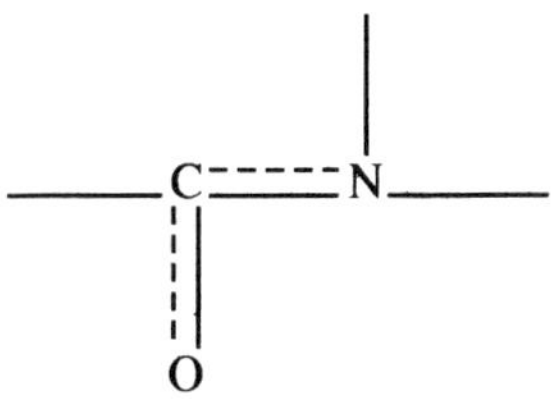

Peptides are named using the N-terminal amino acid as the prefix, and the C-terminal amino acid as the suffix. Glycylglycine is a *homodipeptide*, while glycylalanine is a *heterodipeptide*. When the degree of polymerization exceeds 50, the peptide is considered to be a protein. Peptides are also zwitterionic substances, and are neutral, and uncharged at their isoelectric point. With respect to their acidity constants, the peptides are much less zwitterionic than the amino acids. Glycylglycine (pK_1, 3·1) has only one-fifth the acidity of glycine (pK_1, 2·4).

Fibrous proteins serve in a structural capacity. Water soluble globular proteins serve in catalytic (enzyme), regulatory, and biochemical recognition capacities. Several globular proteins taste sweet. As components of taste cell membranes, globular proteins are likely candidates to be the receptor substance for recognition chemical reactions prerequisite for taste.

7.2.2 Primary, Secondary and Tertiary Protein Structure

The *primary* structure of a peptide or protein refers to the nature and sequence of the amino acids linked through peptide bonds. A sequence such as ser • tyr • met • glu • hist is an example of a primary structure for a peptide with five amino acid residues. The *secondary* structure of a peptide or protein refers to the character of the chain of amino acids. It may be an extended chain of amino acids, a pleated sheet, or a helical pattern.

7.2.2.1 *The α-Helix*

In helical structure, amino acid side chains point away from the major helical axis, and the carbonyl groups of one turn of the helix are hydrogen-bonded to the H–N group of the next turn. For the L-series of amino acids, the most stable helical form is right-handed, as the coils describe a right-handed screw pattern. It is known as the *α-helix*. Every complete turn of the helix, or for every 5·4 Å of pitch, as shown in Fig. 7.5, requires only 3·6 amino acid residues. Therefore, with respect to the number of amino acids per turn, the coils of the helix are rather compact. By convention, a right-handed screw pattern is one that will move away from the operator when turned to the right.

Not all amino acids and amino acid sequences are amenable to α-helix formation. Depending on pH, and the nature of the side chains, portions of a globular protein may prevail in random coil form. Side groups with the same charge inhibit helix formation, as do large bulky side groups. Because the nitrogen atom of proline is a component of a ring structure (Fig. 7.1b), rotation about the C–N bond is not possible. As a result, when proline occurs in a helical sequence, the helical format is broken. However, proline at position N-cap plus one, is believed to be an important component at the N-terminal position of a protein's helical structure. It is half-in and half-out of helical structure. Asparagine is also an important component of the N-terminal position (Richardson & Richardson, 1988). For globular proteins of known structure, about one-quarter of all amino acid residues are found in helices, and the average protein helix is 12 residues in length (Presta & Rose, 1988).

The side groups of a globular protein govern many of its properties. If

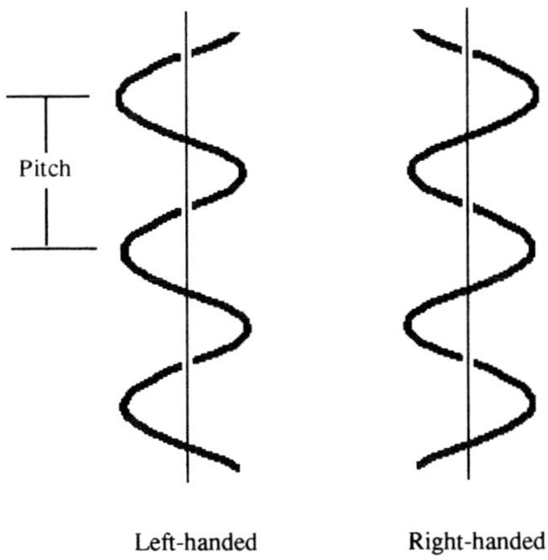

FIG. 7.5. Left- and right-handed helical patterns for the secondary structure of proteins.

they are predominantly linear or aromatic hydrocarbons, they are most stable in an apolar environment. If the environment is aqueous, the apolar side chains are directed toward the inside of a folded globular structure. Charged side groups, such as the carboxyl groups of aspartic and glutamic acids, the amine groups of lysine and arginine, and the uncharged amide groups of asparagine, glutamine and serine, are compatible with a polar aqueous environment. In a nonpolar environment these groups are folded toward the inside of the globular protein structure.

7.2.2.2 *Pleated Sheet Conformation*

The pleated sheet conformation (β-conformation) is shown in Fig. 7.6. In this form, the peptide bonds have a planar orientation perpendicular to the side chains. The association of chains, through intermolecular hydrogen bonds, forms *pleats*. One component of a pleat in an expanded three-dimensional form is shown in Fig. 7.6. Association of several pleats then forms the sheet. An association of helixes (super helix), and pleated sheet conformations, is the form found in structural proteins (keratins, collagen). Those conformations are components of the water soluble globulin structures as well.

7.2.2.3 *Folded and Random Coil Forms*

The usual structure for a globulin is a regular, folded coil (Fig. 7.7). As

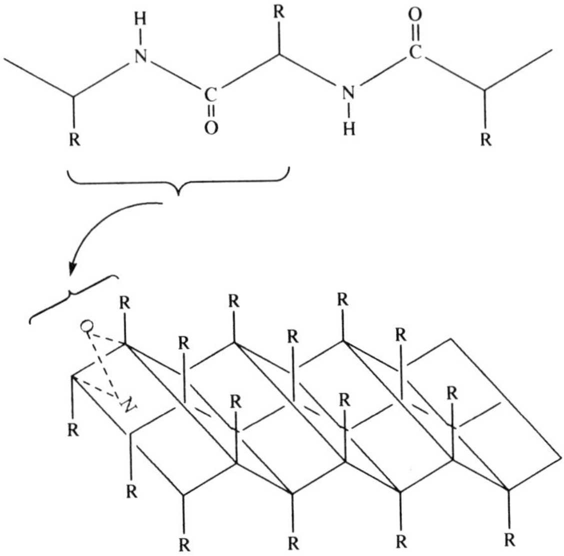

FIG. 7.6. The extended β- or pleated-sheet protein conformation.

globulins are catalysts for biological reactions (enzymes), and also act as carrier proteins, they have distinct folded structures (tertiary structure) characteristic of the active protein form. The native interactive form is held in proper shape by weak chemical interactions. They include: (1) hydrogen bonds between the peptide and other groups; (2) hydrophobic interactions between nonpolar groups; and (3) ionic bonds between anionic and cationic groups of aspartic and glutamic acids, and the epsilon amine group of lysine. Some folding is held in place by interchain disulfide bridges. Hydrogen bonds between the amide groups of

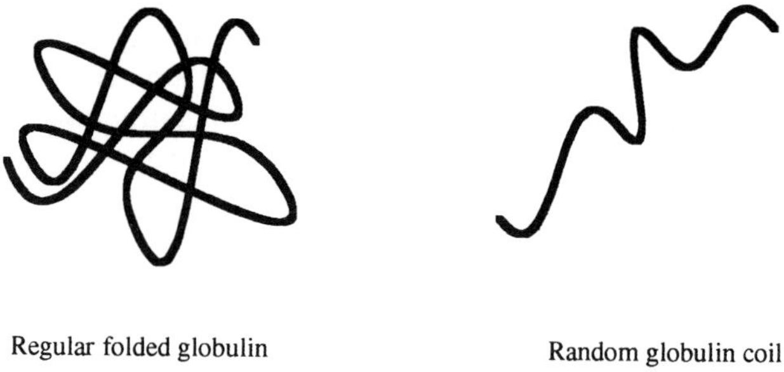

Regular folded globulin   Random globulin coil

FIG. 7.7. The regular folded and random globulin coil.

asparagine and glutamine also play a role. In a polar environment, the hydrophobic interactions seem to be most important. The protein is oriented so that the hydrophobic side amino acid chains are folded to the inside of the structure. There they are attracted to each other by van der Waals dispersion forces. Polar side chains tend to be on the surface of the molecule, and hydrogen-bond to water molecules to impart water solubility. In an apolar environment, hydrogen bonding and ionic interactions are important in stabilizing the molecule, and hydrophobic side chains are directed toward the solvent. An interruption of the forces that hold the protein in its reactive native form leads to denaturation, formation of a random coil, and loss or reduction of specific activity. If denaturation is not too extensive, the random coil may reform, and the characteristic biological activity of the protein returns.

7.2.3 Denaturation

The activity of proteins, either as enzymes, or as sweet-tasting substances, depends on their folded tertiary structure. That structure prevails only within limits of pH and temperature. Most globular proteins unfold, with concomitant decrease in biological activity and solubility, when heated to 60–70°C, or when held at very high or low pH. The process is called *denaturation*. Up to a point, the process is reversible. The favored and reactive protein conformation in solution is the thermodynamically stable form, which is determined by the nature and sequence of amino acids, and their interaction with the solvent.

In solution, the charged groups of the protein are presumed to be at the surface of the molecule, or at the protein–solvent interface. They have the same environment whether in the favored native conformation, or in the unfolded, denatured structure. At the interface, the charged groups are hydrogen-bonded to water. For the most part, the hydrophobic components of the protein molecule are at the interior of the native structure. It is not possible to account for the stability of globular proteins in water unless this assumption is true (Tanford, 1962).

Based on the above assumptions, Tanford calculated the free energy (ΔF) of denaturation from the relation

$$\Delta F = -T\,\Delta S_{\mathrm{conf}} + \Sigma\,\Delta f_{\mathrm{u}}$$

where ΔS_{conf} is the change in conformational entropy of the polypeptide chain, and Δf_{u} is the free energy change for transfer of the small component groups of the molecule from the native form to the unfolded form. Omitting the influence of disulfide cross links, Tanford reasoned that the entropy factor markedly stabilizes the unfolded protein. As there must be a correspondingly large term for the factor $\Sigma \Delta f_u$ that stabilizes the globular form, there must also be Δf_u terms that are large and positive. If solvent interactions of amino acids can be taken as the sum of the contributions from the various parts of the molecule, local interactions of protein molecules are the sum of contributions from peptide groups and amino acid side chains.

From the values given in Table 7.1, Tanford (1962) calculated that $-T\Delta S_{conf}$ and $\Sigma \Delta f_u$ for the contribution to the unfolding of proteins (myoglobin and β-lactoglobin) were about the same (but of opposite sign). This was not true for a third protein (ribonuclease). Elements of Tanford's general procedure, and deliberations, have been applied to derivation of a relation between sweetness and bitterness of amino acids, peptides, and proteins.

7.3 TASTE OF AMINO ACIDS

Amino acids usually taste either bitter or sweet. Those containing a sulfur atom have a sulfury note. Aspartic and glutamic acids taste sour, and their sodium salts taste salty, in addition to a complex taste known as *umami* (Section 1.1.1). The first amino acid to be isolated, by acid hydrolysis of gelatin (1820), was believed to be a sugar (*Sucre de gelatin*) because it tasted sweet (Schmidt, 1938). When found to contain nitrogen, it was named *glycine*, employing the Greek prefix for sweet taste. Because glycine does not contain an asymmetric carbon atom, there are no optical isomers. Among the amino acids that do have enantiomeric forms, about 20% of the L-forms taste either sweet (L-alanine, threonine, solutions of proline, and hydroxy proline), or bitter-sweet (valine, methionine). Others are essentially tasteless (L-glutamine, L-asparagine) or bitter. D-Threonine and L-threonine are about equally sweet, and while both D- and L-alanine are sweet, the D-form appears to be slightly sweeter than the L-form. As the amino acids have application to derivation of taste chemistry principles, their taste attributes are first given as the taste of the crystalline substances (Table 7.2).

Tastes recorded for D- and L-forms of amino acids, in solution, along with the thresholds, and sweetness potency, with reference to the threshold value for sucrose, are listed in Table 7.3. The report by Kaneko (1939), as summarized by Tsuzuki & Yamashita (1968), lists an L-series of dimethyl, di-ethyl-, and di-propyl glycine as tasting bitter, and

TABLE 7.2
Taste assessment of amino acid crystals (Haefeli & Glaser, 1990) with respect to the four basic tastes, and hedonic assessment of their taste attributes (+ = pleasant; – = unpleasant; × = unclassifiable; S = sulfurous).

Amino acid	*Taste of* D-	*Taste of* D,L-	*Taste of* L-
Glycine	—	Sweet (+)	—
Alanine	Sweet (+)	Sweet (+)	Sweet (+)
Isoleucine	Sweet/bitter (+)	Bitter (×)	Bitter (–)
Leucine	Sweet (+)	Sweet (+)	Bitter (–)
Valine	Sweet (+)	Sweet/bitter (–)	Sweet/bitter (–)
Serine	Sweet (+)	Sweet/bitter (+)	Sweet/bitter/sour (+)
Threonine	Sweet (+)	Sweet (+)	Sweet (+)
Aspartic acid	Sour (+)	Sour (+)	Sour (+)
Glutamic acid	Sour (+)	Sour (+)	Sour (+)
Asparagine	Sweet (+)	Sweet (+)	Bitter/sweet (+)
Glutamine	Sweet (+)	Salty/bitter(–)	Salty/bitter (–)
Arginine	Sweet/bitter (+)	Bitter/sweet (+)	Bitter (–)
Lysine	Sweet/bitter (+)	Sweet/bitter (–)	Sweet/bitter (–)
Cysteine	Sweet/bitter/sour (–, S)	Sweet/bitter (–, S)	Bitter/sweet (–, S)
Methionine	Sweet (–, S)	Sweet (–, S)	Bitter/sweet (–, S)
Phenylalanine	Sweet/bitter (+)	Sweet/bitter (–)	Bitter (–)
Tyrosine	Insipid (×)	Insipid (×)	Insipid (×)
Tryptophan	Very sweet (+)	Sweet (+)	Bitter (–)
Histidine	Very sweet (+)	Sweet (+)	Bitter (–)
Proline	Bitter/sour (–)	Bitter/sweet/sour (–)	Sweet/bitter (+)

the D-series as sweet. However, these substances are symmetrical and do not have D- and L-forms. Consequently, the observation that it is the asymmetric L-form of phenylglycine that is sweet, while the D-form is tasteless, is probably in error. Kaneko (1939) concluded that for an amino acid with the general formula $RR'C(NH_2)COOH$, the amino acid will taste sweet only when R is a hydrogen atom, and R′ is smaller than a propyl group. The conclusion is often cited as an example of a relation between stereochemistry and taste, but without specification of the chiral nature of R and R′, the conclusion is not particularly meaningful. In addition to the general D-configurational sweetness requirement for amino acids, the NH_2 group must be in the α-position. β-Alanine and γ-aminobutyric acid are nearly tasteless.

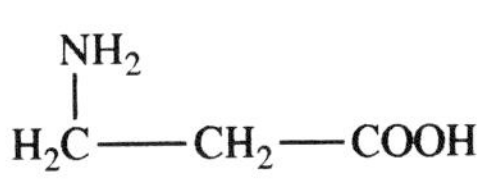

β-Alanine

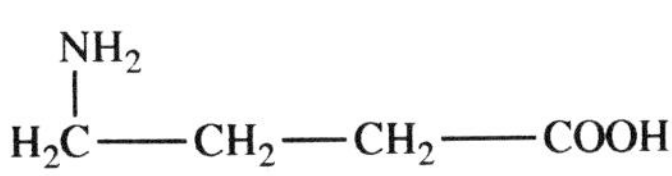

γ-Aminobutyric acid

Substitution on an L- and a D-amino acid can cause unexpected tastes.

TABLE 7.3

Taste, mM recognition threshold, and relative sweetness or bitterness of D- and L-amino acids. Values in parentheses are relative sweetness using 11 mM (0·376% w/v) as the sucrose threshold value (Wieser *et al.*, 1977)

Amino acid	D-	L-	*References*
Alanine	Sweet (0·73×)	Sweet (1·1×)	Heiduschka & Komm (1925)
	Sweet	Sweet	Kaneko (1939)
	Flat	Sweet	Solms *et al.* (1965)
	Sweet, 14·7 (0·71×)	Sweet, 13·8 (0·75×)	Nofre *et al.* (1974)
	Sweet, 10·0 (0·91×)	Sweet, 10·0 (0·91×)	Wieser *et al.* (1975, 77)
	Sweet, 11·2 (0·98×)	Sweet, 16·2 (0·68×)	Schiffman *et al.* (1981*a*)
	Sweet	Sweet	Birch & Kemp (1989)
	Sweet, 7·86 (1·4×)	Sweet, 7·17 (1·5×)	Haefeli & Glaser (1990)
α-NH_2butyric	Sweet, 14·5 (0·78×)	Sweet, 13·3 (0·72×)	Nofre *et al.* (1974)
Arginine	Slightly sweet (D,L)	Flat	Solms *et al.* (1965)
	—	Neutral	Wieser *et al.* (1975, 77)
	—	Bitter	Birch & Kemp (1989)
	Sweet, 2·61 (4·21×)	Bitter, 2·65	Haefeli & Glaser (1990)
Asparagine	Sweet	Neutral	Piutti (1886)
	Sweet, 4·5 (2·4×)	Neutral	Wieser *et al.* (1975, 77)
	—	Tasteless	Nofre *et al.* (1974)
	Sweet, 9·77 (1·1×)	(Bitter/sweet?), 1·62	Schiffman *et al.* (1981*a*)
	Sweet, 4·40 (2·5×)	Bitter/sweet, 67·0 (0·2×)	Haefeli & Glaser (1990)
Aspartic acid	Flat	Flat	Solms *et al.* (1965)
	Sour, neutral	Sour, neutral	Wieser *et al.* (1975, 77)
	Sour, 0·74	Sour, 0·182	Schiffman *et al.* (1981*a*)
	Sour	Sour	Birch & Kemp (1989)
	Sour, 4·0	Sour, 4·0	Haefeli & Glaser (1990)

Glutamine	Sweet, 10·0 (1·1×)	Neutral	Wieser *et al.* (1975, 77)
	Sweet, 3·5 (3·2×)	Sweet/bitter, 9·8 (1·13×)	Schiffman *et al.* (1981*a*)
	Sweet, 5·4 (2·0×)	Sweet-bitter, 43·0 (0·3×)	Haefeli & Glaser (1990)
Glutamic acid	Sour	Sour	Wieser *et al.* (1975, 77)
	Flat	Unique	Solms *et al.* (1965)
	Sour, 0·016	Unique, 0·063	Schiffman *et al.* (1981*a*)
	Sour	Sour	Birch & Kemp (1989)
	Sour, 3·92	Sour, 4·00	Haefeli & Glaser (1990)
Glycine	Sweet (0·64×)	—	Heiduschka & Komm (1925)
	Sweet (1·5×, w/v)	—	Solms *et al.* (1965)
	Sweet, 28·8 (0·36×)	—	Nofre *et al.* (1974)
	Sweet, 30·0 (0·37×)	—	Wieser *et al.* (1975, 77)
	Sweet, 30·9 0·36×)	—	Schiffman *et al.* (1981*a*)
	Sweet	—	Birch & Kemp (1989)
	Sweet, 15·4 (0·71×)	—	Haefeli & Glaser (1990)
Histidine	Sweet	Tasteless, bitter	Abderhalden & Weil (1912)
	Sweet (7·4×, w/v)	Flat	Solms *et al.* (1965)
	Sweet, 3·0 (3·4×)	Bitter, 47·5	Wieser *et al.* (1975, 77)
	Sweet, 1·86 (5·9×)	Flat/bitter, 1·23	Schiffman *et al.* (1981*a*)
	Sweet	Bitter/tasteless	Birch & Kemp (1989)
	Sweet, 1·98 (5·6×)	Bitter, 21·0	Haefeli & Glaser (1990)
Isoleucine	Sweet	Bitter	Ehrlich (1914)
	—	Tasteless	Nofre *et al.* (1974)
	Sweet, 10·0 (0·58×)	Bitter, 11·0	Wieser *et al.* (1975, 77)
	Sweet, 12·5 (0·88)	Bitter, 7·41	Schiffman *et al.* (1981*a*)
	Sweet, 12·0 (0·90×)	Bitter, 13·0	Haefeli & Glaser (1990)

continued

TABLE 7.3
Continued

Amino acid	D-	L-	*References*
Leucine	Sweet	Flat, bitter	Fischer & Warburg (1905)
	Sweet (4·3× w/v)	Bitter	Solms *et al.* (1965)
	Sweet, 2·5 (4·4×)	Bitter, 12·0	Wieser *et al.* (1957, 77)
	Sweet, 5·01 (2·2×)	Flat/bitter, 6·45	Schiffman *et al.* (1981*a*)
	Sweet/bitter	Bitter	Birch & Kemp (1989)
	Sweet, 2·18 (5·0×)	Bitter, 10·60	Haefeli & Glaser (1990)
Methionine	Sweet	Flat	Berg (1953)
	Sweet, 5·0 (2·0×)	Sulfury	Wieser *et al.* (1975, 77)
	Sweet, 5·01 (2·2×)	Bitter, 3·72	Schiffman *et al.* (1981*a*)
	Sweet	Bitter	Birch & Kemp (1989)
	Sweet, 6·50 (1·7×)	Sweet-bitter, 12·7	Haefeli & Glaser (1990)
Norvaline	Sweet	Bitter	Nofre *et al.* (1974)
	Sweet, 6·5 (1·7×)	Bitter	Weisner *et al.* (1975, 77)
Phenylalanine	Sweet	Faintly bitter	Fischer & Mouneyrat (1900)
	Sweet (0·73× w/v)	Bitter (0.23× w/v)	Solms *et al.* (1965)
	Sweet	Bitter	Nofre *et al.* (1974)
	Sweet, 2·0 (5·5×)	Bitter, 57·5	Wieser *et al.* (1975, 77)
	Sweet, 1·55 (7·1×)	Bitter, 6·61	Schiffman *et al.* (1981*a*)
	Sweet, 1·77 (6·2×)	Bitter, 10·0	Haefeli & Glaser (1990)
Proline	(Bitter)	Sweet	Kaneko (1940)
	Neutral	Sweet/bitter, 26·0 (0·4×)	Wieser *et al.* (1975, 77)
	Flat	Flat	Solms *et al.* (1965)
	Bitter	Sweet	Birch & Kemp (1989)
	Bitter, 25·4	Sweet, 16·1 (0·68×)	Haefeli & Glaser (1990)

Hydroxy proline	Tasteless	Sweet	Kaneko (1940)
	Neutral	Sweet, 6·0 (1·83×)	Wieser *et al.* (1975, 77)
	Tasteless	Sweet	Birch & Kemp (1989)
Serine	—	Sweet	Nofre *et al.* (1974)
	Sweet (0·31×)	Sweet (0·37×)	Wieser *et al.* (1975, 77)
	Sweet (0·2×)	Sweet (0·52×)	Schiffman *et al.* (1981*a*)
	Sweet ((0·5×)	Sweet (0·96×)	Haefeli & Glaser (1990)
Threonine	Sweet	Fairly sweet	Berg (1953)
	Sweet (D > L)	Sweet (L < D)	Inui & Kaneko (1961)
	Sweet (0·24×)	Sweet (0·27×)	Wieser *et al.* (1975, 77)
Tryptophane	Very sweet	Flat	Berg (1953)
	Sweet (35× w/v)	Bitter (0·54× w/v)	Solms *et al.* (1965)
	Sweet (37×)	Bitter	Wieser *et al.* (1975, 77)
	Sweet	Flat/bitter	Schiffman *et al.* (1981*a*)
	Sweet/bitter	Bitter	Birch & Kemp (1989)
Tyrosine	Sweet	Bitter	Kaneko (1938)
	Sweet (5·5× w/v)	Bitter (0·05× w/v)	Solms *et al.* (1965)
	Sweet (5·5×)	Bitter	Wieser *et al.* (1975, 77)
Valine	Very sweet	Slight sweet, bitter	Fischer (1906)
	Sweet	Bitter	Wieser *et al.* (1975, 77)
	Flat/bitter/sweet	Tasteless	Schiffman *et al.* (1981*a*)
	Sweet	Bitter/sweet	Birch & Kemp (1989)

[a]Calculated from the racemic mixture (0·92×).

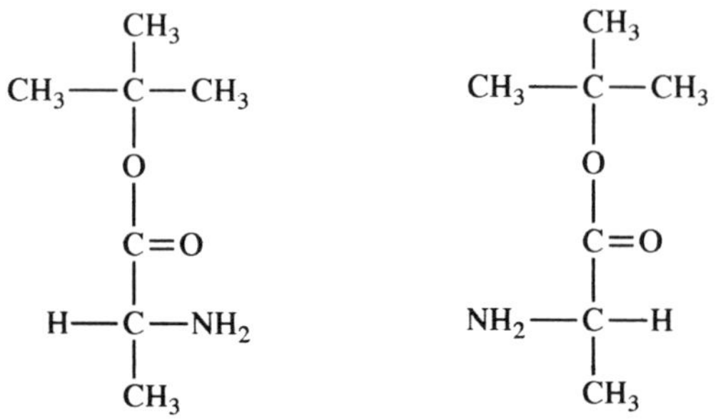

D-Alanine tertiary butyl ester (Bitter) L-Alanine tertiary butyl ester (Sweet)

While both D- and L-alanine, and D- and L-α-aminobutyric acid are sweet, the tertiary butyl esters of the D-compounds are bitter, while the tertiary butyl esters of the L-compounds are sweet (Nofre *et al.*, 1974). Moreover, the L-alanine derivative (L-alanine *tert*-butyl ester hydrochloride) is 40 times as sweet as sucrose on a molar basis (75× on a weight basis), but the methyl and ethyl alcohol analogues are tasteless or bitter. The observation serves as a caveat to those who attempt to derive information on the relation between structure and taste by substituting atoms or groups for other atoms or groups.

Many of the tastes reported for amino acids, such as those summarized in Tables 7.2 and 7.3, have been criticized (cf. Birch & Kemp, 1989). Amino acid acyclic structures vary with pH, and unbuffered solutions of amino acids have a pH corresponding to the isoelectric point. Attempts to control pH with NaOH and HCl have been reported (Solms *et al.*, 1965), but the adjustment also alters taste, as does the use of buffer. In the past, the usual crystalline forms of amino acids were hydrochloric acid salts, and taste assessments of amino acids are complicated by sour and salt taste attributes. This may account for the saltiness attribute assigned (Tables 7.2 and 7.3) to the taste of crystalline L-glutamine, and for the sourness assigned to D-proline.

There is a relation between the sweetness intensity and bitterness intensity of enantiomeric amino acids. Lehmann (1978), using normalized data for the sweetness and bitterness of amino acids, found that the two are directly related, as shown in Fig. 7.8. Haefeli & Glaser (1990) pointed out that compounds with approximately equal D- and L-thresholds (and D,L-thresholds) for various tastes (Table 7.3) are

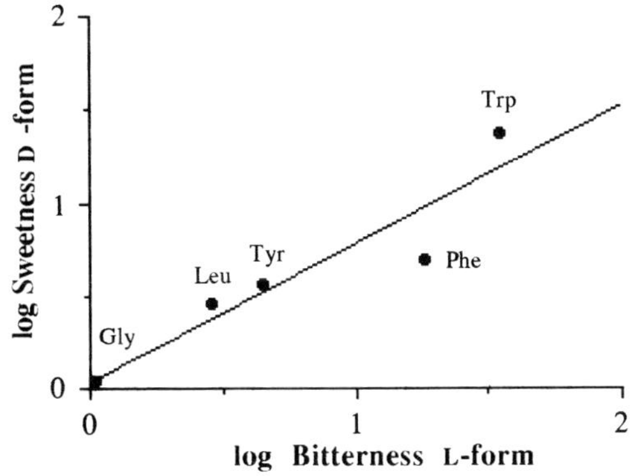

FIG. 7.8. Relation between sweetness and bitterness of D- and L-amino acids (Lehmann, 1978).

alanine, isoleucine, valine, threonine, aspartic and glutamic acids, arginine, lysine, and cysteine. Some compounds (serine, methionine, phenylalanine, tryptophan, and histidine) had similar D- and D,L-thresholds, and still others (leucine, asparagine, glutamine, and tyrosine) had widely different thresholds for D-, D,L-, and L-forms. The reasons for these observations are probably varied, such as the nature of the salts, different taste reaction mechanisms, etc. Of particular importance to the chemistry of sweetness and bitterness is the observation (Belitz *et al.*, 1979) that the sweet and bitter taste potency of antipodal amino acids is related to the hydrophobicity of the amino acid side chains.

7.4 TASTE OF PEPTIDES

The L-amino acids are usually either bitter tasting, or sometimes sweet. The methyl esters of simple dipeptides containing two otherwise bitter-tasting or tasteless amino acids may taste intensely sweet. The reason for the apparent chiral anomaly contained in the sweetness of dipeptides is given in detail in Chapter 11.

7.4.1 Peptides with Sweet Taste

With the discovery (Mazur *et al.*, 1969) that L-aspartyl-L-phenylalanine methyl ester (aspartame) tasted sweet, a plethora of studies were initiated to determine the source of the sweetness, and how the sweetness is affected by substitution. Sucrose is often said to be the organic substance that leads all others with respect to the number of scientific articles devoted to it, but aspartame is the most intensively studied substance with the objective of deducing the molecular features responsible for its high sweetness potency. Until recently, very few articles addressed the reason for sucrose sweetness.

7.4.1.1 *L-Aspartic Acid Peptides*

Aspartame is the methyl ester of the dipeptide L-aspartyl-L-phenylalanine. It is amphoteric (pK_a 3·1 and 7·9), with an isoelectric point at pH 5·2. Solubility at 25°C is only about 1%, but at pH 2·2 a 10% solution can be obtained (Mazur & Ripper, 1979). Depiction of its formation from the component amino acids is shown in Fig. 7.9.

Mazur *et al.* (1969) determined that only L-aspartic acid serves as a source of sweetness as the substitution of any other amino acid, D- or L-, for aspartic acid, results in loss of sweetness. Not even the dipeptide with an amino acid that is the next highest homologue of aspartic acid (glutamic acid) yields a sweet dipeptide. Another critical feature is that the zwitterionic group must be free (unsubstituted).

Only the L,L-aspartame isomer is sweet, and the D-L-, L-D-, and D-D-isomers are bitter, indicating a high degree of configurational specificity for the component amino acids, and for the peptide as a whole. With reference to the Fischer projection structure (Fig. 7.9), transposition of configuration at either the upper asymmetric carbon atom, or at the lower asymmetric carbon atom, eliminates sweetness.

An interesting result of substitution reactions at the carbon atom alpha to the bottommost asymmetric carbon atom is obtained. Substitution of a small alkyl group, and omitting the phenyl group, converts the structure to a D-amino acid ester. Several of these (e.g., L-aspartyl-D-alanine methyl ester) are sweet. The structures of L-aspartyl-D-alanine methyl ester and L-aspartyl-L-alanine methyl ester are shown in Fig. 7.10.

In cases of multiple configuration, seemingly degenerate results are a consequence of the conventions used for specifying configuration. Here, the L-D-isomer is a diastereoisomer of the L-L-form. The enantiomer is the D-D-form. With L-aspartyl-D-alanine esters, sweetness potency increases with the size of the ester group until the *n*-propyl methyl ester is obtained. From then on, increasing the size of the ester group leads to decreasing sweetness potency (Mazur *et al.*, 1973).

Ariyoshi (1984) investigated the effect of both the configuration and substituent size of aspartic acid di-, tri-, and tetra-peptides. He concluded that when there is a small and a large substituent at the asymmetric

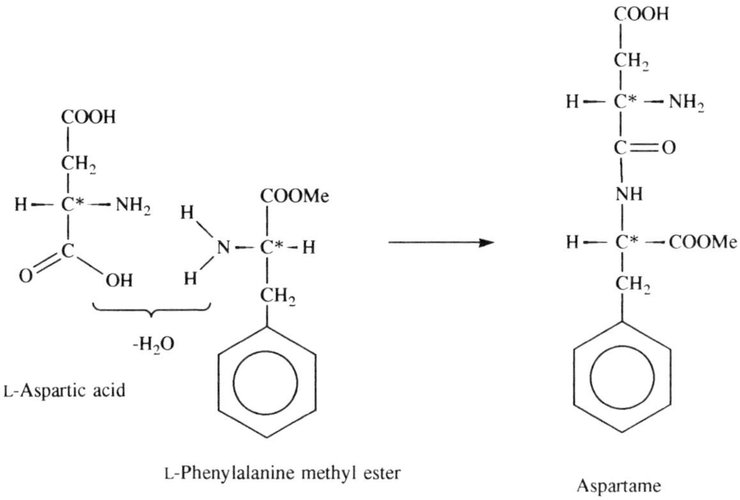

FIG. 7.9. Formation of aspartame.

carbon atom for the second amino acid of the aspartyl dipeptide, taste varies with configuration, as shown in Fig. 7.11. The result is analogous to that observed (cf. Culbertsen & Walters, 1991) for the aspartyl-alanyl methyl esters by Mazur *et al.* (1969, 1973). With special reference to the O–R side chain, where R is an alkyl group, it appeared, as a first approximation (Mazur *et al.*, 1969), that the side chain is essential for sweetness. Among the L-aspartyl-D-alanyl substances, aspartyl phenylalanine is tasteless, while aspartyl-phenylalanine-OMe has more than 100

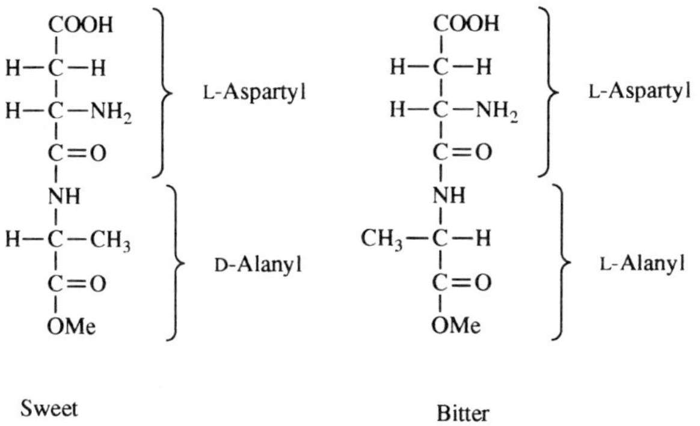

FIG. 7.10. D- and L-alanyl methylesters of L-aspartyl-alanyl dipeptides.

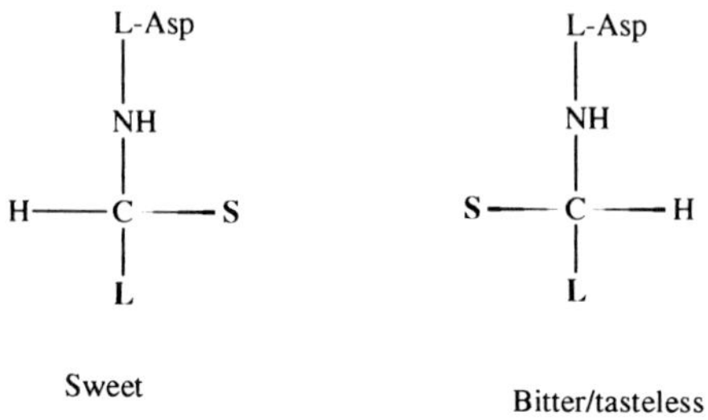

FIG. 7.11. Effect of configuration and size of substituents on the non-aspartyl amino acid of aspartyl-dipeptides. S = 'small,' L = 'large' (Ariyoshi, 1984).

times the sweetness of sucrose. Aspartyl-phenylalanyl-OEt is 10 times sweeter, and the O-Pr and O-Bu derivatives have only about the same sweetness potency as sucrose. Aspartyl-phenylalanyl-*O*-cyclohexane is also tasteless (Fujino *et al.*, 1973), as are the L-histidyl- and L-tryptophyl *O*-methyl esters. These large groups do not possess either the proper lipophilic-hydrophobic balance, or adversely affect the overall conformation of the molecule (Ariyoshi, 1984).

With respect to the nature of the large and the small groups, various substitutions lead to varying sweetness potency. Among the L-aspartyl-aminomalonic acid diesters, Fujino *et al.* (1973) found that when R in the general diester structure shown in Fig. 7.12 was cyclopentyl, sweetness increased to 300–600 times. When R was cyclohexyl, it was 556–880 times sweeter. Finally, when R was fenchyl, sweetness intensity increased

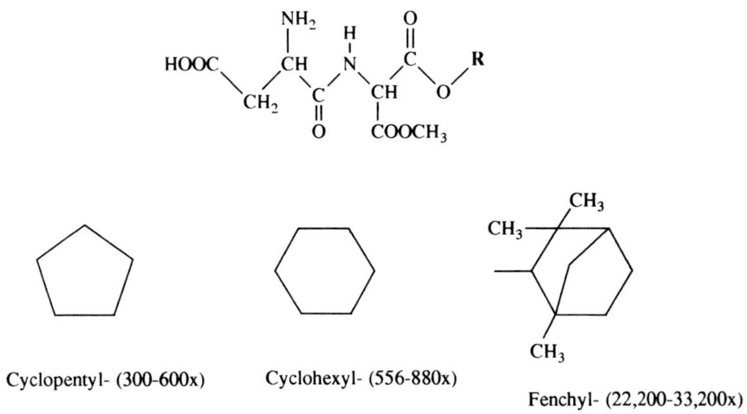

FIG. 7.12. Effect of substitution for substituent R in L-Asp-aminomalonic diesters on sweetness intensity (Fujino *et al.*, 1973).

to 22 200–33 200 times. The number of derivatives and variations on the aspartame structure that have been studied is enormous. van der Wel *et al.* (1987) list the structure and taste potency for over 100 substances.

The favored conformation of aspartame has proved difficult to pin down. It is presumed that it is the so-called favored conformation that tastes sweet. Lelj *et al.* (1976) proposed the $F_I D_{II}$ form as the favored conformation, as shown in Fig. 7.13. van der Heijden *et al.* (1979) proposed a conformation whereby the zwitterionic group and the phenylalanine ring were about in the same plane with the remainder of the molecule. Their conformation is also shown in Fig. 7.13. Iwamura (1981) proposed conformations for aspartyl dipeptide analogues that are more in line with those of Leji *et al.* (1976). Subsequently, Goodman *et al.* (1987) proposed a conformation whereby both the zwitterionic group and the phenylalanine ring are in the same plane, with the phenyl group located at a 90° angle to the backbone for the remainder of the molecule. The arrangement is described as the 'L' peptide conformation. Examination of molecular models (Belitz *et al.*, 1988) support the Goodman proposal, and establish that the ester group is in direct line with the phenyl group, and perpendicular to the remainder of the molecule. The position of the

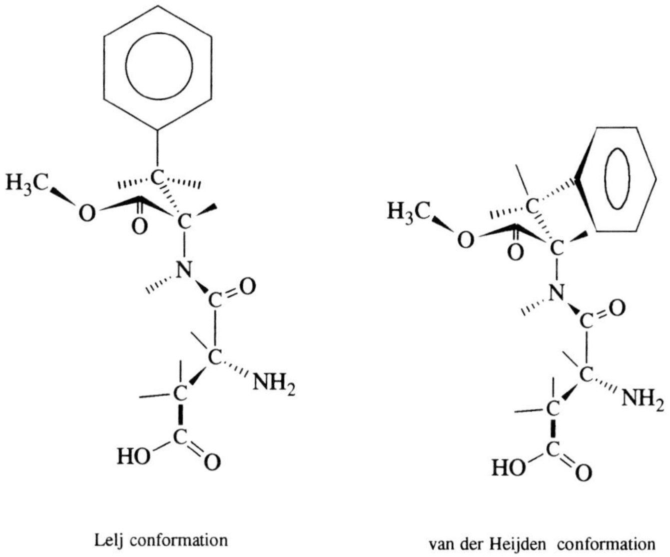

FIG. 7.13. Proposed favored conformations of aspartame (Lelj *et al.*, 1976; van der Heijden *et al.*, 1979).

zwitterionic groups in the Goodman model is opposite to that proposed by van der Heijden *et al.* (1979), but in agreement with that proposed by Lelj *et al.* From free energy calculations, Taylor *et al.* (1991) point out that both conformations (Leji, van der Heijden/Goodman) can be construed to be favored conformations.

Alitame is another novel dipeptide sweetener, somewhat analogous to aspartame, except that the second amino acid is alanine and a terminal amide group serves to replace the methyl ester. It is also much sweeter than aspartame (2000×), and is resistant to acid hydrolysis. The effects of substitution, degree of side chain branching, etc., on sweetness indicate that they are not unlike those reported for aspartame (Glowaky *et al.*, 1991). The effect of altering the nature of aspartyl alanylamide side chain is particularly significant. When S is replaced by a carbon atom, sweetness is only 300 times. When the sulfur atom is replaced by a SO group, sweetness is 300 times, and when replaced by SO_2, to 805 times.

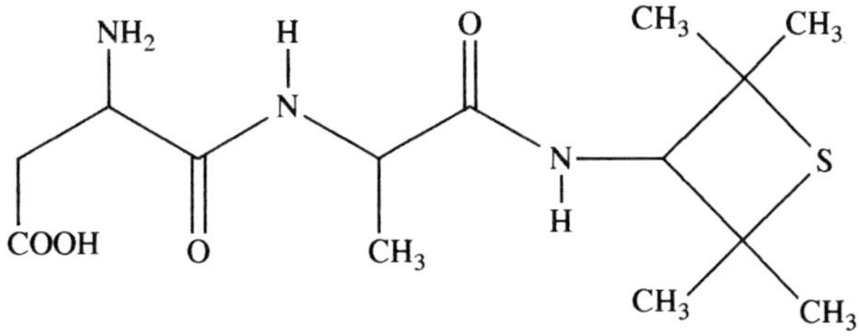

L-α-Aspartyl-N-(2,2,4,4-tetramethyl-3-thietanyl-D-alaninamid
(Alitame)

7.4.1.2 *Retro-Inverso Peptides and Amides*

Dipeptide substances have proved to be valuable for deriving the role of configuration and conformation on sweet taste. Of particular importance are the retro-inverso modifications of dipeptide amides (Fuller *et al.*, 1985). L-Aspartyl-D-alanine amide, known to be sweet, is converted to *N*-(L-aspartyl)-1,1-diaminoalkane compounds (Fig. 7.14), where the C-terminal amide bond of II is the reverse of that for I. A series of *N*-(L-aspartyl)-1,1-diaminoalkane compounds, obtained by varying the nature of R′, were found to have sucrose-like sweetness ranging from 50–75

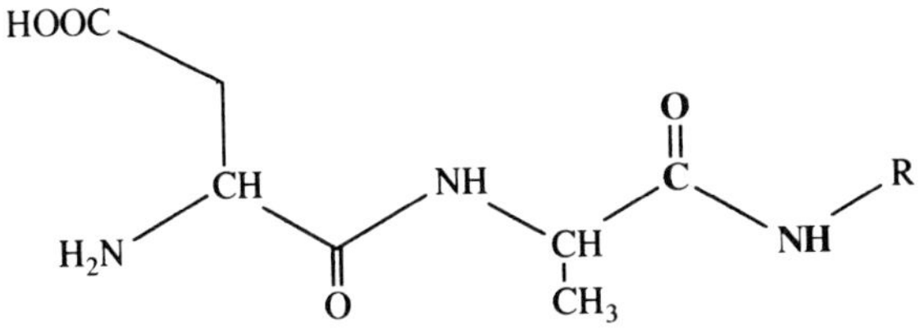

I. L,D-form

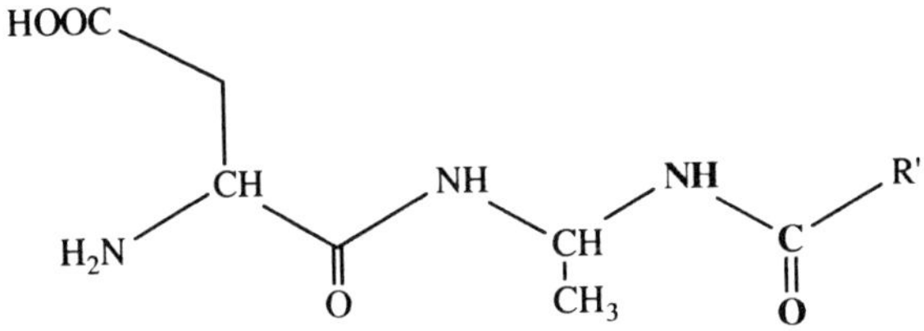

II. L,R-form

FIG. 7.14. L-Aspartyl-D-alanine amide (I) and the retro-inverso compound *N*-(L-aspartyl)-1,1 (*R*)-diaminoalkane (II) (Fuller *et al.*, 1985).

times (R′ = cyclohexyl) to 800–1000 times (R′ = tetramethylcyclopentyl). For comparison with aminomalonic acid diesters (Fig. 7.12), several groups, used to substitute for R′, are given in Fig. 7.15.

Subsequently, Goodman *et al.* (1987) synthesized four configurational isomers of the dipeptide and retro-inverso aspartyl-alanyl tetramethylcyclopentyl substances (Fig. 7.16). Of great significance to configurational and conformational requirements for sweetness, it was found that the L,L-amide is bitter, while the L,D-amide and the two retro-inverso analogues L-R and L-S are sweet. Conformational analysis showed the tetramethylcyclopentanyl ring, and the zwitterionic chelate-ring of the aspartyl moiety, are coplanar. They describe an angle of about 90° in the sweet compounds, and the sweet substances had a common conformation significantly different from the conformation of the bitter substances. These structures are shown in Fig. 7.17, with the aspartyl group in fixed position. Sweet molecules have a small rectangular coordinate z-component, while that for the bitter substance is large. The planar orientation, is designated as the L-conformation. Enantiomers then have a reverse 'L'-conformation. In the bitter L,L-amide, the tetramethylcyclopentanyl ring

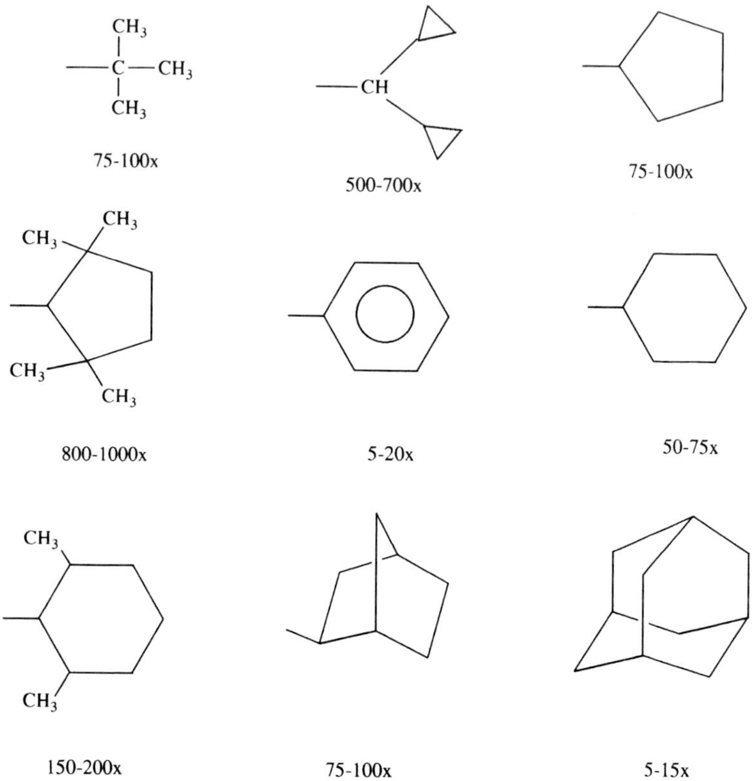

FIG. 7.15. Relative sweetness of compounds formed by substitution for R′ in the *N*-(L-aspartyl)-1,1 (*R*)-diaminoalkane (II) structure in Fig. 7.14 (Fuller *et al.*, 1985).

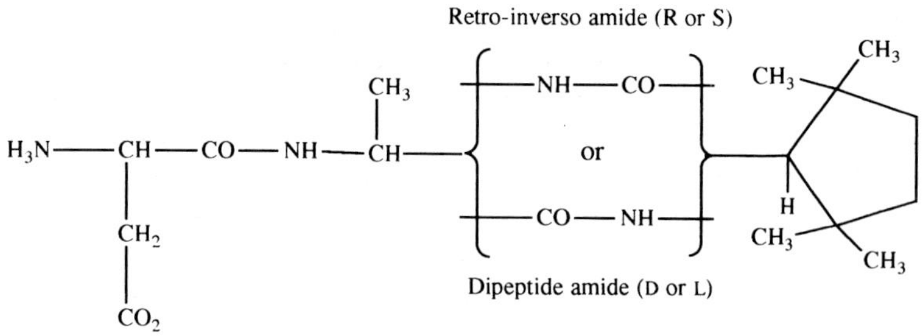

FIG. 7.16. Retro-inverso (R, S) analogues of L-aspartyl-L- or D-alanyl amides with a tetramethylcyclopentanyl group (Goodman *et al.*, 1987).

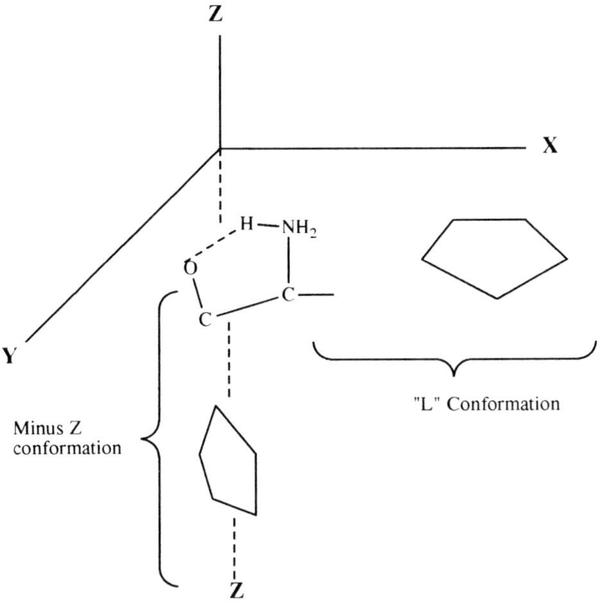

FIG. 7.17. Positioning of sweet and bitter retro-inverso analogues of aspartyl-alanyl amides in rectangular coordinates (Douglas & Goodman, 1991).

is out of the xy plane, and twisted through an angle of about 60°. It is also perpendicular to the rectangular coordinate xy axis, and has a large minus-z component.

The Goodman school established that the L-conformation prevails when dipeptide substances taste sweet, and possess a minus-z component when bitter. A particularly intriguing relation between conformation and taste occurs with the compounds *N*-(L-aspartyl)-*N'*[2,2,5,5-tetramethylcyclopentanal)carbonyl]-(*R*)-1,1-diaminoethane (the L-R-isomer), and *N*-(L-aspartyl)-*N'*-[2,2,5,5-tetramethylcyclopentanyl)carbonyl]-*S*-1,1-diaminoethane (L-S-isomer), as shown in Fig. 7.18. Both reside in the same crystal (Benedetti *et al.*, 1990), and have enantiomeric L-conformations. The L-R-diastereoisomer has the same crystalline conformation as found for L-aspartyl-L-phenylanine methyl ester in solution. There, the ϕ torsion bond of L-aspartyl-L-phenylanine methyl ester is rotated by 40°. In solution, while the L-R-isomer retains the crystalline structure conformation, there is rotation by 180° about the ϕ_{Phe} torsion angle for the L-S-isomer, as a crystalline structure intramolecular hydrogen bond is severed in solution. The conformation of the L-S-isomer is then transformed from

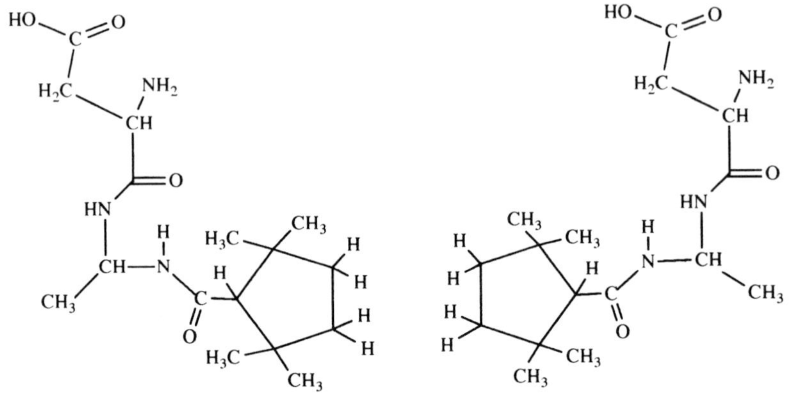

I II

FIG. 7.18. The L-conformation for the L-R-isomer of *N*-(L-aspartyl)-*N'*[2,2,5,5-tetramethylcyclopentanal)carbonyl]-(*R*)-1,1-diaminoethane (I), and the mirror image reverse-L conformational isomer of the L-S isomer of *N*-(L-aspartyl)-*N'*-[2,2,5,5-tetramethylcyclopentanyl)carbonyl]-*S*-1,1-diaminoethane (II) (Benedetti *et al.*, 1990).

mirror-image 'L' to that of the normal 'L' conformation. Both compounds taste sweet.

For aspartylglycyl stereoisomeric analogues of the above compounds (the L-aspartyl-D-alanine amides, and the *N*-(L-aspartyl)-1,1-diamino-alkane retro-inversion compounds), it was established (Goodman *et al.*, 1988) that the retro-inverso analogue is 220 times as sweet as sucrose. It has an L-conformation that is highly favored over the conformation with a minus-z component. On the other hand, the glycine amide analogues, with similar conformations, are about energetically equal. In agreement with an approximate 50/50 distribution of these conformational forms in solution, the glycine amide analogue tastes sweet and bitter.

It was predicted (Yamazaki *et al.*, 1991) that those peptide compounds possessing 2-aminocyclopentane carboxylic acids will taste sweet. The prediction was based on the observation that aspartyl-(1*R*, 2*R*)-2-Ac^{5}c-OMe and aspartyl-(1*S*, 2*R*)-2-Ac^{5}c-OMe, which possess the L-conformation, taste sweet. However, because aspartyl-(1*S*, 2*S*)-2-Ac^{5}c-OMe has a large negative z-component, it was predicted that this substance will be bitter. As aspartyl-(1*R*, 2*S*)-2-Ac^{5}c-OMe has a reverse L-conformation, that compound was predicted to be tasteless. The predictions were subsequently found to be true (Douglas & Goodman, 1991).

In a review of their studies, Goodman *et al.* (1991) noted that a +z projection for the peptide substance hydrophobic group is also equated with tastelessness, and the L-shape for sweetness in peptide systems appears to be general, as alitame also has the L-shape, and tastes sweet.

7.4.1.3 *Extraordinarily sweet organic substances*

Tinti & Nofre (cf. Tinti & Nofre, 1991*a*, *b*; Nofre & Tinti, 1993) embarked on a series of studies that led to preparation of a series of peptide substances that are extraordinarily sweet. The authors describe them as hyperpotent sweeteners that have sweetness greater than 40 000 times on a weight basis when compared to 2% sucrose solutions. Their studies began with a study of a cyano derivative of suosan (Section 8.1.7). It was found that substitution of a sulfur group for the carbonyl oxygen atom in cyanosuosan (650×) increased sweetness to 2900 times, although cyanosuosan itself is not as sweet as suosan (900×).

Recognizing that the aspartame aspartyl residue is also a 3-aminopropanoate moiety, and that cyanosuosan also possessed that group, the authors combined the two substances so that the 3-aminopropanoate group was common to both. The resulting compound (superaspartame) was much sweeter (8000×) than either aspartame or cyano-suosan. Concentrating on the structure of superaspartame, substitutions were made for A, Y, and Z in the reference structure shown on page 245. When X in the dipeptide reference structure is CN, A is S, NCN, O or NH, and with Y as $COOCH_3$, and Z as $C_6H_5CH_2$, sweetness potencies of up to 40 000 times, on a weight basis, were obtained. One example is thiosuperaspartame, where only the carbonyl oxygen atom of superaspartame is substituted with a sulfur atom.

Subsequently, the authors discovered the guanidine sweeteners. For the guanidine reference structure shown on page 246, compounds with a sweetness potency of 200 000 times (sucrononic acid) were obtained. This was especially true when X_1 is CN (when X_2 and X_3 are H), and X_2 and X_3 are Cl (when X_1 is H), and in which R_1 is H or CH_3, and R_2 is a hydrophobic group, such as a cyclononyl group (c-C_9H_{17}). Reference to other compounds significant for their composition and structure is given by Culbertson & Walters (1991), and Tinti & Nofre (1991*a*, *b*). Some of them are shown in Fig. 7.19. As with aspartyl dipeptides, Muller *et al.* (1991) point out that while R-isomer of the 3-amino-3-phenylpropionic acid analogue of *N*-(4-cyanophenyl)-*N'*-carboxyethyl urea is tasteless

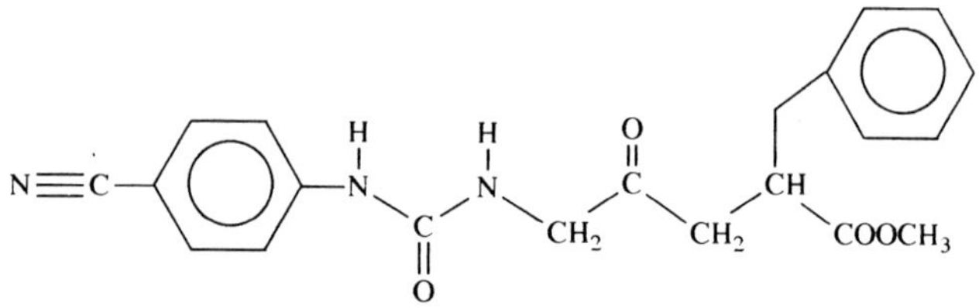

Cyanoarylurea aspartame (7,800x)

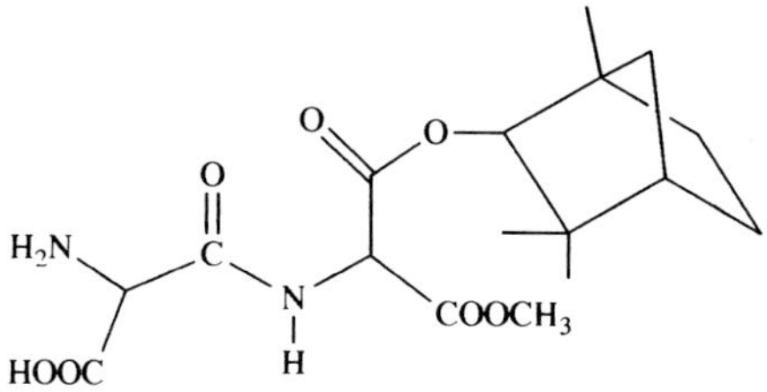

An aspartic acid fenchyl ester (50,000x)

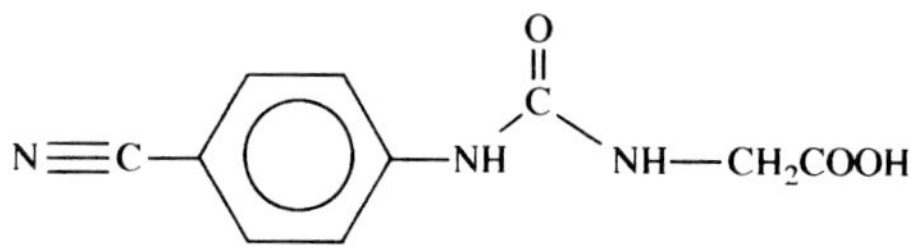

Glycine analog of cyano-suosan (7000x)

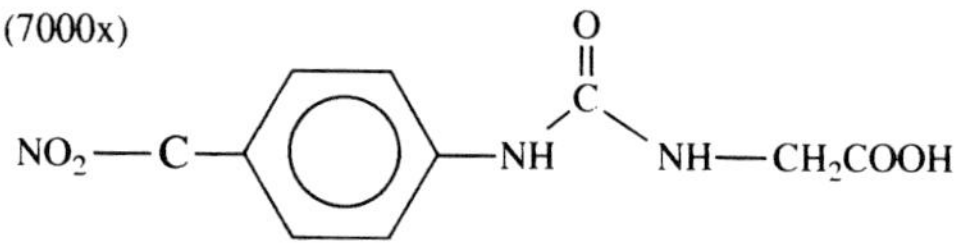

Glycine analog of suosan (10x)

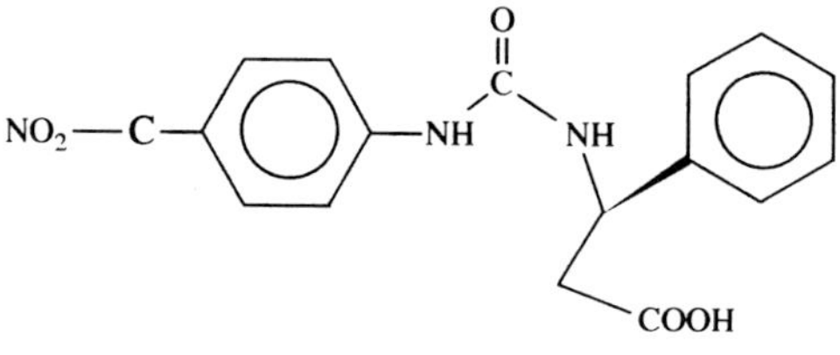

(R)- 3-amino-3-phenylpropionic acid analog of N-(4-cyanophenyl)-N' carboxyethyl urea (Tasteless)

FIG. 7.19. Selected extraordinarily sweet compounds.

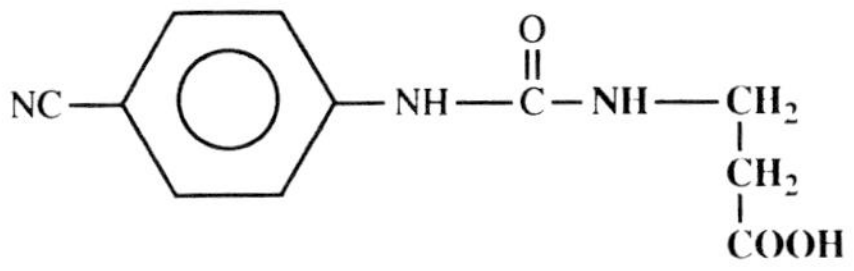

Cyanosuosan (2900x)

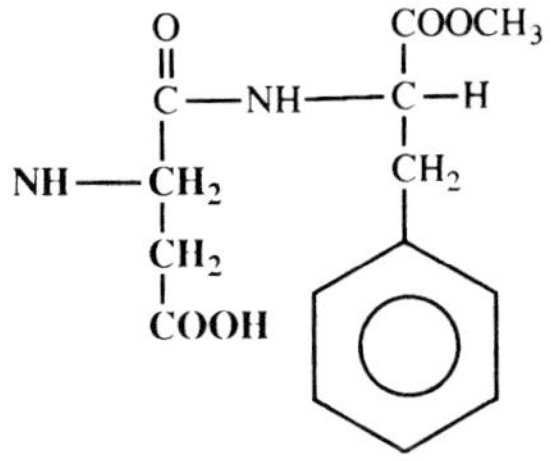

Aspartame (180x)

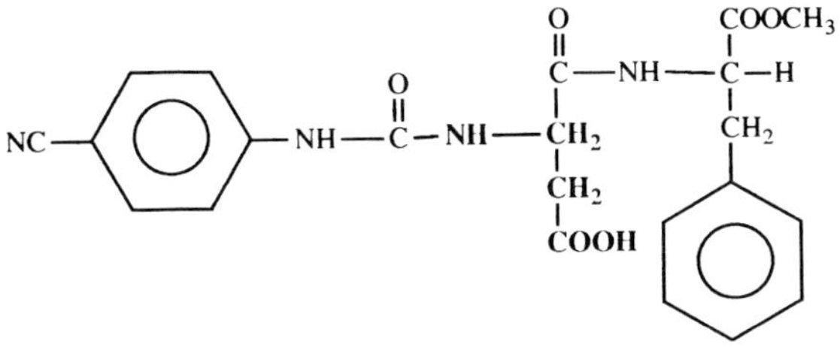

Super aspartame (8,000x)

(Fig. 7.19), the S-form (comparable to L-aspartic acid in the structure of aspartame), is quite sweet.

Structures of additional extraordinarily sweet substances, e.g. sucrononic acid, are presented in Chapter 9. A common feature among them is the presence of single substituents (the nitro group and the cyano group) with strong inductive effects. In the glycine analogues of suosan (Fig. 7.19), which contains a β-alanine residue, the glycine substituent led to a marked decrease

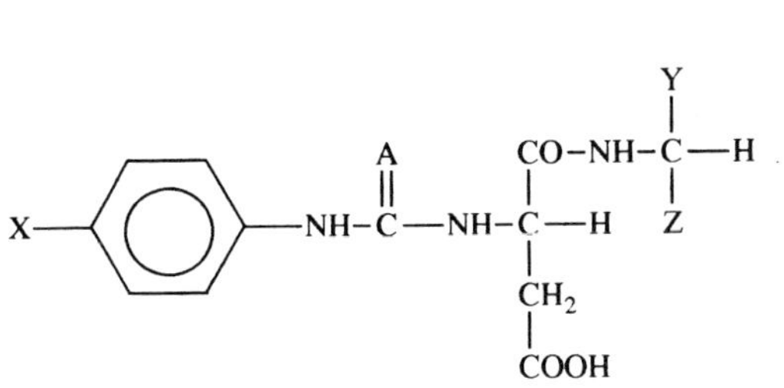

Dipeptide family of high-potency sweeteners

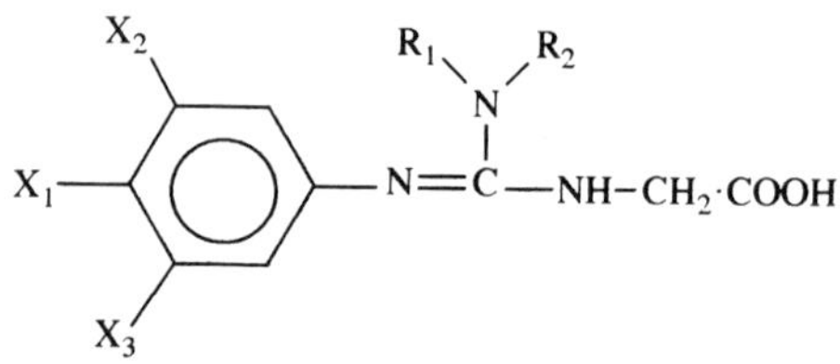

High-potency guanidine sweeteners

in the sweetness of suosan (from 700 to 10). On the other hand, the sweetness of cyanosuosan increased from 900 to 7000.

7.4.2 Peptides with Bitter Taste

The majority of peptides taste bitter. Matoba & Hata (1972) found that tasteless proteins may yield bitter peptides, regardless of the amino acid sequence. On the other hand, Shiraishi *et al.* (1973) found that the proportion of hydrophobic amino acids is generally higher in bitter peptides, and that their sequence does appear to play a role. Phe–Pro bitterness is more intense than Pro–Phe, and Gly–Phe–Pro is more intensely bitter than Phe–Pro–Gly.

According to Nishimura & Kato (1988), nearly all bitter peptides contain the bitter-tasting, hydrophobic amino acids phenylalanine, tyrosine, tryptophan, leucine, isoleucine, and valine. Ishibashi *et al.* (1987) (cf. Nishimura & Kato, 1988) pointed out that when the hydrophobic amino acid leucine was located at the C-terminus of peptides, the peptide was invariably bitter. Ney (1971) proposed that the bitter intensity of protein hydrolysates is related to a hydrophobicity constant (Q) obtained by averaging the Δf values of Tanford (1962) for individual amino acids. In contrast, Adler-Nissan (1988) found that a Q value for a protein does not predict protein hydrolyzate bitterness, as the primary component responsible for bitterness may be only a minor component in the hydrolyzate. The author concludes that the bitterness of peptides is not always a simple function of average hydrophobicity.

7.4.3 Peptides with Sour Taste

As with the amino acids, there is little reason to believe that the structure (component) responsible for the sourness of peptides is not a preponderance of glutamic and aspartic acid residues. The notion was proposed by Tada *et al.* (1984) and Nishimura & Kato (1988). As an example, Ishibashi *et al.* (1987) showed that whenever aspartic or glutamic acid appeared in the structure of the otherwise bitter leucyldipeptides, the principal taste is sour. Interestingly, whenever the γ-carboxyl group of glutamic acid is disposed to bond an NH_2 group in another amino acid in the structure of a peptide, the dipeptide assumes both a sour and an astringent taste (Kirimura *et al.*, 1969).

7.4.4 Peptides with Salty Taste

In attempting to elucidate the role of arginine in peptide bitterness, Tada *et al.* (1984) prepared H-L-Orn-Gly-OH · HCl and H-L-Lys-Gly-OH · HCl. In addition to umami taste, the dipeptides also tasted salty (to many people, the predominant characteristic of the umami taste is saltiness). The retro-inverso-analogs of the aforementioned compounds possessed sour and sweet taste. The compound H-L-Orn-β-Ala-OH · HCl was also found to taste salty, and the retro-inverso compound was sweet/sour (Fig. 7.20). Upon preparing H-L-Orn-Tau · HCl and H-L-Orn-g-Abu-OH · HCl, it was found that little umami taste remained, and the latter two compounds were stated to be twice as salty as NaCl on a weight basis, and equally salty on a molar basis.

7.5 TASTE OF PROTEINS

Several proteins that are soluble in water (globulins) taste sweet. One that is essentially tasteless imparts sweet taste to acid (sour) solutions. Sweetness attributes of the proteins are characterized by slow onset time, high potency, and prolonged duration. This section begins with the unique taste properties of a glycoprotein that does not, by itself, taste sweet, and continues on with discussion of the two most intensely studied protein substances. Other sweet proteins are known, e.g. pentadin (van der Wel *et al.*, 1989), but their properties have not been studied.

7.5.1 Miraculin

Fruits of the West African shrub, *Richardella dulcifica* (*Synepalum dulcifica*), contain a glycoprotein with a molecular weight of about 42 000 that can change the taste of acids from sour to sweet. The substance contains about 373 amino acids (Giroux & Henkin, 1974). The isoelectric point of the protein is at pH 9, and it is basic. The carbohydrate content is reported (cf. Higginbotham, 1979) by Kurihara & Beidler (1968) to be 6·7% (arabinose and xylose). Brouwer *et al.* (1968) reported 7·5–21% (glucose, ribose, arabinose, galactose, rhamnose), and Giroux & Henkin (1974) reported 6·3% (fucose, xylose, mannose, galactose). About 0·1 mg protein is

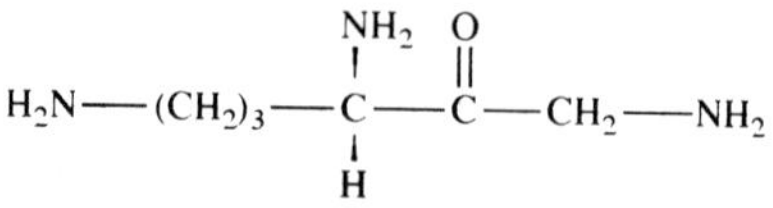

L-Ornithylglycine
(Salty, 0.6x; Umami, 0.8x)

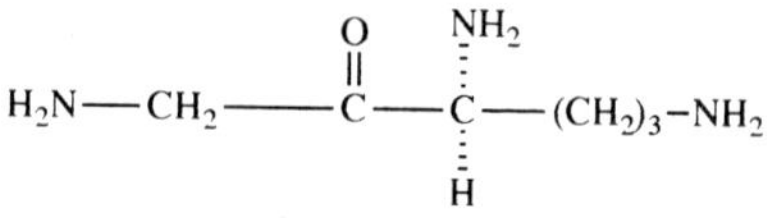

Glycyl-L-ornithine
(Sour, sweet)

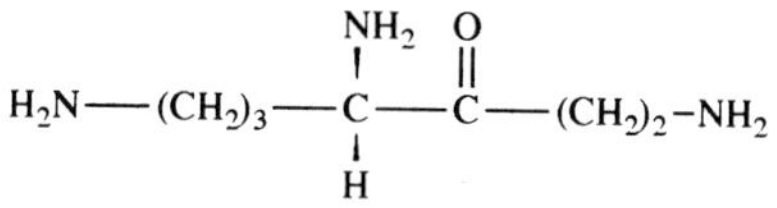

L-Ornithyl -β-alanine
(Salty, 0.8x; Umami, 0.2x)

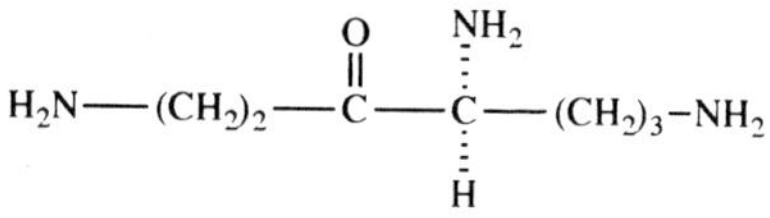

β-Alanyl-L-ornithine
(Sour, sweet)

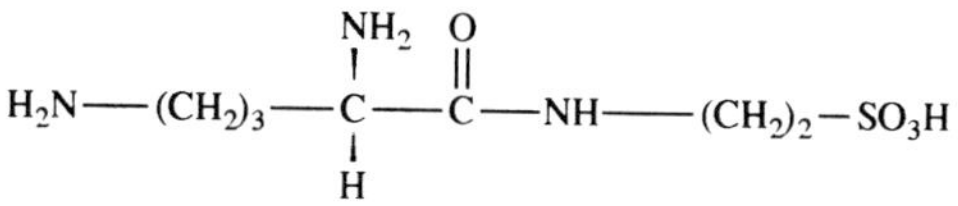

L-Ornithyltaurine (Salty, 1x)

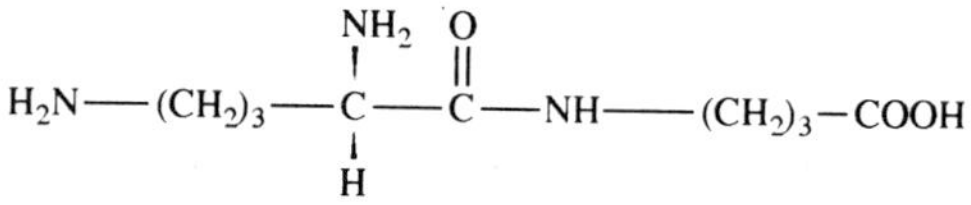

L-Ornithyl-γ-aminobutyric acid (Salty, 1x)

FIG. 7.20. Saltiness-umami taste on a molar basis of dipeptides containing basic-amino acids. × for salty = NaCl saltiness. × for umami = monosodiumglutamate taste (Tada *et al.*, 1984).

sufficient to provide a long-lasting effect. As the substance is inactivated below pH 3, and above pH 12, its action is conformation dependent.

7.5.2 Monellin

The berries of the West African plant *Dioscoroephyllum cumminsii* are intensely sweet. The active principle was initially believed to be a carbohydrate, with a molecular weight of about 10 000 (Inglett & May, 1968, 1969), but Morris & Cagan (1972) and van der Wel (1972) established that it is a protein that is devoid of carbohydrate components. The sweetness on a weight basis is about 1000–2250 times that of sucrose (cf. Higginbotham, 1979), and the molecular weight is about 10 700–11 069. On a molar basis the sweetness is about 100 000–130 000× (Ariyoshi *et al.*, 1991; Kim *et al.*, 1991).

As with the action of Miraculin, sweetness is pH dependent, and the substance is tasteless below pH 2 and above pH 9. The isoelectric point is 9·03 (van der Wel, 1972), and monellin is also a basic protein. It is made up of 94 amino acids arranged in two non-identical tasteless and non-covalently linked subunits (Bohak & Li, 1976). Denaturation leads to loss of sweetness. Kim *et al.* (1991) redesigned monellin, using genetic engineering techniques, and combined the two subunits into one molecule having equal sweetness with the natural substance. This improved thermal stability. Monellin has been crystallized, and the amino acid sequence of the subunits (Frank & Zuber, 1976), correspond to a molecular weight of 11 069. The amino acid sequence of the two subunits, as obtained by Frank & Zuber (1976), is shown in Fig. 7.21. They have

Chain A.

0 5 10 15
Phe-Arg-Glu-Ile-Lys-Gly-Tyr-Glu-Tyr-Gln-Leu-Tyr-Val-Ala-Ser-

20 25 30
Asp-Lys-Leu-Phe-Arg-Ala-Asp-Ile-Ser-Glu-Asp-Tyr-Lys-Thr-Arg-

35 40 44
Gly-Arg-Lys-Leu-Leu-Arg-Phe-Asn-Gly-Pro-Val-Pro-Pro-Pro-

Chain B.

5 10 15
Gly-Glu-Trp-Glu-Ile-Ile-Asp-Ile-Gly- Pro-Phe Thr-Gln-Asn-Leu-Gly-Lys

20 25 30
Phe-Ala-Val-Asp-Glu-Glu-Asn-Lys-Ile-Gly-Gln-Tyr-Gly-Arg-Leu-Thr-Phe

35 40 45 50
Asn-Lys-Val-Ile-Arg-Pro-Cys-Met-Lys-Lys-Thr-Ile-Tyr-Glu-Asn-Glu

FIG. 7.21. Amino acid sequence for the two sub-units of monellin (Frank & Zuber, 1976).

been synthesized by Ariyoshi *et al.* (1991). About 10% of the molecules are believed to possess an extra phenylalanine group at the N-terminal of the A chain. There are nine basic amino acids and six acidic amino acid residues in the A chain. Overall, it is a basic component. On the other hand, there are seven basic amino acids and eight acidic amino acid residues in the B chain, and it is an acidic unit. Collectively, there is an excess of basic over acidic residues, which accounts for the relatively high isoelectric point (Higginbotham, 1979).

7.5.3 Thaumatins

A second series of proteins with an intense sweet taste was isolated from the fruits of *Thaumatococcus danielli* by van der Wel & Loeve (1972). Thaumatins I, II, etc., have about the same amino acid composition, and a molecular weight of 22 000 (207 amino acid residues). The isoelectric point is between 11·5 and 12·5 (van der Wel & Ledeboer, 1989). Thus, they have a strong basic reaction, and like monellin, the tertiary structure is needed for sweet taste. On a weight basis, thaumatin is 5500 times as sweet as 0·6% sucrose. On a molar basis it is 100 000 times sweeter, or about the same as for monellin. As with other intensely sweet substances, the relative sweetness diminishes rapidly with increasing concentration, as shown in Fig. 7.22.

The complete amino acid sequence of thaumatin I was determined by Iyengar *et al.* (1979). The authors also established that the primary structure was a single polypeptide chain with eight intramolecular

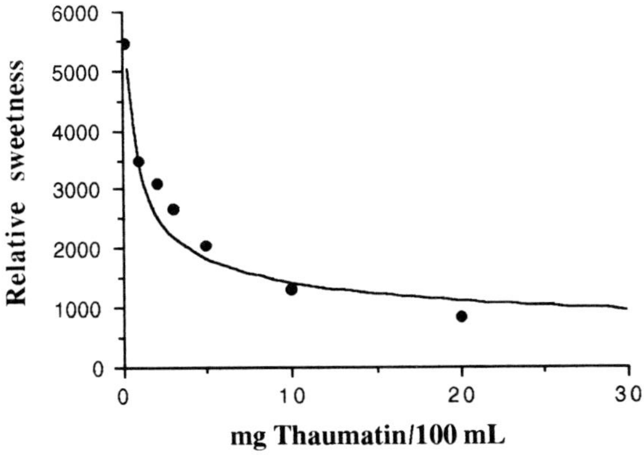

FIG. 7.22. Relative sweetness of thaumatin on a weight basis (Higginbotham, 1979).

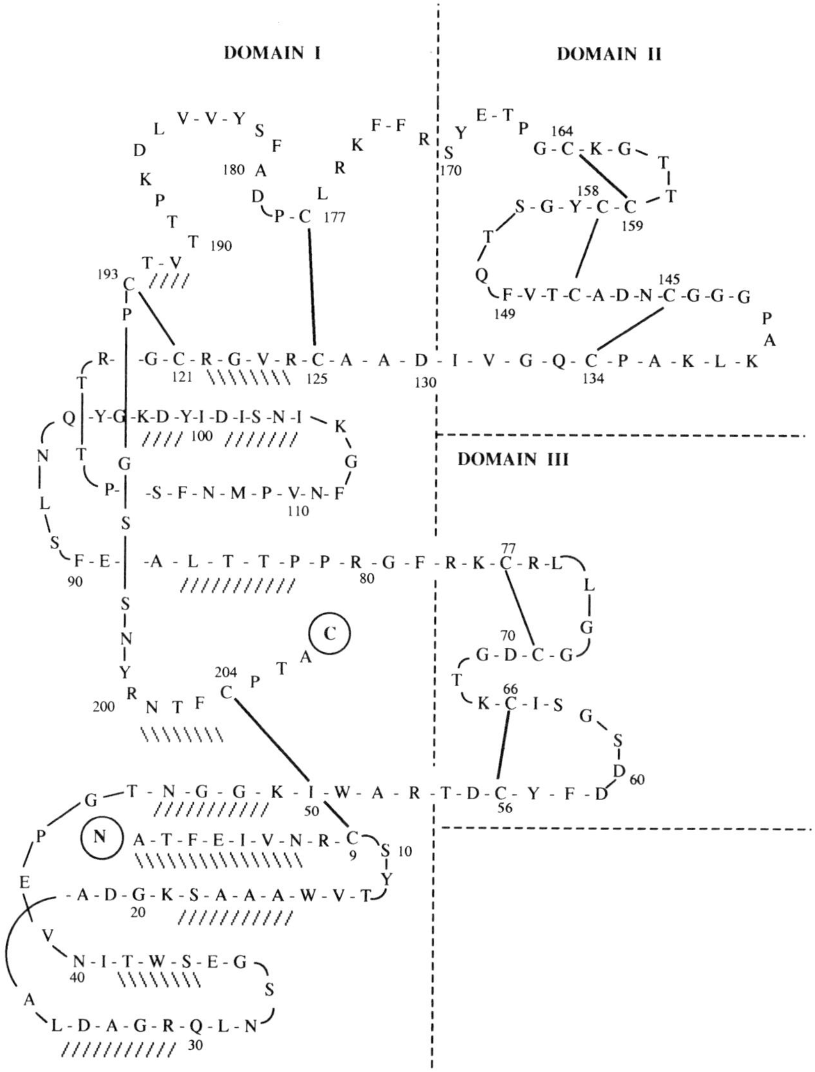

FIG. 7.23. Diagram of Thaumatin I. Heavy bars are the disulfide bonds. The β-strands of the top β-sheet are marked ///, and the bottom strands are marked \\\. N and C indicate the N- and C-termini (van der Wel & Ledeboer, 1989). The code to the amino acids (van der Wel *et al.*, 1984) is Asx, (D/N); Thr, T; Ser, S; Glx, (E/Q); Pro, P; Gly, G; Ala, A; Cys, C; val, V; Ile, I; Leu, L; Tyr, Y; Phe, F; Lys, K; Arg, R.

disulfide bonds. Circular dichroism studies (Edens & van der Wel, 1985) showed very few α-helices, but many β-pleated sheet strands and bends. An extremely flexible region of 28 sequential amino acids was suggested, on the basis of their homology with monellin, as the probable sweet site.

As indicated by X-ray analysis (DeVos *et al.*, 1985), the tertiary structure of thaumatin I has nine antiparallel strands that make up, by far, the largest domain, and it contains the flexible sequence. Because this sequence has antiparallel chains, it is possible that the antiparallel chains mimic the two subunits of monellin. Domain I is formed by a long β-sheet folded in such a way that the strands of one planar part are nearly parallel to, and positioned above those of the remainder of the sheet, as shown in Fig. 7.23.

When subjected to boiling in deionized water, thaumatins show very little loss of sweetness, even after several hours (Higginbotham, 1979). Acetylation of only four of the ten ϵ-amino groups of lysine reduced sweetness to zero. van der Wel & Bel (1976, 1980) believe that as acetylation causes only slight change in the stereochemical nature of the group substituted, but decreases the positive charge on the protein, it is the decrease in charge that led to the reduction in sweetness. When those groups were methylated, the dissociation constant of the ϵ-amino groups was not affected, and there was no change in the taste of either thaumatin or monellin (van der Wel, 1981).

By substituting the guanidino group of arginine by reaction with 1,2-cyclohexanedione to yield an N^7, N^8-(1,2-dihydroxy-cyclohex-1,2-xylene)-L-arginine derivative, no change in thaumatin sweetness potency occurred, even when six of the eleven arginine residues were so modified. As the net effect of this procedure is to lower net charge, it was concluded that some of the lysine residues are probably involved in the initiation of sweet taste. Upon iodination, which also lowers the isoelectric point, it was found that two iodine atoms can be introduced into two of the seven residues without changing sweetness intensity. Introduction of three iodine atoms destroyed the sweetness, as the tertiary structure of the molecule had been altered. In the case of monellin, iodination split the molecule into A and B subunits, with loss of taste.

CHAPTER 8

Organic Compounds

The purpose of this chapter is to present selections from the vast number of saporous organic chemical compounds, other than sugars, amino acids, peptides and proteins that have application to taste theory. Where possible, homologous series of substances, or extensive numbers of derivatives for a single compound, are given as these have more direct application to development of taste-structure theory. However, classification of the substances turned out to be difficult. Compounds listed under the heading 'Ureas,' for example, include some that might very well be considered to be nitroaniline (suosan) compounds, or even substituted amino acids. Most of these substances are taken from the original references, but others are from comprehensive compilations of substances and their taste attributes, such as that by van der Wel *et al.* (1987).

8.1 SWEET ORGANIC COMPOUNDS

8.1.1 Nitroaniline Compounds

The substituted 2-amino-4-nitrobenzene series of substances has been extensively used to derive mathematical models for sweetness-structure relationships. The first report on the sweetness of *meta*-nitroaniline derivatives was made by Blanksma & van der Weyden (1940). To establish relative sweetness scores, water solutions of the compounds were tasted, and sweetness compared to the sweetness of 1% aqueous solutions of sucrose. These procedures were also employed in later reports (Blanksma *et al.*, 1946; Blanksma & Hoegen, 1946), but the unpublished details are contained in a thesis by van der Weyden, Unversity of Leiden (1939).

Verkade *et al.* (1946) examined the relative sweetness of many of the same compounds, and obtained relative sweetness scores that are somewhat different. Part of the reason given for the discrepancy is that the taste procedure of van der Weyden was 'rather summary'. However, the Verkade taste procedures are also contained in an unpublished 1946 University of Delft thesis by C. P. van Dijk. Verkade & Witjens (1946) used 1, 0·5, and 0·25% solutions of sucrose as well as 0·02 and 0·01% solutions of *m*-nitroanaline as standards. Relative scores were derived from the ratio of the weights of substances that gave the same sweetness

TABLE 8.1

Relative sweetness (RS) scores on both a w/v and also on a molar basis for the substituted 2-amino-4-nitrobenzene series of compounds. From data of Blanksma & Hoegen (1946), Verkade *et al.* (1946) and van der Heijden *et al.* (1985*a*)

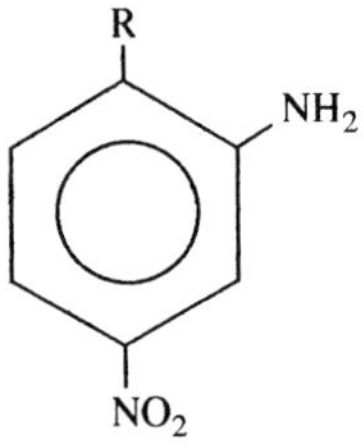

R=H=Nitroaniline

R	*Name*	*Blanksma*		*Verkade*		*van der Heijden*	
		w/v	*Molar*	*w/v*	*Molar*	*w/v*	*Molar*
H	Hydrogen	40	16	—	—	—	—
OMe	Methoxy-	330	162	220	108	167	—
						330	—
OEt	Ethoxy-	1 400	745	950	506	703	—
						1 400	—
O-nPr	*n*-Propoxy-	5 000	2 865	4 100	2 350	3 300	—
O-nBu	*n*-Butoxy-	—	128	1000	614	—	—
O-n-Pent	*n*-Pentoxy-	—	—	—	—	0	0
F	Fluoro-	40	18	—	—		
Cl	Chloro-	400	202	—	—		
Br	Bromo-	800	507	—	—		
I	Iodo-	1 250	965	—	—		
Me	Methyl-	330	147	—	—		
$OCH_2{=}CH_2$	Allyl oxy	—	—	2 000	1 058		
$OCH(CH_3)_2$	iso Propoxy-	—	—	600	344		
OH	Hydroxy-	200	90	120	54		

(Verkade *et al.*, 1946). Presumably, the ratios were obtained by diluting solutions of the test substances. Tabulation of the relative sweetness scores of both schools is given in Table 8.1, along with those determined by van der Heijden *et al.* (1985*a*).

8.1.2 Saccharine

Saccharine (2,3-dihydro-3-oxobenzisosulfazole) is a white crystalline substance formed from toluene and chlorosulfonic acid. Crystals of the acid form prevail as dimers formed by intermolecular hydrogen bonding between the imide hydrogen and the carbonyl (*keto*-) oxygen atoms. Saccharine is a moderately strong acid with a pK_a of 2·3. Of significance to sweet taste theory, it has been established (cf. Lee, 1987) that the imino and the SO bonds describe a 60° projected torsion angle. The salts (Na, K, Ag, Cu, Li) of saccharine are very sweet (Walter & Mitchell, 1986), although the free acid is reported to be a better taste stimulant than the sodium salt (Jakinovich, 1982). Moncrieff (1967) states that the pure acid is about twice as sweet as the commercial product, presumably sodium saccharate. However, the ammonium salt is reported to be even sweeter than saccharine (Moncrieff, 1967). The variation among the salts may be due to an auxiliary taste attribute of the cation. The sweet form of the substance(s) is purported to be the anion (Täufel & Wagner, 1925), or a hydrated form thereof. In addition to sweet taste, saccharine has a bitter (metallic) taste. Although long thought to be due to an impurity, the bitter taste has been conclusively shown to be intrinsic (Rader *et al.*, 1967).

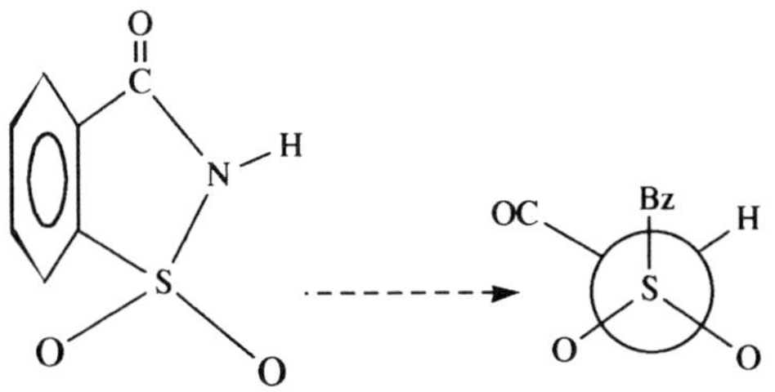

Saccharine conformation

The relative sweetness of saccharine varies with concentration, and is generally considered to be about 200–700 times as sweet as sucrose. A plot of the changing relative sweetness, with concentration, is shown in

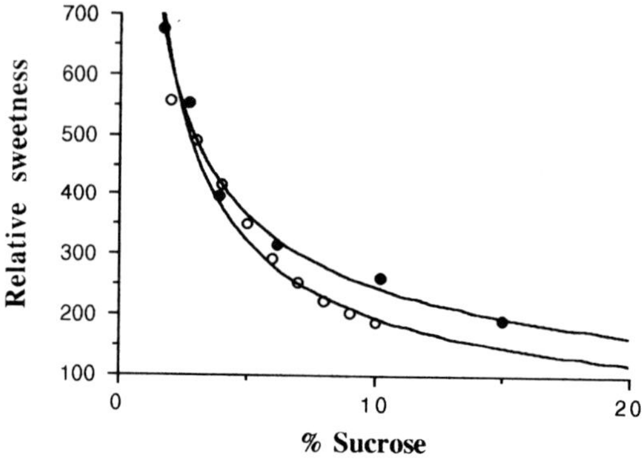

FIG. 8.1. Changing relative sweetness of saccharine with increasing concentration. Solid circles are data of Magidson & Gorbatschow (1923). Open circles are the data of Täufel & Klemm (1925).

TABLE 8.2
Effect of substitution at various positions on the sweetness of saccharine (Hamor, 1961)

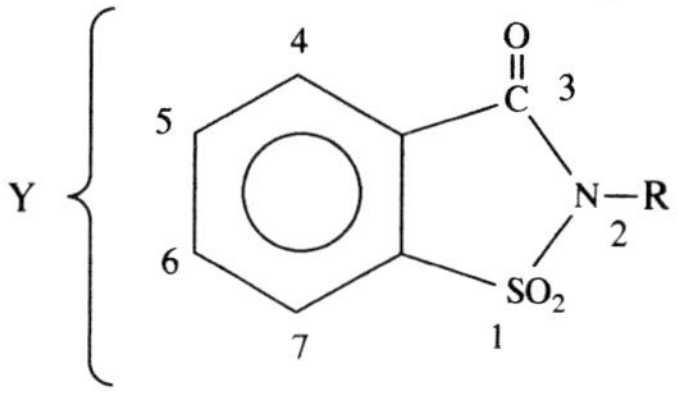

Y	*R*	*Taste*
7-NO_2	H	Slightly sweet, not bitter
6-NO_2	H	Moderately bitter, not sweet
5-NO_2	H	Slightly sweet, then very bitter
4-NO_2	H	Moderately bitter, not sweet
7-NH_2	H	Sweet
6-NH_2	H	Tasteless
H	CH_3	Tasteless
H	C_2H_5	Tasteless
6-NO_2	CH_3	Tasteless
7-NO_2	C_2H_5	Tasteless
6-NO_2	C_2H_5	Tasteless
5-NO_2	C_2H_5	Tasteless

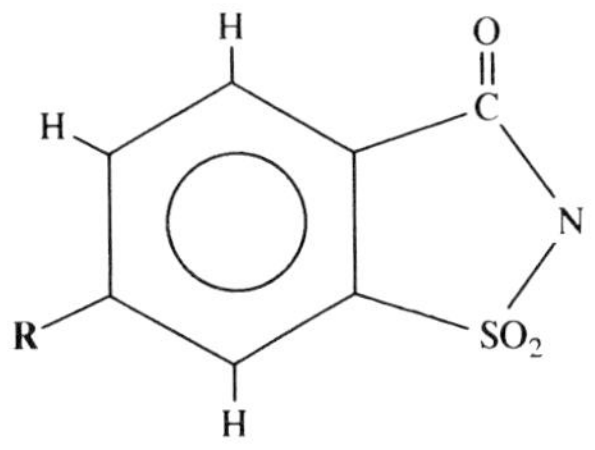

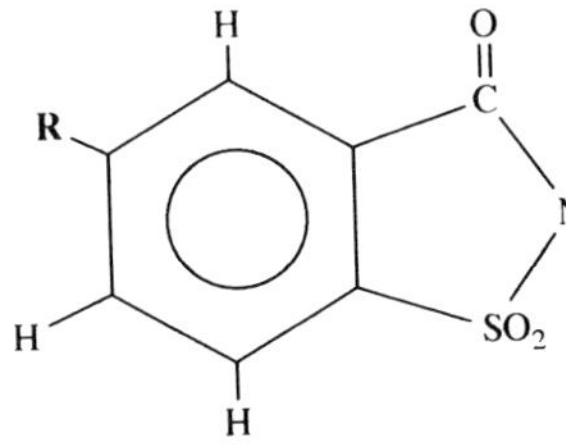

R = F, Sweet (450x).
R = Cl, Sweet, bitter.
R = Br, Very sweet, very bitter.
R = I, Slightly bitter only.

R = NH_2, Sweet.
R = OH, Sweet
R = OCH_3, Tasteless
R = OC_2H_5, Tasteless.

R = NO_2, Bitter.
R = Cl, Slightly sweet, bitter.
R = NH_2, Tasteless.
R = OCH_3, Bitter.

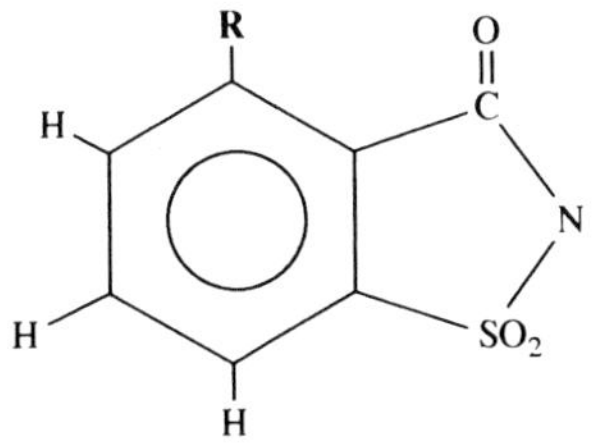

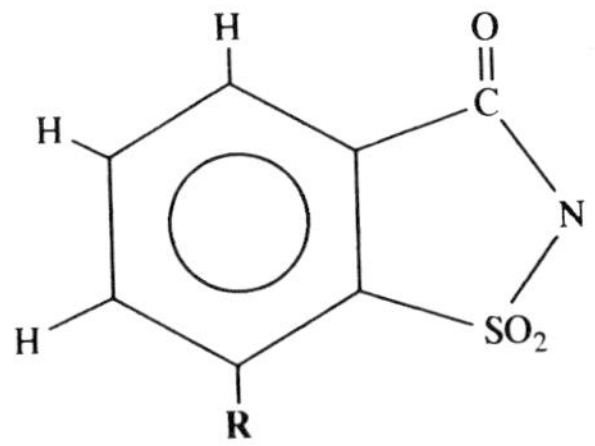

R = Cl, Sweet (225x).
R = NO_2, Slightly sweet.
R = NH_2, Sweet.

R = NO_2, Bitter.
R = OCH_3, Bitter.
R = Cl, Very sweet, bitter.

FIG. 8.2. Taste of saccharine derivatives.

Fig. 8.1 using the data of Magidson & Gorbatschow (1923), and the data of Täufel & Klemm (1925).

A number of 'saccharine derivatives' have been explored to determine their sweetness. Any substitution at the imino group destroys sweetness (Hamor, 1961). Therefore that group has a special role in determining the ability of saccharine to elicit sweetness. Complexing the acid anion with a metal obviously cannot be construed to be a substitution. Other substitutions examined by Hamor (1961) are given in Table 8.2, and have application to the role of inductive effects of substituents, especially those of the NO_2 group. Other substitutions on the benzene ring have varied effect on the taste of saccharine, and those observations are summarized in Fig. 8.2. Of particular interest is substitution on the benzene ring with the chlorine atom. There is an optimum position (six) for the sweetness of the chloro-derivatives. Among the halogeno-series, sweetness is lost when the substituent at that position is iodine. Also of significance, placing a methoxy group at position six leads to tastelessness. Changing the composition of the heterocyclic saccharine ring to two carbonyl groups results in tastelessness, while two sulfoxide

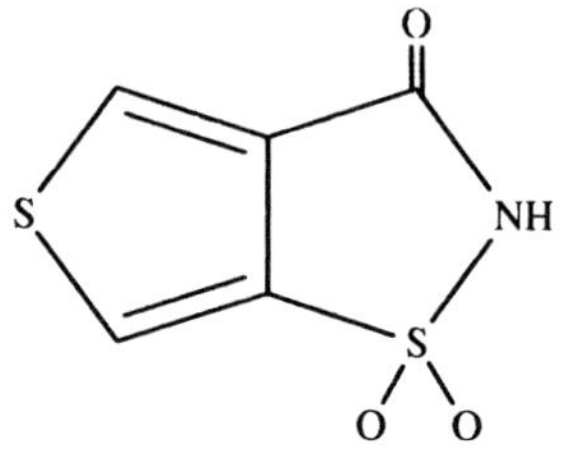

Thiophenesaccharine (1000x)

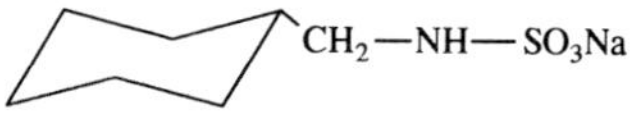

Sodium cyclamate (30x)

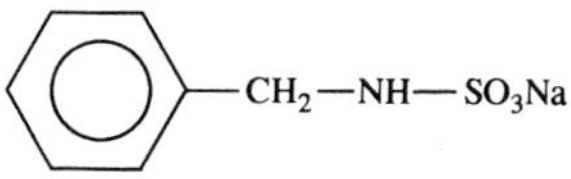

Sodium N-phenylsulfamate
(Sweetish after-taste)

$(C_6H_{11})—CH_2—NH—SO_3Na$

N-n-Hexylsulfamate (Not sweet)

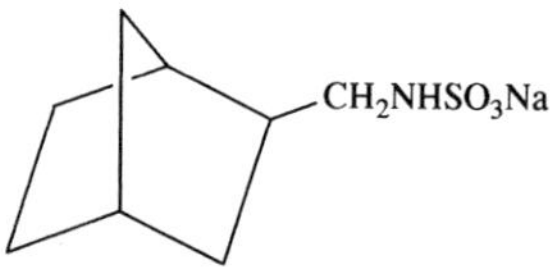

(-?)-Exo-8,9,10-Tribornylsulfamate
(Very sweet)

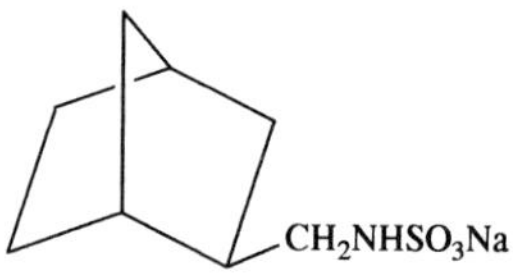

(+?)-Exo-8,9,10-Tribornylsulfamate
(Very sweet)

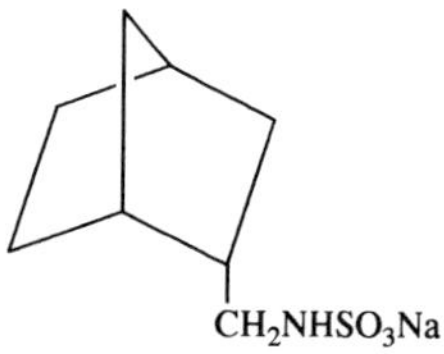

Endo-8,9,10-Tribornylsulfamate
(Either not or faintly sweet)

$Me_3C(CH_2)_2—CH_2NHSO_3Na$

3,3,3-Trimethylpropyl carbosulfamate
(Sweet)

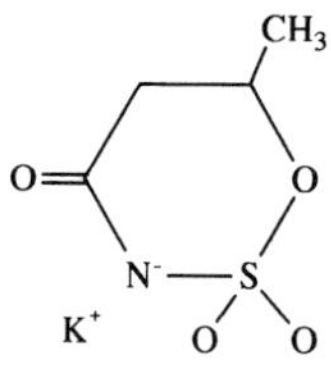

Acesulfame-K (200x)

FIG. 8.3. Sweet-tasting sulfamates.

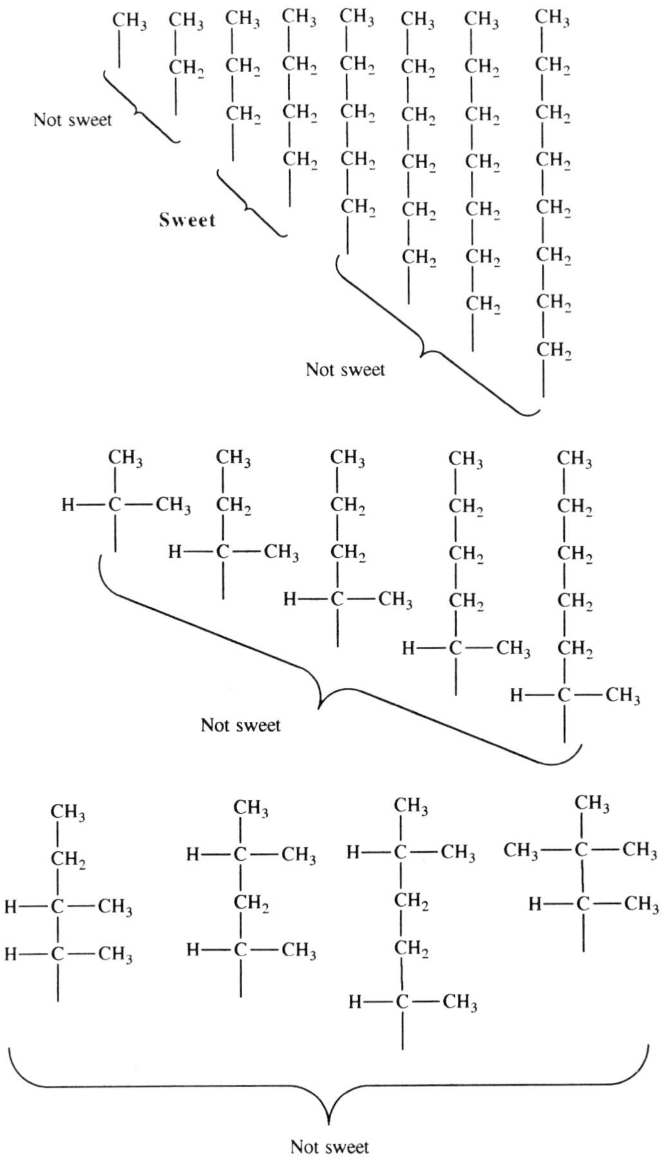

FIG. 8.4a. Effect of the structure and size of the side chain (R) on the sweetness of sulfamates ($R—NHSO_3Na$) (Pautet & Nofre, 1978).

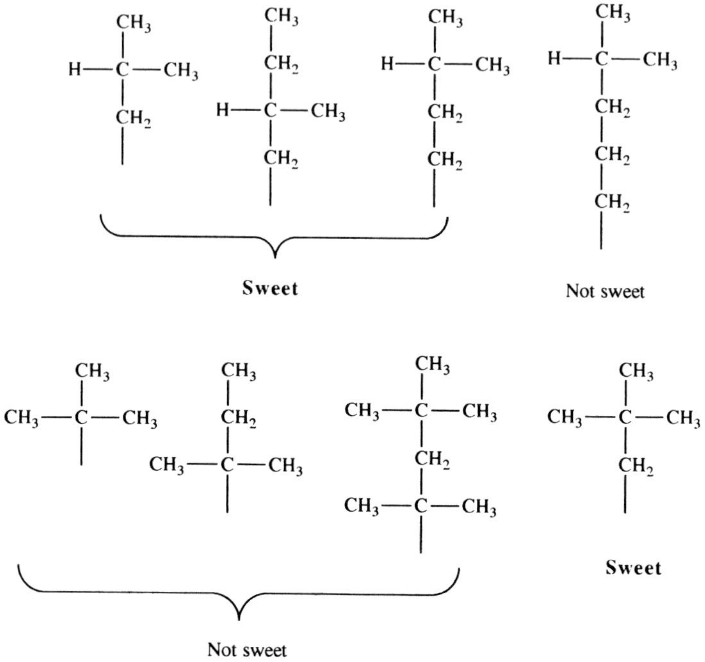

FIG. 8.4b. Effect of the structure and size of the side chain (R) on the sweetness of sulfamates (R—$NHSO_3Na$) (Pautet & Nofre, 1978).

groups retain the sweetness attribute, indicating a special role or function for the sulfonate group.

A more recent report on the taste of saccharine derivatives is directed toward the thiophene saccharines (Rossy *et al.*, 1980). Thiophenesaccharine [2,3-dihydro-3-oxothienol(2,3-*d*)isothiazole]1,1-dioxide is 1000 times sweeter than sucrose.

8.1.3 Sulfamates

Sulfamates contain the grouping —C—NH—SO_2—, and the first compound of this nature found to taste sweet was cyclohexylsulfamic acid (cyclamic acid). It is a crystalline substance with a melting point of 169–170°C. The neutral equivalent end point lies between pH 7·3 and 7·4, indicating that cyclohexylsulfamic acid is a fairly strong acid. The crystals taste lemon-sour sweet (Audrieth & Sveda, 1944). The sodium salt is reported to taste pleasantly sweet, and is about 30 times as sweet as sucrose.

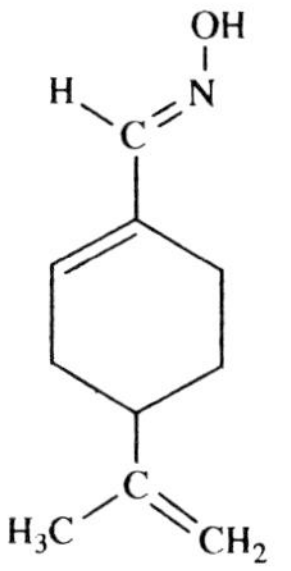

syn-Perillartine (2294/558)

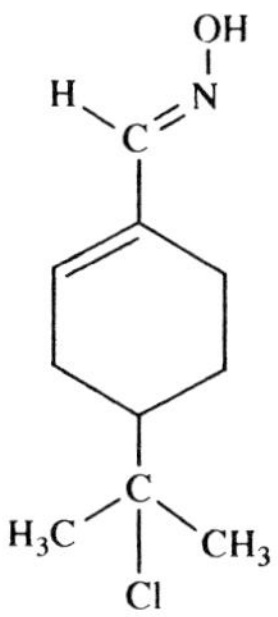

DL-4-(2-Chloro-2-propyl)-
1-carboxaldehyde (Tasteless)

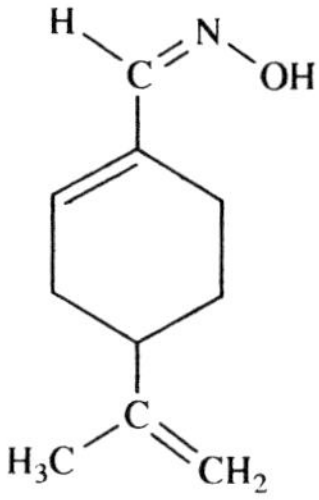

anti-Pellartine
(unobtainable)

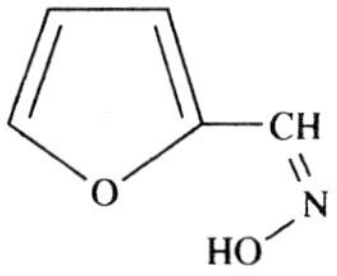

anti-Furfuraldehyde oxime
(Sweet)

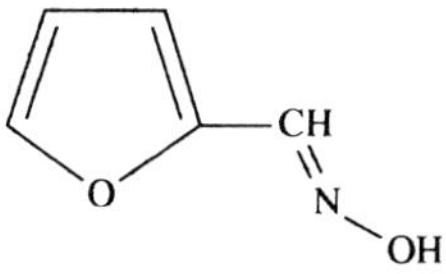

syn-Furfuraldehyde oxime
(Sweet)

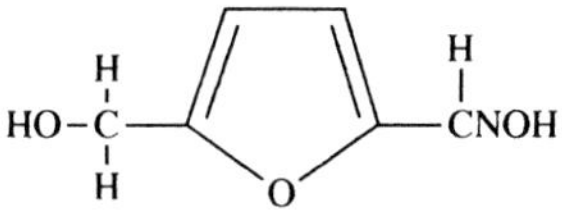

syn- or *anti*-hydroxymethyl
Furfuraldehyde oxime
(Tasteless)

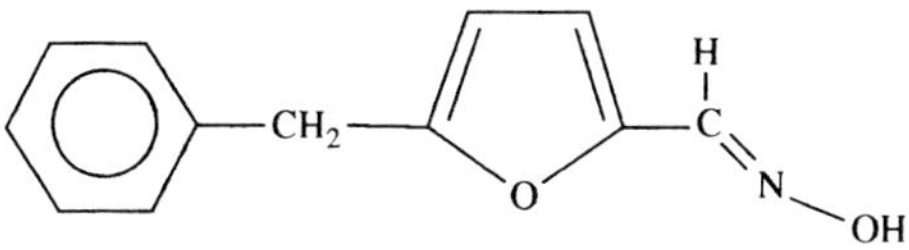

syn-5-Benzyl-2-Furfuraldehyde oxime
(690)

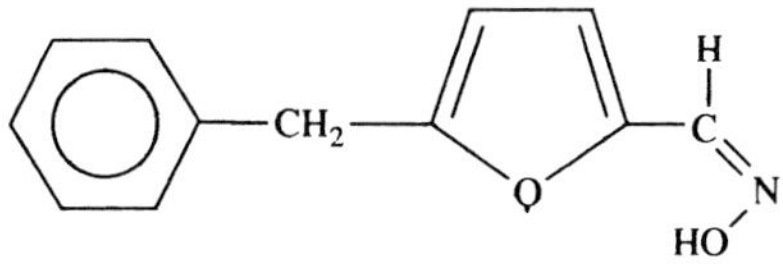

anti-5-Benzyl-2-Furfuraldehyde oxime
(100)

FIG. 8.5a. Structure and relative sweetness/bitterness of selected oximes. The sweetness/bitterness scores are relative to 0·5 M glucose and the single scores refer to sweetness (Acton *et al.*, 1970*a*).

The structure of sodium cyclamate, and several related compounds are shown in Fig. 8.3. The normal straight chain and benzene analogues are essentially tasteless. Also shown in Fig. 8.3 are representatives of a series of carbo- and heterosulfamate sweeteners studied by Spillane & Sheahan (1989). The ones shown are the majority of those reported to taste sweet. Two of the *exo*-tribornylsulfamates are stated to be enantiomers, with apparent equal sweetness, while the third (*endo*-) compound received mixed reports.

Acesulfame K (5,6-dimethyl-1,2,3-oxathiazin(3*H*)-one 2,2-dioxide) is a white crystalline powder without definite melting point. The free acid, however, melts sharply at 123°C. It is reported to be about 200 times as sweet as sucrose (Rymon Lipinski, 1986). Like all high-potency sweeteners, relative sweetness diminishes rapidly with increasing concentration. A series of derivatives is known, with varying sweetness, and their taste and structure are presented in Section 9.3.11.3.

A series of sulfamate side chains, compiled by Pautet & Nofre (1978), along with their taste attributes, is shown in Figs 8.4a and 8.4b. These are subsequently employed (Section 9.3.11.3) to show their application to the relation between structure and the size of R in the general sulfamate structure R—$NHSO_3Na$.

8.1.4 Oximes

Oximes are the product of the condensation of a carbonyl group with various amines. The condensation of acetaldehyde with hydroxyl amine yields acetaldoxime.

$$CH_3CH{=}O + H_2NOH \longrightarrow \underset{\text{Acetaldoxime}}{CH_3CH{=}NOH} + H_2O$$

According to Acton *et al.* (1970*a*) oximes have three common structural features. They are conjugated olefinic unsaturation, alkyl substitution (sometimes as part of a ring on the α-carbon atom), and *trans* orientation of the oxime olefinic bond with respect to the β-substituent. As a consequence, the oximes are *syn*-isomers with little or no tendency to isomerize to *anti*-isomers.

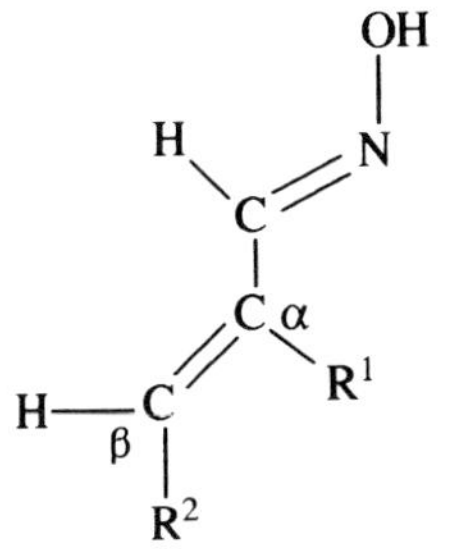

syn-Oxime

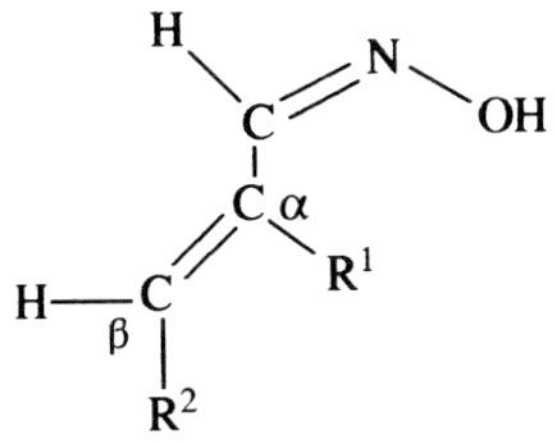

anti-Oxime

As a result of variable nomenclature, care needs to be exercised when comparing taste attributes assigned to oxime structures. The sweetening agent perillartine (perillaldehyde oxime) is the α-*syn*-oxime shown in Fig. 8.5a, but it is also commonly called 'α-*anti-oxime*' in commerce, and in the review literature (Acton *et al.*, 1970*b*). The latter term was assigned at a time when the structure of the oximes was uncertain. The correct *syn*-structure may be given along with the use of the name 'α-*anti*-,' but at other times the *anti*-structure is given. To add to the problem, it is claimed in the early literature that the β-*syn*-isomer of perillartine is tasteless, when it is an entirely different compound that is tasteless (the *tert*-chloride of the *syn*-form above). The comparison has been used (Ferguson & Lawrence, 1958; Kulka, 1967) to emphasize the different tastes of isomers, but the *anti*-form is not obtainable (Acton *et al.*, 1970*b*). The α-*anti*-perilla aldoxime stated by Moncrieff (1967) to be 2000 times as sweet as sugar is the *syn*-structure, and the α-*anti*-anisaldehyde oxime stated to be the sweet form also appears to be the *syn*-isomer.

In a study of the furfuraldoximes, Gilman & Dickey (1930) reported that the two oximes of furfural are equally sweet. While stating that *syn*-5-benzyl-2-furfuraldoxime is 690 times sweeter than sucrose, and the *anti*-form 100 times sweeter, they did not indicate *syn-anti*-structures. The authors recognized that, 'What is called the *syn*-form of some geometric isomers may in reality be the *trans*-form and *vice versa*.' Nevertheless, these are shown in Fig. 8.5a using the convention stated above. Of particular interest among the compounds in Fig. 8.5a, neither the *syn*-nor the *anti*-oxime of 'hydroxymethyl furfural' possesses sweet taste, although the parent compound, furfural, is sweet. Furfural also has a pleasant bread-like odor.

One consequence of the early confusion with oxime nomenclature is

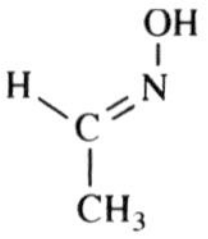

Acetaldehyde oxime (>0)

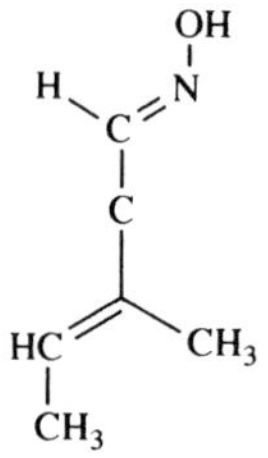

Tiglic aldehyde oxime (44/24)

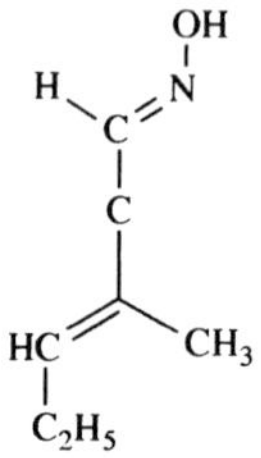

trans-2-Me-2-Pentenal oxime (36/48)

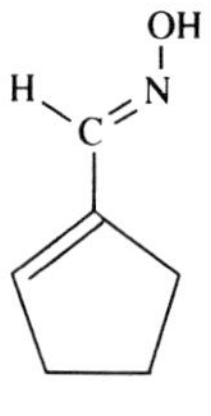

1-Cyclopentene -1- carb-
oxaldehyde oxime (70/30)

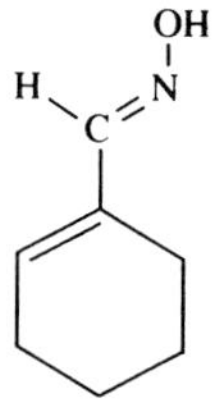

1-Cyclohexane-1-carb-
oxaldehyde oxime (68/38)

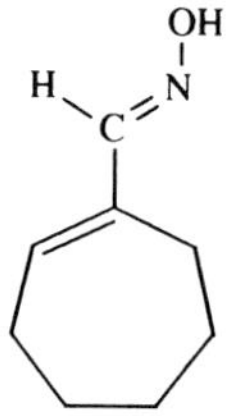

1-Cycloheptene-1-carb
oxaldehyde oxime (61/205)

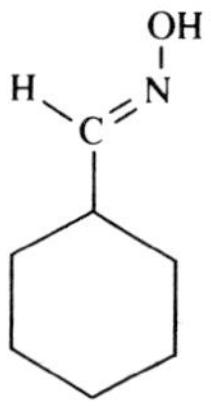

syn-Cyclohexane
carboxyldehyde (26/29)

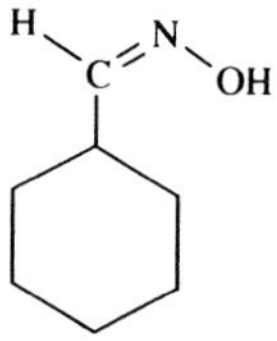

anti-Cyclohexane
carboxaldehyde (0/92)

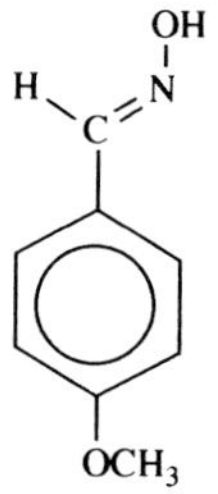

syn-*p*-anisaldehyde oxime (54/180)

FIG. 8.5b. Representative *syn*- and *anti*-oximes, along with sweetness/bitterness scores relative to a glucose standard (Acton *et al.*, 1970; van der Heijden *et al.*, 1985*a*).

that some authors select the wrong oxime structure to emphasize a specific point with respect to the taste of isomers, e.g., Shallenberger & Acree (1971), and Temussi *et al.* (1978). Acton *et al.* (1970*a*) report that it is the *syn*-isomer that is sweet/bitter (cf. Fig. 8.5b), and no data can be given for the *anti*-anisaldehyde oxime, as it is insoluble. Interestingly, in the original article on the taste of the anisaldehyde oximes, Goldschmidt

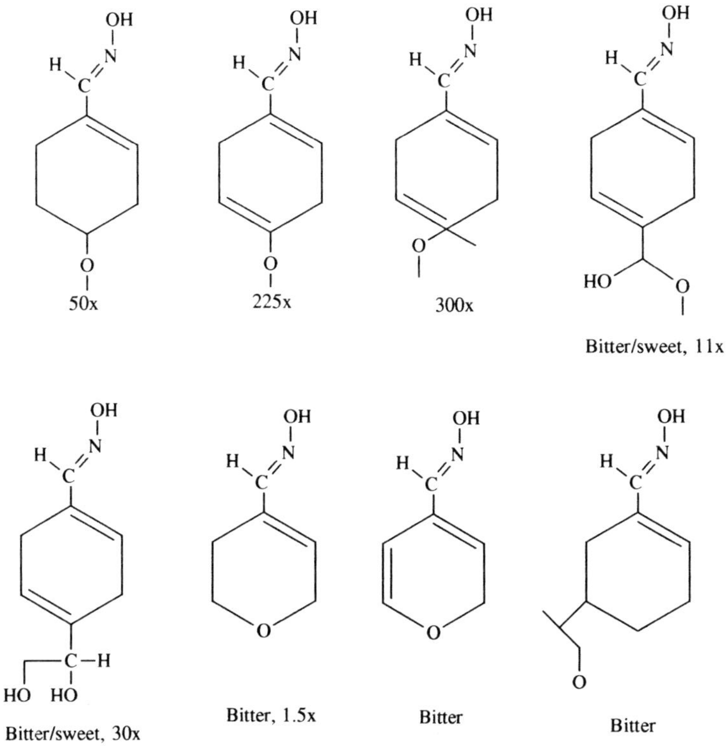

FIG. 8.5c. Taste of cyclohexylaldoximes (Kier, 1980). Bitterness intensities alone are relative to sucrose sweetness intensity.

(1890) indicated, by use of molecular structures, that the intensely sweet-tasting isoanisaldoxime that he had discovered is that structure currently described as *syn*-, which is correct.

A number of additional oxime structures are given in Fig. 8.5b using the nomenclature described above. The relative sweetness-bitterness scores given in Figs 8.5a,b are calculated from total relative taste intensity based on the sweetness of 0·5 M glucose solutions. As an example, pellartine (4-isopropyl-1-cyclohexene-1-carboxaldehyde) has a total taste intensity score of 3100. However, 71% of the total were attributed to sweetness, 18% to bitterness, and the remainder to sour, gingery, and minty notes. van der Wel *et al.* (1987) report that the compound designated SRI oxime, or 4-(methoxy-methyl)-cyclohexa-1,4-carboxyaldehyde synoxime, is 450 times as sweet as sucrose, and does not have the aftertaste of saccharine. If so desired, the sweetness scores relative to sucrose shown for substances in Figs 8.5a,b may be approximated by increasing the reported values by 33%. Finally, most of the substances shown in Figs 8.5a,b had odorous character of even greater significance than that reported for pellartine. The tastes of a number of additional cyclohexylaldoximes (Fig. 8.5c) were used by Kier (1980) to

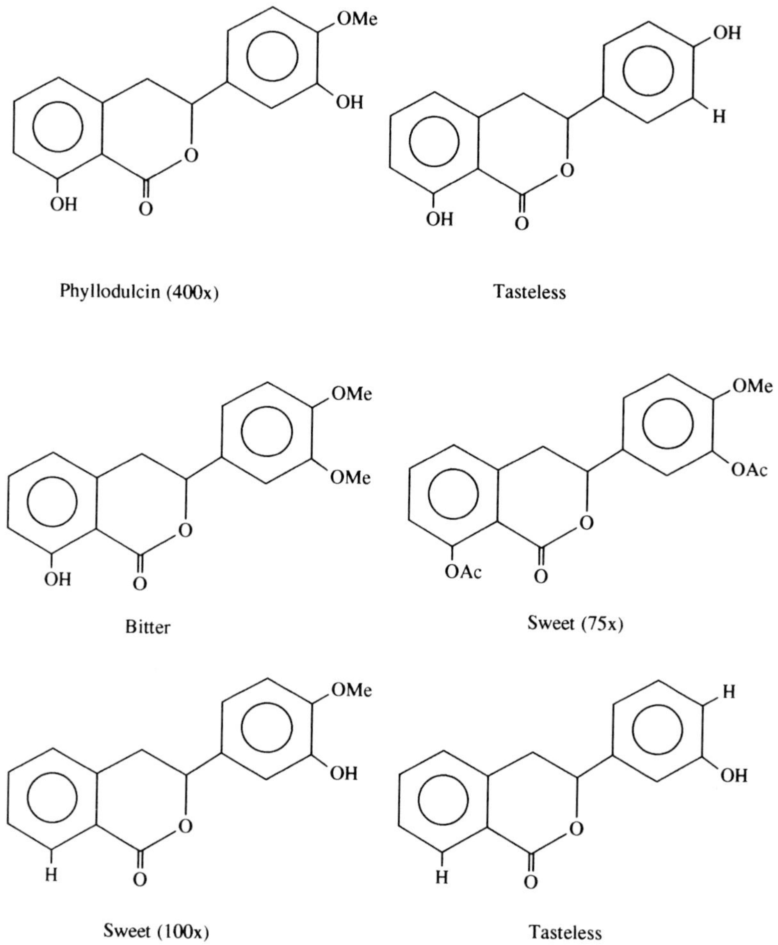

FIG. 8.6. Structure and taste of phyllodulcin and selected derivatives. Sweetness scores indicate the number of dilutions required to equal the sweetness of 3% sucrose (Yamato & Hashigaki, 1979).

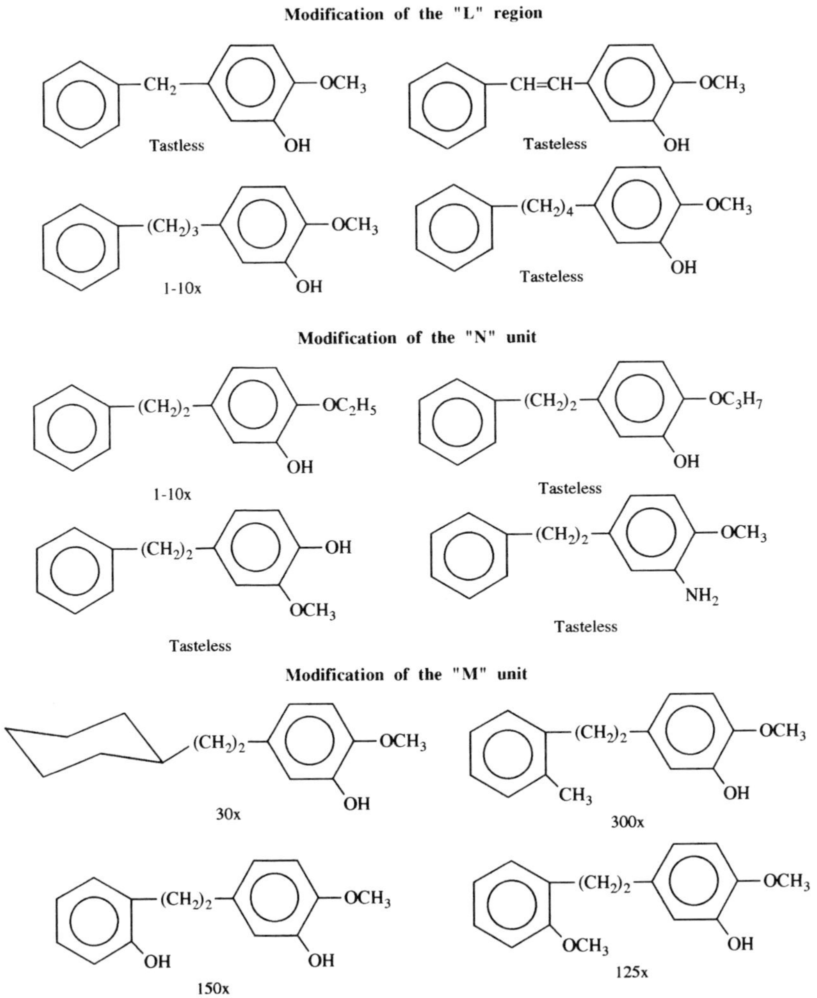

FIG. 8.7. Various isocoumarins and their sweetness relative to sucrose (weight basis) (Yamato & Hashigaki, 1979).

examine the source for the sweet and bitter taste variation of these compounds. These are discussed in Section 9.3.11.2.

8.1.5 Diphenyl Substances

In this section, a great number of interrelated substances (isocoumarins, flavanones, dihydrochalcones, etc.) are grouped under the diphenyl heading, even though some are not truly diphenyl substances.

8.1.5.1 *Isocoumarins*

A diphenyl compound, phyllodulcin, isolated from the leaves of *Hydrangea macrophylla seringe*, and with a structure that categorizes it as a derivative of 3,4-dihydroisocoumarin, is the only known naturally occurring isocoumarin derivative with a sweet taste. Coumarin (1,2-benzopyrone) has a fragrant vanilla-like odor, but has a burning taste. On a weight basis, van der Wel *et al.* (1987) report that phyllodulcin is 200–300 times as sweet as sucrose, but Yamato & Hashigaki (1979) give a value of 400 times the sweetness of 3% sucrose. Phyllodulcin is also reported to have an intense, clean sweetness that is both slow in onset, and lingers for some time (DuBois *et al.*, 1977*a*). Its structure, along with the taste of related substances, is given in Fig. 8.6. In a study directed toward the relation between the taste and structure of isocoumarins, Yamato and others (see Yamato & Hashigaki, 1979) studied many isocoumarine-like

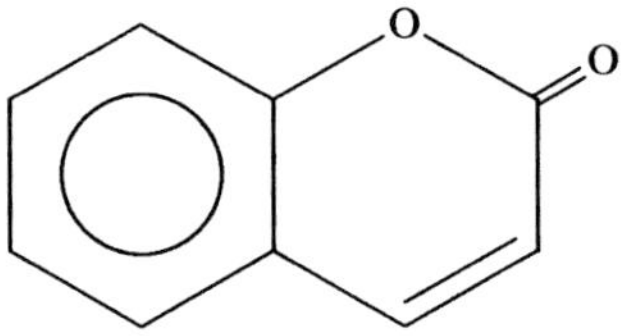

Coumarin

substances. The structure of the reference compound, with a sweetness score of 300×, was divided into L, M, and N sections. Substances with substitutions on the L, M, and N units were prepared to ascertain the effect on taste. Examples are given in Fig. 8.7.

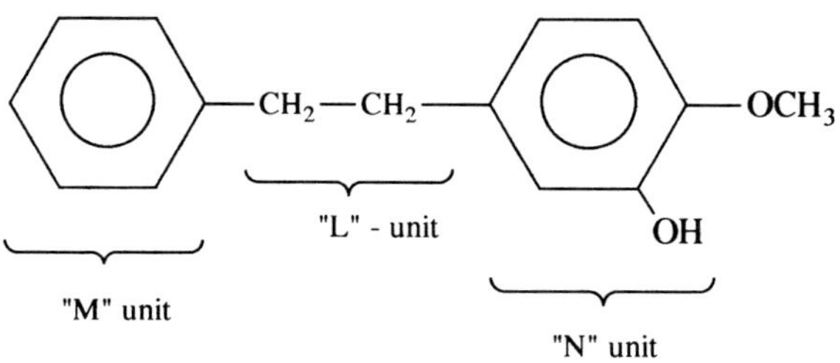

Reference isocoumarin structure

8.1.5.2 *Flavone Derivatives*

Flavones are wide-spread in nature, especially as the nucleus of many pigments (anthocyanins, etc.), as are compounds with the flavanone

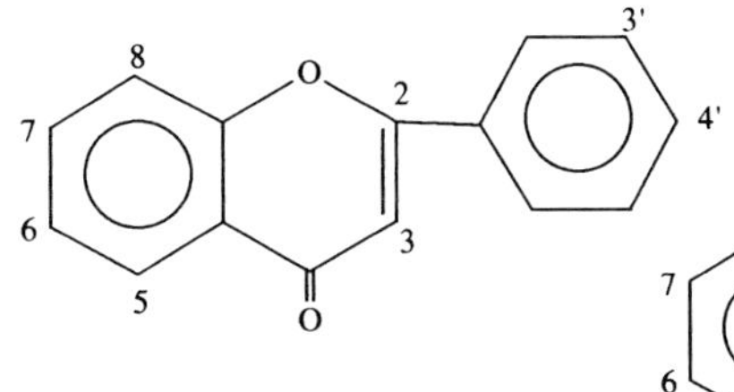

Flavone

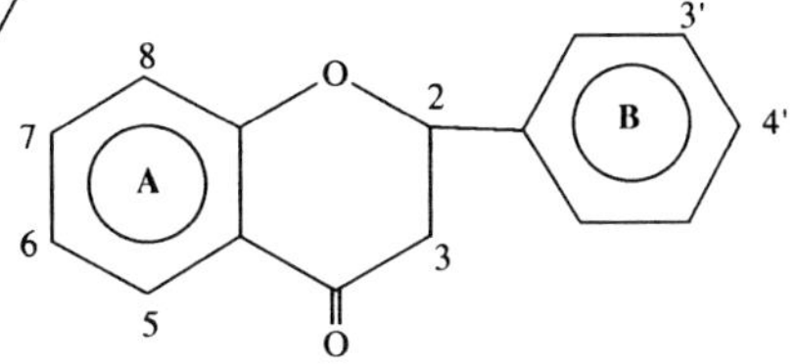

Flavanone

nucleus (usually glycosides). The structure of several flavanones, derivatives (flavanols), and their sweetness, is shown in Fig. 8.8. A number of relations between structure and taste are evident in the taste of the different flavanones. Apparently, the B ring must have an hydroxyl group at position 5, and transposing the OMe and OH groups on the B ring alters taste from sweet to tasteless and *vice versa*.

Among the naturally occurring dihydroflavanols, (+)-dihydroquercetin 3-acetate, from *Tessaria dodoneifolia*, is reported to taste about 80 times as sweet as sucrose (Kinghorn & Soejarto, 1991). When the 4′ methyl ether derivative was prepared, it was found to be 400 times as sweet as sucrose. For comparison with the compounds in Fig. 8.8, both compounds have the R configuration at positions 2 and 3.

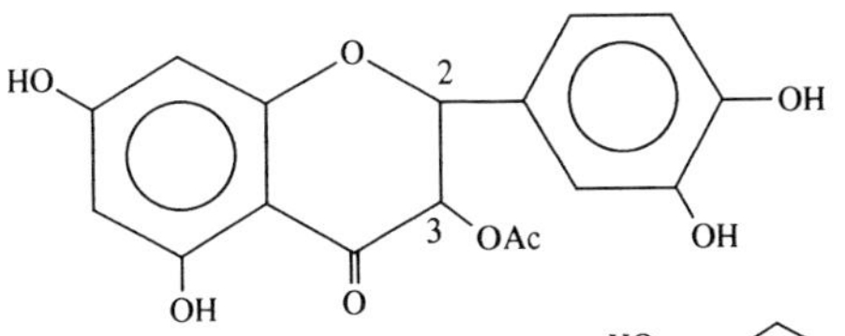

(+) Dihydroquercetin 3-acetate (80x)

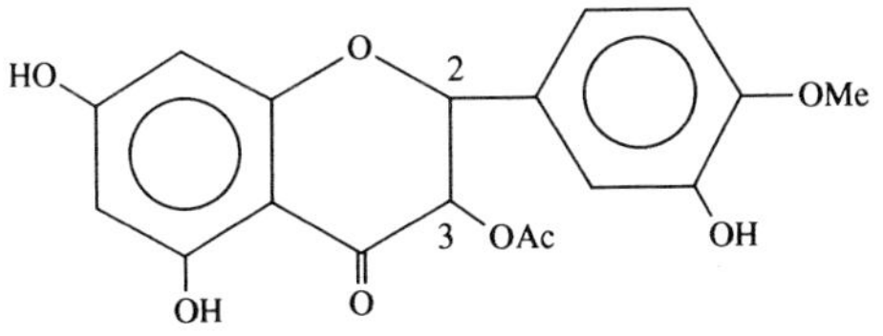

(+) Dihydroquercetin 3-acetate 4′ methyl ether (400x)

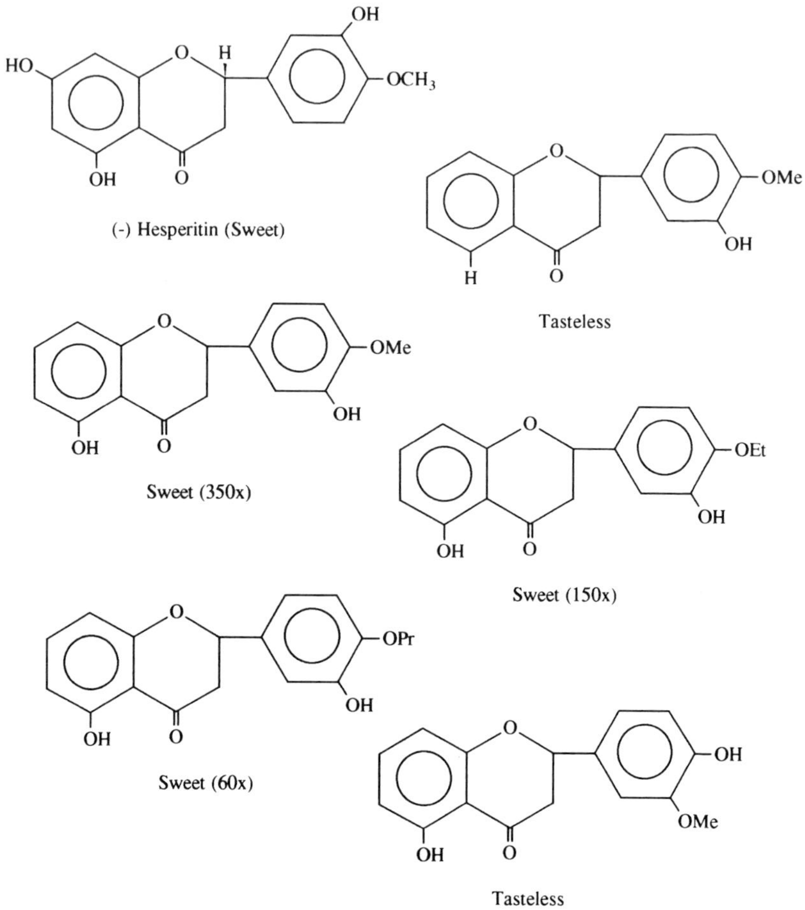

FIG. 8.8. Taste of selected flavanones (Yamato & Hashigaki, 1979).

Neoastilbin, a sweet dihydroflavanol from *Englelhardtia chrysolepis* differs from (+)-dihydroquercitin 3-acetate in that the acetate group at position 3 is replaced by an α-L-rhamnosyl group. In this case, it is the 2S,3S-diastereoisomer that is sweet, while the 2R,3R compound (astilbin) is not at all sweet. Dick (1981) established that the favored conformation for flavans, flavanones, and phyllodulcin (Fig. 8.6) is semiplanar.

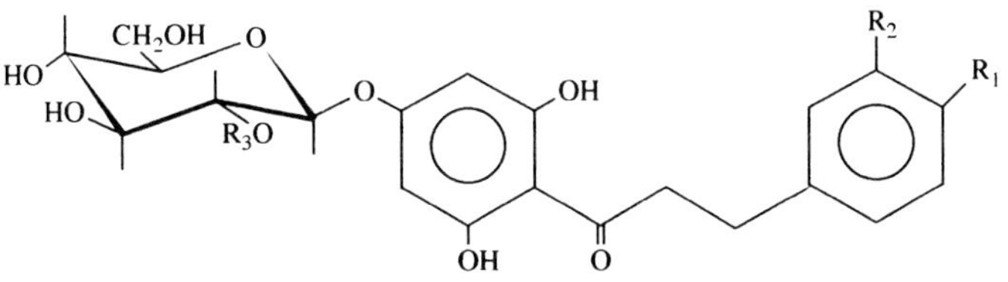

Naringin DHC, R_1 =OH; R_2 = H; R_3 = α-L-Rhamnosyl

Neohesperidin DHC, R_1 = OCH_3; R_2 = OH; R_3 = α-L-Rhamnosyl

Hesperetin DHC, R_1 = OCH_3; R_2 = OH; R_3 = H

(α-L-Rhamnosyl =

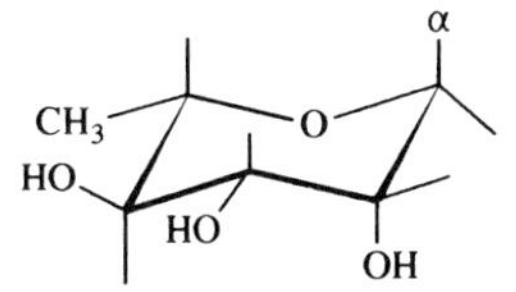

)

FIG. 8.9. Structure of three dihydrochalcones to emphasize the glycoside group(s) (Horowitz & Gentili, 1986).

8.1.5.3 *Dihydrochalcones*

The intensely sweet dihydrochalcones are prepared from naturally occurring, but bitter-tasting, flavanones, from citrus fruit (Krbechek *et al.*, 1968; Horowitz & Gentili, 1969; Horowitz & Gentili, 1986). The compounds have been extensively studied, and results obtained (Horowitz & Gentili, 1969, 1971; Guadagni *et al.*, 1974; DuBois *et al.*, 1977*a*, *b*) add significantly to the knowledge of taste chemistry. The flavanones may or may not be glycosides, and the dihydrochalcones prepared from them may or may not be glycosides. Three of the original dihydrochalcones, which have been intensely studied, are glycosides prepared from naringin, neohesperidin, and hesperidin, respectively, and are shown in Fig. 8.9. To emphasize the true size of the molecule for neohesperidin, it is redrawn (Fig. 8.10) to show the full structure of the neohesperidosyl sugar group. The terminally substituted benzene ring group has also been rotated about the connecting carbon–carbon single covalent bonds to bring the entire molecule into a planar depiction.

The taste properties of the dihydrochalcones are particularly interesting (Guadagni *et al.*, 1974). Although their relative sweetness

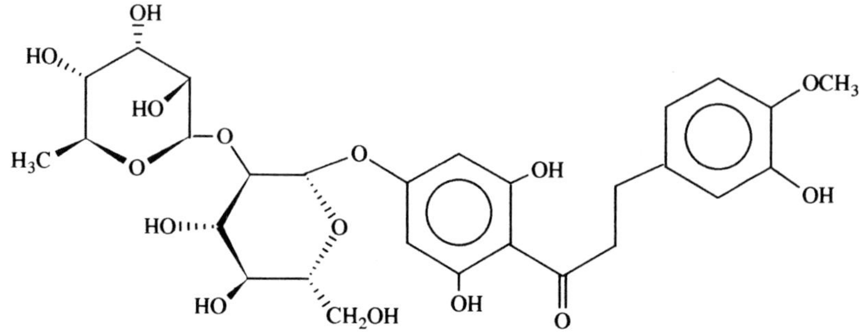

Neohesperidin dihydrochalcone

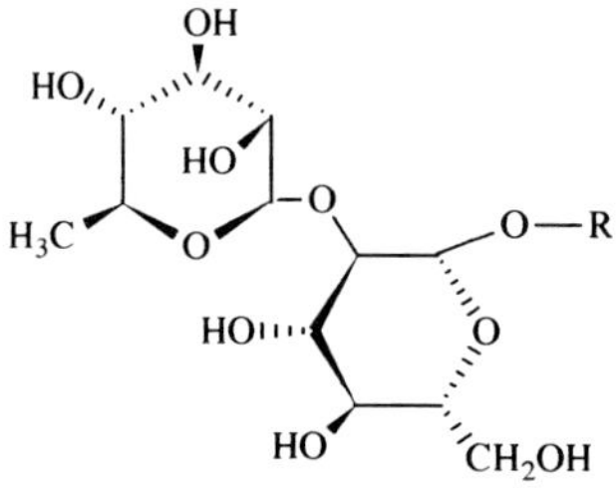

Neohesperidose
(2-*O*-α-D-Glucopyranosyl-β-D-glucopyranoside)
(Sweet)

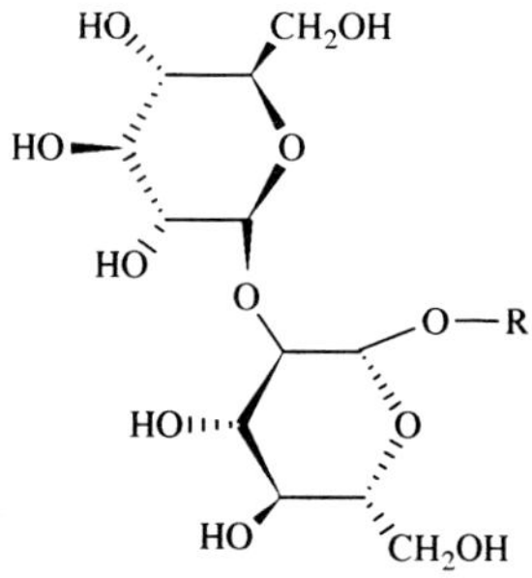

2-*O*-β-D-Glucopyranosyl-β-D-glucopyranoside
(Tasteless)

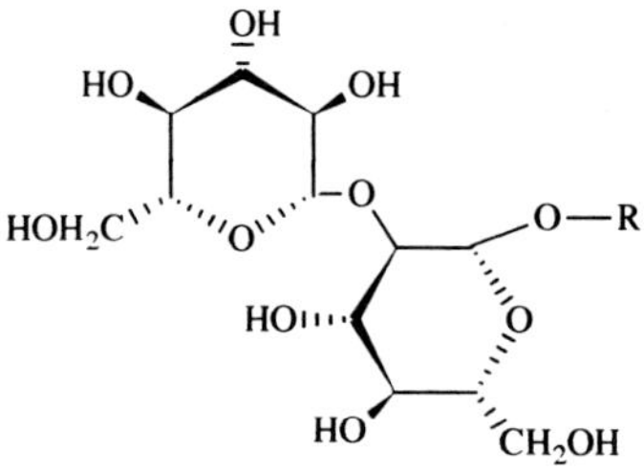

2-*O*-β-D-Glucopyranosyl-β-D-glucopyranoside
(Tasteless)

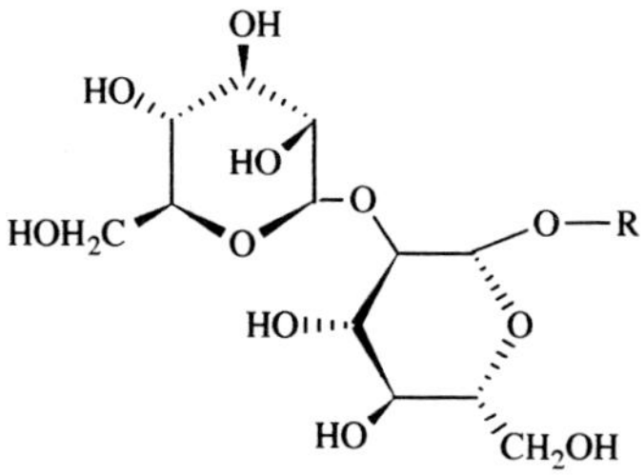

2-*O*-α-L-Mannoopyranosyl-β-D-glucopyranoside
(Sweet)

FIG. 8.10. Effect of varying the composition of the glycoside on the sweetness of neohesperidin dihydrochalcone (Esaki *et al.*, 1975; Kamiya *et al.*, 1976*a*, *b*).

diminishes rapidly with increasing concentration, at the sweetness level of 5% sucrose, neohesperidin DHC is about 250 times as sweet as sucrose. At threshold for both compounds, it is 1875 times as sweet. Values for naringin DHC are 78 times and 289 times, respectively, and those for hespertin DHC are 78 times and 289 times, respectively. The sweet dihydrochalcones also have a menthol-like aftertaste (Krbechek *et al.*, 1968).

The onset time of detection of neohesperidin sweetness is slower than for sucrose. In addition, the taste persists about three times as long (DuBois *et al.*, 1981; DuBois & Lee, 1983), or about 49 s for neohesperidin DHC vs about 19 s for sucrose. While the substances have an aftertaste, they do not have a bitter aftertaste, and distilled water acquires a sweet taste (cf. Section 3.1.4) after tasting a dihydrochalcone

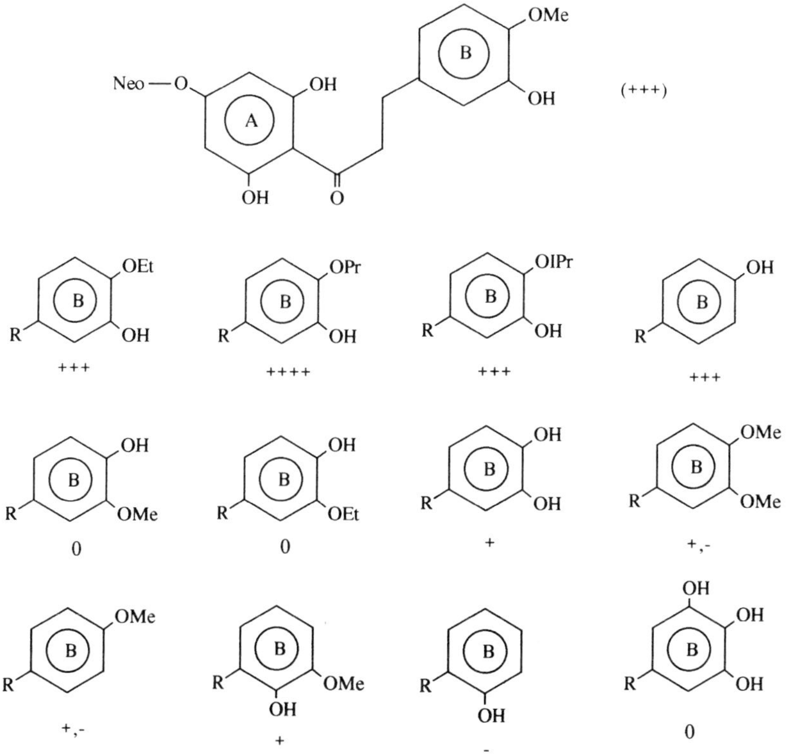

FIG. 8.11. Effect of substitution on ring B on the taste of neohesperidin dihydrochalcone (Krbechek *et al.*, 1968; Horowitz & Gentili, 1971).

(Horowitz & Gentilli, 1986). The aftertaste has been described as licorice-like. A final unusual aspect to the taste of the dihydrochalcones is that the sweet taste is perceived on the roof of the oral cavity, as well as at the back of the tongue (Crosby *et al.*, 1979).

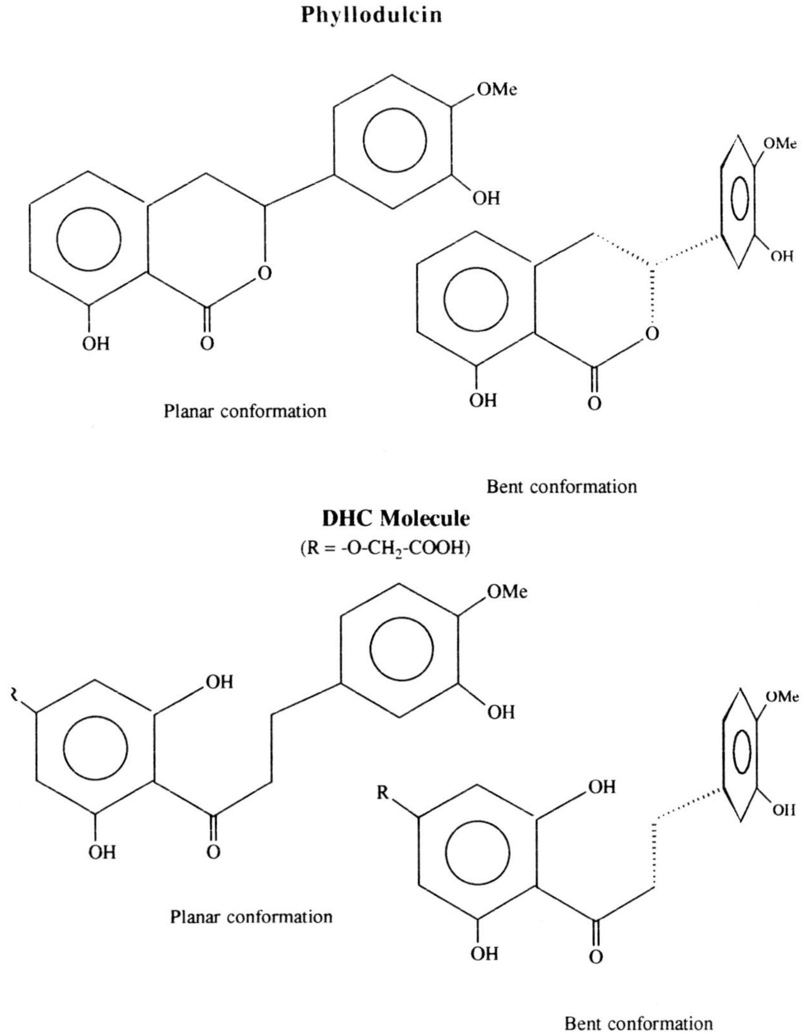

FIG. 8.12. Conformations of phyllodulcin and 4-carboxymethyl hesperitin dihydrochalcone (DuBois *et al.*, 1977*b*).

Variation in composition of the sugars effects the sweetness of neohesperidin dihydrochalcone. Several of the structures investigated by Esaki *et al.* (1975) and Kamiya *et al.* (1976*a*, *b*) are shown in Fig. 8.10. The sugars are positioned to occupy the same plane as that for the aglycone (R).

Substitution on the B ring has marked effect on the sweetness of neohesperidin dihydrochalcone, and the observations of Krbechek *et al.* (1968) and Horowitz & Gentili (1971) are summarized in Fig. 8.11. To deduce the probable conformation of the dihydrochalcones that elicit sweetness, DuBois *et al.* (1977*b*) examined phyllodulcin and the —CH_2CO_2Na salt of hesperitin DHC, as the compounds have similar sweet taste attributes (slow in onset, lingering). They proposed that phyllodulcin exists in a conformational equilibrium between a planar form, where the B ring is *pseudo*-equatorial, and a bent form, where the B-ring is *pseudo*-axial. Although both conformations (Fig. 8.12) appeared to be relatively stable (Crosby *et al.*, 1979), the bent form was selected for both compounds. In contrast, Dick (1981) chose the planar conformation.

8.1.5.4 *Isovanillyl Compounds*

A final group of substances related to phyllodulcin and the dihydrochalcones are viewed as being isovanillyl substances (Arnoldi *et al.*, 1991). In this context, the parent compound, 4-hydroxy-3-methoxybenzaldehyde, has a pleasant aromatic odor and taste, but the properties of isovanillin (3-hydroxy-4-methoxybenzaldehyde) are not recorded. Directing their attention to the B ring of substances with the general isovanillyl structure shown in Fig. 8.13, Arnoldi *et al.* (1986) observed that the 1,4-benzodioxan derivative was 450 times as sweet as sucrose. Other substances, as shown in Fig. 8.13 had varying tastes. To elicit taste, Arnoldi *et al.* (1991) determined that the B ring of the general isovanillyl formula must not be planar. This was also observed for dihydroisocoumarins and flavones. It was further determined that sweetness is optimum when the distance from the point of attachment of the isovanillyl group to the B ring has an angle of 30·7°C and a distance of 3·62 Å.

In a subsequent contribution, Arnoldi *et al.* (1993) introduced sulfur atoms in place of the B ring oxygen atoms of selected isovanillyl compounds. In all cases, sweetness was either maintained, or significantly enhanced. For the 1-S analogue of the 1,3-benzodioxan isovanillyl compound (3000×), sweetness was increased to 9000 times. For the 1-S,3-S benzodioxin isovanillyl compound, sweetness was increased to 5000 times.

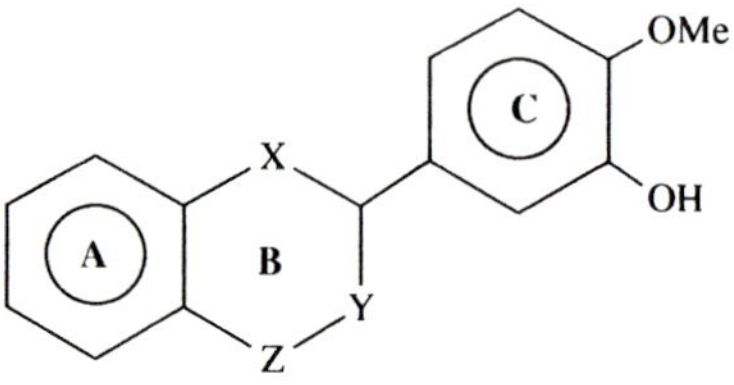

General formula for isovanillyl substances

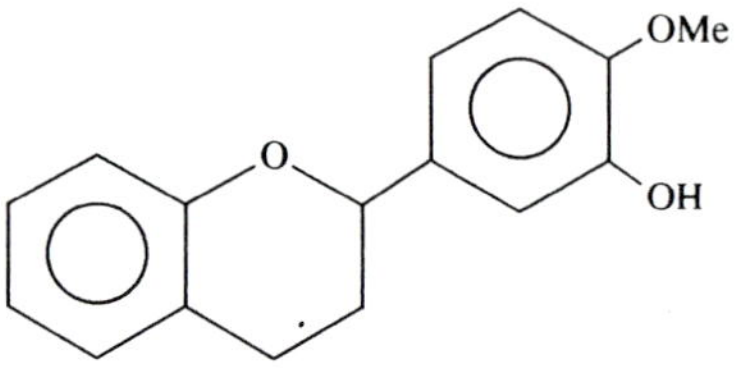

Benzopyran derivative (350x)

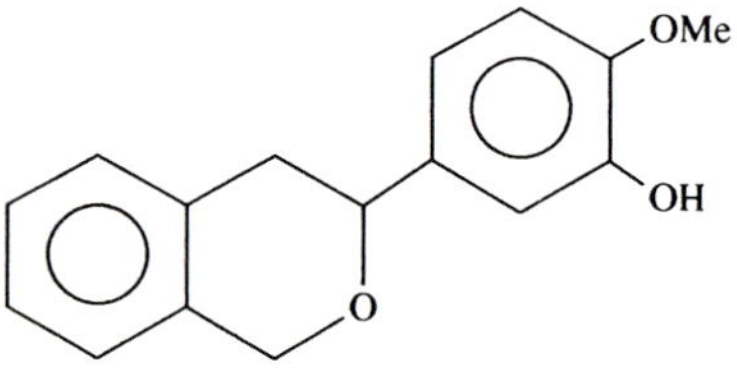

Benzopyran derivative (350x)

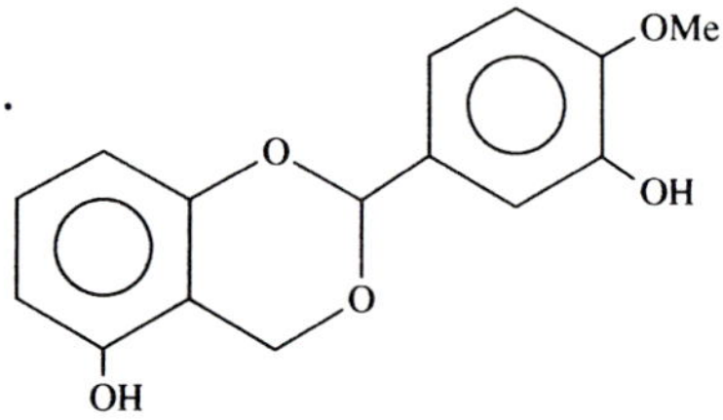

1,3-Benzodioxan derivative (3000x)

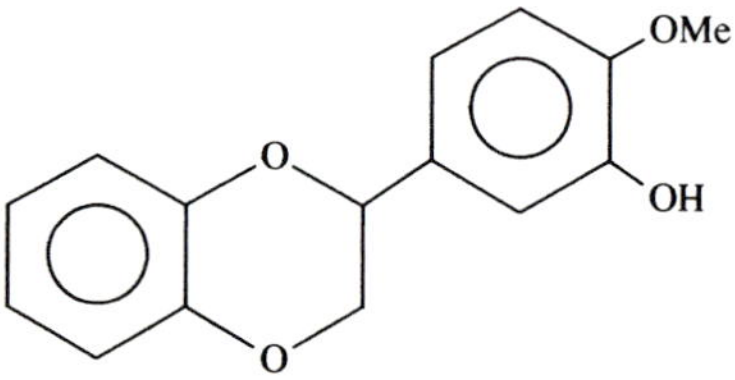

1,4-Benzodioxan derivative (450x)

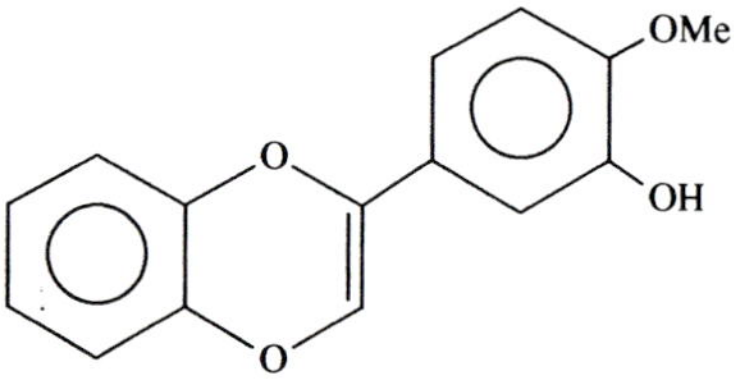

1,4-Benzodioxin derivative (Tasteless)

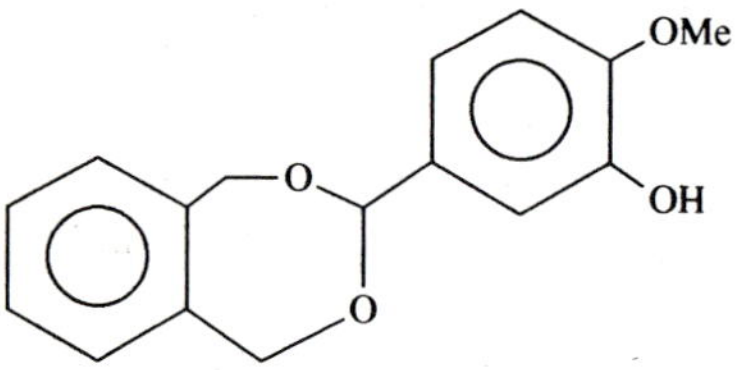

2,4-Benzodioxepine derivative (Tasteless)

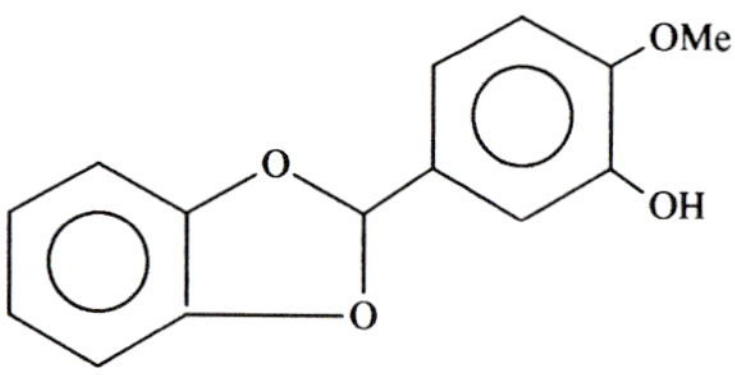

1,3-Benzodioxole derivative (150x)

FIG. 8.13. Structure and taste of isovanillyl substances (Arnoldi *et al.*, 1991).

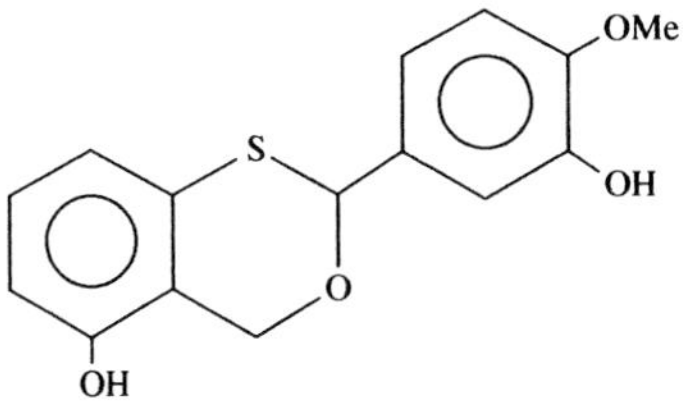

9000x

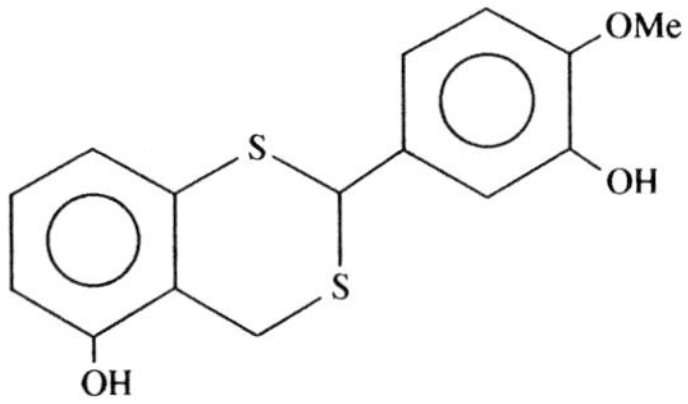

5000x

Sulfur analogues of the 1,3 benzodioxin isovanillyl substances

8.1.6 Sesquiterpenes

In a search of the Mexican ethnobotanical literature (cf. Kinghorn & Soejarto, 1991), a passage in a 1570–1576 monograph entitled 'Natural History of New Spain,' described a plant found by Aztecs to have a sweet taste. It was subsequently identified as *Lippia dulcis* Trev., and a sesquiterpene sweetener (hernandulcin), isolated from it. Upon synthesis, a mixture of hernandulcin and the epimer (epihernandulcin) was obtained. The natural (+)-isomer was judged to be about 1000 times as sweet as sucrose (molar basis), while the epimer was devoid of sweetness. It was also established (Compadre *et al.*, 1988) that the C-4′–C-5′ double bond of hernandulcin is necessary for sweetness.

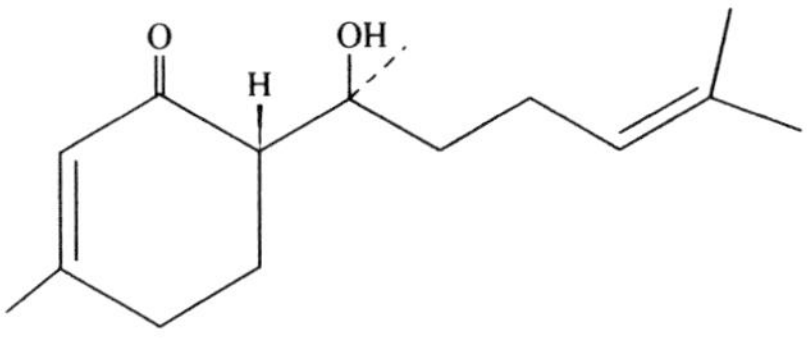

(+) 6S, 1'S- Hernandulcin (2000x)

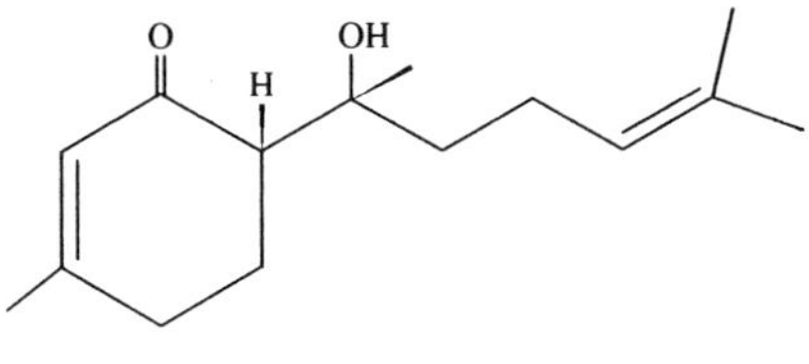

(+) 6S, 1'R- Epihernandulcin (Not sweet)

8.1.7 Ureas

A number of compounds that have a urea grouping taste sweet. They can also be considered to be derivatives of nitroanilines and amino acid and/or peptide compounds. Generally, if a compound has a urea grouping it is placed under this heading. Unlike the amino acid group, a urea group is uncharged and nonionizable. Crystalline urea itself has a cool, bitter taste.

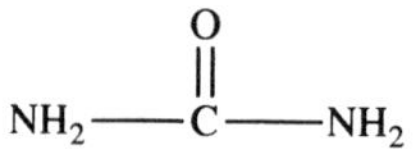

Urea (bitter)

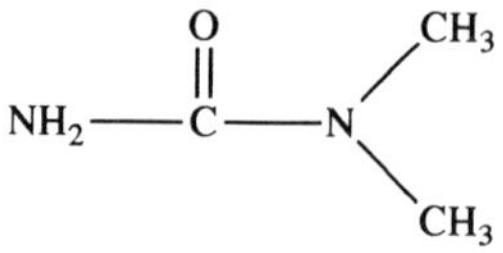

α-Dimethyl urea
(sweet)

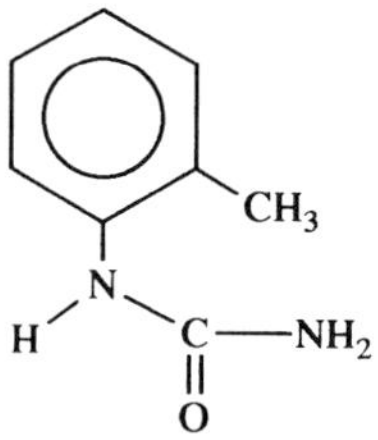

ortho-Tolylurea (tasteless)

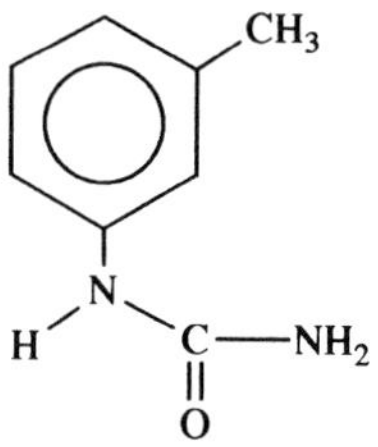

meta-Tolylurea (bitter)

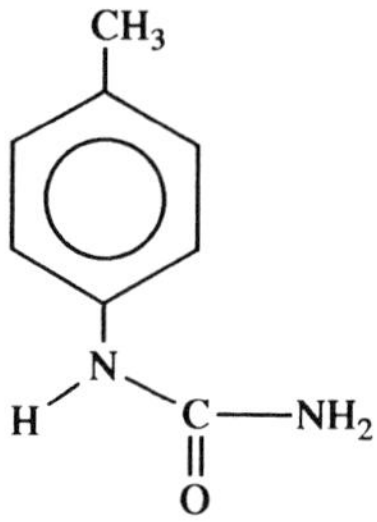

para-tolylurea (sweetish)

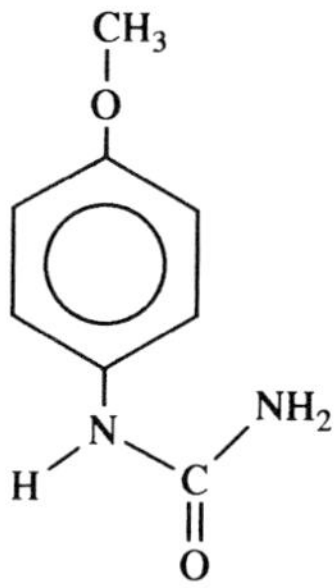

4-Methoxyphenyl urea (<250x)

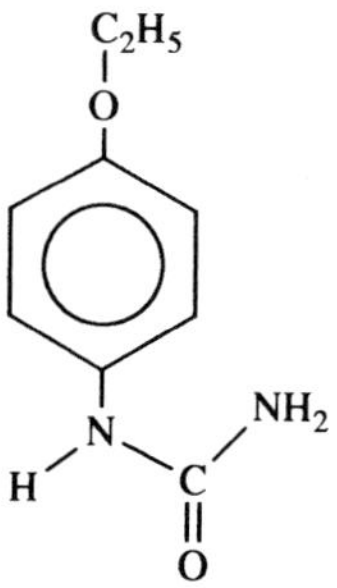

4-Ethoxyphenyl urea,
Dulcin (250x)

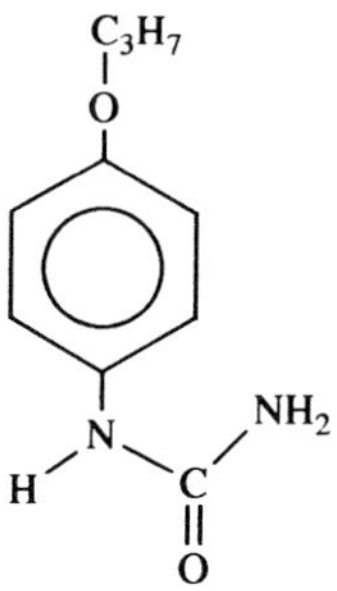

4-Propoxyphenyl urea,
(Sweet)

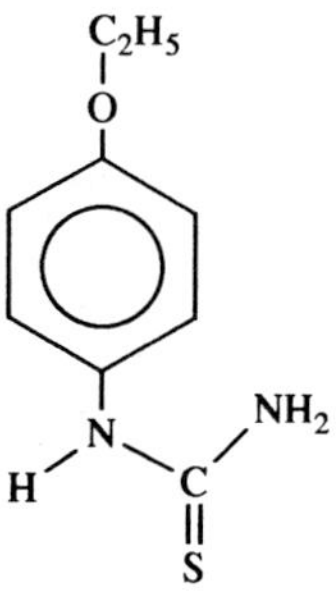

4-Ethoxyphenylthiourea
(bitter)

FIG. 8.14. Structure and taste of urea derivatives relative to that of sucrose.

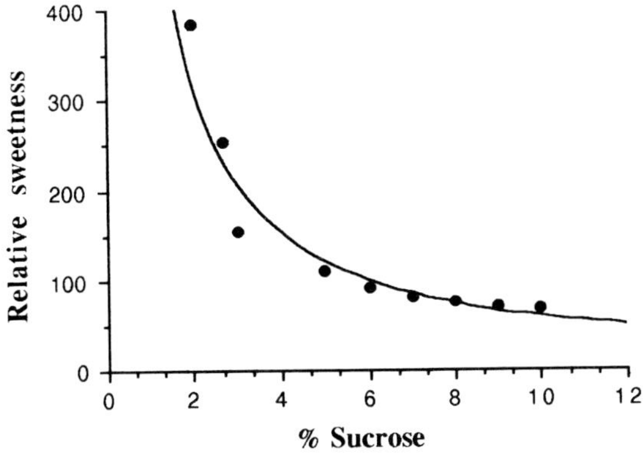

FIG. 8.15. Decrease in the relative sweetness of dulcin with increasing concentrations of sucrose (Täufel & Klemm, 1925).

The first group of sweet urea substances to be considered are the phenyl ureas, and are characterized by dulcin. A number of different urea derivatives and their taste, relative to sucrose, are shown in Fig. 8.14. Among the tolylureas, the position of the ring substituent methyl group influences the taste of the compound (Moncrieff, 1967). The *ortho* compound is tasteless, *meta* substitution leads to bitterness, and *para* substitution leads to potential for sweetness.

Although dulcin (4-ethoxyphenylurea) is 250 times as sweet as sucrose, the 4-methoxy compound is not as sweet. In these substances, the position of the alkoxy group is again important, as placing the ethoxy group in an *ortho* position leads to tastelessness. A plot of the data of Täufel & Klemm (1925) for dulcin relative sweetness at equi-sweet concentrations of sucrose is shown in Fig. 8.15. The second group of substances are a suosan alkane-chain analogue and the thio-analogue of suosan. The structure of suosan and its thio-derivative are shown below. Petersen & Müller (1948) found that the sodium salt of 1-(2′-carboxyethyl)-3-(4′nitrophenyl) urea (suosan) is 350 times (weight basis) as sweet as sucrose while the lower (methoxy) analogue is less sweet. Tinti *et al.* (1982) assigned a score of 10× to this compound (molar basis). Tinti *et al.* (1982) also report that suosan is 700 times as sweet as sucrose on a molar basis, but Crosby *et al.* (1979) report that the taste potency of the sodium salt is 173×. They also report that it possessed significant bitterness (29% of the total taste potency).

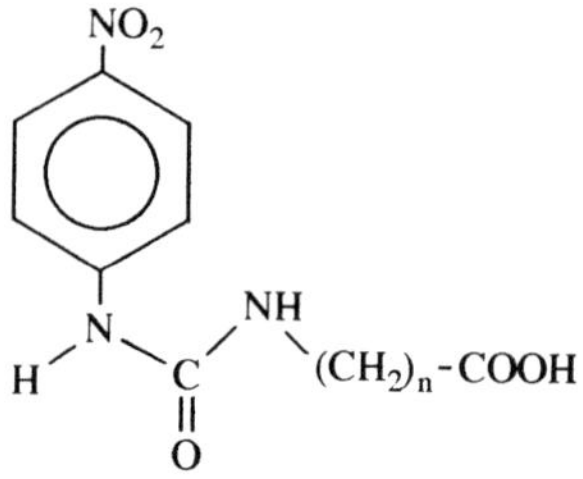

n = 1. (10x)
n = 2. Suosan (350x, 173x, 700x)

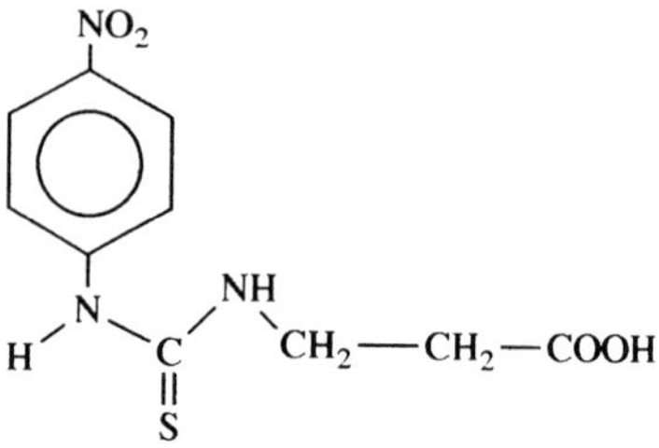

(2400x)

Unlike the thio-analogue of dulcin, the thio-analogue of suosan is reported (Tinti *et al.*, 1982) to be very sweet (2400×, molar basis). However, Crosby *et al.* (1979), assigned sweetness scores on a weight basis to the potassium salt and free acid of 552 times and 602 times, respectively. They also point out that these substances are even more bitter than the dulcin thio-analogue. Indeed, the total taste intensity of the free acid thio-analogue of suosan is 37% sweet and 62% bitter.

Of special significance is the taste of suosan and its derivatives, and although the structures of several were previously shown, the tabulation, prepared by Tinti *et al.* (1982), is given in Table 8.3.

TABLE 8.3
Taste of suosan derivatives (Tinti *et al.*, 1982)

Compound	*Taste*
1. $NO_2—C_6H_4—NH—CO—NH—CH_2—CH_2—COOH$	Sweet (700×)
2. $NO_2—C_6H_4—NH—CO—NH—CH_2—CH_2—CH_3$	Tasteless
3. $NO_2—C_6H_4—NH—CO—NH_2$	Tasteless
4. $NO_2—C_6H_4—NH—CO—NH—CH_2—CH_2—COOCH_2$	Tasteless
5. $NO_2—C_6H_4—NH—CO—NH—CH_2—COOH$	Sweet (10×)
6. $NO_2—C_6H_4—NH—CO—NH—CH_2—CH_2—CH_2—COOH$	Tasteless
7. $NO_2—C_6H_4—NH—CO—NH—CH_2—CH_2—COOSO_3$	Tasteless
8. $NO_2—C_6H_4—NH—CO—NH—CH{=}CH—COOH$	Tasteless
9. $NO_2—C_6H_4—NH—CO—O—CH_2—CH_2—COOH$	Tasteless
10. $NO_2—C_6H_4—CO—NH—CH_2—CH_2—CH_2—COOH$	Tasteless
11. $NO_2—C_6H_4—NH—CS—NH—CH_2—CH_2—COOH$	Sweet (2400×)
12. $NO_2—C_6H_4—CH_2—CO—NH—CH_2—CH_2—COOH$	Bitter
13. $NO_2—C_6H_4—O—CO—NH—CH_2—CH_2—COOH$	Sweet (20×)
14. $NO_2—C_6H_4—NH—NH—CO—CH_2—CH_2—COOH$	Tasteless
15. $NO_2—C_6H_4—CO—NH—NH—CH_2—CH_2—COOH$	Tasteless
16. $CN—C_6H_4—NH—CO—NH—CH_2—CH_2—COOH$	Sweet (450×)
17. $C_6H_4—NH—CO—NH—CH_2—CH_2—COOH$	Tasteless
18. $NH_2—CO—NH—CH_2—CH_2—COOH$	Tasteless

8.1.8 Sulfones

A recent addition to the list of substances that taste sweet is a series of sulfones, prepared by Ratajczak & Polański (1991*a*, *b*) after observing that 2,2-dimethyl-1-phenylsulfonylcyclopropanecarboxylic acid tasted

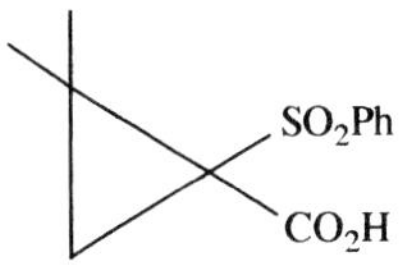

2,2 Dimethyl-1-phenylsulfonyl-cyclopropane carboxylic acid

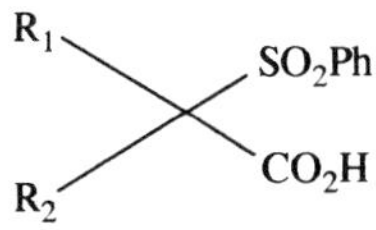

Arylsulphonylalkanoic acids

sweet. Among the significant observations made are that both 2,2-dimethyl-1-phenylsulfonylcyclopropane-carboxylic acid enantiomers taste sweet, and unbranched cyclic modifications of the parent compound

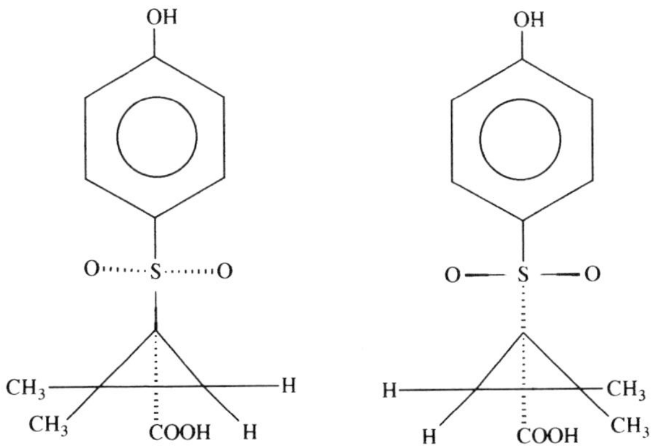

2,2-dimethyl-1-phenylsulfonylcyclopropanecarboxylic acid enantiomers

are tasteless. The sweetest sulfone (450×) was obtained when R_1 = isopropyl, R_2 = H and Ph = EtPh. When the carboxyl group is esterified (ethyl ester), sweetness is quenched. However, the sodium salts of the active acids taste sweet. For the arylsulphonylalkanoic acid acyclic derivatives, incorporation of an electron-withdrawing atom in the benzene ring induced bitterness.

8.2 BITTER ORGANIC SUBSTANCES

Belitz & Wieser (1985) reported that the number of known bitter

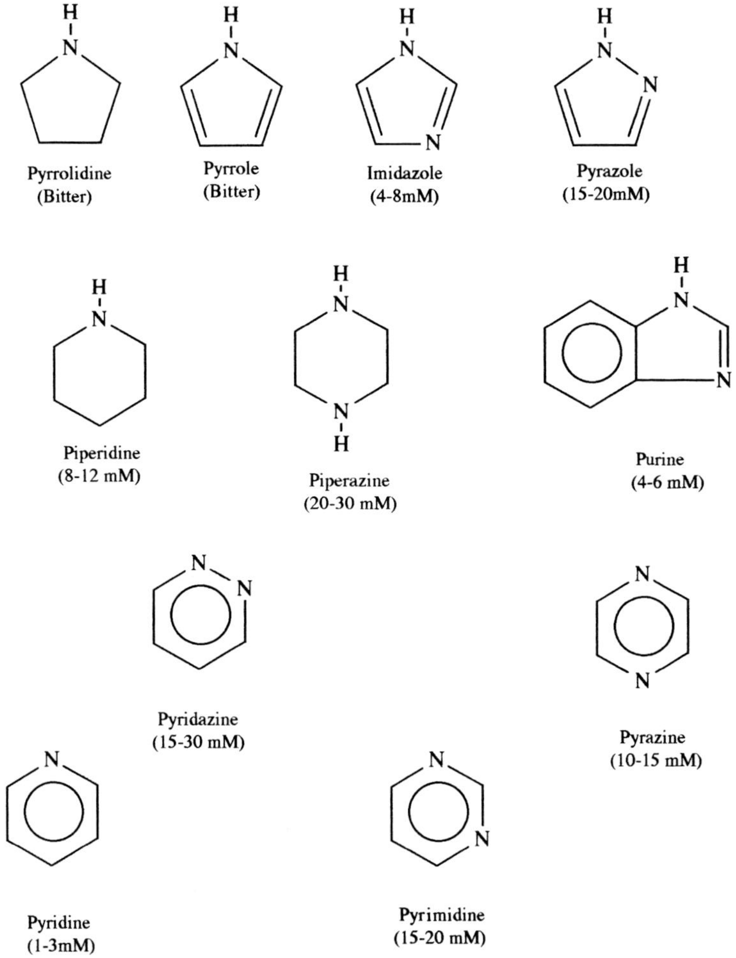

FIG. 8.16. Structure and bitterness threshold for the simplest members of the alkaloids (Kuninaka *et al.*, 1980). See also Belitz & Wieser (1985).

compounds far exceeds the number of compounds known to be sweet. The observation suggests that the structural and functional group requirements for bitterness are not as rigid as for sweetness. Many bitter tasting amino acids and peptides were presented in the previous chapter, and while they will not be discussed here, it is pointed out that primary amines (R—NH_2) are uniformly more bitter than the corresponding amino acids. The threshold for bitterness of amines is only one-half to one-tenth that of the corresponding bitter amino acids.

One problem with routine study of bitter substances is that bitterness scores are generally the recognition threshold concentration only. A reference standard for relative bitterness has not yet been agreed upon, although van der Heijden (1992) used picric acid (threshold 1×10^{-3} M) for comparing the relative bitterness of substances.

8.2.1 Alkaloids

Alkaloids are base-reacting organic substances that occur as salts of organic acids (malic, citric, etc.). They usually contain a nitrogen atom, but not always. The simplest nitrogen heterocyclic members, shown in Fig. 8.16, along with their bitterness thresholds, are strong bases that are freely soluble in water. They may also have penetrating, unpleasant odors (pyridine).

There seems to be only one simple relation between structure and bitterness for these compounds. A decrease in the overall hydrophobicity, by introducing a second hetero-atom into the structure, decreases bitterness. The further removed a hydrophobic substituent is from the ring hetero-atom, the greater the bitterness, as observed for substituted piperidine derivatives (Belitz & Wieser, 1985). Some of the more commonly known higher molecular weight alkaloids with potent bitterness are shown in Fig. 8.17. A brief discussion of each of them follows.

8.2.1.1 *Strychnine*

Strychnine belongs to the isoquinoline group of alkaloids, along with morphine (the opium alkaloids), and is obtained from the seeds of the plant genus *Strychnos*. The pH of a saturated solution is 9·5. The bitterness threshold, given as 1 : 130 000 computes to be $2{\cdot}3 \times 10^{-5}\,\mathrm{mol\,l^{-1}}$ on a molar basis (mol. wt = 334·4). Pfaffmann *et al.* (1971) give the molar threshold as $1{\cdot}6 \times 10^{-6}$.

8.2.1.2 *Brucine* (10,11-dimethoxystrichnine)

Brucine, like strychnine, is also found in *Strychnos* seeds. Its threshold is given as 1 : 220 000, which gives $1{\cdot}2 \times 10^{-5}\,\mathrm{mol\,l^{-1}}$ for the molar basis threshold (mol. wt = 394·5).

8.2.1.3 *Piperine*

Piperine belongs to the pyridine group of alkaloids, which includes nicotine. It is also the alkaloid found in black pepper (*Piper nigrum*). The alkaloid is neutral to litmus, and tasteless at first, but has a burning aftertaste. The pungent taste of pepper is due to an essential oil, and not to the alkaloids.

8.2.1.4 *Solanine*

Solanine belongs to the tropane group of alkaloids found in *Solanaceae* and cocoa. Other tropane alkaloids are atropine, and cocaine. Solanine is found in the sprouts (0·04%) of potato tubers. It also occurs in tubers exposed to light during growth, hence its name. This high molecular weight (868·04) alkaloid is a glycoside containing a molecule each of glucose, galactose and rhamnose as a discreet trisaccharide. Although exceedingly bitter, its threshold is not known.

8.2.1.5 *Quinine*

Quinine is extracted from the bark of the Cinchona tree (*Chinona officinalis*). Its structure is given in Section 1.5.1. The pH of saturated solutions is about 8.8. It is also known as the chloride salt ($C_{20}H_{24}N_2O_2{\cdot}HCl$), which is neutral to litmus. Its bitterness threshold is given as 1 : 30 000, or 1×10^{-4} on a molar basis (mol. wt 360·88) in the Merck Index, and $3{\cdot}0 \times 10^{-5}$ on a molar basis by Pfaffmann *et al.* (1971). Quinine is the classic antimalarial drug. It is also known as the sulfate ($2C_{20}H_{24}N_2O_2{\cdot}H_2SO_4$), with a bitter threshold of $8{\cdot}0 \times 10^{-6}$ M (Pfaffmann *et al.*, 1971), and its solutions are also neutral.

According to Belitz & Wieser (1985) there is only low stereospecificity for quinine and its derivatives for the bitter receptor. To taste bitter, the hydroxyl group needs to be above the plane of the quinoline nucleus, and the quinuclidine nucleus behind it. Reduction, esterification, or oxidation of that hydroxyl group decreases bitterness, and saturation of the vinyl group increases bitterness.

8.2.1.6 *Caffeine*

Caffeine is the common alkaloid of coffee and tea, and its structure is given in Section 1.5.1. Water solutions are neutral, and the molar threshold (Pfaffmann *et al.*, 1971) is $7{\cdot}0 \times 10^{-4}$.

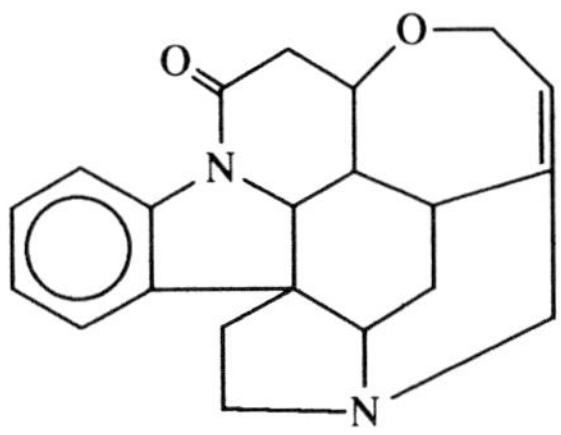

Strichnine (1:130,000)

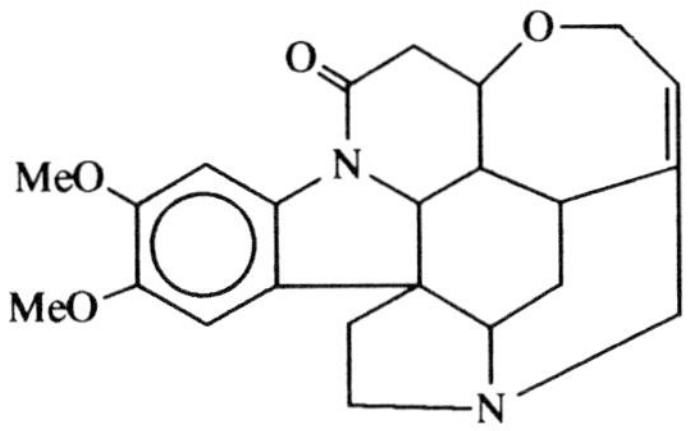

Brucine (1:220,000)

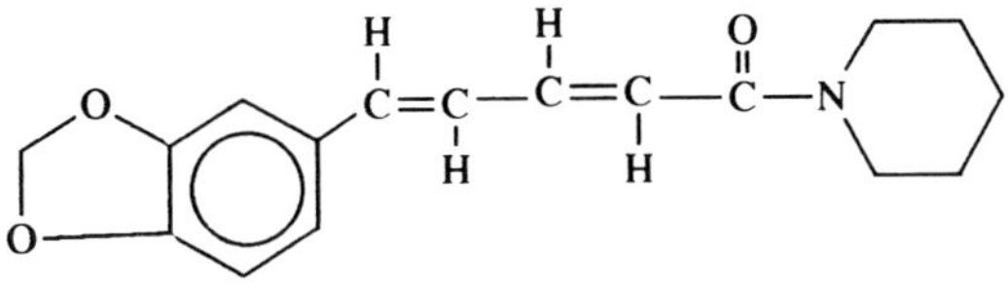

Piperine (Tasteless)

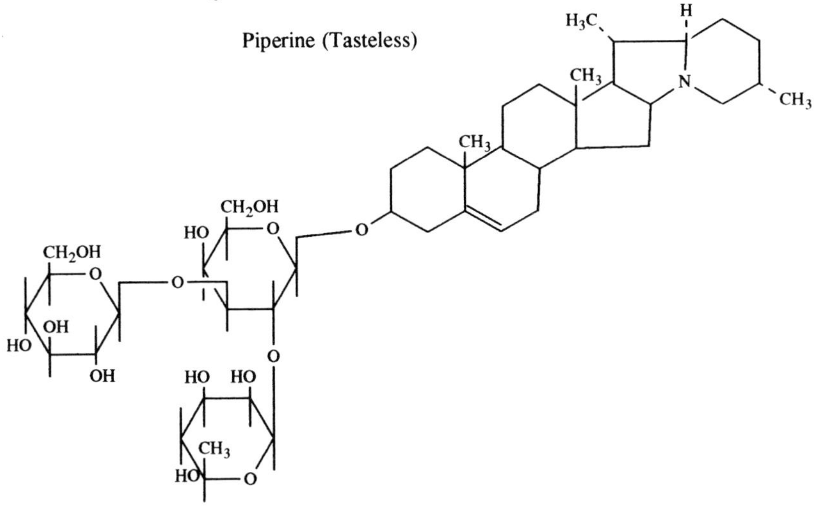

Solanine (Bitter)

FIG. 8.17. Selected structures of complex alkaloids.

8.2.2 Amides and Thioureas (Thioamides)

Many amides are bitter (Wieser & Belitz, 1975). Compounds with the general formula $RCONH_2$ show bitterness increase, up to a point, with an increase in the size of R (Belitz & Wieser, 1985). The bitterness thresholds (mM) are R = methyl, 100; ethyl, 50–55; propyl, 17·5–22·5; butyl, 17·5–22·5. When R = C_6H_5, the threshold is only 0·8–1·0 mM. The authors also list increasing bitterness scores for the formula H and CH_3 for R_1 in the structure R_1CONHR_2. The dimethyl compound (R_1, R_2 = CH_3) alone is fairly bitter (10–15 mM), and bitterness increases again with increasing the size of R_2.

Thioamides are also bitter, and thioureas present a special case in the annals of taste chemistry (Section 1.1.5), particularly phenylthiourea (phenylthiocarbamide, PTC). About one in four persons have limited ability to discern the bitterness of these substances, unless the concentration is quite high. To 'tasters,' the bitter threshold for PTC is 2×10^{-5} M, but to 'nontasters' it is about $8{\cdot}0 \times 10^{-3}$ M.

8.2.3 Diterpenes

One series of homologous substances with bitter taste that was studied in detail for a relation between structure and bitter taste is the diterpenes (Kubota & Kubo, 1969). A reference structure for 16 additional bitter-tasting terpene reference structures, along with the first example of 10 tasteless reference compounds, is given in Fig. 8.18.

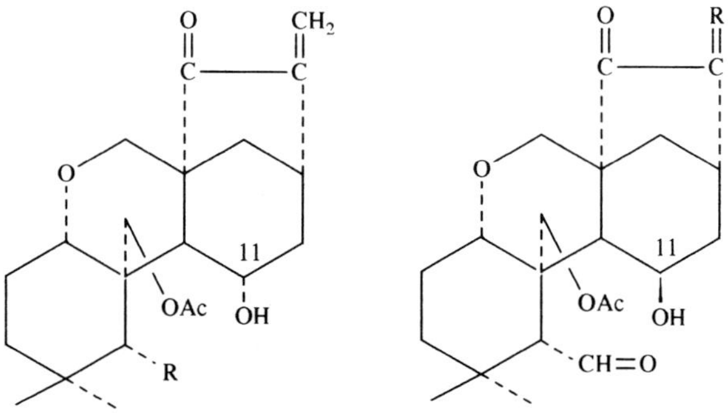

Bitter diterpenes
R = CHO, R' = > CH_2
R = CHO, R' = α-CH_3, β - H
R = COOH, R' = α-CH_3, β - H
R = COOCH3, R' = α-CH_3, β - H

Tasteless diterpenes
R = > CH_2
R = α-CH3, β - H

FIG. 8.18. Selected structures of bitter-tasting and tasteless diterpene substances (Kubota & Kubo, 1969).

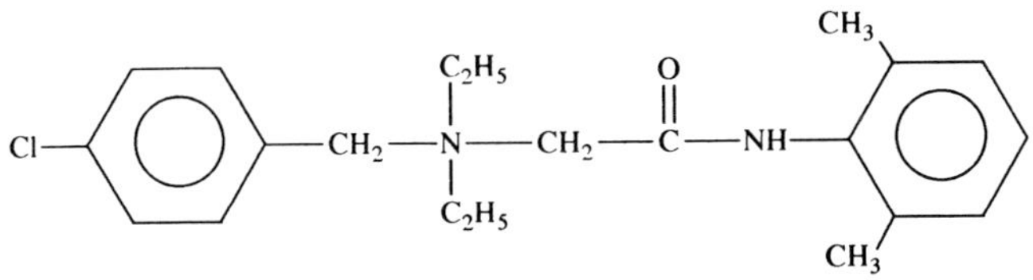

Denatonium chloride

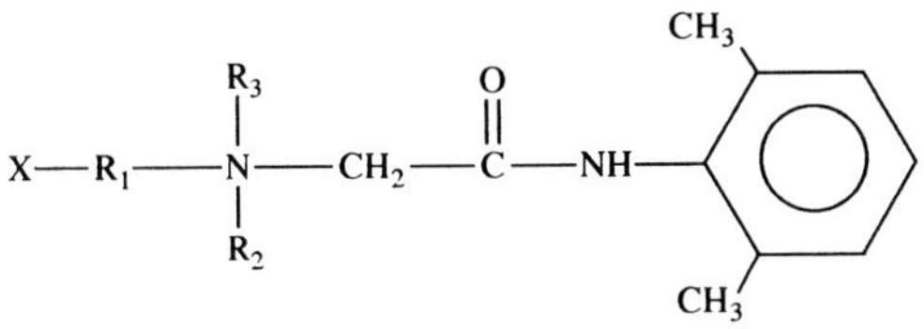

Reference structure I

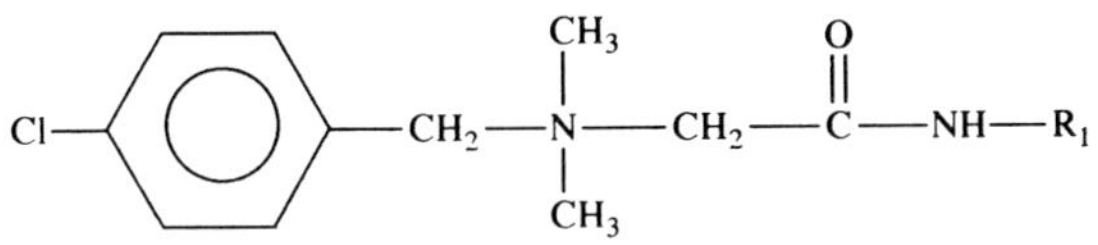

Reference structure II

FIG. 8.19. The extraordinarily bitter substance denatonium chloride and reference structures for Table 8.4.

8.2.4 Extraordinarily Bitter Organic Substances

Several compounds are now known that have extraordinarily high bitterness potency. Perhaps the most bitter compound known to date is denatonium chloride. Its structure is shown in Fig. 8.19. Because of its extreme bitterness, Saroli (1984, 1985) studied the effect of various substitutions on the basic structure to obtain insight into the relation between structure and bitterness. The results obtained by substituting on reference structure I in Fig. 8.19 are summarized in Table 8.4.

Saroli observed a rapid increase in bitterness when the ammonium nitrogen atom substituents were increased from methyl to butyl (Nos. 1–3). Ignoring the possible effect of the halogen substituent, it appeared that each addition of one carbon atom to the size of the alkyl group chain increased bitterness by a factor of ten, until a 4-carbon atom chain was attained. From then on, bitterness increased only slightly (Nos. 5–7).

TABLE 8.4
Effect of substitution on reference structure I (Fig. 8.19) on the bitterness threshold ($mol\,l^{-1}$) of denatonium chloride derivatives (Saroli, 1984)

No.	R_1	R_2	R_3	*X*	*Threshold*
1	CH_3	CH_3	CH_3	Cl	$1–2 \times 10^{-3}$
2	C_2H_5	CH_3	CH_3	Br	$2–4 \times 10^{-4}$
3	C_4H_9	CH_3	CH_3	I	$2–4 \times 10^{-5}$
4	C_5H_{11}	CH_3	CH_3	Cl	$2–4 \times 10^{-5}$
5	C_6H_{13}	CH_3	CH_3	Cl	$4–8 \times 10^{-5}$
6	C_8H_{17}	CH_3	CH_3	Cl	$4–8 \times 10^{-5}$
7	$C_{10}H_{21}$	CH_3	CH_3	Cl	$4–8 \times 10^{-5}$
8	$C_6H_5CH_2$	CH_3	CH_3	Cl	$1–2 \times 10^{-7}$
9	$C_6H_5CH_2CH_2$	CH_3	CH_3	Cl	$2–4 \times 10^{-5}$
10	$C_6H_5CH_2CH_2CH_2$	CH_3	CH_3	Cl	$2–4 \times 10^{-5}$
11	$C_6H_5CH_2$	C_2H_5	CH_3	Cl	$3–6 \times 10^{-8}$
12	$C_6H_5CH_2$	C_4H_9	CH_3	Cl	$3–6 \times 10^{-8}$
13	$C_6H_5CH_2$	$C_6H_5CH_2$	CH_3	Cl	$2–4 \times 10^{-7}$
14	$C_6H_5CH_2$	C_2H_5	C_2H_5	Cl	$1–2 \times 10^{-8}$

Upon addition of a benzyl group, bitterness again increased exponentially (No. 8).

The optimum group for hydrophobicity was found to be contained in the structure $C_6H_5CH_2$, as bitterness drops with increasing the distance of the phenyl group (Nos. 9 and 10). When R_1 is a benzyl group, either an ethyl group or a butyl group at R_2 again increases bitterness (compounds 11 and 12), as does a benzyl group at that position (No. 13). Compound 14 in the table is denatonium chloride.

Saroli (1985) next directed his attention to modification of reference structure II in Fig. 8.19. Only the modifications for R_1 and their effects on the bitter threshold are given here. When R_1 in reference structure II was increased in size from methyl through hexyl, the bitterness threshold decreased from 2 to $4 \times 10^{-5}\,mol\,l^{-1}$ until the size was propyl ($4–8 \times 10^{-6}$). Neither an increase in the length of R, up to hexyl, nor an increase in branching, up to and including pentyl derivatives, significantly altered the taste threshold.

Even substances with a phenyl group at R_1 did not change the bitterness threshold. Increasing the distance of the phenyl group from the peptide oxygen atom, by insertion of CH_2 groups up to $(CH_2)_2$, increased the threshold only slightly. Substituting the aromatic ring with methyl groups again increased the bitterness, and maximum bitterness was

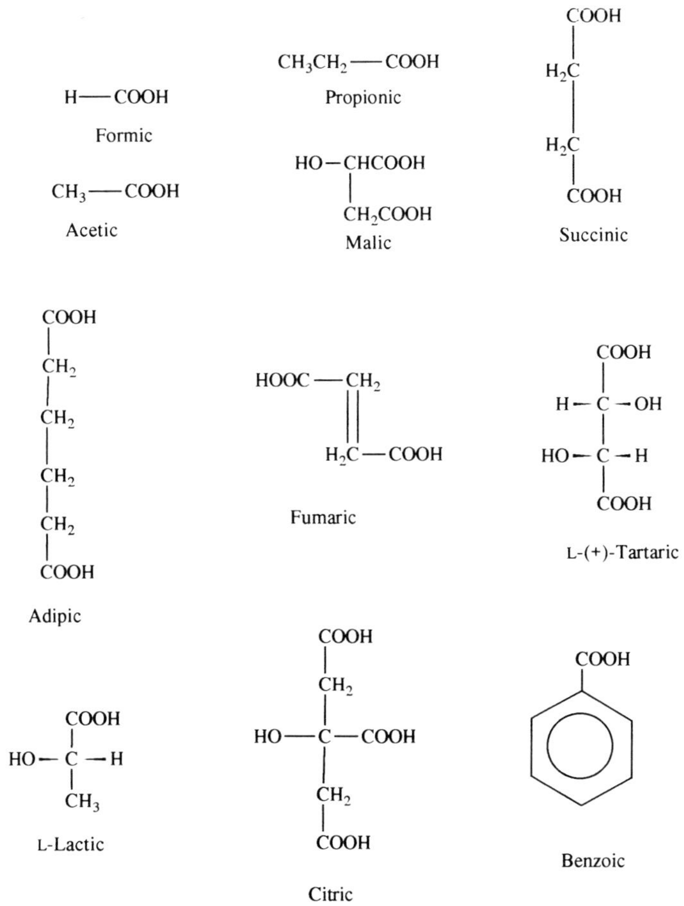

FIG. 8.20. Structure of selected organic acids.

obtained with the 2,6-dimethylbenzyl compound. This was also the case found for denatonium chloride.

8.3 SOUR AND SALTY ORGANIC SUBSTANCES

Sourness and saltiness of organic substances are considered together for obvious reasons. Their structures have little bearing on the threshold for sourness, and their salts, particularly sodium salts, are always salty, but have additional tastes. A large number of organic acids are employed in foods as acidulants. The structures of representative examples are shown in Fig. 8.20.

Formic acid. Formic acid is obtained from the distillation of ants. It is a colorless liquid with a pungent odor. It is also found in plants (nettle) which prick and irritate the skin.

Acetic acid. Acetic acid is a weak organic acid produced in pure form (glacial acetic acid) as a clear liquid with a pungent odor. The acid function is the carboxyl group. Acetic acid is also obtained by fermentation of carbohydrates (vinegar). The pK_a is $1{\cdot}8 \times 10^{-5}$, and the pH of a 1·0 M solution is 2·4. That of a 0·1 M solution is 2·9, and an 0·01 M solution has pH 3·4. In addition to sourness its solutions have an aromatic note.

Adipic acid. Adipic acid (1,4-butanedicarboxylic acid) is found in beet juice, and the melting point is 152°C. pK_{a1} is $3{\cdot}9 \times 10^{-5}$, and pK_{a2} is $5{\cdot}3 \times 10^{-6}$. The pH of a 0·1% solution is 3·2.

Benzoic acid. Benzoic acid (phenylformic acid) is a common component (up to 0·05%) of berries. Its crystals are odorless and sublime at 100°C. p$K_a = 6{\cdot}4 \times 10^{-5}$, and the pH of a saturated solution is 2·8. It is only slightly soluble (0·2%), and the sodium salt, whose solutions are slightly alkaline (pH 8), is reported to be sweetish and astringent.

Citric acid. Citric acid (2-hydroxy-1,2,3-propantricarboxylic acid) is found in both plants (up to 8% in citrus juices) and animals. It is prepared on an industrial scale by fermentation of sugars. The anhydrous crystals melt at 153°, and the monohydrate melts at 100°. The dissociation constants for the three carboxyl groups are p$K_1 = 3{\cdot}128$, p$K_2 = 4{\cdot}761$, and p$K_3 = 6{\cdot}396$. The crystals are reported to have a pleasant, sour taste.

Lactic acids. L-Lactic acid (*S*-2-hydroxypropanoic acid, L-milk acid) is present in animal tissues and blood. Its crystals melt at 53°C and the specific rotation is −2·6°. The pK_a is 3·83. D-Lactic acid (*R*-2-hydroxypropanoic acid, D-milk acid) has a specific rotation of +2·6°, and the pK_a is reported to be 3·79. The racemic form is also known.

Malic acid. Malic acid (hydroxysuccinic acid, apple acid) is found in many fruits as the L-form. It is a solid with a melting point of 100°C and a specific rotation of −2·3°. The pK_{a1} is 4×10^{-4}, and pK_{a2} is 9×10^{-6}. It has a smooth and tart taste.

Fumaric acid. Fumaric acid (*trans*-1,2-ethylenedicarboxylic acid) occurs in many plants. Its crystals sublime, and it is used in foods. The p$K_{a1} = 9{\cdot}3 \times 10^{-4}$, and p$K_{a2} = 2{\cdot}9 \times 10^{-5}$.

Propionic acid. Propionic acid (propanoic acid) is obtained from wood pulp by fermentation. It is an oily liquid with a disagreeable pungent odor. The pK_a is $1{\cdot}34 \times 10^{-5}$. The calcium salt is a mold inhibitor.

Succinic acid. Succinic acid (butanedioic acid) occurs free in fossils, fungi, and lichens, but is prepared commercially from acetic acid. It is also a coenzyme factor in metabolic pathways. The crystals melt at 185–187°, and are reported to have a very acid taste. The pH of a 0·1 M solution is 2·7.

Tartaric acid. Tartaric acid (L-tartaric, *d*-tartaric acid) occurs in many plants, particularly grapes. The dissociation constants are $pK_{a1} = 1{\cdot}04 \times 10^{-3}$, and $pK_{a2} = 4{\cdot}6 \times 10^{-5}$. The pH of a 0·1N solution is 2·2. The taste is that of a strong acid, but said to be refreshing in dilute solution. It is also stated to taste bitter. The potassium salt of this acid precipitates during grape juice fermentation. The D-form is also known, and an optically inert *meso*-form has $pK_{a1} = 3{\cdot}11$, $pK_{a2} = 4{\cdot}80$, which is significantly different from that for the L-form (2·93 and 4·23, respectively).

8.3.1 Sourness of Organic Acids

The sourness of organic acids, as with the inorganic acids, is due to the hydrogen ion, even though the organic acids do not completely dissociate in solution. They may be mono-, di-, or tri-protic, and their dissociation constants and equivalency need to be considered when assigning sourness scores. For total taste attributes, the nature of the anion is important, and may contribute sweet, bitter, odoriferous, and astringent notes. With citric acid on a weight basis taken to be unity with respect to sourness, the relative sournesses of selected acids (Sausville, 1974) are fumaric (0·67–0·72), tartaric (0·80–0·85), malic (0·89–0·94), adipic (1·10–1·15), phosphoric (0·55–0·60).

On a molar basis, as on a weight basis, it might be expected that the relative sourness of the organic acids would vary appreciably due to their different normality. This is the case. Fabian & Blum (1943) found the following molar concentrations of the organic acids to be equally sour: lactic, $8{\cdot}5 \times 10^{-4}$; malic, $7{\cdot}5 \times 10^{-4}$; tartaric, $7{\cdot}0 \times 10^{-4}$; acetic, $2{\cdot}1 \times 10^{-3}$; and citric, $7{\cdot}0 \times 10^{-4}$. Presumably, then, the ordering of the acids in terms of decreasing sourness would be citric and tartaric > malic > lactic > acetic. When the normality of the acid solutions is considered, setting aside the ability of the anion to contribute various taste and flavor notes, the difference in relative sourness is minimal. All organic acids have a threshold around or near 0·001–0·002 N (Pfaffmann, 1959).

CHAPTER 9

The Common Saporous Units for Taste

One heuristic principle of taste chemistry is that there is a common chemical feature (saporous group, saporophoric group) among substances with the same taste quality. Presumably then, those groups are isosteric (iso-functional) for a given taste (Section 2.9). As a consequence, different tastes may have different modes of chemical interaction with the taste receptor. This chapter explores the possible existence of an acidophore for sourness, a halophore for saltiness, a glycophore for sweetness, and picrophores for bitterness. There is an imbalance with respect to the extent with which each is treated. This is because the saporous groups increase in complexity in the order sour > salt > (sweet, bitter). It is also due to the disproportionate amount of study devoted to sweetness because of its economic importance.

9.1 THE COMMON ACIDOPHORE

There seems to be little doubt that the common denominator among compounds with sour taste is the hydrogen (hydronium) ion (Sections 3.6 and 8.3.1). Therefore, the hydrogen ion is the simplest saporous group, i.e. it is the acidophore for sour taste. The hydronium ion (H_3O^+), when fully hydrated, has long been suspected to be complexed with three additional molecules of water, as shown in Fig. 9.1. However, it is the incompletely hydrated hydrogen ion that is probably the true acidophore. The ion is so small it is difficult to envisage it as possessing, for any length of time, four water molecules of hydration. Of particular significance to sourness, molar solutions of acids greater than one normal (H_2SO_4 and citric acid) are proportionately more sour than those that are one normal, for example HCl. If the hydrogen ion is the acid-taste saporous group, it follows that a

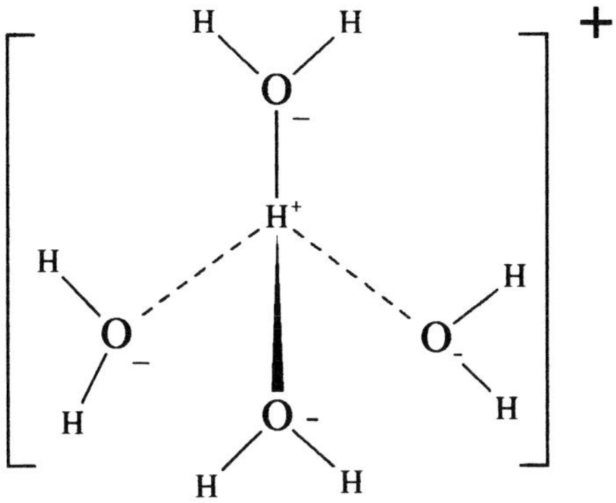

The hydronium ion $(H_9O_4)^+$, i.e., "H_3O^+"

FIG. 9.1. The completely hydrated hydrogen ion.

compound may have an acidophore equivalency greater than one. In addition, weak (incompletely dissociated) acids, such as acetic are, on a normal basis, just as sour as HCl. Therefore, principles of chemical equivalency also apply (Section 3.6) as sourness intensity is directly related to the hydrogen ion concentration *and* the total acidity of weak acids.

9.2 THE COMMON HALOPHORE

Although all acids are characterized by sourness, and have a common acidophore, saltiness is unique as a taste in that it is truly characteristic of only one compound, sodium chloride. Moreover, the taste is due to both the anion and the cation (Section 3.4). Other inorganic and organic salts taste somewhat salty, of course, but the taste is admixed with other taste attributes, usually described as bitter. If sourness is due primarily to an acidophore, it would seem to follow that true saltiness is due to a very special halophore consisting of the ion pair Na^+ and Cl^-. If only the sodium and the chlorine ions are capable of eliciting a true salt taste, an obvious question is, 'What is so special about those antipolar ions that cannot be duplicated by any other antipolar ion pair?'

Part of the answer seems to be that solutions of sodium chloride are neutral (pH 7), and have little acid- and alkaline-taste features. Moreover, the crystalline structure coordination number is six. In the crystal, each sodium ion is symmetrically coordinated with six chlorine atoms, and *vice versa*. In addition, among the group I elements (Li, Na, K, Cs), it is most difficult to subtract an electron from sodium (Section 3.2.1), but among the Group VII elements (Fl, Cl, I, Br), it is easiest to add an electron to chlorine (Section 3.2.2).

Finally, the majority of salts with a predominant, as opposed to true, salty taste have a solution equivalency of one, although sodium sulfate may be an exception. Therefore, the halophore for true saltiness is herein assigned to the sodium and chlorine ions. It is tentatively shown in Fig. 9.2 as incompletely hydrated ions. The reason is because prior to the onset of saltiness, dilute solutions of NaCl taste sweet. Fully hydrated ions are therefore reserved to explain the glycophore attribute of sodium chloride ions (Section 9.3.9.1), and also the anomalous conductivity exhibited by dilute NaCl solutions (Section 3.2.4). For strong saltiness, cluster ionic structures (Fig. 3.13) with reversed charge are herein proposed. At high concentrations, their existence is mandated by the loss in molecular conductivity with increasing concentrations of salt. Interestingly, at high salt concentrations, the bipolar identities of the halophore are transposed (Section 3.2.4), but with little apparent effect on the perception of the saltiness modality.

The saltiness profile and taste spectrum of salts vary, of course, with the nature of the cation and the anion, but the feature that inorganic salts have in common is that they ionize to yield a metal cation (M^+) and a nonmetal anion (X^-). As with inorganic salts, saltiness among salts of the organic acids is generally restricted to the sodium ion. One example is monosodium glutamate, but it, like organic acids, has taste attributes other than saltiness. In this case it is proposed that the saporous unit for

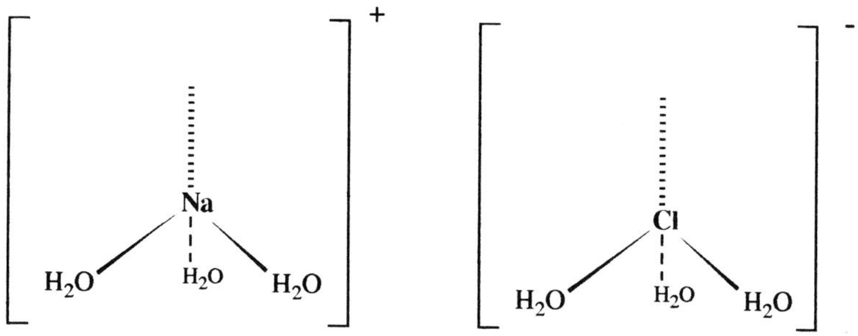

FIG. 9.2. Incompletely hydrated sodium and chlorine ions as the halophore for saltiness.

the salty/umami taste is the sodium cation, in combination with various nucleotide base (Section 1.1.6.2) and/or basic peptide anions. Here, however, some complications arise.

Upon examining 21 dipeptides, Tada *et al.* (1984) proposed that whenever n = 1 and m varied among 1–3 in the general structure of the basic amino acid peptides (Fig. 9.3), the taste of the dipeptides was predominately umami. The umami taste was not affected by the length of the basic amino acid side chain. The salty taste predominated when m = 1 and n varied among 1–4. The salty taste also occurs in sulfonyl and *p*-toluenesulfonyl salts, but the latter compounds are also bitter. It was concluded that the distance between the amino group of the branched chain (shown as an extension from the peptide bond), and the C-terminal carboxyl group is important for governing both the umami and the salty tastes. As H-L-Orn-Tau · HCl and H-L-Lys-Tau · HCl was believed to possess only salty taste, at least equal to or greater than NaCl, it was proposed that the salt-taste quality is influenced by the degree of dissociation. It was also pointed out that salty peptides do not contain the sodium ion, but it is now known that the substances are not true NaCl taste mimics, either.

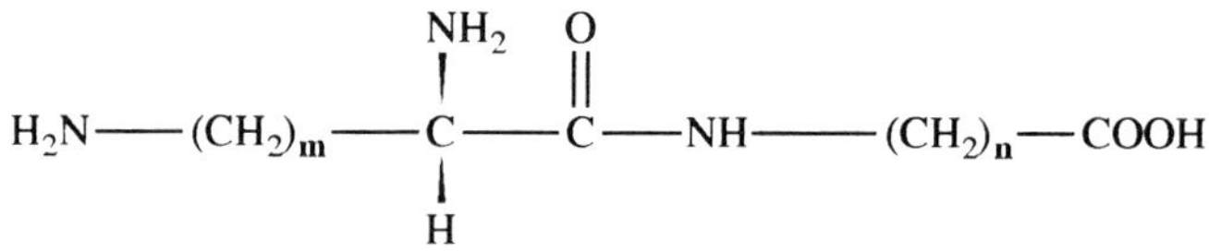

m = 1-4; **n** = 1-3

FIG. 9.3. General structure for salty-umami taste (Tada *et al.*, 1984).

9.3 THE SEARCH FOR THE COMMON GLYCOPHORE

Salient features of the search for the common glycophore are shown in Table 9.1. It can be observed that progress in this endeavor, up to a point, parallels progress in development of general chemical theory.

9.3.1 The Glucogene

In a compilation entitled *Die Organischen Geschmackstoffe* Cohn (1914) recognized that specific chemical saporous groups, such as $(-OH)_n$, NH, NH_2, NOH and NO_2 were associated with sweetness. Cohn also observed

TABLE 9.1
The search for the common glycophore

Structural examples	*Common denominator*
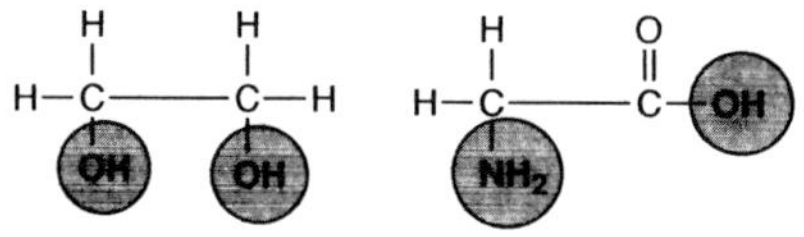 Ethylene glycol Glycine	Cohn (1914). Saporous groups, comprising a 'glucogene'
 Ethylene glycol Chloroform	Oertly & Myers (1919), Glucophores, auxoglucs.
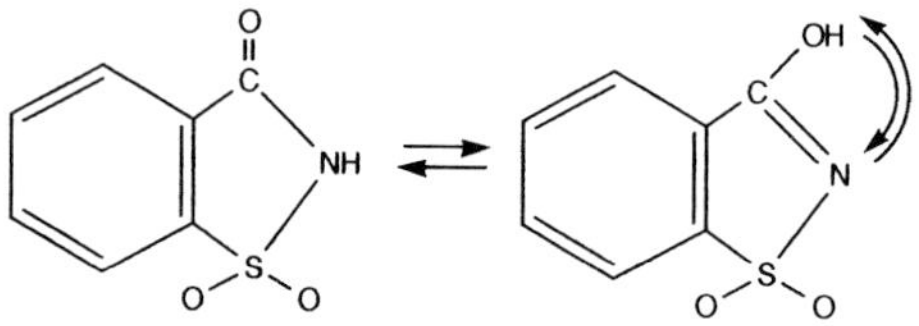	Kodama (1920). Vibratory hydrogen. Tsuzuki *et al.* (1954). Resonance energy.
$\left\{ -H^{+},\ -X:^{-} \right\}$	Warfield (1954). A taste couple consisting of a proton and unshared electron pair. Belitz *et al.* (1979). A nucleophilic, electrophilic unit.
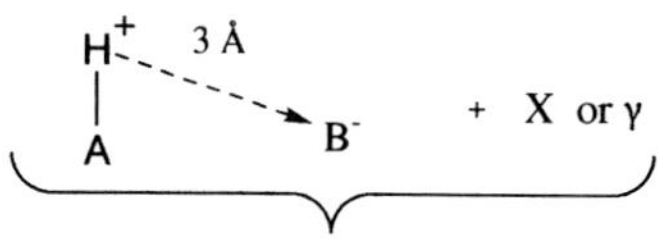 A hydrogen bonding unit plus a hydrophobic bonding site.	Shallenberger & Acree (1967), AH,B Kier (1972), AH,B,X Shallenberger & Lindley (1977), AH,B,γ

that these groups usually occurred in pairs, and described those pairs as a *glucogene*. Examples (Table 9.1) are multiple hydroxyl groups (ethylene glycol, sugars, and sugar alcohols), and NH or NH_2 plus a carbonyl oxygen atom (saccharine and the D-amino acids). The oximes are given as an example of N and OH. Nitrobenzene is given as an example of NO_2, although the second group was not specified.

9.3.2 The Auxogluc and Glucophore

Oertly & Myers (1919) coined the idea that Cohn's saporous groups had different chemical functions, and that the glucogene was composed of two groups with different chemical functionality. Through analogy to the need for an auxochrome and a chromophore in order for a compound (dye) to possess colorant properties, they proposed that a *glucophore* and an *auxogluc* were needed for sweetness. Auxoglucs for ethylene glycol and an amino acid are illustrated in bold print in Table 9.1. The glucophores make up the remainder of the molecule. Together, they comprise Cohn's notion of a glucogene. Oertly & Myers recognized six different types of glucophores, and four different types of auxoglucs. Each became progressively more complex as the structure of the sweet tasting compound became more complex. For hexoses, the auxogluc was assigned to $C_nH_{2n-1}O_n$ and the glucophore was assigned to the –COC–HOH–(H) group More often than not, however, the auxogluc was assigned to a lone hydrogen atom (ethylene glycol, glyceraldehyde, glycine, and halogenated hydrocarbons).

9.3.3 Vibratory and Resonating Hydrogen Atoms

In 1920, Kodama pointed out that all substances that taste sweet contain 'vibratory' hydrogen that interacts intramolecularly with a proton-accepting group to form a hybrid structure. To Kodama's way of thinking, if a compound possessed chemical groupings that lead to hydrogen vibration, that system probably was responsible for, or at least prerequisite to, sweet taste. Tsuzuki *et al.* (1954) subsequently described this idea in terms of 'resonating forms.' It now seems clear that Kodama had attempted to describe, using the best chemical theory available, the current notion of the intramolecular hydrogen bond.

9.3.4 The Taste Couple

In 1954, R. B. Warfield presented a paper at the 126th Meeting of the American Chemical Society's Division of Agricultural and Food Chemistry entitled '*Taste and Molecular Structure*.' The full manuscript

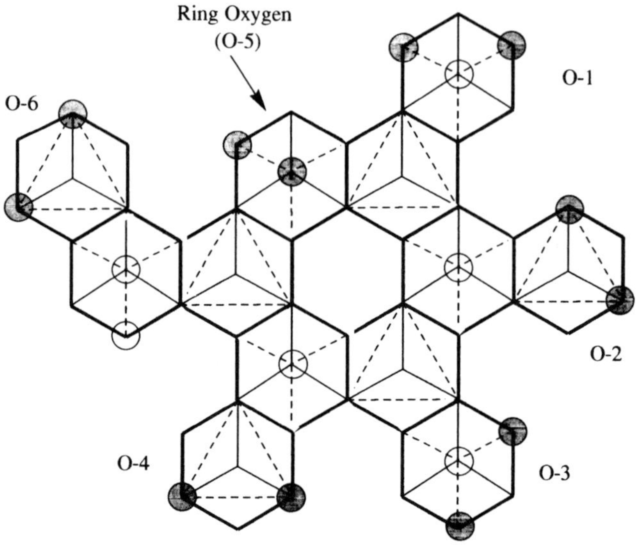

FIG. 9.4. The Reinicke (1943) structure for β-D-glucopyranose indicating a series of two oxygen tetrahedra, separated by two carbon tetrahedra, as the unit responsible for sweet taste.

was never published, but in the program abstract of the paper, the molecular feature common to compounds that have the same taste quality was defined as a *taste couple*. The couple was viewed as being composed of a proton, and a neighboring unshared electron pair. For several classes of organic compounds it is stated that sweet, bitter, and tasteless substances could be associated with specific differences in electrostatic patterns arising from the taste couples. When the taste qualities of inorganic salts were compared with factors such as ionic radius and hydration, it was found that the same 'qualitative difference' between sweet and bitter electrostatic patterns arose. The ideas of Warfield have many parallel ideas in current thinking on the chemical nature of taste.

9.3.5 Derivation of the Sugar Glycophore as AH,B

The identity of the glycophore common to the sugars has also been sought for many years, as there is need to understand the reasons for the varying sweetness of the sugars. A common theme running through sugar sweetness studies is that if the sugar glycophore could be identified, the result might shed light on the nature of the component(s) responsible for sweetness observed for many different chemical substances. Many investigators subscribe to the notion (Shallenberger & Acree, 1967) that the primary glycophore necessary for the taste modality described as sweetness is an AH,B unit in the proton donor–acceptor sense with which AH,B is used to define and describe the hydrogen bond (Section 2.7.3.1). Others, of course, are not all that certain that the tenet applies to all substances that taste sweet. The following sections, therefore, describe the trail of deliberation and conjecture that led to development of the theory. In the development, and while the Greek prefixes *gluc-* and *glyc-* both mean sweet, the term *glycophore* is employed to indicate a two- or three-component saporous unit with different chemical functions responsible for the perceived sweetness modality. The term glucophore is reserved for Oertly & Myers (1919) notion of a two-component glucophore and auxogluc system, i.e., the glucophore is only one component of a sweet taste grouping.

9.3.6 Reinicke's Oxygen Tetrahedra in *Para* Position

The monosaccharides form a complete set of configurational isomers, ring isomers, and optical antipodes (enantiomers). Therefore, they are ideal for studying relations between stereochemical structure and sweetness. Reinicke (1943) deduced that the glycophore for the sugars 'was two oxygen tetrahedra in *para* position with two intervening carbon tetrahedra.' The structure for β-D-glucopyranose is shown in Fig. 9.4 as it appeared in a 1943 edition of *Der Zukerindustrie*. The structure is a three-dimensional representation of β-D-glucopyranose in the plane of the paper, and as depicted in Fig. 9.4, is viewed from above. After superimposing the two possible orientations for the tetrahedra, the apexes of the resulting six-pointed stars were connected to form a

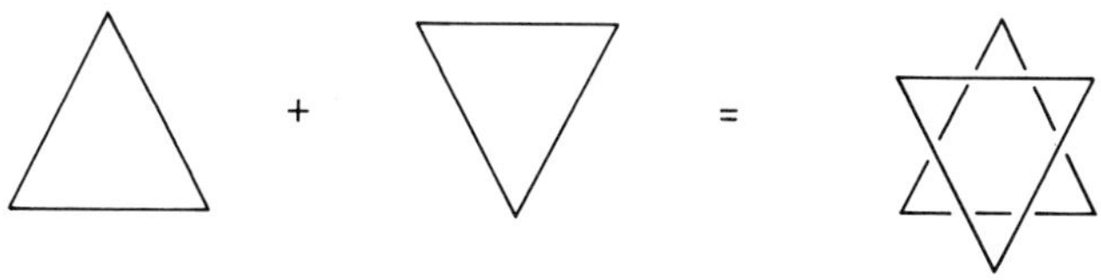

Superimposing two triangles to form a star

hexagon. Each carbon or oxygen atom tetrahedron is drawn with an apex pointing either up or down. Of the two possible orientations for each tetrahedron, those with the apex projected toward the viewer carry a central open circle. Those without the open circle have the apex oriented down. The structure shown in Fig. 9.4 is β-D-glucopyranose in the $^4C_1$ conformation (Section 5.1.2.2). The author seemed to indicate, through analogy to the structure of graphite, that the glycophore for this sugar was either all or one of the three possible glycol groupings with a diequatorial conformation. Reinicke's structures represent, in superimposed form, the two possible chair forms for a hexose sugar proposed by Haworth (1929), and he chose between them. In the drawings below, the two forms are isolated, and the possible glycophores proposed for β-D-glucopyranose in the $^4C_1$ conformation are shown in braces. The

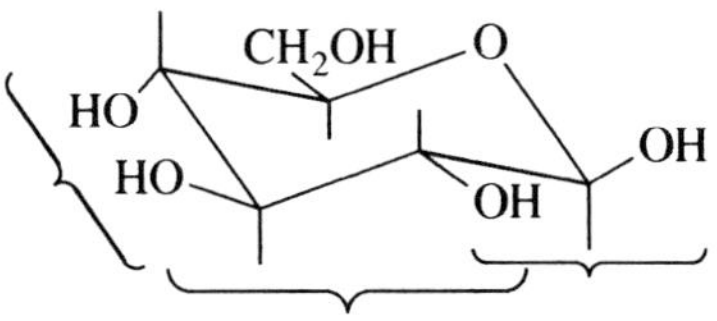

β-D-Glucopyranose $^4C_1$

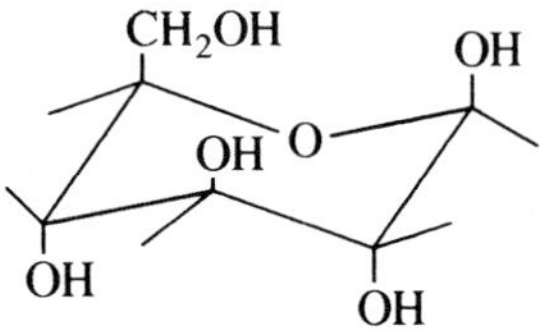

β-D-Glucopyranose $^1C_4$

sugar glycophores chosen are consistent with the knowledge that ethylene glycol (dihydroxyethane) is sweet, whereas ethanol (monohydroxyethane) is not. The way with which the ethylene glycol moiety might be viewed as a bifunctional group was not, however, mentioned.

9.3.7 Rationalization of the Varying Sweetness of the Sugars

Although hexoses have the same molecular formula ($C_6H_{12}O_6$), some are

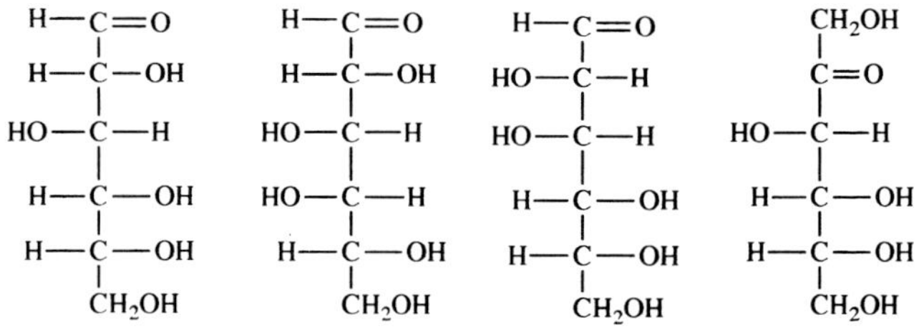

Glucose (1) Galactose (0.5) Mannose (0.5) Fructose (3)

FIG. 9.5. Comparison of the configurational structures for selected hexoses. The relative sweetness scores shown are for crystalline α-D-gluco- and galactopyranose, and β-D-manno- and β-D-fructopyranose, respectively.

sweet (e.g., glucose), others are very sweet (e.g., fructose), still others are mildly sweet (galactose and α-D-mannose), and one (β-D-mannose) is bitter. On the assumption that the glycol unit is the glycophore for the sugars, it remained to be established what the special features of that glycophore responsible for the varying sweetness of the sugars might be. Examination of the structural attributes of the various sugar glycophores commenced with consideration of the various forms available for depicting the structure(s) of the sugars.

9.3.7.1 *Fischer Configurational Structures*

Fischer projection structures for the monosaccharides afford an opportunity to examine sugar sweetness in the light of variable asymmetric carbon atom configurations. The configurational structure of glucose is 2+, 3−, 4+, 5+ pentahydroxyhexanal. The structure of this hexose, and the structure of other selected hexoses, along with estimates of the relative sweetness assigned to their crystalline forms are shown in Fig. 9.5. Structurally, crystalline galactose and mannose are the 4-epimer and 2-epimer of glucose, respectively. Transposing the configuration of either of those latter two carbon atoms of glucose diminishes relative sweetness by a factor of two. That observation leads immediately to the question as to what those two hydroxyl groups might have in common. The answer does not seem to be related to the change in their *cis*, *trans* relationship to other hydroxyl groups because the hydroxyl group on C-2 of fructose is absent, and the remainder of the molecule has a configuration identical to the configuration of glucose. In essence, glucose and fructose are the *aldehydo*- and the *keto*-analogues for an identical C-3 to C-5 configuration. As fructose in crystalline form is three times as sweet as glucose, the Haworth ring structures of these sugars were examined.

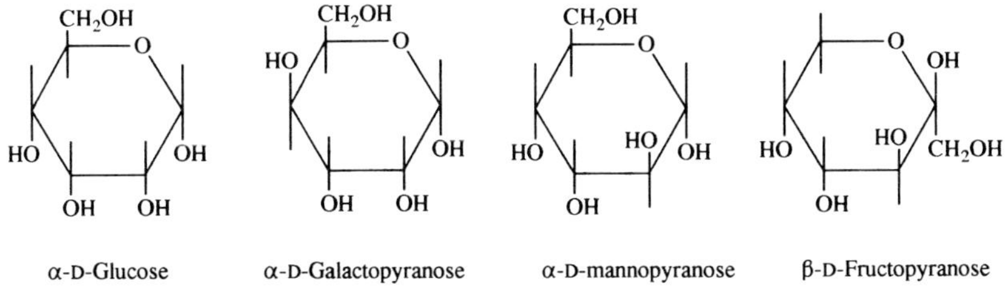

FIG. 9.6. Haworth structures for crystalline forms of hexoses of varying sweetness.

9.3.7.2 *Haworth Ring Structures*

Upon examination of the Haworth ring structures of the sugars, it is still not apparent what the significance of changing the configuration at C-4 in glucose to yield galactose and at C-2 in glucose to yield mannose might be (Fig. 9.6). Both the OH-4 group of galactose and the OH-2 group are on the top face of the Haworth ring, and are two carbon atoms removed from the ring oxygen atom, but what the taste significance of that common feature might be yet remains obscure.

Indeed, in considering the ring forms of the sugars, the relation between structure and varying sweetness became more complicated. Instead of one sugar for the glucose, galactose, and mannose configurational structures, there are now two. They are known as the *alpha*- and *beta*-anomers, and are in equilibrium in water solution (Section 5.3.3). While the two anomers of both glucose and galactose possess about the same relative sweetness, and the α-D-anomer of mannose is also sweet, the β-D-anomer is bitter. In addition, when crystalline fructose (β-D-fructopyranose) is dissolved in water, it establishes an equilibrium with other tautomeric forms, primarily the furanose ring forms, and its sweetness markedly diminishes.

9.3.7.3 *Conformational Structures*

Ultimately, sugar ring forms are puckered and not planar as depicted in Haworth formulas. As a result, any hydroxyl group written either above or below the ring in the Haworth formula now occupies one of two possible positions in the favored puckered-ring form. Whether above or below the ring, it can now also be *axial* or *equatorial* (Section 5.1.2.2). It

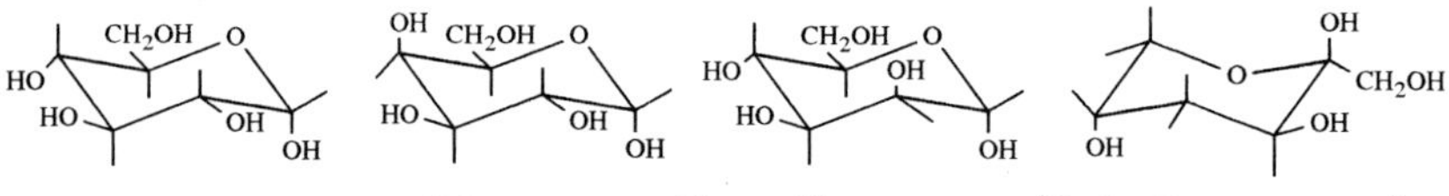

α-D-Glucopyranose $^4C_1$ α-D-Galactopyranose $^4C_1$ α-D-Mannopyranose $^4C_1$ β-D-Fructopyranose $^2C_5$

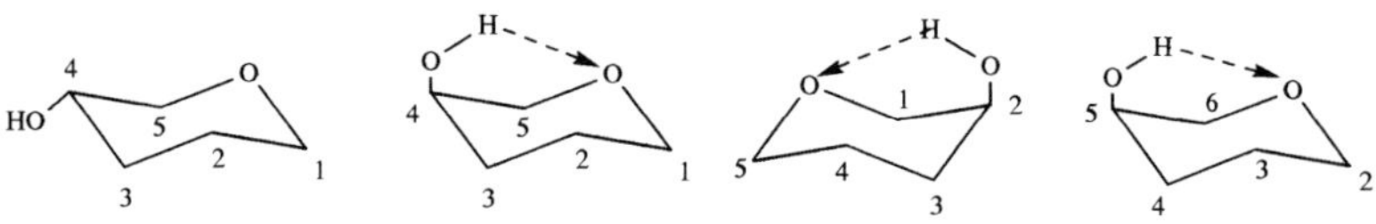

α-D-Glucopyranose $^4C_1$ α-D-Galctopyranose $^4C_1$ α-D-Mannopyranose $^4C_1$ β-D-Fructopyranose $^5C_2$

FIG. 9.7. Favored conformations of selected hexose sugars illustrating the steric relation of hydroxyl groups to the ring oxygen atom.

was here that a common structural denominator between OH-4 of galactose and OH-2 of mannose was found. While both were above the ring and two carbon atoms removed from the ring oxygen atom in the Haworth formulas, it was now observed that both OH groups are axial, i.e. perpendicular to an average plane for the ring. The analogous OH groups in glucose occupy an equatorial position, and are in about the same average plane as that for the six-membered ring.

Figure 9.7 depicts the favored conformations for the hexoses, followed by reorientation of the favored conformations for mannose and fructose. This emphasizes the steric relations between those ring hydroxyl group substituents. The feature that galactose and mannose have in common with each other, but not with glucose, is that the distinguishing axial hydroxyl groups bear the same, albeit mirror-image, disposition with respect to the position of the ring oxygen atom.

Two possibilities now arise as to why galactose and mannose are only one-half as sweet as glucose. The first is that they do not fit the sweetness receptor as well as glucose, although space-filling (CPK) models of galactose, mannose and glucose indicate that their shapes are only slightly different. In addition, there are many facets of sweetness-structure phenomenon that a 'lock and key' tenet do not satisfactorily answer. Fructose has the same C-3 to C-5 configuration as D-glucose. The steric disposition for OH-5, with respect to the ring oxygen atom, is also the same as that as for OH-2 and OH-4 of mannose and galactose. This feature is apparent when the stripped structure for fructopyranose, employing the $^5C_2$ conformation, is shown in Fig. 9.7. Nevertheless, fructose is three times as sweet as glucose.

Although an axial/equatorial and equatorial/equatorial pair of hydroxyl groups are the same distance apart, and can equally serve as Reinicke's (1943) glycophore, a second reason was postulated for the glucose, galactose, and mannose sweetness difference. It was based on the observation that an axial hydroxyl group at a given position in diastereoisomers is not necessarily the same distance as an equatorial group from a fixed point elsewhere in a sugar molecule.

Thus, both galactose and mannose have axial OH groups the same distance from the ring oxygen atom, and they are closer to it than is an equatorial group at the same position. Presumably then, they might interact with the ring oxygen in some manner that diminishes ability of at least one glycol grouping to elicit sweet taste. This thought served as the basis for exploring the possible role of hydroxyl group hydrogen bonding as a reason for the varying sweetness of the sugars.

9.3.7.4 *Intramolecular Hydrogen Bonding in Model Compounds*
Studies on the propensity of hydroxyl groups on six-membered rings to hydrogen bond intramolecularly were first initiated by Brimacombe *et al.* (1958) and Barker *et al.* (1959). Upon examining infrared spectra of selected alcohols in dry carbon tetrachloride, at a concentration where intermolecular hydrogen bonding was not a factor (< 0·005 M), single and multiple hydroxyl absorption bands were observed. Following principles set forth by Kuhn (1952), a sharp band occurring at 3630 cm^{-1} was assigned to free OH absorption. A second broader band occurring at 3580 cm^{-1} was assigned to intramolecularly hydrogen-bonded hydroxyl groups. Results obtained with cyclohexane and cyclopentane diols, and derivatives of tetrahydropyran and tetrahydrofuran, are given in Table 9.2. Data (Barker *et al.*, 1959) to demonstrate than an axial hydroxyl group two carbon atoms removed from a six-membered ring oxygen atom will bond to that atom are shown in Fig. 9.8. The compound in the figure is tetrahydropyran 3-ol. In Table 9.2 it can be seen that *trans*-cyclohexane diol shows an absorption spectrum for both free and hydrogen-bonded hydroxyl groups, and the compound probably exists as an equilibrium between two forms. Either hydroxyl group can be the

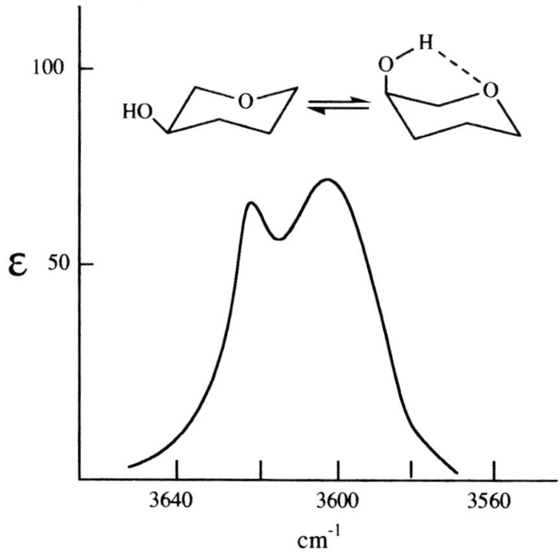

FIG. 9.8. Infrared absorption for the hydroxyl region of tetrahydropyran 3-ol (Barker *et al.*, 1959).

TABLE 9.2
Absorption bands (cm^{-1}) for free and intramolecularly hydrogen-bonded hydroxyl groups in selected compounds, the estimated strength of the bond (Δv), and the probable structure as indicated by the bonding. Data of Brimacombe *et al.* (1958) and Barker *et al.* (1959)

Compound	*Free OH*	*Bonded OH*	Δv	*Structure*
Cyclohexane *trans*-1:2-diol	3633	3602	31	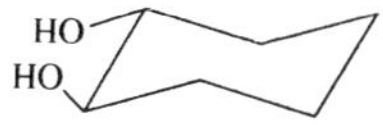
Cyclohexane *cis*-1:2-diol	3632	3592	40	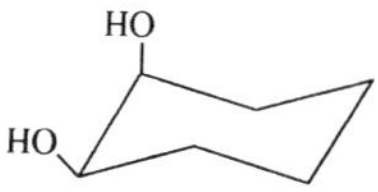
Cyclopentane *trans*-1:2-diol	3624	—	—	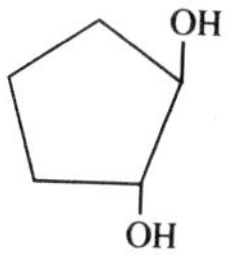
Cyclopentane *cis*-1:2-diol	3624	3579	45	
Tetrahydropyran-2-ol	3620	—	—	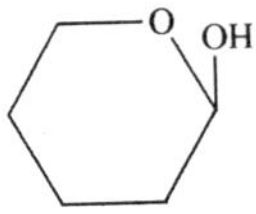
Tetrahydropyran-3-ol	3620	3604	16	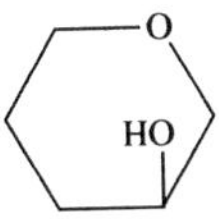
Tetrahydropyran-4-ol	3622	—	—	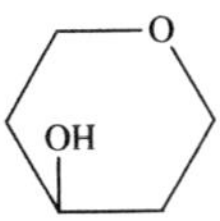
2-hydroxymethyl-tetrahydropyran	—	3597	—	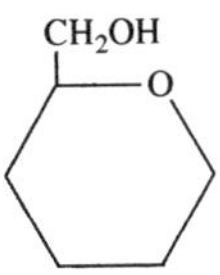

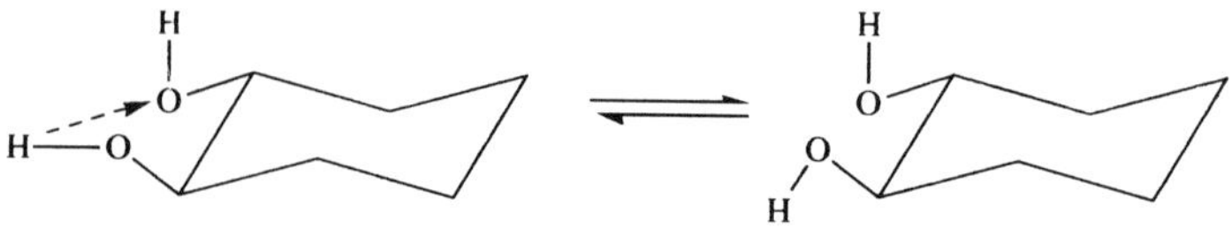

Equilibrium between hydrogen bonded and free *trans*-cyclohexane diol OH groups.

donor or the acceptor in the equilibrium. For *cis*-cyclohexane diol, results on the same order were obtained, but the intensity of the intramolecular hydrogen bonding is stronger, as indicated by a large value for Δv. This result probably has its origin in the observation that *cis* hydroxyl groups on a six-membered ring can approach each other more closely than can *trans* hydroxyl groups. The closest approach for *trans* OH groups is 3·0 Å; the closest approach for *cis* OH groups is 2·5 Å. The equilibrium for the *cis*-cyclohexane diol compound is probably between the free hydroxyl form and a combined chair and boat conformation, with the latter stabilized by a strong asymmetrically bifurcated intramolecular hydrogen bond.

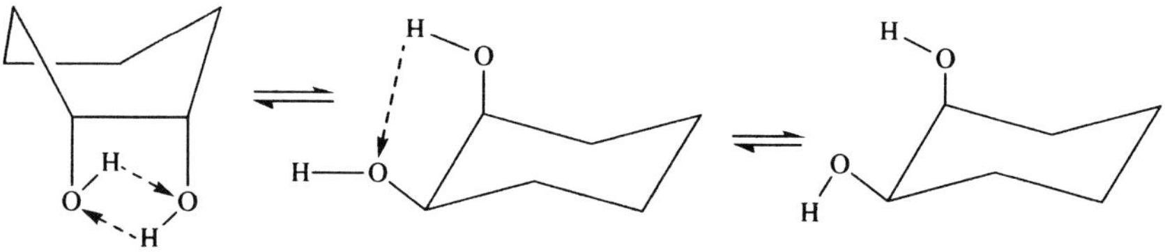

Equilibria between hydrogen bonded and free *cis*-cyclohexane diol OH groups

For the cyclopentane diols shown in Table 9.2, the *trans* configuration showed free hydroxyl group absorption only. The *cis*-cyclopentane diol showed significant intramolecular hydrogen bonding (large Δv). This indicates that the bifurcated hydrogen bond is particularly strong in five-membered ring compounds. This difference is probably due to the relatively small magnitude for five-membered ring puckering (Section 5.1.2.2), which holds the vicinal hydroxyl groups in a position that favors bifurcated bonding.

The monohydroxy tetrahydropyran and furan compounds, shown in

Table 9.2, serve as models for the propensity of sugar hydroxyl groups to hydrogen bond to the sugar ring oxygen atom. As shown in Table 9.2, this arises for the six-membered ring when a hydroxyl group is two carbon atoms removed from the ring oxygen atom, but not when it is at the first or third carbon atom. Similarly, the hydroxyl group of a methylene carbon atom, external to the ring, at the position of the first carbon atom exists only in the intramolecularly bonded form. This is also the case for the hydroxyl group of an extracyclic methylene carbon atom on the tetrahydropyran ring. However, in these models, a hydroxyl group at either the first or second position is not capable of bonding to the ring oxygen atom.

The aforementioned observations and assumptions derived from them would seem to preclude the ability of at least one of the sugar ethylene glycol groups from having any capability to elicit sweet taste. To explore the possibility that intramolecular hydrogen bonding might be responsible for the varying sweetness of the sugars, the infrared spectra of the sugars were examined. It became clear (Shallenberger, 1963) that the sugar glycophore is the glycol moiety, and the glycol sugar grouping(s), and glycol itself, are rendered bifunctional by virtue of intramolecular hydrogen bonding. Moreover, the sweetness of the sugars seemed to vary with the propensity of sugar hydroxyl groups to intramolecularly hydrogen bond.

9.3.7.5 *Hydrogen Bonding in Sugars*

At the time the sugar varying sweetness studies commenced, the only procedure available for deducing hydrogen bonding in substances was analysis of hydroxyl group absorption bands in the near infrared region of the electromagnetic spectrum. For sugars, a serious deficiency in the procedure was that it required working with solid materials (nujol mull technique), as the hydroxyl groups of water obliterate hydroxyl absorptions from substances dissolved in water. Nevertheless, the near infrared spectra for a number of sugars with varying sweetness were recorded (Shallenberger, 1963), and an example obtained for crystalline β-D-fructopyranose is shown in Fig. 9.9. As for alcohols in apolar solvents (Kuhn, 1952), and cyclohexane diols (Cole & Jefferies, 1956), sharp bands in the region of 3400–3600 cm^{-1} were assigned to free hydroxyl groups. Broad composite bands were assigned to hydrogen-bonded hydroxyl groups.

A crude measure of the approximate strength of the intermolecular hydrogen bonding in the crystal was obtained from the arithmetic

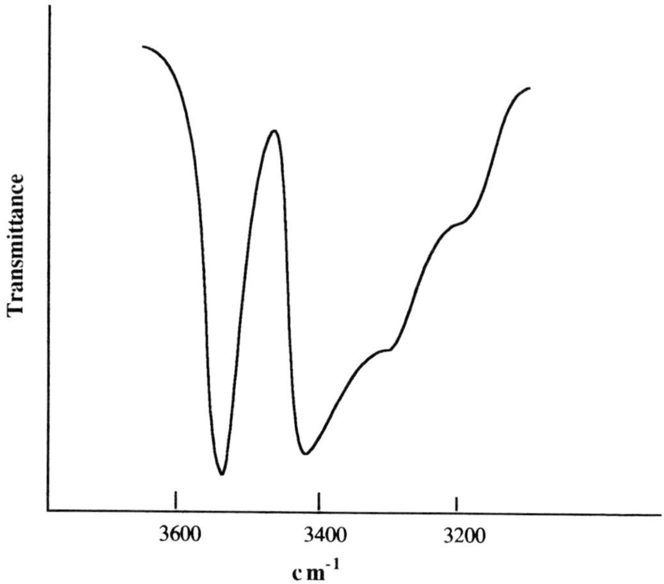

FIG. 9.9. Infrared spectrum of the hydroxyl absorption region of crystalline β-D-fructopyranose (Shallenberger, 1963).

difference (Δv in reciprocal centimeters) for the position (v, cm^{-1}) for the free hydroxyl, and for the bonded hydroxyl groups. Generally, among sugars that could be grouped in empirical series (fructose and fructose-containing oligosaccharides; aldohexoses; lactose anomers), Δv was largest for those sugars with the least sweetness. As hydrogen bonding in crystalline materials is almost entirely intermolecular, extrapolation of the data to indicate intramolecular hydrogen bonding propensity in solution was believed to be tenuous. Nevertheless, the hydrogen bonding parameters seemed to relate to other factors that are governed, in turn, by the magnitude of crystalline bonding and packing. Among them are water solubility, the ease of solubilization in water, and even the hygroscopicity of the crystals. Those factors seemed to be inversely related to any propensity for sugar intramolecular hydrogen bonding.

At the time, one widely accepted tenet maintained that intramolecular hydrogen bonding in any compound ought to be completely quenched in a polar environment, particularly in water. On the other hand, the polarity of the environment in which the sugar sweetness reaction occurred was not known with any degree of certainty either. The

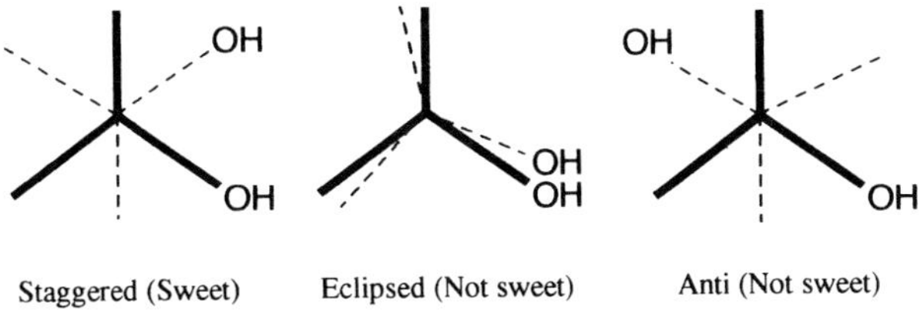

FIG. 9.10. Relation between the disposition of adjacent (vicinal) hydroxyl groups, in modified Newman projection, and their propensity to intramolecularly hydrogen bond (staggered, weak; *eclipsed*, strong; *anti*, none). Their proposed ability to elicit sweet taste is also given.

prevailing heuristic notion, especially among physiologists, was that the environment was very probably lipophilic and hydrophilic, or biphasic. Presumably the initial chemistry of the sweet taste sensation takes place in a lipophilic/hydrophilic environment analogous to the postulated nature of biological membrane surfaces. In that case, intramolecular hydrogen bonding among the sugars might very well be a factor governing their relative sweetness.

It is now recognized that when a hydroxyl group is near an inherent hydrophobic function within the same compound, it has little recourse

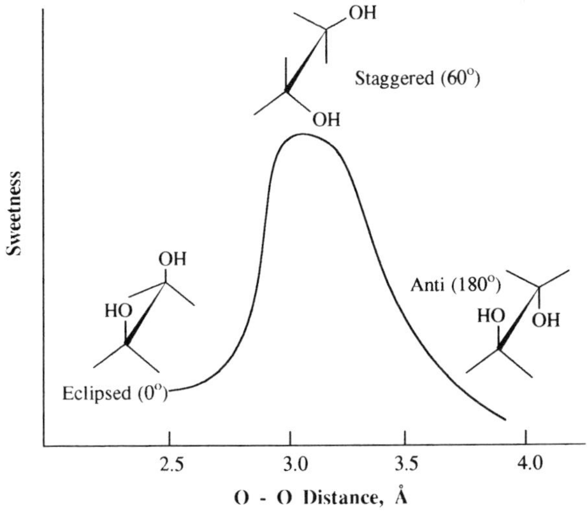

FIG. 9.11. *Pseudo* energy-like diagram to illustrate that the optimum glycol conformation for sweetness is staggered, with a projected hydroxyl group angle of 60°.

but to hydrogen bond to a neighboring polar function, regardless of the nature of the solvent. The methylene hydroxyl group of the sugars is one example. Because of the relative hydrophobicity of its surroundings (Section 5.4.4), it hydrogen bonds intramolecularly to the neighboring group (Lemieux & Brewer, 1973). In an apolar environment, a chain of cooperative hydrogen bonds (Section 2.7.3.4) is established. Two examples are shown below for β-D-glucopyranose. Following the

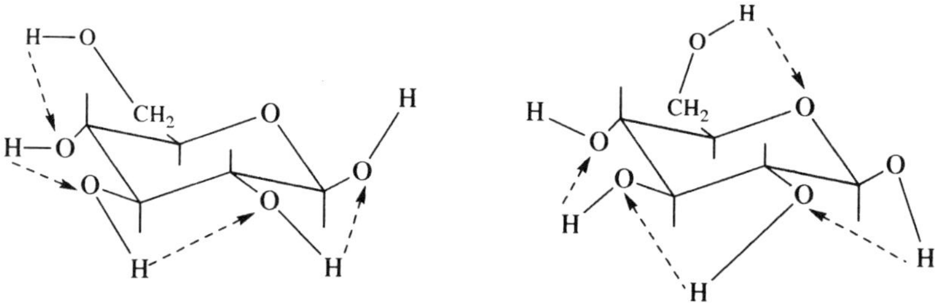

Cooperative hydrogen bonding pattern examples in β-D-glucopyranose

suggestion of Lemieux & Brewer (1973), it was proposed (Shallenberger & Lindley, 1977) that such intramolecular hydrogen bonding would in effect make OH-4 strongly disposed to bond to the ring oxygen atom if it occupied an axial position (e.g., galactose). However, it has been shown (Nishida *et al.*, 1984) that in water solution the primary alcohol hydroxyl group at C-6 is associated most of the time with the ring oxygen atom. Presumably then, the cooperative chain-like hydrogen bonding would proceed in counterclockwise fashion, both in water solution and in a relatively apolar environment. In either case an axial OH-4 is a special steric case with multiple potential for bonding.

Upon examining many sugars and sugar derivatives, it became clear that when any hydroxyl group was sterically disposed to hydrogen bond, either to another hydroxyl group in an *eclipsed* conformation, or to the ring oxygen atom, sugar sweetness was diminished. In addition (Shallenberger, 1963), it was observed that sweetness was diminished when glycol hydroxyl groups were too far apart, as in the *anti* glycol conformation (Fig. 9.10). This proposal, fundamental to explaining the varying sweetness of the sugars, is shown as a *pseudo*-energy diagram in Fig. 9.11. Sweetness then seemed to be at a maximum when the glycol OH groups had a proton-to-neighboring oxygen atom distance of about 3 Å. The empirical plot shown in Fig. 9.11 is derived from considering compounds in which all hydroxyl groups appended to a ring form of the sugars are *eclipsed* (fructofuranose structures), or *anti* (crystalline levoglucosan, or solution fructofuranoses), or *staggered* (glucopyranose anomers).

An immediate corollary to this deliberation arose. If the sugar glycol moiety was viewed as an AH,B system, as normally used to define and describe the hydrogen bond (Section 2.7.3.1), it seemed that this unit might be the common denominator to explain sweetness in compounds regardless of chemical class. Then, if the sweetness of sugars varies inversely with their propensity to hydrogen bond intramolecularly, a second corollary is that the initial chemistry of the sweet taste response is due to intermolecular hydrogen bonding with the receptor site. The glycol moiety, as an AH,B unit must, then, both sterically and functionally, be isomorphic with the receptor site.

9.3.8 The Glycol Group as an AH,B Hydrogen Bonding Unit

That hydrogen bonding played a role in the varying sweetness of the sugars, and that the glycol unit is the glycophore for the sugars, receives support from the realization that the phenomenon is probably the reason why ethylene glycol tastes sweet. The required bifunctionality is induced by intramolecular hydrogen bonding. Thus, while sugar glycol groups,

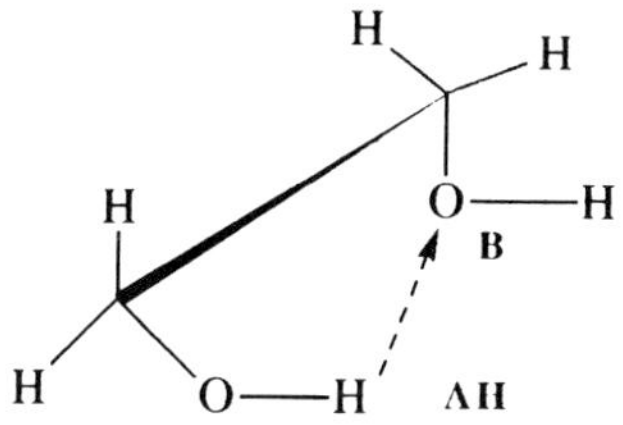

Ethylene glycol

and glycol itself, obtain necessary bifunctionality through intramolecular hydrogen bonding, the intramolecular bonding might also serve as a weak hydrophobic probe for access to the receptor site (Section 9.3.8.3). In any event, it seemed that if sugar glycol hydroxyl groups are too strongly bonded intramolecularly, their ability to act as a glycophore is restricted. The rationalization of the varying sweetness of the sugars, on the basis of hydrogen bonding, could then be examined from the viewpoint of the glycol unit as an AH,B hydrogen bonding system.

9.3.8.1 *Cyclitols and Aldohexoses*

The notion that α-glycol groups with the staggered conformation were the AH,B unit of the sugars was carefully examined by Birch and others (cf. Birch, 1976). Those studies included compounds such as cyclohexane polyols (Birch & Lee, 1971), deoxy sugars (Birch & Lee, 1974), and sugar methyl ethers (Lindley *et al.*, 1975). It was found that cyclohexane 1,2/4,5-tetrol was tasteless even though it contained two AH,B units. Parenthetically, cyclohexane 1,2/4,5-tetrol is a symmetrical molecule.

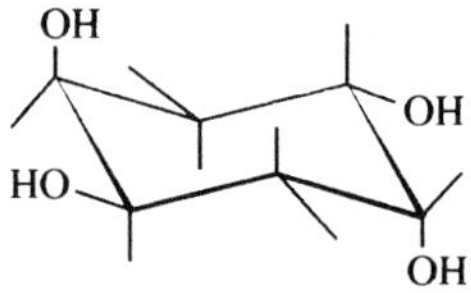

1,2/4,5 Cyclohexane tetrol

It was also found that D-protoquercitol, with a configuration analogous, except for the ring oxygen atom, to that of α-D-mannose was tasteless. On the other hand, (+)-viboquercitol, with a configuration analogous to that of α-D-glucose was sweet, and although the enantiomer is listed as having trace sweetness, it too is probably sweet. Neither of the enantiomers have yet been isolated in pure form, and the taste assessment applies to a racemic mixture. Directing their attention to the inositols, the Birch group found that only *allo*-inositol, and the two cyclohexanehexols (+)- and (−)-*chiro*-inositol were sweet. The latter two substances are configurationally related to D- and L-glucose. All other inositols, including the *epi*-, *neo*-, *myo*-, and *muco*-inositols, had only trace sweetness (Fig. 4.4).

It therefore seemed that the hemiacetal heterocyclic oxygen ring structure was a primary factor for sweetness in the sugar series, in contrast to the cyclohexane ring for the cyclitols. That assessment is probably true. The reason for it seems to be that substitution of the two different types of rings with axially or equatorially disposed hydroxyl groups leads to the presence or absence of symmetry in the two series of substances. Cyclitols that are sweet have, without exception, a dissymmetric distribution of axial/

equatorial ring hydroxyl substituents. They are (+)- and (−)-*chiro*-inositol, *allo*-inositol, and (+)- and (−)-viboquercitol (Fig. 4.4). With the exception of (+)-protoquercitol (trace sweet), all other cyclitols with only trace sweetness are symmetrical. An asymmetric electron distribution pattern seems to be prerequisite for sweetness among the cyclitols, as it also is for ethylene glycol. In the case of cyclitols, the dissymmetry is imposed by the hydroxyl group substitution pattern. In this latter sense, ethylene glycol has an asymmetric electron distribution pattern, but 1,2/4,5 cyclopentane diol does not.

From examination of the sweetness of a number of sugar derivatives, Birch & Lee (1971) proposed that OH-4 of the glucopyranosyl structure is of primary importance in the sweetness of glycosides containing glucose. Upon examining α,α-trehalose and α,α-*galacto*-trehalose Birch *et al.* (1975) and Birch (1976) noted that, in accord with the proposal by Shallenberger (1963) for glucose and galactose, the fourth hydroxyl group of those substances possessed special hydrogen bonding features. The group participates in a particularly strong hydrogen bond in the α,α-trehalose crystalline lattice structure. Upon comparing the sweetness of methyl α-D-glucopyranoside and α,α-trehalose, it was found that isomolar solutions are equally sweet. This indicates that there is but one common group for the sweetness of these substances, even though one is a monosaccharide and the other a disaccharide. Apparently, dual or multiple AH,B units (multiple glycophore equivalency), does not necessarily ensure enhanced sugar sweetness. Possibly, the solution conformation of trehalose leads to only one functional AH,B glycophore.

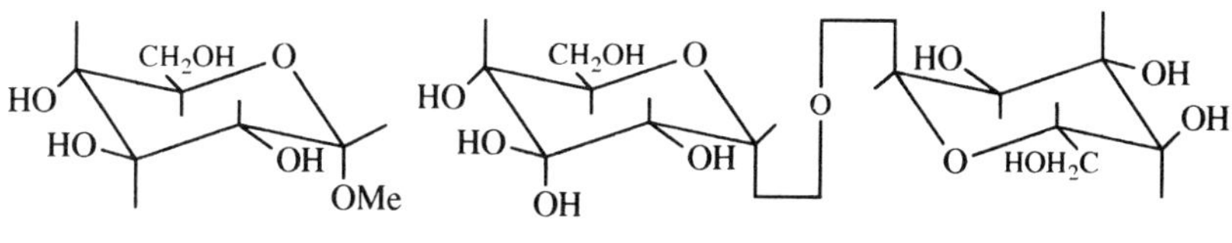

Methyl-α-D-Glucopyranoside

"α,α-Trehalose"

The latter observation also suggested to the Birch group that the fructofuranoside moiety of sucrose must be involved in eliciting sweet taste. On a molar basis, sucrose is the sweetest sugar, and the AH,B equivalency of sucrose is probably greater than one.

With OH-4 tentatively established as AH, it would seem to follow that O-3 must be B in *aldo*-hexopyranose structures. The search for a primary glycophore for the sugars, employing substitution and other reactions, is fraught with uncertainty. Minor changes in the electronic nature of a given compound may have major effect on the electronic nature of the glycophore. Nevertheless, the assignment of AH to OH-4 receives support from the observation that cyclohexane 1,2/4,5 tetrol, an analogue of galactose, is tasteless (Section 4.3).

The primary AH,B tenet also gains support from consideration of the structure and sweetness of deoxy sugars. Birch & Lee (1974) noted that sweetness reduction occurs in monodeoxy sugar derivatives only when the oxygen atom of the third hydroxyl group is eliminated. While this finding does not seem to be immediately in accord with the assignment of OH-4 to AH, it is not clear from results obtained with deoxy sugars whether OH-3 is AH or B in the parent compound. In fact, O-3 could be B if OH-4 is AH, and if OH-3 is AH, then O-2 might serve as B. In any event, it was found that while monodeoxy sugars are always either sweet, trace sweet, and only sometimes bitter, dideoxy sugars are never sweet, and always bitter.

Upon examining the methyl glycosides of glucose, it was found that when the proton of an additional hydroxyl group was substituted with a methyl group (At C-2 through C-6), the resulting substances were still sweet, although they also had a trace of bitterness. This result seemingly contradicts that obtained with deoxy sugars where removal of the third oxygen atom caused marked reduction in sweetness. However, in the case of the methyl glycosides, the third oxygen atom is not removed, only the proton. For the most part, then, O-3 must be B in the primary AH,B system, and OH-4 must be AH. Elimination of either of those positions

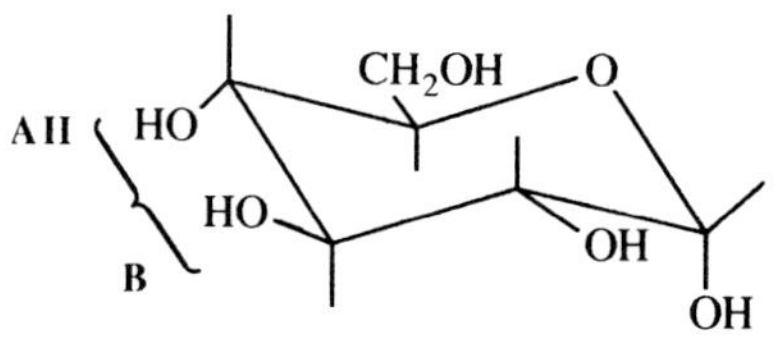

Primary AH,B in α-D-glucopyranose

by substitution does not necessarily eliminate sweetness. Subsidiary AH,B units may yet remain to elicit sweetness. The assignments gain credence with recognition that carbohydrate substances, where OH-4 is substituted (maltodextrins, starch, lactose), are either tasteless, or sweetness is diminished. There are cases that arise through substitution

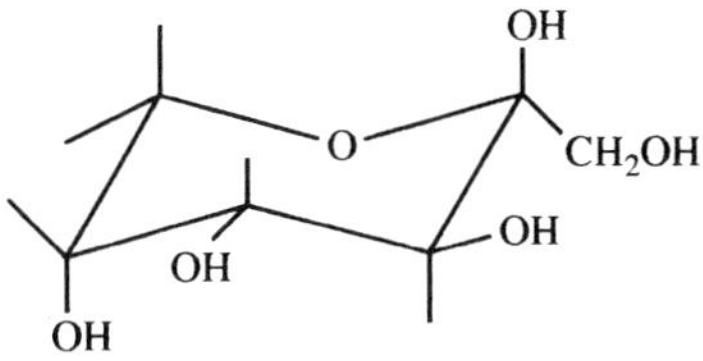

β-D-Fructopyranose

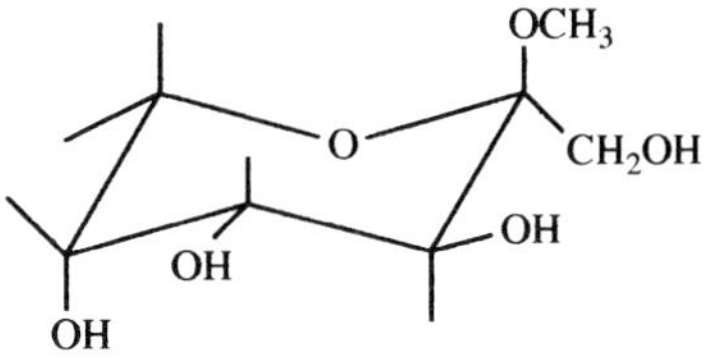

methyl-β-D-Fructopyranoside

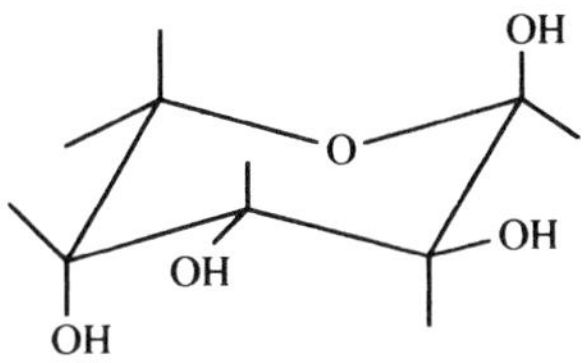

β-D-Arabinopyranose

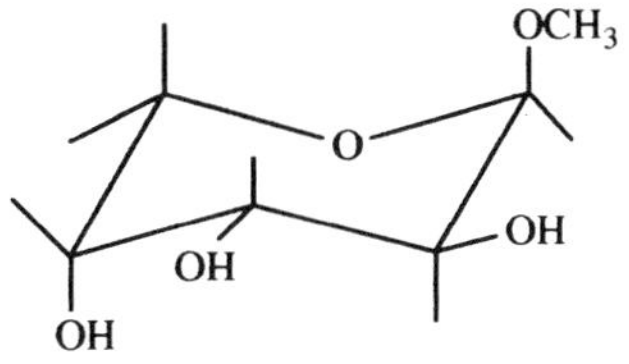

methyl β-D-Arabinopyranoside

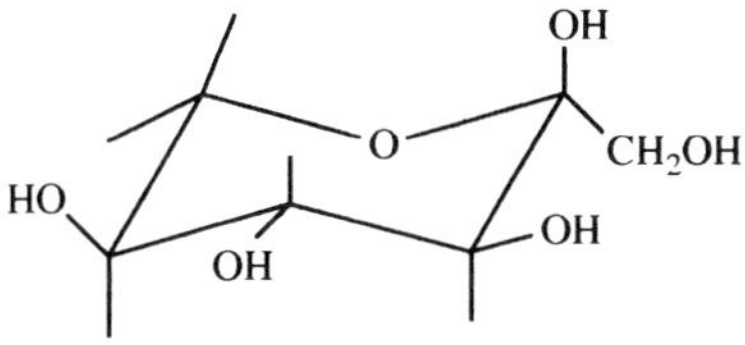

α-L-Sorbopyranose

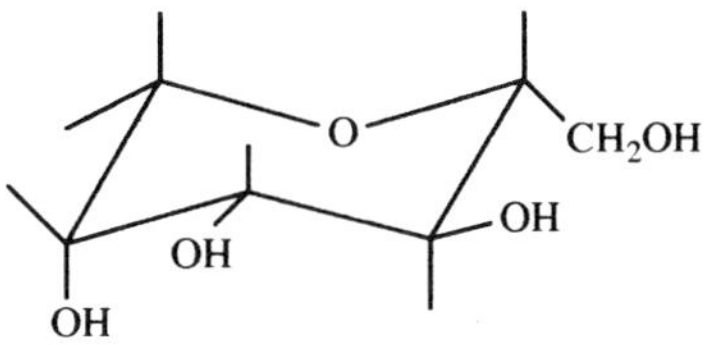

2-deoxy-D-Fructopyranose

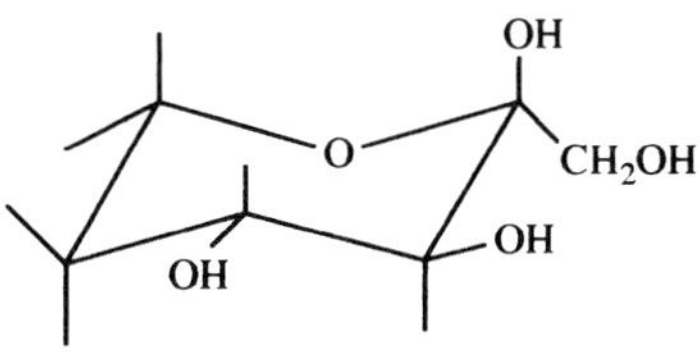

5-deoxy-D-*threo*-Hexulose

FIG. 9.12. The structure of various sweet ketose sugars, derivatives, and structurally related aldopentose analogues.

that suggest, at times, OH-3 acts as AH, and O-2 as B, and *vice versa*. With multiple AH,B systems, the sugars exhibit great versatility and variation with respect to primary AH,B systems.

By applying observations derived from intramolecular hydrogen bonding phenomena, infrared studies, and the effect of hydroxyl group substitution or elimination, a rather complete picture of the reasons for the varying sweetness of the aldohexoses begins to emerge. For glucose, the identity of OH-4 as AH and O-3 as B is commensurate with the decrease in sweetness when OH-4 is substituted (lactose, maltose). That OH-4 is a primary aldose AH is commensurate with the observation that, when it is disposed to intramolecularly hydrogen bond, sweetness is diminished (galactose). While OH-2 of mannose is not necessarily a primary AH, the bonding of that group to the ring oxygen atom diminishes sweetness by virtue of eliminating one set of available glycol groups. Alternatively, or perhaps in addition, an axial OH-2 group sterically hinders primary AH,B from interacting with the receptor.

9.3.8.2 *Ketohexoses*

The identification of primary AH in β-D-fructopyranose, and the identity of B, varies with interpretation of infrared spectra, and the significance of substitution reactions. The structure of a number of reference compounds employed to evaluate the AH,B units of selected ketoses are given in Fig. 9.12. As an example of an AH,B unit in the sugars, Shallenberger & Acree (1967), selected the OH-2 group β-D-fructopyranose as AH, and the OH-1 oxygen atom was proposed as B. The assignment was based on the knowledge that di-D-fructose dianhydride I, which does not contain a free anomeric center, is tasteless (Section 5.3.6). It was also based on the notion that the fructopyranose OH-2 group is the most 'acidic' hydroxyl group. If OH-2/O-1 was indeed the primary AH,B unit, Lindley & Birch (1975) reasoned that α-L-sorbose should also be quite sweet. After consideration of the free hydroxyl, and the hydrogen-bonded absorption bands of a series of ketose sugars (2-deoxy-D-fructopyranose, methyl β-D-fructopyranoside, β-D-arabinopyranose, methyl-β-D-arabinopyranose, and α-L-sorbopyranose), Lindley & Birch (1975) proposed that the primary AH,B unit of β-D-fructopyranose was the same as that assigned by Shallenberger & Acree (1967). The authors pointed out that β-D-fructopyranose and β-D-arabinopyranose are sweet. They also show strong free hydroxyl absorption bands at 3520 and 3530 cm^{-1}, respectively. On the other hand, 2-deoxyfructose and methyl β-D-fructopyranoside are much less sweet. While β-D-arabinopyranose is not as sweet as fructose, methyl-β-D-arabinopyranose is even less sweet. As α-L-sorbopyranose is also less sweet than fructose, and since deoxy fructose and the methylated derivatives show no free hydroxyl absorption band, the authors concluded that OH-2 is AH and O-1 B for the primary fructopyranose glycophore.

The above reasoning was based on the premise that OH-5 in fructose is bonded to the ring oxygen atom, allowing OH-2 to remain relatively unbonded. However, OH-5 in α-L-sorbose is incapable of bonding to the ring oxygen atom, and OH-2 of sorbose does bond. In the intramolecularly bonded form, it is incapable of partaking in the intermolecular hydrogen bonding phenomena believed to be prerequisite to sweetness among the sugars.

Shortly after the Lindley–Birch proposal, Martin *et al.* (1982) reexamined the reason given for the low sweetness of a α-L-sorbopyranose. They synthesized 5-deoxy-D-*threo* hexulose(5-deoxy-'β-D-fructose,' 5-deoxy-'α-L-sorbose'). Taste tests indicated that 5-deoxy-D-*threo*-hexulose is much sweeter than L-sorbose, and nearly as sweet as D-fructose. It was concluded that the proton releasing tenet of Lindley & Birch (1975) is open to question. Martin *et al.* (1982) concluded that the large difference in sweetness between the ketohexoses epimeric at C-5 is probably steric in nature. This precludes correct alignment of the sugar (sorbose) on the taste receptor. The authors do not consider the fact that 5-deoxy-D-*threo* hexulose is far more lipophilic than L-sorbose, and one reason for its higher sweetness score may be due to more rapid access to the receptor (Section 10.3.4).

A final possible reason for the varying sweetness of these substances is that the axial hydroxyl group at C-5 for fructose imparts an element of

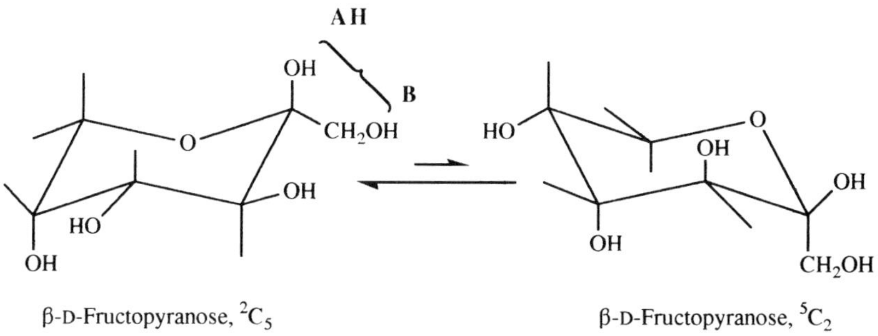

conformational stability to the fructose conformation. It maintains the axial OH-2 proton to the O-1 oxygen atom distance of 3 Å. The two axial hydroxyl groups in the fructopyranose $^2C_5$ conformation counterbalance each other, and this conformational stability element is lost in the sorbose and arabinose structures.

Additional attempts to identify the primary AH unit of fructose, through interpretation of infrared absorption spectra, gave mixed results. Examination of the hydroxyl absorption bands for the sugars using infrared and Raman laser techniques (Szarek *et al.*, 1984) resulted in assignment of the absorption bands of sugars to specific hydroxyl groups. Through correlation with crystallographic patterns from neutron diffraction and X-ray studies, estimates were made with respect to the distance between the donor group's proton and the receptor hydroxyl group's oxygen atom. This led to an estimate of bonding strength. Results obtained for β-D-fructopyranose were as follows:

The broad band in Fig. 9.9 occurring at 3190 cm^{-1} was assigned to the anomeric hydroxyl group (OH-2). It was thought to bond through an infinite cooperative chain to the oxygen atom of the OH-1 hydroxyl group contained in a second fructose molecule in the crystal. This bond was the shortest (1·750 Å), and strongest. The next strongest bonds were between OH-1 and O-3 (1.965 Å), and OH-3 and O-5 (1.977 Å). The OH-1 and OH-3 groups were also observed to partake in an infinite chain structure. In addition, each of these hydroxyl groups was observed to participate in an asymmetric bifurcated hydrogen bond system involving the ring oxygen atom. The bifurcated structure is believed to be energetically comparable to, or more favorable (by 1–2 kcal mol^{-1}), than a normal, linear hydrogen bond. The OH-5 group was assigned a bonding function with O-2 with an absorption band at 3415 cm^{-1}, and a length of 1·869 Å. Lastly, the OH-4 group was found to belong to the finite chain class of bonding (i.e., as a donor group only), with an absorption band at 3525 Å. As the distance for the bond is the longest (2·065 Å), this bond is the weakest, and yields the sharpest absorption band (Fig. 9.9). Szarek *et al.* (1984) then proposed that for fructose, the decreasing order of hydrogen bond strength is HO-2 > HO-1 > HO-3 > HO-5 > HO-4.

Szarek *et al.* (1984) further proposed that the strongly bonded hydroxyl group at 3200 cm^{-1}, assigned to OH-2 for both fructose and sorbose, is not as strongly bonded in the crystal for sorbose as it is for fructose. As it is not as capable of intermolecular hydrogen bonding with the receptor either, L-sorbose is not as sweet as D-fructose. The primary AH of fructose must then be assigned to OH-2.

Alternatively, Kanters *et al.* (1977) showed that the sharp absorption band at 3525 cm^{-1} is due to the hydroxyl group attached to carbon atom number one, and Mathlouthi & Portman (1990), proposed that the weakest bond in the crystal is probably the first to bind to the receptor.

AH is then probably OH-1 and B is O-2, which is in accord with the assignment proposed by Hough (1989). However, through analogy with the glucose glycophore, Birch & Shamil (1988), proposed that the primary glycophore of fructose is the 3,4-α-glycol unit. Clearly, the identity of primary AH,B for β-D-fructopyranose is fraught with uncertainty. After considering tripartite glycophore requirements in Section 11.2.2.4, it is concluded that OH-5 is AH and O-3 is B for the fructopyranose glycophore.

Among the chemical evidence cited for OH-2 as AH, assignment of the fructose primary AH to OH-2 can explain the decrease in fructose sweetness that occurs during either conventional or thermal-shift mutarotation. The fructofuranoses retain OH-2 as an AH,B unit during formation of their five-membered rings. OH-2 is now strongly disposed, due to its innate hydrophobicity, and to the steric location of O-6, to hydrogen bond to O-6 to form a *pseudo*-bicyclic compound (Shallenberger *et al.*, 1980). The

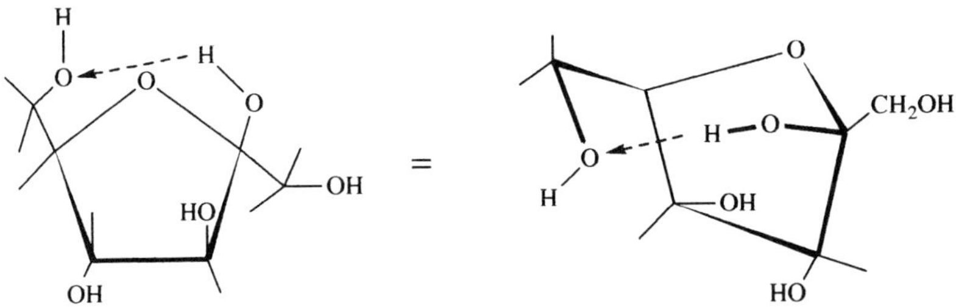

Formation and structure of pseudo bicyclic β-D-fructofuranose

postulated formation of the β-D-fructofuranose chelate ring is supported by the finding that the specific rotation is +64° (Shallenberger *et al.*, 1977), or perhaps +78°, by indirect computation (Shallenberger, 1978). However, Schneider *et al.* (1985) calculated that it is about −11°. The present author favors the direct computation made by Shallenberger *et al.* (1977), or the indirect computation of Shallenberger (1978), as increasing concentrations of fructose cause the equilibrium specific rotation of fructose to become more positive. This indicates that the major furanose form present (β-D-fructofuranose) is dextrorotatory (cf. Browne & Zerban, 1941). It is believed that free β-D-fructofuranose is devoid of sweetness (Section 6.4.1), by virtue of OH-2 being bonded to O-6. On the other hand, the anhydro-sugar analogue, 2,5-anhydro-D-mannitol (2,6-anhydro-(β)-D-fructofuranose) is sweet and bitter (Shamil *et al.*, 1989).

The observation that methyl β-D-fructofuranoside is devoid of sweetness is evidence for the notion that OH-2 in β-D-fructofuranose is intramolecularly bonded to OH-6 (Shallenberger, 1978). Furthermore, 6-thio-β-D-fructopyranoside is sweeter than the free sugar, and 5-thio-methyl β-D-fructofuranoside is essentially tasteless (Lindley *et al.*, 1976*b*). Upon chlorination to form 6-chloro-6-deoxy-D-fructofuranose, sweetness for a substituted free fructofuranose tautomer is obtained, although the compound is not as sweet as sucrose (Dziedzic & Birch, 1981; Hough, 1989). The finding that the 1,6 dichloro-derivative is even sweeter (Section 6.4.5.3) further supports the contention that the fructofuranoses are not sweet, and are probably bicyclic structures by virtue of intramolecular hydrogen bonding. The case for AH as OH-2 and O-1 as B for β-D-fructofuranose seems to have been settled by Gan *et al.* (1990). Those chemists prepared 1,6-di-*S*-methyl-1,6-dithio-D-fructofuranose, which cannot form a bicyclic chelate ring. It is reported that the compound is 15–20 times as sweet as sucrose, and it would appear that O-3 is the probable B function, and not O-1.

Further discussion of the identity of the fructose AH,B function is contained in the derivation of a tripartite glycophore in Section 9.3.10.2 and Section 11.2.2.4. In the latter section, it is concluded that for β-D-fructopyranose, OH-5 is AH, and O-4 is B. In that case, the fructofuranoses would not possess a primary AH function as OH-5 is involved in ring closure.

9.3.8.3 *Sucrose*

The glucose OH-4, O-3 groupings are one possible AH,B unit for sucrose. Sucrose, on a molar basis, is about three times as sweet as glucose, suggesting that the AH,B equivalency for this disaccharide is greater than one (Shallenberger, 1983*b*). Confirmation of that assessment was obtained by consideration of the sweetness of a sucrose configurational isomer, and partially methylated derivatives of sucrose. Lindley *et al.* (1975) found that while sucrose is sweet, *galacto*-sucrose has only trace sweetness. This indicates that the OH-4 group of glucose has a special role, not only in the sweetness of glucose, but also sucrose. Examination of the taste of methyl derivatives of sucrose revealed that they varied in sweetness depending upon the specific hydroxyl groups substituted. In the Haworth structure of sucrose, with numbered positions as shown below, it was found that 6′-*O*-methyl sucrose was quite sweet. Therefore, the C-6 hydroxymethylene group of sucrose apparently has little to do with sucrose sweetness.

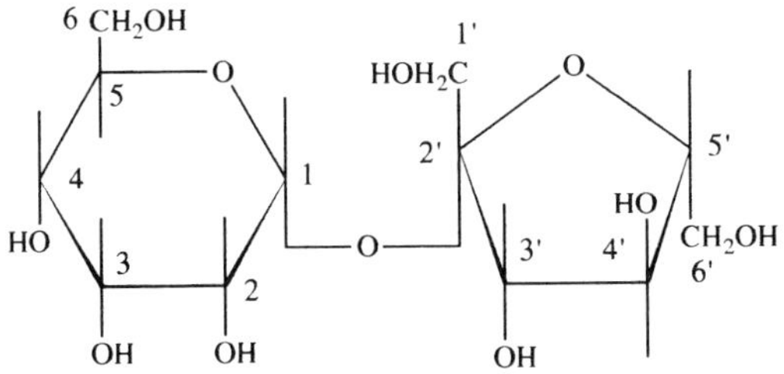

Numbered positions for the Haworth formula of sucrose

Methylation of the glucose No. 4 hydroxyl group diminished sucrose sweetness as 4-*O*-methylsucrose is not as sweet as the parent compound. This finding confirmed the probability that the glucose glycophore is functional in sucrose. Methylation of both the C-6 and C-6′ hydroxyl groups to form 6,6′-di-*O*-methylsucrose also failed to diminish the taste of the parent compound. This eliminated those hydroxyl groups as having any significant role in sucrose sweetness. The 4,6′-, 4,6-, and 4′,6-di-*O*-methyl esters are also not as sweet as sucrose, confirming the results obtained by substituting the 4-position only. However, methylation of the No. 1′ primary alcohol group and the 6′ primary hydroxyl group of the fructose moiety (1′, 6′-di-*O*-methylsucrose) also diminished the sweetness. This is the first concrete indication that the fructose OH-1 group plays some role in governing the sweetness of the parent compound.

Thus, the glucopyranoside AH,B unit is a possible candidate for sucrose AH,B (see also Mathlouthi & Portman, 1990), and the OH-2 group is an additional AH. The O-1 atom of the fructofuranoside moiety is an additional B. These groups were also selected by Boch & Lemieux (1982) as a probable sucrose AH,B unit. Explanation of the basis for the Boch & Lemieux (1982) rationale for the AH,B assignment is particularly significant. Those authors point out, through analogy to the probable initial recognition chemistry in immunological chemical reactions, that the main driving force for the binding of oligosaccharides to proteins is hydrophobic. As the glucopyranoside OH-2 group and the fructofuranoside O′-1 atom of sucrose are intramolecularly hydrogen-bonded, they serve as a hydrophobic probe to access the taste receptor. In concentrated solutions, the O-5 oxygen atom and the OH-6′ hydrogen atom may form a third sucrose AH,B system (cf. Mathlouthi, 1981).

Lichtenthaler *et al.* (1991) examined computer generated hydrophobicity potentials, and the hydrogen bonding network for sucrose. The authors concluded that the hydrogen bond between O-2 and OH-1′

of sucrose, or between O-2 and OH-3′, plays a role in enhancing the H-donor (AH) capacity of the glucosyl OH-2. Then, the glucosyl O-3 atom serves as B, as shown below, for the favored conformations of sucrose.

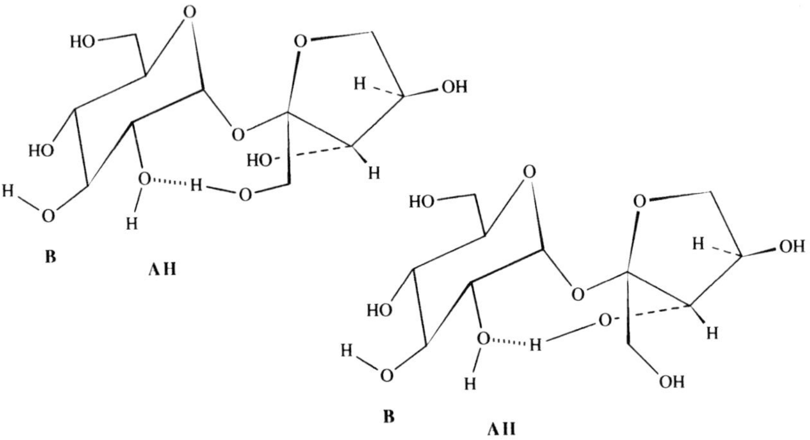

AH,B in two proposed hydrogen bonded structures of sucrose

Presumably, it is the bonding between OH-2 and O-3′ that serves as the hydrophobic probe for interaction with the receptor. The assignment is also favored by Rohse & Belitz (1993). Tran & Brady (1990) consider the sucrose molecule to be relatively rigid. What the implication might be with respect to the reactive groups for sucrose sweetness is open to question. According to Pérez *et al.* (1993), the description of sucrose as a rigid molecule under all conditions has to be mistrusted. Upon binding to

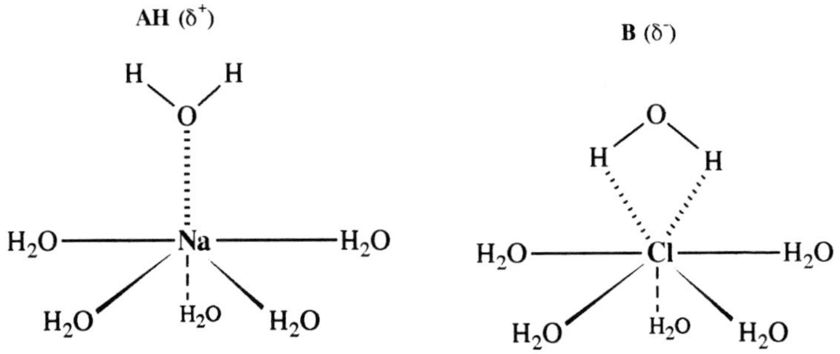

FIG. 9.13. Proposed AH,B units in weakly ionic, but fully hydrated, sodium and chlorine atoms in dilute solution.

a protein, carbohydrates can adopt conformations that correspond to higher energy local minima in response to local conditions.

The idea that hydrogen-bonded units (hydroxyl groups) of the sugars serve as a hydrophobic probe has interesting application and implication. The receptor can be considered to be hydrogen-bonded and to serve initially as a weak hydrophobic site. The thought is not unlike Kodama's (1920) notion of vibratory hydrogen (Section 9.3.3). The first stage for the initial chemistry of sweetness is hydroxyl group hydrophobic interactions, followed by stronger intermolecular hydrogen bonding interactions (Section 10.3.1). To document their observations, Boch & Lemieux (1982) point out that the saccharine molecule shows near isomorphous relationship with the hydrophobic region of sucrose. The NH grouping of saccharine occupies a region near that of the OH-2 group of sucrose.

An intriguing aspect of the analogy between sweetness and immunochemical reactions is that Hough & Edwardson (1978) found strong correlation between relative sweetness and formation of antibodies to sweet-tasting substances. In support of this latter tenet, Boch & Lemieux (1982) pointed out that sucrose binds only weakly with proteins. The tenet is also in accord with the notion that the described hydrophobic bonding region of the sugar should be wedge-shaped to penetrate satisfactorily the hydrophobic cleft of the receptor site.

To conclude Section 9.3.8 it can be stated that the difficulty associated with assigning AH,B, with confidence, to various sugar structures is about equal to the difficulty of understanding their seemingly intractable stereostructures. Nevertheless, elucidation of the primary sugar glycophores has application for deducing the nature of sweetness recognition chemistry.

9.3.9 AH,B in other Sweet Compounds

With establishment of the sugar glycophore as the fundamental unit necessary for sugar sweetness, the question arose how the glycophore might be related to the sweetness of nonsugars. As sugars varied in sweetness with propensity for intramolecular hydrogen bonding, it was clear that the sugar glycophore was simply an AH,B unit as used to define and describe the hydrogen bond. If that was the case, then it was possible that the chemical function common to all substances that taste sweet is an isosteric AH,B unit.

9.3.9.1 *Metal Salts*

The AH component of the glycophore in dilute solutions of lithium, sodium, and potassium chloride (Section 3.4.1) can be attributed to the fully hydrated cation ($M^+ \cdot OH_2$). The B component then is the fully hydrated anion ($Cl^- \cdot H_2O$). The identity of AH and B for dilute solutions of lithium, potassium, and sodium chloride, are probably those shown for NaCl in Fig. 9.13. In this sense, full hydration of certain metal ions yields AH and B forms capable of participating in hydrogen bonding phenomena, as opposed to the ionic structures of Fig. 9.2. Incomplete hydration at higher concentrations renders them more capable of partaking in electrostatic (ionic) interactions. At high dilution, increasing the NaCl concentration causes molecular conductivity to decrease slowly until a critical concentration is attained (Section 3.4). Thereafter conductivity decreases rapidly. It is at this point that taste changes from sweet to salty. In a system wherein the discrete cations and anions with a coordination number of six (Section 3.2.4) can be envisaged to be fully hydrated (Fig. 9.13), those complexes (with lowered conductance ability and lowered ionic nature) can possibly act as AH and B (Shallenberger, 1993). Incompletely hydrated anions and cations and/or ion clusters at higher concentrations then elicit saltiness (Section 9.2).

Beryllium chloride and lead acetate do not ionize in water solution in a conventional manner. Instead, they also form complex hydrated ions. The AH,B units of these substances are again assigned to water of hydration, as with dilute solutions of the lithium, sodium, and potassium salts. In the former case, however, AH,B resides in the structure of a single complex ion. Consequently, because of inductive effects, the sweet taste intensity of these inorganic substances is enhanced. One possible example of a discrete AH,B unit is shown below for the trimeric form of beryllium ion hydrate (see also Section 3.3.1), and beryllium chloride fits the category of a rather sweet substance (> 50×). The arguments for AH,B for hydrated lead acetate ions are approximately the same as those offered for beryllium chloride solutions.

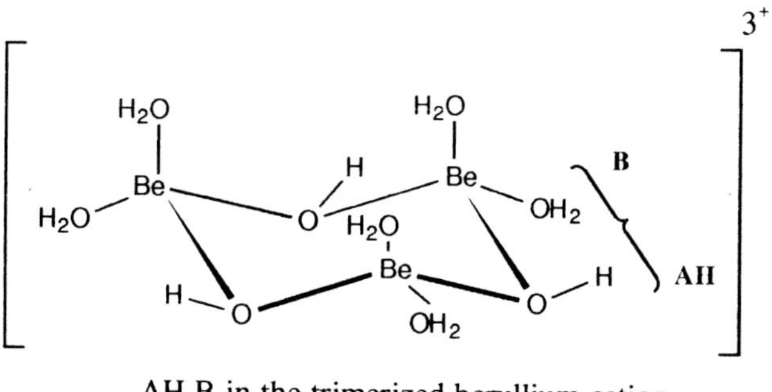

AH,B in the trimerized beryllium cation

9.3.9.2 *Halogenated Hydrocarbons*

Halogenated hydrocarbons, such as chloroform and tetrachloroethane, taste sweet. The ability of the hydrogen atoms in these compounds to participate in a hydrogen bond is due to the strong inductive effects (electron-withdrawing power) of the chlorine atoms. These, in turn, then serve as B in the AH,B system, as shown for chloroform and tetrachloroethane. The general inductive effects of halogens, NO_2, and CN have an important role in the increased activity of AH,B in high potency sweeteners such as the chlorosucroses and guanidine sweeteners (Chapter 8).

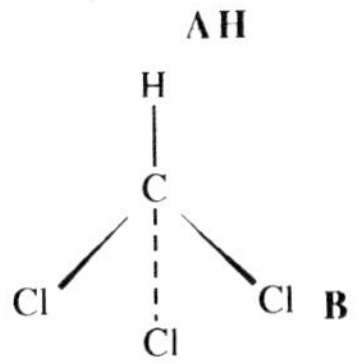

Chloroform

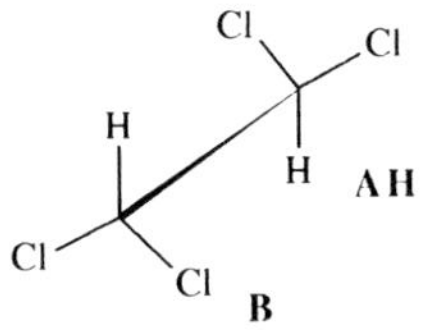

Tetrachloroethane

9.3.9.3 *Polyhydroxy Alcohols*

The formation of an intramolecular hydrogen bond between the two hydroxyl groups of glycerol degrades the electronic symmetry of the molecule. One hydroxyl group is a proton donor in a hydrogen bonding system, and the oxygen atom of the second is a proton acceptor (Section 9.3.8). For unsaturated monohydroxy alcohols, the AH function is apparently the hydroxyl group, and B is the unsaturated center. Alternatively, AH may be an 'acidic' methine proton at the carbon atom alpha to the carbon atom alpha containing the hydroxyl group.

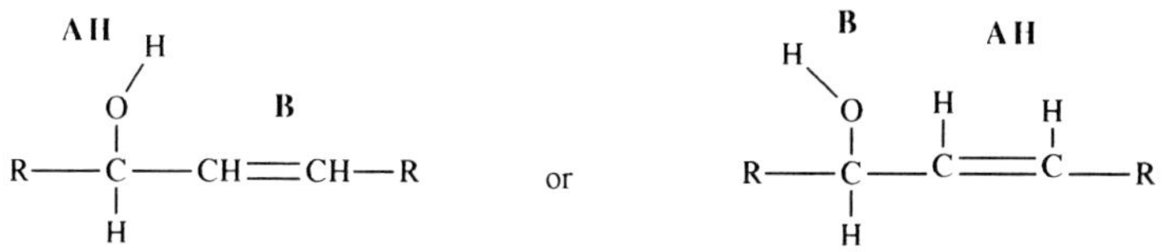

Unsaturated alcohols

The AH,B of diols with the hydroxyl groups separated by more than 3 Å (cf. Goodwin *et al.*, 1981) is not necessarily a pair of hydroxyl groups. It may be an acidic proton at an unsaturated center as AH, and the oxygen atom of a hydroxyl group as B. Thus, in substances such as phloroglucinol (Section 4.0), B is an oxygen atom of a hydroxyl group and AH is an aromatic ring proton. Activation of methine protons to serve as an AH unit is due to the inductive effects of the hydroxyl groups in phloroglucinol. Activation of methylene protons by the NO_2 groups of nitroglycerin (Section 1.5.3) allows them to serve as AH, while an oxygen atom of an NO_2 group serves as B.

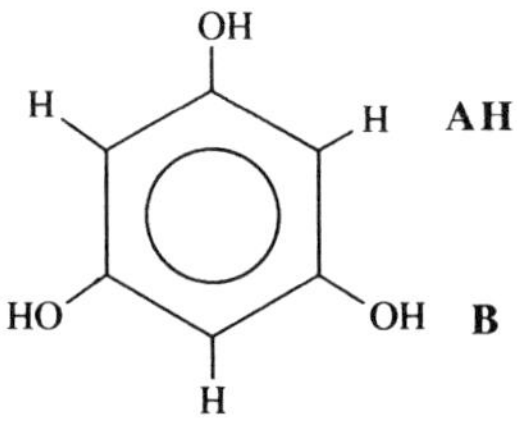

Phloroglucinol

9.3.9.4 *Amino Acids, Amides, Peptides, and Proteins*

Amino acids in water exist in a dipolar ionic, or zwitterionic form, but at the isoelectric point they have little net charge. For amino acids, amides, peptides, and proteins, the amine and the carboxyl oxygen atoms seem to be an unambiguous AH,B pair. The problem of the general sweetness of D-amino acids and bitterness of L-amino acids is reserved for Chapter 11.

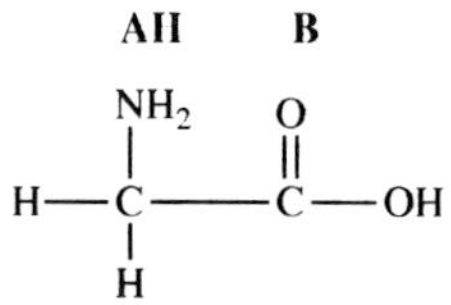

Glycine

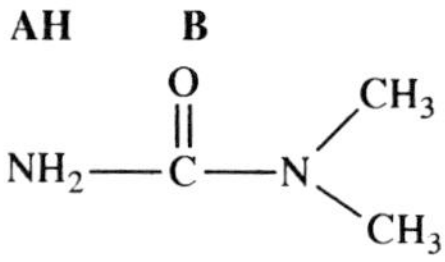

α-Dimethyl urea

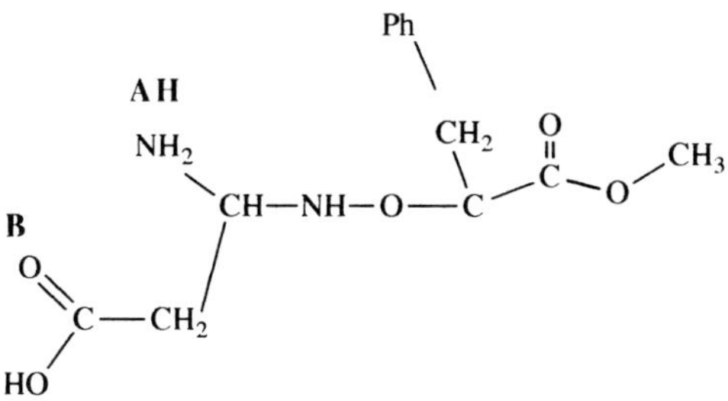

L-Aspartyl-L-phenylalanine methylester

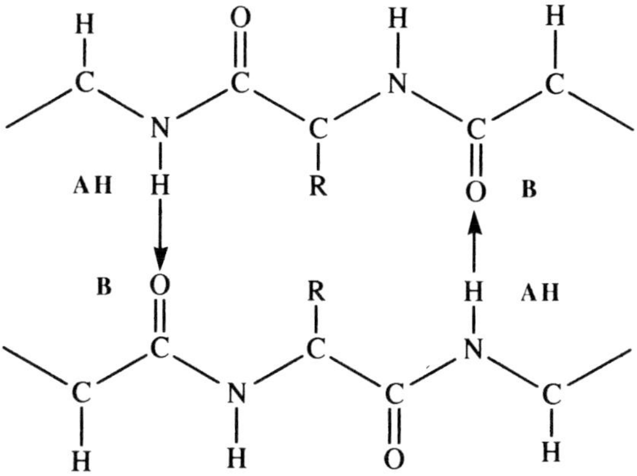

FIG. 9.14. Potential AH,B units in two subunits of a sweet tasting protein.

Apparently, zwitterionic functions, as such, are not necessary for sweetness, and may not even be responsible for sweetness. A substituted urea that cannot take on a zwitterionic form, such as α-dimethyl urea, is sweet, but the parent compound (urea) is bitter.

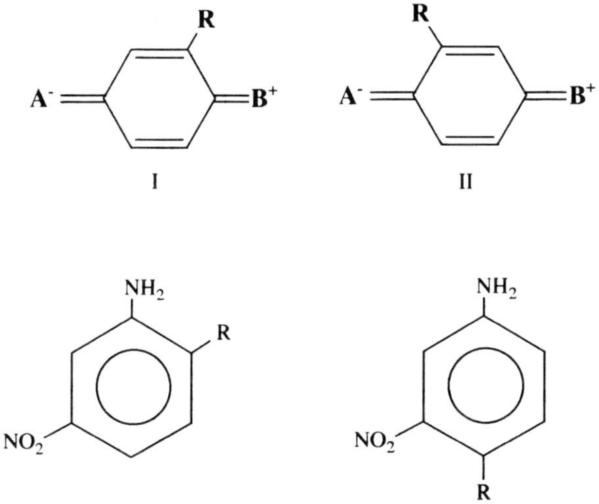

I II

III IV

FIG. 9.15. Effect of the nature and position of substitution on the sweetness of nitroaniline analogues (Ferguson & Childers, 1960).

A particularly significant point that can be made with respect to amino acid and peptide sweetness is that a series of compounds that have a common glycophoric unit as an absolute requirement for sweetness (the L-aspartyl residue) vary in their sweetness potency over a range of five log units. Some of them mentioned in Chapter 7, along with the common logarithm of their sweetness potency, are: Asparagine (1), Aspartame (2·25), Cyanosuosan (3·46), Superaspartame (3·90), Aspartic acid fenchyl ester (4·69), and Sucrononic acid (5·30).

For sweet-tasting proteins, a number of AH,B units are possible. Some carboxylic acid residues in their amide form have been implicated in the ability of thaumatin to elicit sweet taste. The lysine residues also seem to be involved in a proton–donor mechanism (van der Wel, 1981). As Monellin is made up of sub-units that are independently tasteless, the tertiary structure of the thaumatins is also essential for sweetness (van der Wel & Ledeboer, 1989). Additional AH,B units can be assigned to those

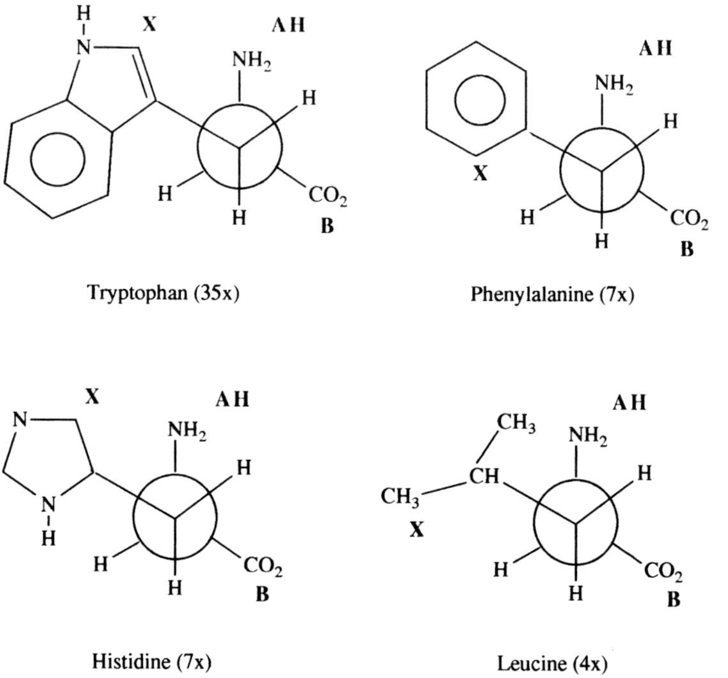

Tryptophan (35x) Phenylalanine (7x)

Histidine (7x) Leucine (4x)

FIG. 9.16. Favored conformation of selected amino acids demonstrating the common position (X) for the high electron density side chain (Kier, 1972).

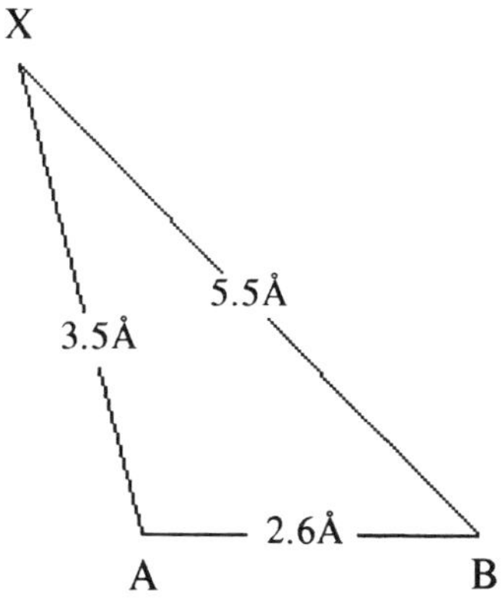

FIG. 9.17. The Kier (1972) tripartite arrangement for the glycophore of sweet tasting compounds.

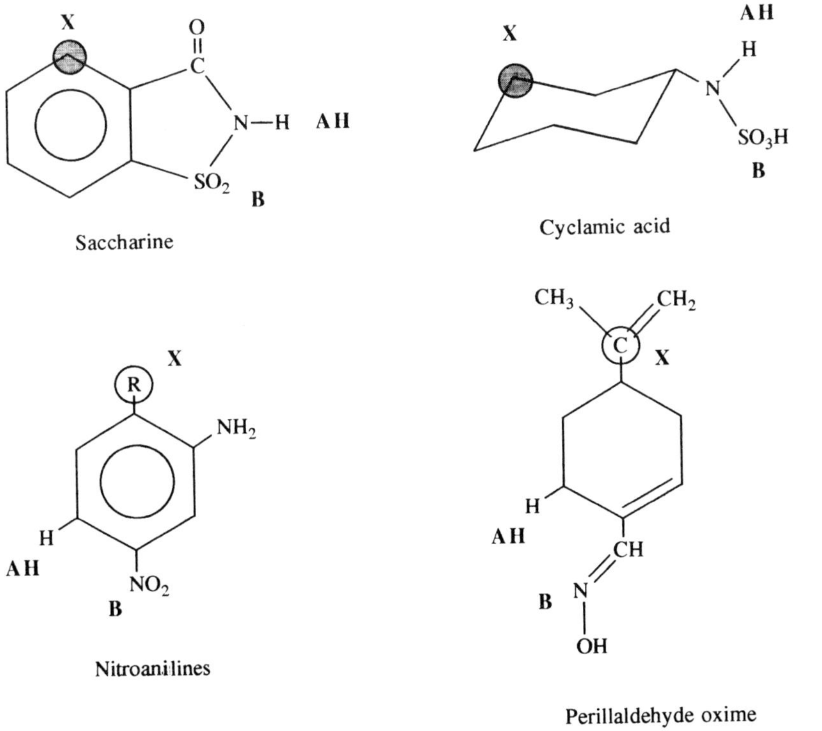

FIG. 9.18. The tripartite AH,B,X groups in selected high-intensity sweet substances (Kier, 1972).

groups that play a role, through hydrogen bonding, in forming the tertiary structure of proteins. They are also capable of participating in an intermolecular hydrogen bond with the receptor, and serve as candidates for pleated sheet AH,B units, as shown in Fig. 9.14.

9.3.10 The Tripartite Glycophore for Sweetness

As an introduction to the manner with which substitutions, or the function of a third glycophore component, may have an effect on the sweetness potency of a glycophore group, several observations recorded by Ferguson & Childers (1960) are presented. In reviewing the nature of substitutions on the sweetness of the 2-substituted 5-nitroaniline series of compounds, they noted that the substances possess both an electron donating (B+) and an electron withdrawing group (A−). In addition, they possessed a third group (R) located in an off-center position (Fig. 9.15). It was pointed out that the sweetness potency of a nitroaniline molecule depends on the degree with which it approximates either structure (I) or (II). For example, *m*-nitroaniline has a sweetness of 40×, but lacks an R group (Table 8.1). However, whenever R in either (III) or (IV) in Fig. 9.15 is an electron donating group, I is the sweet isomer, and II is not. From the observations of Ferguson & Childers, it can be concluded that, depending on their position in a molecule, and their general chemical nature, substituents can have either a positive or negative inductive effect on a glycophore.

9.3.10.1 *Chemical Nature of the Third Glycophore Component* (γ)

To account for the stereochemical specificity shown toward the enantiomeric amino acids Kier (1972) proposed that there is a third component in the sweetness glycophore, designated as X. Upon examining the structural features of the D-amino acids, Kier noted that a group high in electron density occurred in the same position with respect to the zwitterionic functions (AH,B). The finding was based on conformational considerations and molecular orbital calculations, and examples are shown in Fig. 9.16. The particular point in each side-chain position is especially rich in electrons. In all cases, the tripartite arrangement of the groups A, B, and X approximate the scalene arrangement shown in Fig. 9.17. Kier employed the distance from atom A to B in his calculations, whereas Shallenberger & Acree used the distance from the AH proton to B. Upon examining sweet compounds belonging to other groups of substances, Kier noted that the X site and triangle also occur in saccharine, cyclamic acid, nitroanilines, and

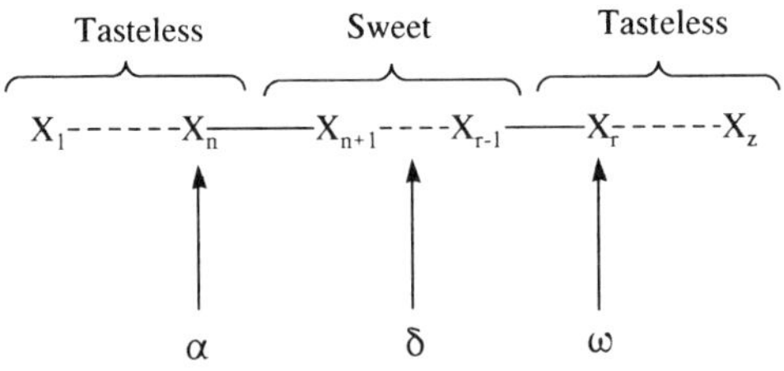

FIG. 9.19. The position of the area in the side chain of a homologous series of sweeteners for optimum (δ) sweetness (van der Heijden, 1992).

perillaldehyde oxime (Fig. 9.18.). Because X can have functions other than action as a dispersion site, such as lipophilic and inductive effects, it is hereafter designated as γ. Liberty is taken to do this because although X and γ can have the same properties and reactions, the properties and functions may also be different. Thus, the bipartite AH,B glycophore becomes tripartite AH,B,C, but the 'C' is represented by the small Greek letter γ.

9.3.10.2 *AH,B,γ in the Sugars*

According to the example set by Kier (1972), the sugars were examined for a possible electron-rich center. They were found to have an unambiguous γ function that bears near identical geometric relation to the probable primary sugar AH,B of Birch, as well as for the sweet amino acids (Shallenberger & Lindley, 1977). For glucose, γ is the C-6 methylene carbon atom, and for fructose it appeared to be the ring methylene carbon atom, although the C-1 methylene group of fructose is a second possible site.

Although compounds such as chloroform and ethylene glycol lack a specific γ site, they do have an element of lipophilicity. To emphasize the hydrophobic character of the third site (cf. Höltje & Kier, 1974), Shallenberger & Lindley (1977) chose to designate it as γ (greasy, oily). The third electron-rich site possesses hydrophobicity, and an element of lipophilicity, in addition to being capable of dispersion bonding. One final innate feature that γ may possess is that it has potential for an inductive effect on the electronic character of AH,B. It would now seem that it is in this sense that γ serves to enhance the sweetness potency of substances. The degree of enhancement is governed by the electron withdrawing power of γ, as evident from observations on its optimum *distance* and *position* from AH,B.

9.3.10.3 *Optimum Position of* γ

Within several homologous series of compounds, van der Heijden *et al.* (1985*a*, *b*) found that there is an optimum position for the location of γ, and the optimum position is partly a function of the nature, or isosteric identity, of AH,B. For any given side-chain complement to AH,B, the γ function occupies a position (point, small area) that represents minimum (α), optimum (δ), and maximum (ω) hydrophobic site distance from AH,B. A fourth parameter (S), also introduced by the van der Heijden group, is the distance between an atom belonging to α, δ, and ω, and the *plane* described by A to H to B. That observation has application to receptor topology. With reference to Fig. 9.19, the α (ω) position is the terminal *n*th (*r*th) atom in a side chain of a tasteless compound in a homologous series of sweeteners (van der Heijden, 1992). A side chain

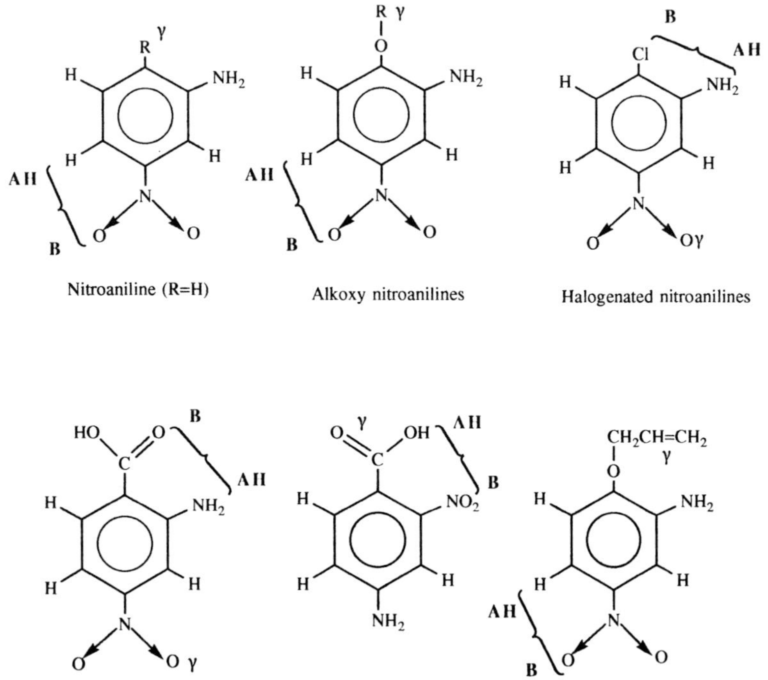

Nitroaniline (R=H) | Alkoxy nitroanilines | Halogenated nitroanilines

2-Amino-4-nitrobenzoic acid | 2-Nitro-4-amino-benzoic acid | 1-allyloxy-2-amino-4-nitrobenzene

FIG. 9.20. Assignment of AH,B and γ functions to the nitroaniline series of substances.

with a terminal $n + 1$th ($r - 1$th) atom, definitely tastes sweet. The δ position denotes the center (or intermediate point) of side chains in molecules originating from a series of sweeteners with the *highest* sweetness intensities. Selected compounds considered by van der Heijden *et al.* (1985*a*, *b*) are reproduced in Chapter 8 under the appropriate headings (Ureas, nitroanilines, etc.). Those tables also served to emphasize the basis for the conclusions the authors made with respect to the significance of γ, and subsequently serves as the basis for continued discussion in Section 9.3.12.1. Those observations are used below to identify AH,B,γ in a number of selected organic substances to serve as cases for a common isosteric glycophore.

9.3.11 General Occurrence of AH,B,γ in Sweet Organic Compounds

With the premise that there is a third glycophore component related to enhanced sweetness, its rather specific directional relations to AH,B yield a triangulation principle for identification of AH,B, and γ in sweet substances. The principle serves to reduce the uncertainty contained in attempts to identify AH,B by substitution reactions that may alter a parent compound's electronic character.

9.3.11.1 *Nitroaniline Compounds*

For several substituted nitroaniline substances, Shallenberger & Acree (1967) selected the *ortho* aromatic hydrogen atom as AH, and an NO_2 oxygen atom as B. Nitroaniline (nitrobenzene), which is quite sweet, only has an NO_2 substituent appended to the benzene ring; hence AH must be located on the ring (Fig. 9.20). Several investigators, aware of the effect of electron withdrawing substituents on the properties of benzene ring methine hydrogen atoms, have accepted the proposal (Verkade, 1968; Kier, 1972; van der Heijden *et al.*, 1985*a*). Others are not that certain of the AH assignment. It has been reported that three strong electron-withdrawing groups are required before an aromatic hydrogen atom can participate in hydrogen bonding (Allerhand & Schleyer, 1963). Consequently, Crosby *et al.* (1979) selected the *ortho* amino group as AH, and the alkoxy group oxygen atom to be B. The reassignment may seem reasonable in view of the Allerhand & Schleyer thesis, but the original assignment gains credence when the third component of the glycophore is considered (van der Heijden *et al.*, 1985*a*).

For the 2-propoxy-5-nitroaniline group of substances, in which the nitro group is replaced by CN or CO_2^-, Tinti *et al.* (1980, 1981) found that while the cyano derivative was sweet (about one-third that of the nitro derivative), the carboxyl derivative was not. This suggests that the NO_2 group of the alkoxy nitroanilines is not the B function, at least for these *derivatives*, and that the identity of AH, and B for the nitroaniline substances can vary. Moreover, as the identity of AH,B varies among the nitroaniline compounds, so does the identity of γ. The probable identity of AH,B and γ in several nitroaniline substances is shown in Fig. 9.20.

For the O—R side chain, and with increasing length of the side chain, the loss of sweetness just beyond an optimum position is characterized by abruptness. In considering these derivatives, it would appear that the amine group is not AH, and the alkoxy oxygen is not B, as indicated by Crosby *et al.* (1979) and others, as there is an increase in sweetness with increasing chain length, with the optimum γ position (δ) occurring with the propoxy derivative. Consequently, the assignment of both AH,B and γ for the alkoxy nitroanilines also seems certain. Another name for the propoxy compound is 'P-4000.' As the pentoxy derivative is tasteless, the terminal carbon atom is designated as ω (van der Heijden *et al.*, 1985*b*).

The assignment of γ in the nitroanilines as shown in Fig. 9.20 serves as an example of how consideration of it, along with AH,B, either confirms or amends empirical AH,B assignments. Now, however, the halogenated nitrobenzene compounds pose a problem. It is not clear if the effect of the halogen substituent is to create a new AH,B system with an oxygen atom of the NO_2 group as γ, or whether AH,B is the same as for the parent compound, and the halogen substituent itself serves as γ. The problem is even more complicated by the fact that a halogen substituent may serve as either B in an AH,B system, or as γ in a tripartite system. Help in

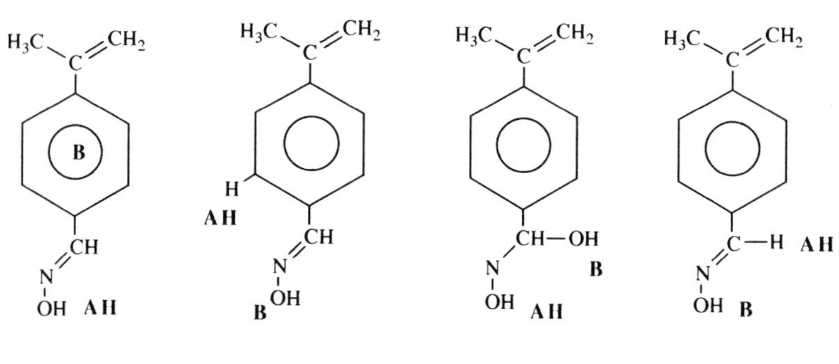

I II III IV

FIG. 9.21. Suggestions for the identity of AH and B in the structures of oximes.

deciding between these assignments comes from consideration of the other compounds in Fig. 9.20.

For the 2-amino-4-nitrobenzoic acid substance, B is assigned to the carbonyl function of the carboxyl group, and AH is assigned to the amino group. If the assignment is correct, then γ is again an NO_2 oxygen atom for 2-amino-4-nitrobenzoic acid. If the carboxyl hydroxyl group of 4-amino-2-benzoic acid is AH, and an NO_2 group is B, the position of γ is uncertain. The γ site might be assigned to the carbonyl oxygen atom for 2-nitro-4-aminobenzoic acid or it might also be a position on the benzene ring. The final compound in Fig. 9.20 (1-allyloxy-2-amino-4-nitrobenzene) retains the initial assignment of AH,B in nitroaniline, and places γ in the unsaturated —$CH_2CH{=}CH_2$ side chain.

9.3.11.2 *Oximes*

As a group, the oximes possess various sensory attributes (sweet and bitter taste, characteristic odor). With respect to the relation between structure and those sensory attributes, the oximes are also a seemingly intractable group of substances. As a consequence they have been the object of considerable study. The very uncertainty of the identity of AH led Temussi *et al.* (1984) to conclude that the identity of AH is not important, provided the fit of the compound with the sweet receptor is good. Perillaldehyde oxime (perillartine) is an example of a compound with several possibilities for the AH,B unit, but many of the possible selections have unusual distance parameters. For this reason Shallenberger & Acree (1971), although employing an initially inappropriate oxime structure, selected the hydroxyl function of the oxime group as AH, and the π-bonding cloud of the aromatic ring as B, as now shown (I) in Fig. 9.21. Kier (1972) chose an aromatic ring hydrogen *ortho* to the oxime group as AH, and selected the hydroxyl oxygen of the oxime group as B (II). That assignment is in accord with the observation (Acton *et al.*, 1970*a*) that a double bond, conjugated with the oxime, is essential for sweet taste in oxime derivatives of perillartine. It is also in accord with the subsequent analysis of a series of cyclohexylaldoximes, using molecular connectivity calculations (Kier, 1980). These are presented in Section 10.6.4.3. Finally, the assignment is in accord (Section 9.3.13.4) with the molecular electrostatic potential recognition pattern (Venanzi & Venanzi, 1988*a*).

Beets (1978) proposed that hydration of the perillaldehyde oxime function (III in Fig. 9.21) would convert it into an α-glycol-like group with the freedom to assume the favored staggered AH,B conformation. Although the suggestion is favored by consideration of the sweetness of oxime analogues, and the lack of sweetness of the *anti*-oxime, it is doubtful that hydration is capable of quenching a double bond. Parenthetically, the lone hydroxyl hydrogen atom of the *anti*-oxime is sterically disposed to intramolecularly hydrogen bond to the π cloud of the benzene ring, and serves to explain why the *anti*-compound is not sweet. Alternatively, the lack of sweetness may be due to steric hindrance by the benzene ring. In any event, based on the observations that an oxime group is not hydrated in water solution (as indicated by n.m.r spectra), and that acetaldehyde oxime is slightly sweet (R=H in place of the benzene ring), van der Heijden *et al.* (1985*a*) proposed that AH is C—H of the oxime, and the oxime hydroxyl oxygen is B (IV in Fig. 9.21). As noted in Section 8.1.4, there is confusion for the nomenclature of the oximes, and while *anti*-anisealdehyde oxime tastes equally sweet and bitter, the *syn*-oxime is bitter only (Acton *et al.*, 1970*b*). Taste is only one of the oxime sensory attributes, and when a bitter note is the predominant taste attribute, the odorous note is usually described, albeit inconsistently, as sour, sharp, and/or burning. When sweetness is the predominant taste, the odorous component is usually described as methol-, mint-, ginger-, and even bread-like.

van der Heijden (1985*a*) felt certain that AH,B is located in the oxime portion of the molecule, and the optimum position (δ) for γ was assigned to the terminal carbon atom of the vinyl group, which is also consistent with electrostatic potential calculations. The initial position (α) was

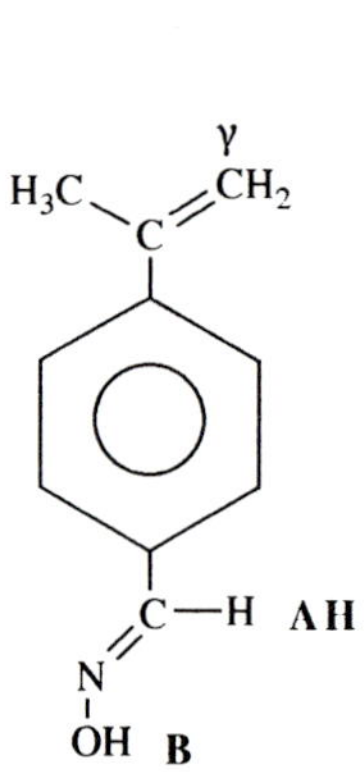

assigned to the carbon atom attached to the oxime group. In view of the previous discussion on the location of AH,B in these substances, the most probable assignment of the AH,B,γ glycophore to the oximes is that proposed by van der Heijden *et al.* (1985*a*), while the α position is probably more commensurate with the assignment of (X) by Kier (1972) to the first carbon of the vinyl group. At this point, the difference seems trivial, although the assignment for No. II in Fig. 9.21 seems most likely. The relation of the AH,B function to the oxime group is further indicated by the taste of furfuraldehyde oximes. Furfural oximes are sweet, and the 5-hydroxymethyl furfural oximes are not. The hydrophobic function, γ, is at position 5 in those substances.

By invoking triangulation, the CH group of the oxime turns out to be AH, but the initial γ position (α) assigned to the carbon atom attached to the oxime group by van der Heijden does not hold for the furfural oximes unless the furfural oxime C-5 and the C-2 γ function are extinguished by the 5-hydroxymethyl group. The furfuraldehyde oximes also seem to be a case where AH,B is not functional as a glycophore unless accompanied by a suitable hydrophobic site (Section 10.3.4.1), an assessment that probably also applies to oximes.

In a study of the influence of structure on either the sweet or bitter tastes of the oximes, Kier (1980) used a discriminate analysis technique to find the best structural representation for both tastes. Using $^1\chi$ and $^4\chi_p$ connectivity indices (Section 10.6.4.3), it was found that a two-variable linear discriminant function correctly assigned the proper taste (primarily bitter or primarily sweet) to 17 out of 20 oximes (examples are shown in Fig. 8.5b,c). If sweetness/bitterness intensity is set aside, the predictive ability of the procedure had 85% accuracy. Kier concluded that larger, more branched substituents on cyclohexyaldoximes increases the tendency for the molecule to be bitter, while smaller substituents, with

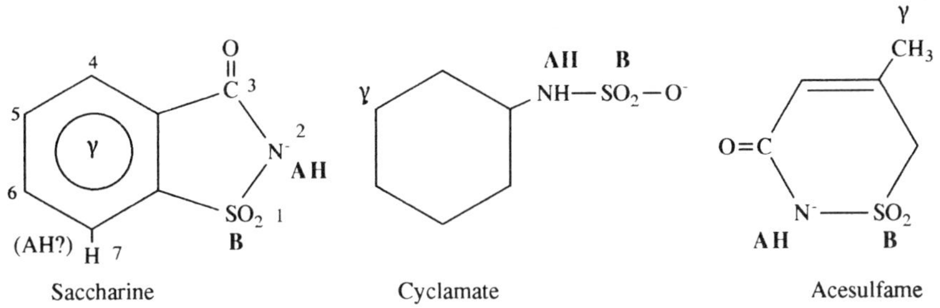

FIG. 9.22. Assignment of AH,B and γ to saccharine and selected sulfamates.

four or fewer atoms increased sweet taste. Kier also noted that the presence of unsaturation or heteroatoms is not a critical influence on either sweet or bitter taste.

9.3.11.3 *Saccharine, Sulfamates*

In contrast to difficulty in identifying AH and B in the oximes, numerous investigators concur with the identity of A and B in saccharine and the sulfamates, but it is the identity of H that is open to deliberation. For the acid form of saccharine, AH is assigned to the imino hydrogen atom and B is assigned to a sulfoxide oxygen atom (Shallenberger & Acree, 1967; Kier, 1972; Beets, 1978; van der Heijden *et al.*, 1985*b*). The fact that the hydrogen atom of the imino group is doubly active due to the strong electron withdrawing power of the adjacent sulfonate and carbonyl groups lends credence to the AH assignment, as does the observation that any substitution at that position eliminates sweetness.

The assignment of B to the sulfoxide oxygen atom gains credence from the observation (cf. Lee, 1987) that one of the sulfonate oxygen atoms describes a 60° torsion angle with the imino group, thus placing that proton at a distance of 2·8 Å from the oxygen atom (Section 8.1.2). Replacing the sulfoxide group with a carbonyl group eliminates sweetness, suggesting that it is the heterocyclic ring sulphonate group that is important for sweet taste, and not the carbonyl group.

For sulfamates, Pautet & Nofre (1978) assigned NH to AH, and SO_3^- to B. They also pointed out that the synclinal conformation ($\phi = 60°$) is most appropriate for that AH,B assignment, and the three oxygen atoms of the SO_3^- group are equivalent in their action as a proton accepting group. The same functional groups are also contained in cyclohexylsulfamic acid (cyclamate), and 3,4-dihydro-6-methyl-1,2,3-oxathiazin-4-one 2,2-dioxide (acetosulfame). The anionic form of these compounds is shown in Fig. 9.22 because there may not be sufficient concentration of the undissociated acids for the AH assignment to elicit sweetness through straightforward hydrogen bonding.

All the substances have high pK_a (that of saccharine is 2·32). They, and their salts, persist in water solution almost exclusively in ionized form (cf. Crosby *et al.*, 1979; Kasperson & Primack, 1986; Rymon Lipinski, 1986; Venanzi & Venanzi, 1988*b*). The usual sodium, calcium and other salts of saccharine and cyclamate, and the potassium salt of acesulfame (acesulfame K) would seem to lack an AH function, *per se*. The assignment of AH to the imino group would also seem to require participation of an acidic hydrogen in a noncovalent (hydrogen) bond.

This is consistent with the observation that saccharine (strongly acidic), the dihydrochalcones (moderately acidic) and the alkoxynitroanilines (weakly acidic) have similarly strong sweetness magnitude. This led Crosby *et al.* (1979) to point out that for saccharine, the C-7 aromatic carbon atom is most probably AH, and it is capable of participating in a van der Waals interaction to elicit sweetness in place of hydrogen bonding. Not only is that suggestion not in accord with the function of a required third site for the enhanced sweetness of the compound, but any substitution of the imino group results in loss of sweetness (Hamor, 1961). The concern (Crosby *et al.*, 1979; Jakinovich, 1982; Venanzi & Venanzi, 1991) with the chemistry of the saccharine and sulfamate imino hydrogen atom does not take into account the possibility that the initial chemistry of sweet taste does not necessarily take place in a polar environment. It also fails to take into account the possibility that either a proton scavenged from water, or water of hydration of the anion itself, can serve as AH. The molecular electrostatic potential pattern for the acesulfame anion is not consistent with the pattern of other sweet compounds, which indicates that the active form of the anion is hydrated (Venanzi & Venanzi, 1991). In any event, Audrieth & Sveda (1944) and others (Pautet & Nofre, 1978) recognized that a free hydrogen on the nitrogen of the cyclamates was absolutely necessary for sweetness, *viz.* $RNHSO_3X$, where X is any salt-forming group.

As the potassium salt of 3,4-dihydro-6-methyl-1,2,3-oxathiazin-4-one 2,2-dioxide (200×) is sweeter than the parent compound (cf. Rymon Lipinski, 1986), which is 130×, the assignment of the acesulfame K glycophore AH to hydration water receives support in the premise that the salt is sweeter because the free acid formed by solvation of the parent substance is neutralized. Thus, not only is the sourness depression of sweetness obviated, but the dilute solution of KCl (Section 3.4.2) also contributes to the sweetness.

In a study of a number of cyclamate derivatives, Benson & Spillane (1976) proposed that while methylation of the imino group greatly reduced sweetness, the imino group was not strictly necessary for sweet taste. It was then proposed that AH is CH in the structure —$CHN(R)SO_3$. The proposal receives support from the listing of sulfamate derivatives and their taste (Figs 8.4a,b), and the observation also serves to support the thesis that substitution reactions can alter the identity of a parent compound's glycophore. In addition to an apparent influence related to the size of R in the general structure $RNHSO_3Na$, only those compounds with a methylene (CH_2) carbon atom *alpha* to the nitrogen atom are sweet. Benson &

Spillane also suggest that the more crucial part of the NH—SO_3 group is the SO_3 moiety.

Benson & Spillane (1976) also chose the carbon atom hydrogen substituent on the cyclohexane ring, and adjacent to the sulfamate group, as AH for cyclamate because adamantane, which lacks such an atom, is tasteless. However, insertion of a methylene group between the cyclohexane ring and the sulfamate function also quenches the sweetness of cyclamate (Nofre & Pautet, 1975). One can invoke a steric hindrance tenet to explain the case for adamantane, but the problem generated by the lack of sweetness of *N*-cyclohexylmethylsulfamate is not quite that clear-cut as *N*-cyclopentylmethylsulfamate is reported to be sweet (Bragg *et al.*, 1978).

While arguments, such as that presented for the case of adamantane, are not particularly convincing (cf. Crosby *et al.*, 1979), other problems are generated by assigning AH to a group other than —NH for saccharine and the cyclamates. Methylation of the acesulfame ring nitrogen atom destroys sweetness (Clauss & Jensen, 1973). Therefore, the ring nitrogen atom is obviously A in the AH,B system for saccharine, cyclamate, *and* acesulfame K, and the manner by which it functions as AH is variable and depends on whether the compound is the free acid or a salt.

Addition of sodium ions has been reported (Magidson & Gorbatschow, 1923) to decrease the sweetness of both saccharine and sodium saccharine. The explanation offered for the observation was that the sweetness of

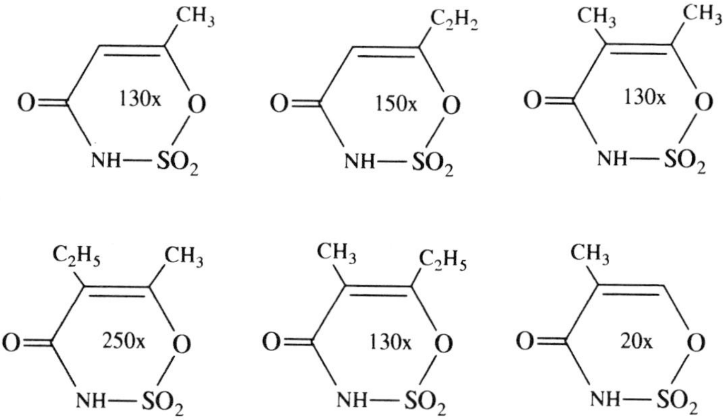

FIG. 9.23. Structure and sweetness of oxathiazinon dioxide substances (Clauss & Jensen, 1973).

sodium saccharine must be due to the undissociated acid, and addition of sodium ions suppresses dissociation. Thus, the sweetness of sodium saccharine was said to be 'conditioned' by the anion. Finally, the acesulfames serve as examples of compounds with multiple AH,B systems, a fact borne out by interpretation of the position and function of the hydrophobic component of the tripartite glycophore.

For the identity of γ in saccharine and the sulfamates, with AH and B as assigned in Fig. 9.22, Kier (1972) assigned X (γ) to the fourth carbon atom of the benzene ring. van der Heijden *et al.* (1985*b*) placed the first (α) position for γ in the middle of the bond to which the benzene ring is attached. The assignment of AH for saccharine to an aromatic hydrogen atom (Crosby *et al.*, 1979) leaves the identity of γ quite uncertain. As derivatives of saccharine, formed by substitution of five-membered sulfur-containing ring compounds joined to the benzene ring at positions 5 and 6 are more than twice as sweet as saccharine (Section 8.1.2), the optimum position of γ was placed at a position more distant from AH,B than that shown for saccharine. The position of γ for cyclamic acid obviously lies in the cyclohexane ring, and the position marked in Fig. 9.22 is that assigned by Kier (1972). van der Heijden *et al.* (1985*a*), from consideration of the homologous series of sulfamates (Section 8.1.3), placed the optimum position for γ in the center of the cyclohexane ring.

The varying sweetness of acesulfame derivatives, with AH,B as assigned in Fig. 9.23, affords opportunity to examine the function not only of γ, but also the effect of substitution on the identity of AH,B. Figure 9.23 shows the relative sweetness of oxathiazinon dioxide substances obtained by Clauss & Jensen (1973). The parent compound of acesulfame K (the potassium salt without an appended methyl group) is 20 times as sweet as sucrose. The acid form of acesulfame K (130×) is only 65% as sweet as the potassium salt, as acesulfame K has a sweetness potency of 200 times. In the acesulfame derivatives shown in Fig. 9.23, it is clear that derivatization not only changes the position of γ, but also the identity of B. With a single methyl group adjacent to the ring oxygen atom, B is the carbonyl oxygen atom. With an ethyl group at a position adjacent to the ring carbonyl group, B becomes SO_2, and the resultant AH,B unit is favored because of the AH proton to B distance. Alternatively, with AH,B assigned to NH and SO_2, the γ site is more appropriate in one position (*ortho* to the carbonyl group) than in the other. The effect of the double transposition of the ethyl and methyl groups in the acesulfame derivatives favors that conclusion. Although acesulfame does not have a tripartite grouping (Fig. 9.22) fully commensurate with the Kier triangle, it would seem that acesulfame, and the derivatives shown in Fig. 9.23, have glycophores with multiple choice for the B function.

9.3.11.4 *Halogenated Organic Compounds*

The halogens in general, but chlorine in particular, also play an important role in the taste of organic compounds. Generally, halogenation leads to either increased sweetness or increased bitterness. Halogen substitution on aminonitrobenzene compounds leads to increasing sweetness. Nitroaniline (Fig. 9.20) has a relative sweetness of 40, and substituting fluorine for the ring hydrogen atom *para* to the NO_2 group does not change relative sweetness (Table 8.1). Sweetness then increases in the order Cl (400×), Br (800×), and I (1250×), however.

Halogenation of saccharine (Fig. 8.2) in the *para* position with fluorine leads to a compound that is sweet, but with bitter aftertaste. Substitution with chlorine leads to a compound that is still sweet, but with accentuated bitterness. Substitution with bromine further enhances the sweetness, but the compound is also extremely bitter. Finally, substitution with iodine

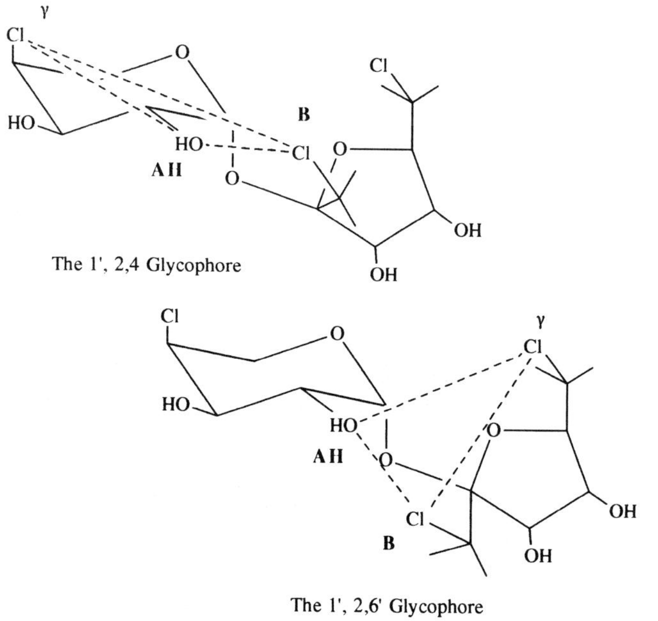

FIG. 9.24. Two possible tripartite glycophores in sucralose.

leads to loss of sweetness, leaving only a slightly bitter taste. While AH,B in 6-chloro-D-tryptophan is probably the amine and carboxyl oxygen atoms, respectively, the compound is 1300 times as sweet as sucrose (van der Heijden *et al.*, 1985*b*). The effect of the chlorine atom, as well as for halogen substitution on saccharine and nitroaniline, is probably inductive, accentuating the activity of AH,B.

For chlorinated sucrose and sucrose derivatives, sweetness may or may not be enhanced manyfold, and one of the results of such substitution is to redirect the identity of the sucrose AH,B, or for the halogen to play the role of B itself. Thus, through reference to the numbered sucrose structure (Section 9.3.8.3) for 1′, 6′ dichlorosucrose, Hough & Phadnis (1976) assigned AH to OH-2 and B to the 1′chloro-substituent. For trichloro-*galacto*-sucrose Kanters *et al.* (1988) proposed three possible sites. For AH, and B they were OH-3′ and O-2, OH-4′ and O-3′, and OH-2 and O-3′, respectively. For the same compound Mathlouthi & Portmann (1990) assigned AH,B to OH-3 and O-2, and to OH-3′ and O-3, respectively. van der Heijden *et al.* (1985*b*) selected OH-3 and OH-4 as AH,B but did not commit themselves to a decision as to which was AH and B. Without doubt, sucralose (trichloro-*galacto*-sucrose) possesses multiple glycophores and those selected by Hough & Khan (1978, 1993*a*, *b*) that include a chlorine atom as B are shown in Fig. 9.24. There, the two glycophores occur on two slightly different sucralose conformations. Finally, based upon the potential energy map for the compound, Hooft *et al.* (1993) located AH,B,γ for trichloro-*galacto*-sucrose (sucralose) at O-2, O-3, and Cl-1′.

From computer generated hydrophobicity potentials, Lichtenthaler *et al.* (1991) concluded that the hydrogen bond between HO-1′ or HO-3′ and 2-O of sucrose and its halogenated derivatives plays an integral role in enhancing the H donor (AH) capacity of the glucosyl-2-OH. In both cases, then, the glucosyl 3-O atom serves as B, and the γ (X) function is located somewhere in the neighborhood of C-1′–CH_2–H-3′. The authors also pointed out that the hydrophobic region can vary within a certain range and reach-over *through* the C-6′–CH_2 to the axial side of the glucosyl C-4 atom.

A unique feature among the chlorosugars is the observation (Lee, 1987) that among the disaccharides, chlorine and other halogen substituents enhance only the sweetness of sucrose. Among monosaccharides, only the fructofuranoses, which are thought to be tasteless as free forms of fructose, are rendered sweet (Hough, 1989) by chlorination (cf. Fig. 6.8). Other halogenated sugars, including methyl α-Dglucopyranosides, α,α-trehalose, maltose, and lactose, are invariably bitter. A final significant point to be made about the varying sweetness of the chlorosugars is that substitutions that affect sweetness invariably take place on what can be considered to be a given *face* of the molecule (Hough, 1989).

9.3.11.5 *Diphenyl Substances (Isocoumarins, Flavanones, and Dihydrochalcones)*

The various isocoumarin derivatives studied by Yamato & Hashigaki (1979) give clear indication as to the identity of AH,B in these substances.

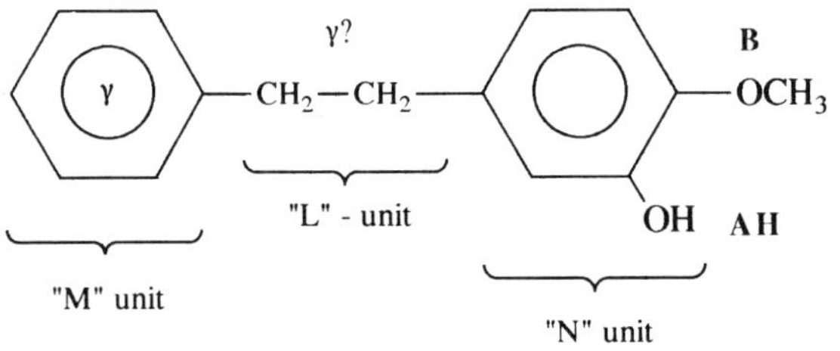

From consideration of the isocoumarin reference structure of Section 8.1.5, the hydroxyl group(s) and methoxyl groups on the 'N' unit appear to be an obvious example of AH,B. The AH to B distance parameters are 2·5–4·0 Å when the dihedral angle between OH and the ring is equal to or greater than 45°. Furthermore, as OH and OCH_3 are hydrogen bonded (Dietrich *et al.*, 1976), the angle does not exceed 90°. This AH,B characteristic is in accord with the premise (Section 9.3.8.3) that the hydrogen-bonded glycol group acts as a hydrophobic probe, and as the driving force for the recognition chemical interaction for sweet taste.

The change in sweetness intensity with relatively minor change in substituents in the above reference structure is dramatic. Substituting the methoxyl group with an alkoxy group drops the relative sweetness intensity from 300× to 1–10×. When the —OH and the methoxy groups are interchanged, the compound is rendered tasteless. Not only is the sweetness of the isocoumarins extremely sensitive to minor changes in the electronic nature of AH,B, they are also sensitive to the position of AH,B in the molecule. Apparently, many compounds have an AH,B, but do not taste sweet because AH,B does not have the appropriate electronic properties, or the proper arrangement with respect to γ.

For the isocoumarins, the second benzene ring ('M') is an obvious γ

function, and was so selected by Yamato & Hashigaki, although van der Heijden *et al.* (1985*a*) assigned δ to the 'L' unit. As with the effect of relatively minor substitution at the 'N' unit on the capacity of AH,B to elicit sweetness in these substances, change in the electronic nature of γ also markedly affects sweetness. When a cyclohexane ring is substituted for the benzene ring, sweetness drops from 300× to 30×, presumably because of the drop in the hydrophobicity of γ, and because of the change in the inductive effect on AH,B.

Substitution in the 'L' unit to increase its size by one CH_2 group lowered sweetness from 300× to 1–10×. Substitution of an NH group for a methane carbon atom in the connecting side chain between the diphenyl rings immediately adjacent to the 'M' group had no effect on sweetness, but placing the NH group adjacent to the 'N' unit decreased sweetness to 250×. Thus, the former had no effect on the hydrophobicity of 'M', but the latter seemed to affect the electronic nature of AH,B. When either an oxygen atom or a sulfur atom was substituted for the 'L' unit CH_2 nearest 'M', the sweetness of the parent compound diminished by a factor of 2, again presumably because of decreased hydrophobicity of the 'M' unit.

While the assignment of AH and B for the flavanones and dihydrochalcones is not particularly difficult, the identity of γ poses a problem. There is some evidence that whenever a glycosidic moiety is present, it resides in the glycoside. For the dihydroflavanols (Section 8.1.5.2) discussed by Kinghorn & Soejarto (1991), the $2R,3R$ form of the nonglycosidic compound (+)-dihydroquercitin 3-acetate is sweet, and while the $2S,3S$ α-L-rhamnosyl analogue is sweet, the $2R,3R$ α-L-rhamnosyl analogue is not.

The sugars in the dihydrochalcone glycoside structure may serve as a source of AH,B units (Hodge & Inglett, 1974). The C-3 and C-4 hydroxyl groups of D-glucose seem to be of fundamental importance in determining the taste properties of the flavanones and dihydrochalcone neohesperidosides (Horowitz & Gentili, 1969; van Nierkerk & Koeppen, 1972). However, they do not appear to be necessary for the high sweetness potency exhibited by these substances. The closely related structure phyllodulcin is devoid of glycosides. When the glycosidic disaccharide component (neohesperidose or, 2-*O*-α-L-Rhamnosyl-β-D-glucopyranose) of neohesperidin dihydrochalcone (Section 8.5.1.3) is modified slightly to form hesperidose (6-*O*-α-L-Rhamnosyl-β-D-glucopyranose), the resulting compound (hesperidin dihydrochalcone) is tasteless.. Yet, removal of the rhamnose unit from the disaccharide

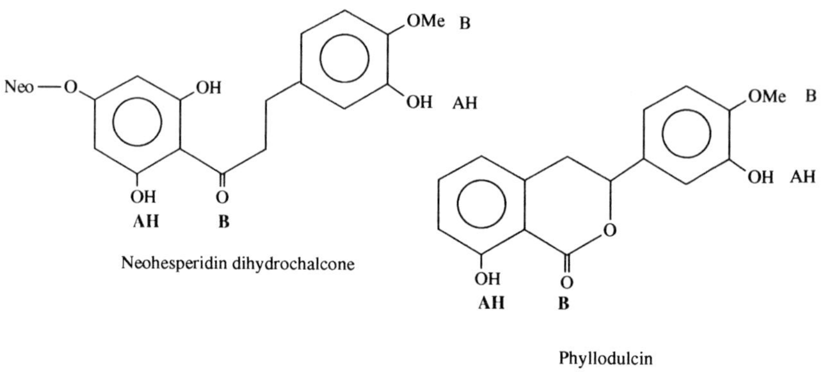

FIG. 9.25. Identity of AH,B in dihydrochalcones.

yields a compound (Hesperitin dihydrochalcone glucoside) that is only somewhat reduced in sweetness (Horowitz & Gentili, 1969). If the rhamnosyl group of neohesperidose is replaced by another hexose that prevails in the $^4C_1$ conformation (α- or β-D-glucose; α-D-galactose or -mannose), sweetness again disappears. Then, when the α-L-Rhamnosyl moiety of neohesperidose is replaced by a sugar that persists in the favored conformation $^1C_4$, sweetness is returned to the modified neohesperidin dihydrochalcone (Esaki *et al.*, 1975; Kamiya *et al.*, 1976*a*, *b*). It would thus appear that the effective glycophore component in the sugar portion of neohesperidin dihydrochalcone is the methyl group, and it must be located in a steric position attainable only in the $^1C_4$ sugar conformation.

Further studies on the substitution of the neohesperidin dihydrochalcone aglycone with groups other than a sugar at the same position leads again to compounds with varying degrees of sweetness. A hydroxyl group alone at that position yields a compound (neohesperitin) with a sweetness of 665× (DuBois *et al.*, 1977*a*, *b*). Obviously then, a sugar at position 4 in the aglycone unit of neohesperidin dihydrochalcone is not a requirement for sweetness, but when it is present rigid stereochemical requirements must be met. The sweetest compounds were obtained by substituting the neohesperitin aglycone at that position with groups that essentially had strong inductive effects.

With respect to substitution in the B ring of neohesperidin dihydrochalcone (Fig. 8.11), great variations in sweetness are again encountered. Unlike the isocoumarins, increasing the size of the alkoxy substituent on the B ring did not significantly affect sweetness, as the compound is reported to be very sweet, but transposing the methoxy and the hydroxy groups leads to tastelessness, as was true with isocoumarins.

DuBois *et al.* (1977*a, b*), after intensive studies of the dihydrochalcones, came to the conclusion that AH and B are probably those groups identified in Fig. 9.25 with the bold AH,B specification. With respect to the identity of γ, DuBois *et al.* (1977*a, b*) observed that there is a second AH,B in the structures of the dihydrochalcones and phyllodulcin in the 'bent' conformation (Section 8.5.1.3). The second B also occupies the same position as Kier's proposed third dispersion site. Therefore, the second B may actually be a form of Kier's (1972) third high electron density glycophore component.

If that is the case, then both dihydrochalcone ring units can alternatively serve as the source AH,B, and both can also serve as γ, particularly when a gycosidic unit is absent to act as γ for the AH,B shown in bold type in Fig. 9.25. All told, then, it is possible that the dihydrochalcones and related compounds have multiple glycophoric equivalency. In addition, the dihydrochalcones are exemplary examples of the observation/tenet that substitution on a parent sweet compound serves to alter the glycophore unit identity, not only with respect to AH and B (Section 9.3.8.1), but also for the identity of γ.

van der Heijden *et al.* (1985*a, b*) did not include the dihydrochalcones and related substances in their report. Nevertheless, van der Wel *et al.* (1987) state that the dihydrochalcones fit a tripartite scheme. Substituting a propyl group, corresponding to δ in the alkoxynitroanilines, for the B-ring methyl group increases the sweetness of the dihydrochalcone from 612× to 2000×. The second glycophoric B is probably now van der Heijden's, δ, and the second AH (OH) serves no direct sweetness function, other than to raise the glycophoric equivalency of the dihydrochalcones from one to two.

9.3.11.6 *Sesquiterpenes*

Although diterpenes are bitter (Section 8.2.3), one of the most obvious and clear-cut cases for the location of AH,B in a substance occurs with the sesquiterpene (+) 6*S*, 1′*S*-hernandulcin, where the side chain C-1 hydroxyl and the ring carbonyl group serve as AH,B, respectively. Comprade *et al.* (1987, 1988) concluded that because the 1′R epimer is tasteless, the bulky side chain must prevent the non-sweet isomer from fitting into the sweet-taste receptor site. Alternatively, as the double bond for sweet taste is necessary for sweetness, rearranging the structure to be commensurate with those shown in Section 8.2.3, and assigning the role of γ to the unsaturated center of the side chain would suggest that the former isomer, which is now congruent with the Kier (1972)

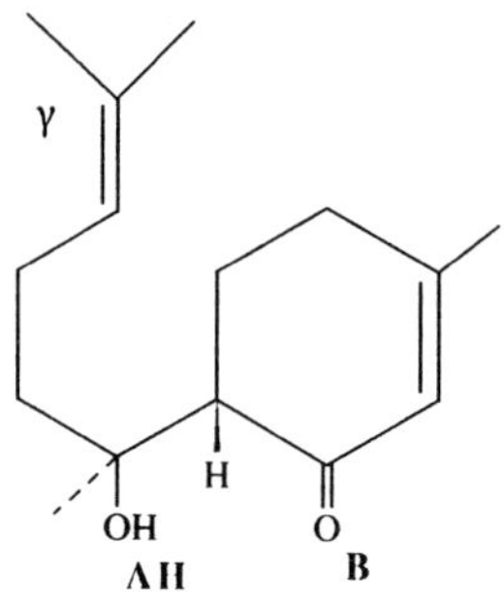

(+) 6S, 1'S- Hernandulcin (1000x)

tripartite AH,B,γ glycophore, is capable of eliciting sweet taste, while the latter is not.

9.3.11.7 *Ureas*

The significance of the ureas to deducing principles of taste chemistry is that they are nonionic substances that may be exceedingly sweet. A simple urea, such as asymmetrical (α-)-dimethyl urea is sweet, although urea itself is bitter. The AH and B functions are probably uncharged

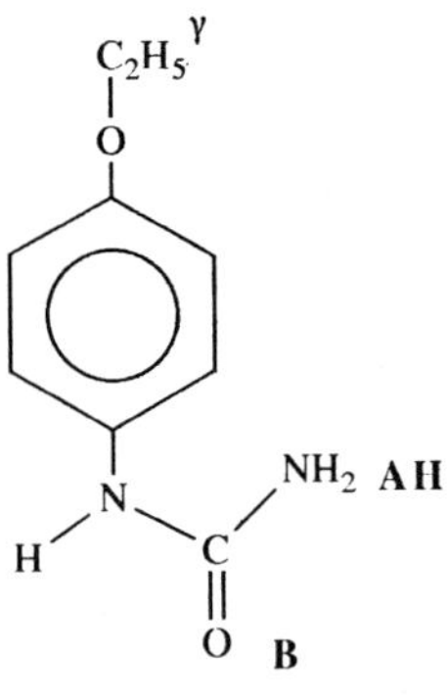

Dulcin

amine and the carbonyl groups, respectively. As *ortho*-tolylurea is tasteless, *meta*-tolylurea bitter, and *para*-tolylurea 'sweetish' (Fig. 8.14), the γ glycophore unit of these compounds is apparently the methylene substituent. The van der Heijden α-position for γ in 4-ethoxyphenylurea (dulcin) is located on the first carbon atom of the alkoxy group, as 4-methoxyphenylurea is not as sweet. As the propoxy derivative is only slightly sweet, the optimum position for γ is at the second carbon atom of the dulcin ethoxy substituent (van der Heijden *et al.*, 1985*b*). The structure shown for dulcin, and the comparable structures in Fig. 8.13, lead to an E and a Z conformation for the urea group. For the phenyl-substituted ureas and thio-ureas, the 'E' conformation is favored because

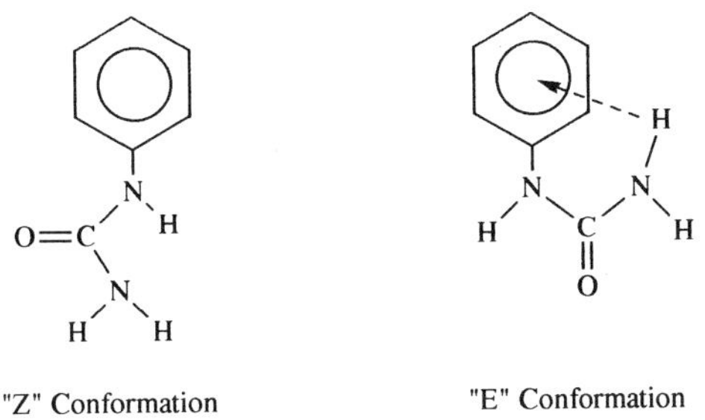

the NH_2 group is hydrogen-bonded to the phenyl ring placing the —NH—CO—NH_2 group perpendicular to the ring, as opposed to the 180° orientation in the 'Z' conformation (Walter & Ruess, 1971). Based on the latter observation, van der Heijden *et al.* (1985*b*) proposed that dulcin also has the 'E' conformation. The argument adds credence to the assignment of the terminal amino group to the AH sweetness function, and to the proposed mechanism for the recognition chemistry for the sweetness of these substances. While the lone hydrogen atom of *anti*-oxime hydroxyl groups is bonded to the aryl ring, and serves to explain the lack of sweetness in those compounds, the bonding here of one hydrogen atom serves to activate the other. Substituting a sulfur atom for the carbonyl carbon of dulcin (4-ethoxyphenylthiourea) leads to a compound that is only bitter (Section 1.5.2).

From a study of the sweetness of suosan (350×) and its derivatives, Tinti *et al.* (1982) determined that the terminal carboxyl group is probably the B component of the AH,B couple as substitution of a methyl group at that position, or methylation of the carboxyl group, led to tastelessness. The second imino nitrogen atom served as AH to complete the AH,B couple. With the benzene ring in a position

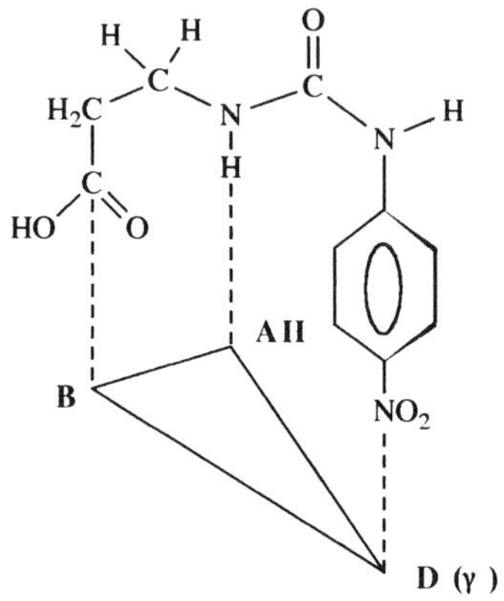

Suosan perspective drawing

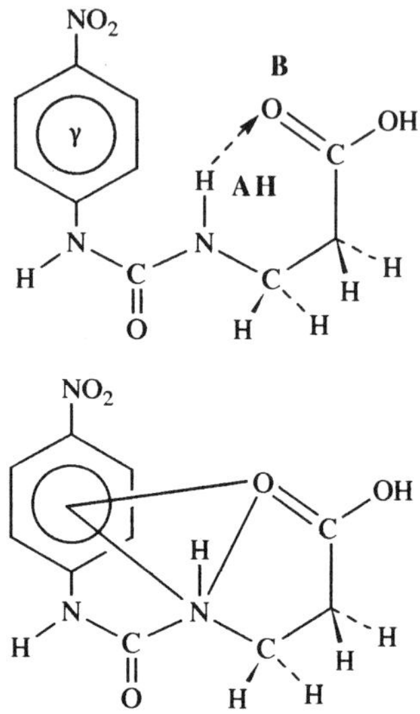

Suosan planar drawing

perpendicular to that primary AH,B unit, they also proposed that the NO_2 group serves as a second B (D), and the glycophore for suosan is essentially an AH,B,D unit. This proposal is shown above with suosan drawn in perspective. Although the idea of a second B is an interesting thought, the authors offer no proof that the conformation that places an NO_2 group in the position of Kier's X site is valid. The NO_2 group probably does not function in the capacity of γ, but as a substituent appended to it. Then, in planar conformation, it would have a strong inductive effect on AH,B as that structure, also shown above, also possesses the Kier triangle. The latter AH,B,γ arrangement is also proposed by van der Heijden *et al.* (1985*b*). The final comparison between the suosan and dulcin series of sweet substances is that substituting an S=O group for the carbonyl group in suosan does not eliminate sweetness, as it did with 4-ethoxyphenylthiourea. With two possible AH,B units in suosan derivatives, such substitution apparently serves to increase the sweetness of suosan from 350× to 2400×.

9.3.11.8 *Sulfones*

For the sulfones, Ratajczak & Polański (1991*a*) assigned AH to the carboxyl group on the basis of the observation that the sodium salts of the acids taste sweet, and presumed that the hydrogen in this case originates, as with salts of saccharine, from water. Another possibility suggested is that AH is the benzene ring *ortho*-hydrogen and B is SO_2. The aliphatic side chains (Section 8.1.8) were assigned to γ. An alternative assignment might be the methine proton of the side chain as AH, an SO_2 oxygen atom as B, and the isopropyl group or the benzene ring as the hydrophobic function.

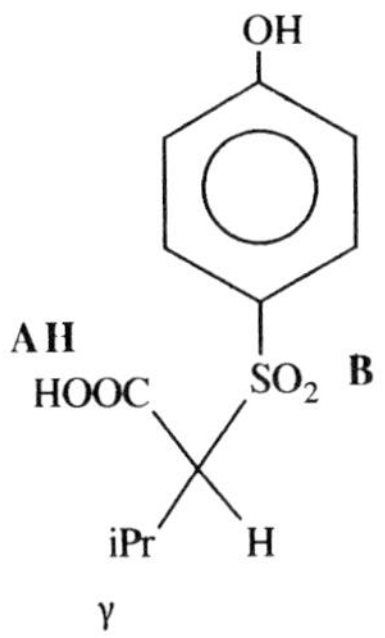

An arylsulfonylcycloalkanecarboxylic acid

9.3.11.9 *Amino Acids and Peptides*

The assignment of AH,B,γ in the amino acids and peptides seems relatively simple for the AH,B grouping, but the position of γ is often uncertain. With the peptides in particular, the position of γ rests with the ability to determine the reactive conformation. That AH,B is the aspartyl amine and carboxyl groups of aspartame is readily deduced from the observation that only the aspartyl group elicits sweetness, and any substitution at the zwitterionic center eliminates sweetness (Mazur *et al.*, 1970).

From the preliminary considerations of Mazur *et al.* (1969), a primary candidate for γ is the methoxy group. Increasing the size of that group leads to loss of sweetness, even though L-histidyl- and L-tryptophyl *O*-methyl esters and Asp-Phe-*O*-cyclohexane are tasteless. Perhaps these latter substances, with large side-chain groups, do not possess either the proper lipophilic/hydrophobic balance, or they adversely affect the overall conformation of the molecule (Ariyoshi, 1984). For alitame derivatives, where the thietanyl group probably serves as γ, it was observed that oxidation of the sulfur atom led to decreased sweetness (Section 7.4.1.1), and the effect was ascribed (Glowaky *et al.*, 1991) to the increased polarity of the dipeptide amides. This is a case of the side chain acting not only as a third (γ) site, but substitution on that third group has strong inductive effect on AH,B.

The lack of agreement as to the favored conformation for aspartame now seems relatively immaterial. Both conformations are at the same energy level (Taylor *et al.*, 1991). However, the position of γ is open to question, and there might be two appropriate hydrophobic functions. van der Heijden *et al.* (1979) assigned the hydrophobic function to the ring, and that assignment is favored by Rohse & Belitz (1991). If the methylene ester group is also γ, then the conformations proposed by either Lelj *et al.* (1976) or Goodman *et al.* (1987) are equally reactive. The two possibilities for γ in the two favored conformations for aspartame are shown below.

An interesting observation was recorded with respect to AH and γ of aspartame by Shinoda & Okai (1985). Those authors obtained H-glycyl-phenylalanyl methyl ester and H-L-alanine-phenylalanyl methyl ester, which taste bitter, and do not have the purported aspartame B grouping.

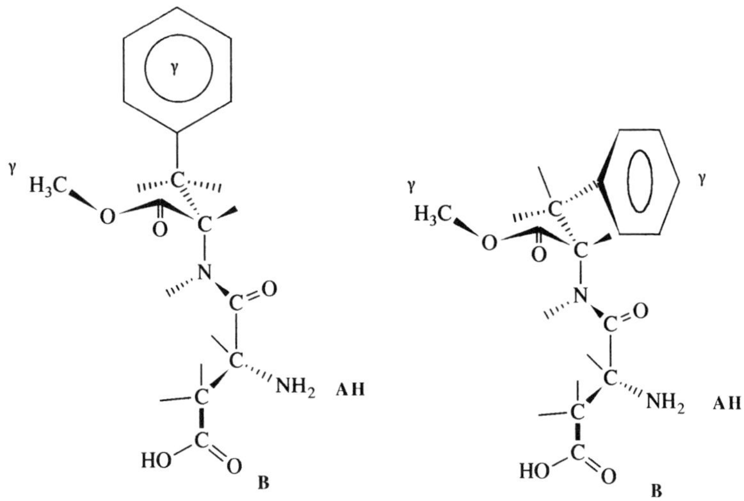

AH,B,γ in two aspartame conformations.

When B, in the form of free acetic or citric acid was added to a solution of those compounds, the mixture was reported to taste about 3 times as sweet as a sucrose standard. Adding aqueous ammonia, presumed to represent free AH, to a solution of succinyl-L-phenylalanyl methyl ester or glutaryl-L-phenylanalylmethyl ester (aspartame without a B function) did not impart sweetness.

More recently, Seki *et al.* (1990), described an 'intermolecular' type of sweetener similar to the Shinoda & Okai (1985) model (Section 12.4.2.2). An 'unsweet' (0·7×) peptide (δ-acetyl-L-ornithyl-β-alanine methyl ester hydrochloride) with AH,B,γ components was rendered more sweet by addition of HCl as 'B' to the solution to complement the function of the original AH,γ. The addition of the acid nearly doubled (1·8×) the sweetness intensity of the original compound. It may also have simply altered the conformation.

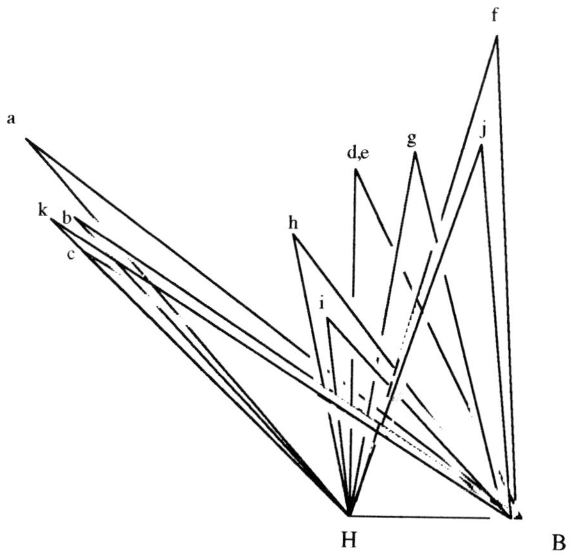

FIG. 9.26. Relative position of AH, B, and γ (δ) for homologous series of sweeteners: a = ureas; b = nitroanilines; c = oximes; d = chlorosugars; e = isocoumarins; f = tryptophan; g = dipeptide esters (other than k); h, i = 5-substituted acesulfams and sulfamates; j = saccharine; k = methyl ester group in L-Asp-L-Phe-OMe (van der Heijden *et al.*, 1985*a*, *b*).

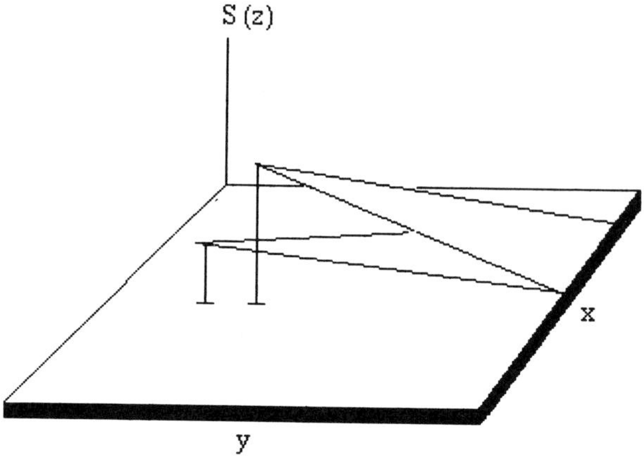

FIG. 9.27. Effect of the S(z) coordinate on the position of δ (i.e., γ).

9.3.12 General Geometry of AH,B,γ

The planar tripartite arrangement of AH,B and γ as first set forth by Kier (Fig. 9.17) for amino acids soon received confirmation for the sugars, and other substances. This is especially true for those substances with high sweetness potency which was, of course, also a Kier premise. For the latter substances, the most comprehensive study is that of van der Heijden *et al.* (1985*a*, *b*). The authors derived a set of coordinates for AH,B and γ, and described the planar variation on the coordinates established by Kier.

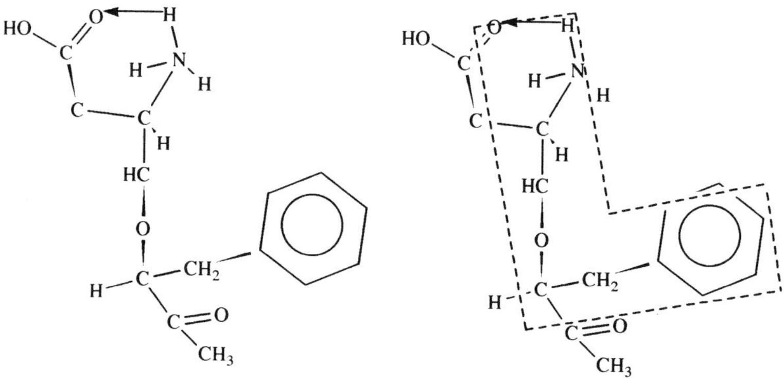

FIG. 9.28. The L-conformation for dipeptides and related compounds (Goodman *et al.*, 1987).

9.3.12.1 *AH,B,γ Coordinates*
The optimum position for γ (δ), as described in Section 9.3.10.3, varies with the identity of AH and B. The elevation (z axis), designated as 'S' for the position of δ, above the plane described by A to H to B, will also vary (van der Heijden *et al.*, 1985*a*, *b*). Subsequent reference to this feature of the topology of the sweet receptor site is presented in Chapter 11.

An x,y plot of these values to show their similarity and dissimilarity is given in Fig. 9.26. When values for S are included in specification of the location of optimum γ (δ), at least three and possibly four types of receptors become evident. They depend not only on the nature of AH,B, but also on the position and elevation of δ. An example of the effect of S on the position of δ is shown in Fig. 9.27. Three distinctly different glycophoric arrangements are: 1. The nitroaniline group (a, b, c, k); 2. The sulfamate group (h, i); and 3. The isocoumarin group (d, e). A possible fourth includes f, g, j and 6-substituted acesulfames.

Consideration of the geometry of AH,B,γ, led to several analogies and predictions. D-Tryptophan and L-aspartyl-L-phenalanine methyl ester (f, g in Fig. 9.26) might be expected to share the same receptor site. That conclusion was also reached by Schiffman *et al.* (1981*b*) on the basis of adaptation data. The earlier literature states that 2-propylthio-5-

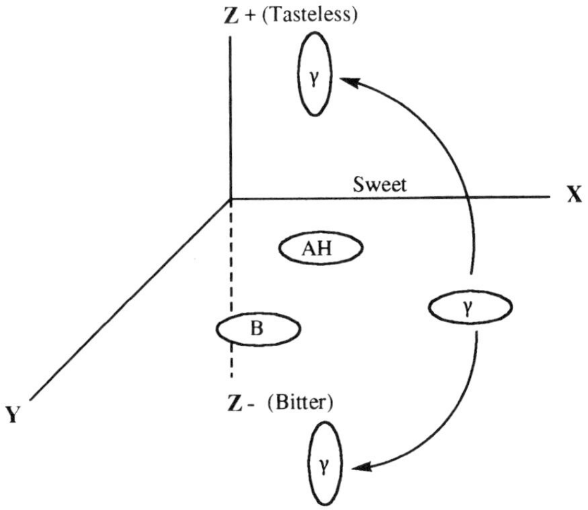

FIG. 9.29. The Goodman coordinates for the taste of dipeptides and retro-inverso substances.

nitroaniline is tasteless, but it was observed to possess an appropriate tripartite glycophore. Subsequently, it was found that the literature assignment had been applied to the wrong isomer, and the compound is in fact quite sweet (van der Heijden *et al.*, 1985*a*).

From the foregoing discussion it is clear that the geometry for the arrangement of the A to H to B, in relation to γ components of the glycophore, is essentially planar. However, γ often skewed above a reference plane (g = z_0), and from the perpendicular. As viewed from above, it is generally skewed to the left. As substances that elicit taste are three-dimensional structures, several investigators elected to examine the nature of the glycophore from that point of view.

9.3.12.2 *Sweet/Bitter/Tasteless AH,B,γ Coordinates*

Goodman and coworkers (cf. Douglas & Goodman, 1991) sought to deduce basic principles of taste perception from consideration of the structure, stereochemical properties (configurational and conformational), and taste of a series of dipeptide isomers and derivatives. It was found that the sweet conformations for sweet-tasting dipeptides and retro-inverso amide analogues have the L-conformation (cf. Fig. 7.17) shown for aspartame in Fig. 9.28. That structure, with the benzene ring as γ is in accord with the tripartite AH,B,γ planar glycophore. A significant feature of the favored conformation for aspartame sweetness proposed by Goodman is the chelate zwitterionic ring. It is assumed *a priori* on the basis of nuclear magnetic resonance characteristics (Douglas & Goodman, 1991). The Cα—Cβ bond of the aspartyl residue possesses a staggered conformation with the carboxyl group and the amino group in the *gauche* minus position, and the sp^2 orbital of the terminal aspartyl carboxylate carbon atom and the Cα—Cβ bond are coplanar. Presumably, then, there is no charge on the zwitterionic group. The latter observation has application to the nature of the initial chemistry of the sweetness sensation as elicited by amino acids and peptides (Section 10.3.5.1).

Upon examining other dipeptide analogues (Section 7.4.1.2) it was found that the position of the benzene ring was also associated with either tastelessness or bitterness. This observation is depicted using rectangular coordinates in Fig. 9.29 as a variation on Fig. 7.17. The conclusion is based on the study of alitame models, and consideration of the taste of aspartyl dipeptides containing β-aminocyclopentanecarboxylic acid methyl ester as the second residue. The compounds (Douglas & Goodman, 1991), with the assigned sweet, bitter, and tasteless coordinates, and corresponding to the coordinate diagram of Fig. 9.29 are:

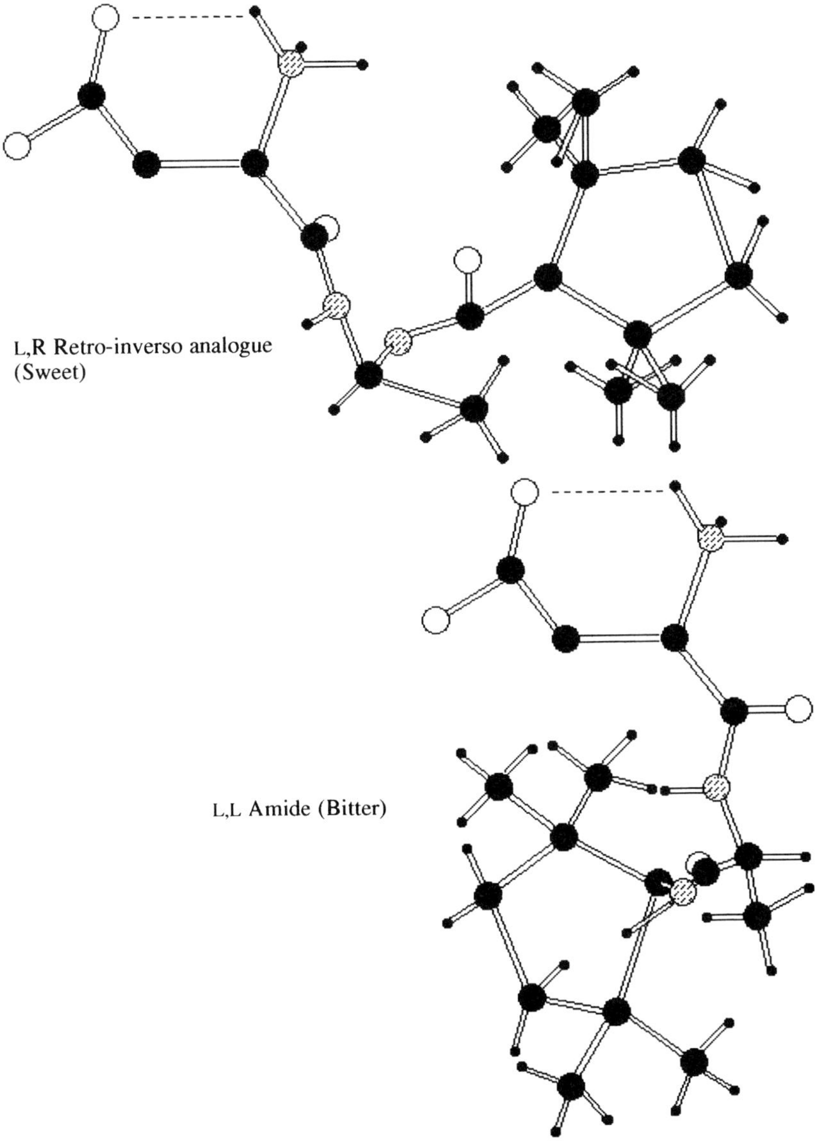

L,R Retro-inverso analogue (Sweet)

L,L Amide (Bitter)

FIG. 9.30. Minimum energy conformations for the sweet retro-reverso compound *N*-(L-aspartyl)-*N*′-[(2,2,5,5-tetramethylcyclopentanyl)carbonyl]-(*R*)-1,1-diamino-ethane and the bitter amide L-aspartyl-L-alanyl-2,2,5,5-tetramethylcyclopentanyl-amide. Open atoms are oxygen, solid large atoms carbon, solid small atoms hydrogen, and hatched atoms, nitrogen (Goodman *et al.*, 1987).

Asp-(1*S*, 2*S*)-Ac[5]-OMe (minus Z), bitter.
Asp-(1*R*, 2*R*)-Ac[5]-OMe (L-conformation, or plus X), very sweet.
Asp-(1*R*, 2*S*)-Ac[5]-OMe (Reverse L-conformation, or minus X), tasteless.
Asp-(1*S*, 2*R*)-Ac[5]-OMe (L-conformation, or plus X), sweet.

The difference in sweetness intensity for the two sweet isomers is probably due either to the inductive differences between the two configurations, or to slightly different positions for γ, as in the van der Heijden *et al.* (1985*a*, *b*) sense. When the remainder of the molecular structures shown for the coordinate positions are included, an interesting result is obtained. For sweetness, the zwitterionic unit is chelated, and presumably uncharged, and AH,B,γ of sweet molecules form a common plane on the surface of the molecule. Bitterness occurs when one component of AH,B, along with γ, lie on a common face of the molecule, but the other component of AH,B does not. Tastelessness results when neither component of bipartite AH,B and γ have a common face on the surface of a molecule. The significance of the Goodman coordinates position for AH,B,γ is the clear depiction of where AH,B and γ occur with respect to the structure of the whole molecule, and in particular, with respect to the different faces than can be envisaged for the entire molecular structure. This is clearly seen in the stereo-drawings of a sweet retro-reverso compound *N*-(L-aspartyl)-*N*′-[(2,2,5,5-tetramethylcyclopentanyl)carbonyl]-(*R*)-1,1-diaminoethane and the bitter amide L-aspartyl-L-alanyl-2,2,5,5,-tetramethylcyclopentanylamide (Goodman *et al.*, 1987) in Fig. 9.30, for comparison with Fig. 7.14. The sweet L,R compound has the L-conformation, and the bitter L,L compound has the minus z-conformation.

The use of coordinates by the Goodman school appears to be the first successful attempt at assigning the various tastes of related compounds with different configurational and conformational attributes, to what are essentially different *chiral point groups*. That sweetness, *per se*, can be related to compounds belonging to a given symmetry point group was pointed out in Section 2.11.1.6.

9.3.13 Alternatives to AH,B,γ as a Common Glycophore

9.3.13.1 *An Electrophilic/Nucleophilic Unit as the Glycophore*

Belitz *et al.* (1979) viewed the AH,B sugar glycol unit as an example of a more general combination of an electrophilic center with a positive polar charge (p^+) and a nucleophilic center with a negative polar charge (p^-). It is essentially an electrophilic/nucleophilic (e/n) unit, on the order of that proposed by Warfield (1954). The dipolar glycophore was then designated (Belitz *et al.*, 1981) as an e_s/n_s system where the subscript refers to the sweet substance (stimulant). The e_s/n_s unit was viewed as a functional entity prerequisite for sweetness in place of AH,B (Belitz *et al.*, 1990). With the e_s/n_s tenet the probable initial chemical interaction for sweetness is unspecified, and may now be construed to be either one of several weak nonbonded interactions. Among them are ionic interactions for amino acids and peptides, and presumably a displacement reaction for other substances (see Section 10.3.5.2). In addition, it is more convenient to *jointly* consider sweetness and bitterness in terms of e/n units.

With reference to the taste of amino acids and amines, Belitz *et al.* (1981) proposed that there are six types of sweet/bitter stimulants. Viewing the AH,B unit as an electrophilic/nucleophilic/ (e/n) system and γ as a general apolar function they are: (1) compounds with e/n only that are only sweet (glycine, D- and L-alanine); (2) compounds with e/n and a third, apolar function that have increased sweetness potency (D-norleucine); (3) compounds that taste sweet and bitter (cycloleucine, i.e. 1-amino-cyclohexane-1-carbocyclic acid); (4) compounds that taste bitter (L-norleucine); (5) compounds that taste very bitter (octane sulfonic acid); (6) compounds that again taste very bitter such as i-pentylamine. Diagrams of the saporous units and the receptors are presented in Section 10.3.5.2.

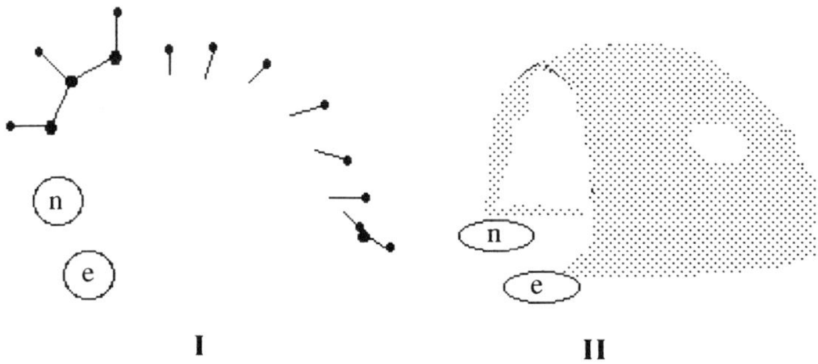

FIG. 9.31. Topological estimation of forbidden zones for the glycophore (Rohse & Belitz, 1991, 1993).

Whether compounds taste sweet (type 2), bitter (type 4) or bitter-sweet (type 3) depends on the steric arrangement of the e_s/n_s, γ system. Among the organic class of substances, bitter taste was found to require suitable apolar and polar groups, and the latter may be either p^+ alone, along with an apolar function (type 5), p^- alone, along with an apolar function (type 6), or both p^+ and p^-, along with an apolar function (types 3 and 4). It was further observed that a distinction between the sweet and bitter taste of the D- and L-amino acids only becomes apparent when R (the side chain) is larger than the propyl group. Equally important, taste threshold values for both sweetness and bitterness were observed to depend on the length of R, and to pass through a minimum for both acyclic and cyclic amino acids. For acyclic compounds, when R = the hexyl group (six-carbon chain), D- and L-unbranched amino acids are tasteless whereas the pentyl group (5-carbon chain), while not imparting sweetness, does impart bitterness.

Belitz *et al.* (1987) were able to determine the positions, in three-dimensional space, that are not commensurate with sweetness, i.e. the positions are 'forbidden' (see also Rohse & Belitz, 1991, 1993). It was found by computer construction that stereostructures of sweet-tasting amino acids (including the tertiary butylester of L-alanine), 2-amino benzoic acid, and derivatives of saccharine and cyclamate describe a common sweetness volume. Derivatives that were not sweet possessed structural components that fell beyond the boundaries described by those that were. When the dipolar sweet system was placed at a common position, it turned out that there are spatial positions for the hydrophobic components of the glycophore, and the position of other substituents, that are also favored. In other words, the procedure assists in forming an impression of the allowable shape of the side chain containing the hydrophobic group.

The deduced primary structure of the glycophore is shown as (I) in Fig. 9.31.

Model (I) is to be viewed along the z axis. The position of γ in this orientation can be construed as lying somewhere within the x, y, and z coordinates, but n and e are in the same plane. This notion of the orientation of n/e is derived from previous attempts to rationalize the combined sweetness and bitterness of the amino acids and other compounds (e.g. Belitz *et al.*, 1979). To accomplish this, the authors placed the third hydrophobic function in two planes, one in the same plane containing n/e and the other in a plane parallel to it. There are no forbidden zones beneath it. The planar tripartite system described a zone

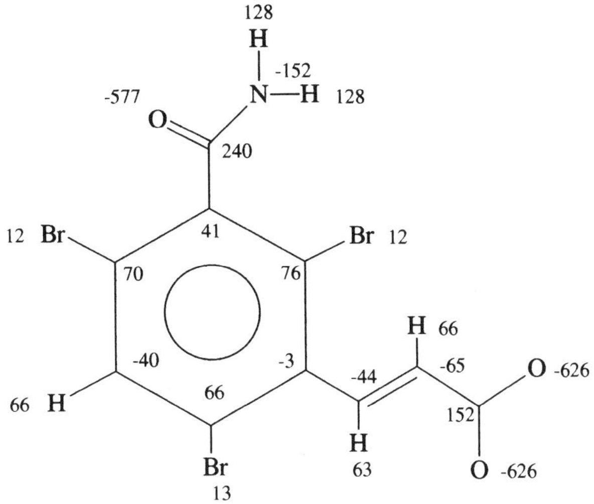

3-Carboxamido-2,4,6-tribromo-cinnamic acid

FIG. 9.32. Charge distribution for 3-carboxamido-2,4,6-tribromo-cinnamic acid (Belitz *et al.*, 1990).

beyond which γ, *contained in the bulk of the remainder of the molecule*, cannot exceed. In addition, bulky groups that extend beyond the forbidden zone are not allowed. In short, the model describes the three-dimensional limits imposed upon the glycophore. It also paints a discrete picture of the probable topology of the receptor site. To bring the forbidden zone proposal of the Belitz school into congruence with the skewed triangular arrangement of AH,B,γ, *n* and *e* in structure I in Fig. 9.31 are placed in the xy plane in structure II. Thus, γ occupies a position inside the space delimited by the forbidden zone.

What is not clear is the extent to which the forbidden zone represents straightforward steric hindrance, loss of inductive effects, and/or loss of hydrophobic effects. Possibly, it is some combination of all three, and the subject receives further treatment in Chapters 10 and 11. The n/e, γ (AH,B,γ) site does not have a bottom side, i.e. it is at or on a surface of either a membrane, or large molecule (the receptor). Provided that the proper conformation and proper n/e functions are selected, a host of substances, in addition to those mentioned above, can be accommodated by the model. Among them are the dipeptide esters, and sucrose (Rohse &

Belitz, 1991), and several other sugars and sugar derivatives (Rohse & Belitz, 1993).

It was also found that several tastants could not be accommodated by the aforementioned glycophore model. They are the oxathiazinone dioxides (Rohse & Belitz, 1988), and the 2,4,6-trihalogenated 3-carboxyalkylbenzamides (Belitz *et al.*, 1990), and dulcin. The n/e distance was nearly double that for the glycophore presented above. As a result, a second glycophore was proposed for these compounds. The validity of the latter assessment depends upon the identity of n/e. When 3-carboxamido-2,4,6-tribromocinnamic acid is examined by means of a charge distribution, a second type of n/e unit does seem necessary.

The probable identity of e_s/n_s for a compound is obtained by calculating the charge distribution for the constituent atoms. Those for the exceedingly sweet substance 3-carboxamido-2,4,6-tribromo-cinnamic acid in units of 10^{-3} e (Belitz *et al.*, 1990) are shown in Fig. 9.32. From the charge distribution values obtained, e_s was assigned to the carbamoyl hydrogen atoms because they are the most positive atoms on the surface of the molecule. The n_s moiety was assigned to the carboxyl oxygen atoms as they are the most negatively charged. The distance between the two groups ranges between 6·60 and 8·30 Å, depending upon the conformation examined. The assignment is supported by the observation that esterification of the carbamoyl group, substitution by CN, or exchange of the carbamoyl group by a carboxyl group abolishes sweet taste. From those observations, it would appear that 3-carboxamido-2,4,6-tribromo-cinnamic acid is a valid case for a second type of glycophore with large distance between the functional groups.

On the other hand, the compound would seem to obtain its high sweetness potency in part from multiple AH,B groups. Substitution of an AH or B moiety of either one of them so disrupts the overall electron distribution that neither is capable of eliciting sweetness. From the viewpoint of the AH,B tenet, it would seem that the lone aromatic hydrogen atom of 3-carboxamido-2,4,6-tribromo-cinnamic acid would serve elegantly as an AH function, and a neighboring halogen atom would serve as a B function. In addition, the carbamoyl NH_2 group might serve in the capacity as AH and a bromine substituent as B. Using dipolar symmetry principles (Chapter 11) to deduce the probable identity of AH,B, the methine proton ($^+$66e) is a candidate for AH, and the unsaturated center in the side chain (−65e) is a candidate for B. A more conventional AH,B assignment for dulcin is presented in Section 9.3.11.7.

Rohse & Belitz (1993) subsequently recognized a series of different e/n systems. They are OH/OH in sugars, NH_3^+/COO^- in peptides and amino acids, NH/SO_2 in oxathiazinone dioxides and isothiazalone dioxides, NH/SO_3 in sulfamic acids, OH/CO in β-hydroxyketones, $CONH_2/COO^-$ in carboxyalkyl benzamides, $CONH_2$/OR in alkoxyphenyl ureas, and NH_2/CO in ureas. The e/n glycophores are about the same as those listed for AH,B glycosteres in Section 9.13.3.3. Finally, Rohse & Belitz recognized two general groups of e/n substances. One group has an e/n distance of about 2·5 Å, and the other has an e/n distance of about 8·0 Å.

9.3.13.2 *General Glycophore Functions and Families*

One consequence of the uncertain nature of AH and/or B in high-potency sweeteners, and the great diversity of substances that taste sweet, is the conclusion that it is unlikely that there is a common glycophore for sweetness (Crosby *et al.*, 1979). The principal example given is the dipeptides, where the glycophore is assigned to the zwitterionic function, rather than the hydrogen bonding outlined in Section 9.3.9.4.

On the other hand, the zwitterionic character of peptides is much weaker than that for their component amino acids (Section 7.2.1), a fact not commensurate with proposals that the initial sweetness chemistry of the exceedingly sweet peptides is ionic in character. The polarity of the environment where the initial chemistry of the sweetness phenomenon takes place is not known. As the environment is a biphasic polar/apolar system, the polarity is almost certainly reduced, which would also tend to reduce the zwitterionic character of substances. If that is the case, the mechanism for sweetness enhancement is made all the more mysterious.

From the observations (Section 7.4.1.3) of Tinti & Nofre (1991*a*, *b*), and Nofre & Tinti (1993), it has become apparent that there are cases for extraordinarily sweet AH,B, and γ groups. These appear to be due to the strong inductive effects of certain groups (NO_2, CN), and/or multiple glycophore equivalency. Several extraordinarily sweet compounds have essentially planar tripartite AH,B,γ glycophores, and a group with strong inductive effects that may also be construed to be a second 'B.' In formulating the chemical nature of the third glycophoric site to explain enhanced sweetness, Kier (1972) pointed out that it is capable of participating in either charge transfer or dispersion bonding. The former interaction is, of course, a relatively strong one, and the latter somewhat weaker. Shallenberger & Lindley (1977) pointed out that the dispersion bonding feature was probably only one attribute of the third component, and preferred to recognize that if the dispersion bonding was to be viewed as hydrophobic bonding capability, the second attribute was that of also imparting an element of lipophilicity to the compound. The latter property, perhaps, better enables a compound to gain access to the receptor site. To those deliberations, it can now be added that the third component can also have a strong inductive effect on the polarity of AH,B, and extraordinary sweetness may result. Tinti & Nofre (1991*a*, *b*), and Nofre & Tinti (1993) place a different interpretation of the reason for high sweetness potency for the compounds they discovered (Section 7.4.1.3). Their conclusions, and the reasons for them follow. Alternative reasons, in compliance with the aforementioned rationalization of variable sweetness potency are given in italic print.

Acknowledging the apparent general occurrence of AH,B Tinti & Nofre pointed out that the carboxyl group is a logical choice for B in amino acids and peptides, and NO_2 is a logical choice for B in certain nitroaniline substances. Attention was then drawn (Tinti & Nofre, 1991*a*) to the nonequivalence of NO_2 and COOH groups in taste, as shown by the compounds below. From examination of those substances

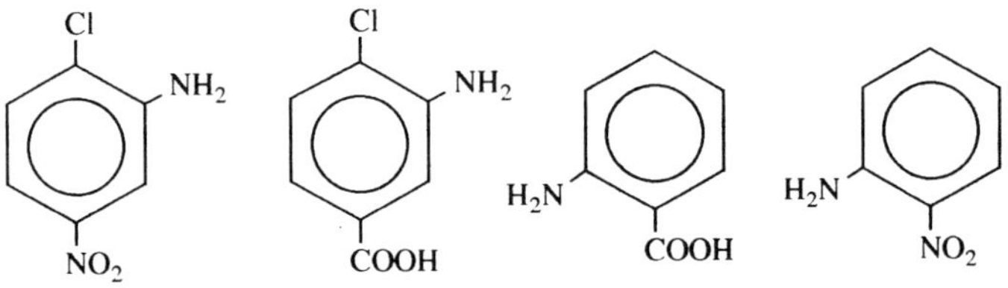

Sweet (400x) Bitter Sweet (120x) Bitter

it appeared that NO_2, Cl, and COOH have varying roles, not only in the nitroaniline substances, but also in the sweetness properties of *N*-carbamoyl β-alanine derivatives shown below. When X and X′ were NO_2 and COOH, respectively, the compound is sweet (suosan, 700×). When CN was substituted for the NO_2 group (cyanosuosan), sweetness was still present (450×). When a carboxyl group was placed at X instead, sweetness was lost, and when a CN group was placed at the X′ position in suosan, sweetness again disappeared. The authors concluded that specific sites are involved for sweetness, one still named B, specific for the anionic CO_2^- group, and a second site named D, specific for the NO_2/CN groups. Thus, sweeteners containing only a CO_2^- group, such as amino acids or aspartame, and sweeteners containing only an NO_2 group, such as P-4000 (the *n*-propoxy-derivative in Table 8.1) possess different glycophores. *Alternatively, the principle that substitution* *reactions may either alter the identity of AH,B, or has profound inductive amplification or attenuation effect on AH,B can be applied. For the case of the N-carbamoyl β-alanine derivatives, it appears that CN,NO_2 and COOH simply have opposite inductive effect on the polarity of a common AH,B glycophore.*

N-Carbamoyl derivatives

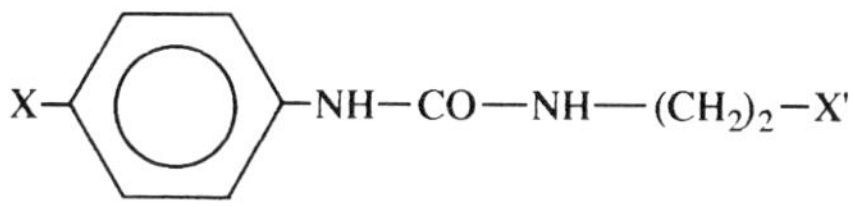

X	X'	Taste
NO_2	COOH	700x
CN	COOH	450x
COOH	COOH	Tasteless
NO_2	CN	Tasteless

The two B groups of suosan and cyanosuosan were next viewed with particular interest, and an attempt was made to mimic the distance between them by synthesizing *N*-trifluroacetyl glutamic acid. The sweetness (3000×) of the substance obtained was surprising, as glutamic acid, as a substitute for aspartic acid in the structure of aspartame-like dipeptides, does not have sweet taste. As a consequence of the latter observation, it was concluded that NO_2 and CN are special B functions as a glycophore component, and that the traditional B functional assignment should be viewed, in the case of the carboxyl group, not as a hydrogen bonding component of an AH,B couple, but as a B group with an ionic bonding function. As the distance between ionic B and D is critical, and while simultaneous functioning is not required for sweetness, their occurrence in one structure leads to extraordinarily high sweetness. Only two groups are now stated to be needed for sweet taste, and one does not even need to be an AH function (Tinti & Nofre, 1991*a*). The conclusion is attractive from the viewpoint that it obviates the need to identify an AH in those cases (e.g., the oximes) where it is difficult to identify AH. However, the high-potency sweeteners have a readily identifiable AH, and those compounds with uncertain AH have one that is identifiable by other means (Section 9.3.13.4). *It seems somewhat premature to abandon the 100 years of methodical study leading to the* *notion that there is a common chemical feature among compounds with sweet taste. Although high-potency sweetness is not easily explained, multiple glycophores and the presence of groups with strong inductive effects on the polarity of the glycophore seem reasonable. A chlorine atom has strong inductive effects on the sweetness of the sugars, but the identity of the glycophore is essentially unaltered.*

Continuing on, the authors recognized that aspartame, which does not contain a D-group and suosan, which does, also possess a common 3-aminopropanoate group. By constructing a compound containing a suosan moiety (with its attendant D group) and an aspartame moiety with a single 3-aminopropanoate group common to each, the author's synthesized superaspartame (8000×) (Section 7.4.1.3). *Another way to view this result is to acknowledge that D is a special case of Kier's electron-rich X* (γ) *site for both the suosan and the cyanosuosan moieties of superaspartame. In the conformation described for suosan in Section 9.3.11.7, the* NO_2 *group occupies the specified steric position. Superaspartame is unique in that it has an AH,B unit accompanied by two* γ *units, either or both of which have strong inductive effect on the polarity of AH,B.*

In a sweetness potency/structure search *tour de force*, Tinti & Nofre (cf. 1991*a*) employed the *N*-carbamoyl dipeptides and cyano suosan to gain further knowledge on the nature of glycophoric functional groups. Using the observation that substituting a ureido (NHCONH) with a thioureido (NHCSNH) group in cyanosuosan increased sweetness more than threefold, the authors prepared a compound with the ureido group of

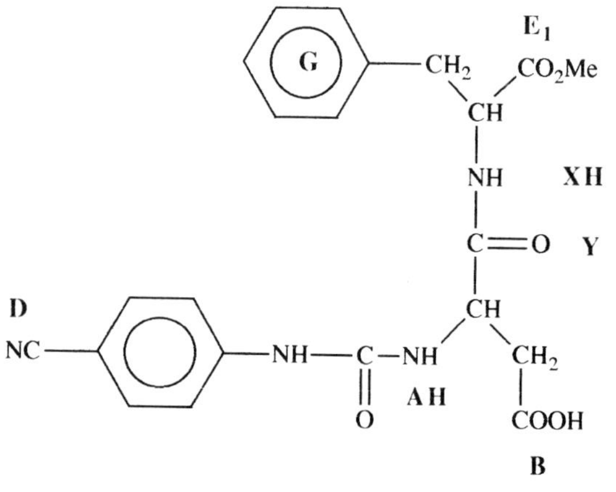

FIG. 9.33. Multiple glycophoric units in superaspartame (Tinti & Nofre, 1991*b*).

TABLE 9.3
Glycophoric groups in selected compounds (Tinti & Nofre, 1991b)

Compound	*Sweetness*	*AH*	*B*	*G*	*D*	*Y*	*XH*	E_1	E_2
CN · PhNH$^+$=C(NHc-Nonyl)Gly (Sucrononic acid)	2×10^5	NH(Ph)	CO_2^-	c-Nonyl	CN				
diCl · PhNH$^+$=C(NHCHMePh)Gly	$1{\cdot}2 \times 10^5$	NH(Ph)	CO_2^-	CHMePh	Cl				
(NO_2 · PhNHCS) Aspartame	$5{\cdot}5 \times 10^4$	NH(Ph)	CO_2^-	Bz	NO_2	CO	NH	CO(OMe)	
L-Asp-NHCH(CO_2Me)CO_2Fn	$5{\cdot}0 \times 10^4$	NH_3^+	CO_2^-	Fn		CO	NH	CO(OMe)	
Suosan	$7{\cdot}0 \times 10^2$	NH(CH_2)	CO_2^-		NO_2				
Saccharin	3×10^2			Bz		CO	NH	SO	SO
Aspartame	2×10^2	NH_3^+	CO_2^-	Ph		CO	NH	CO(OMe)	
Cyclamic acid	50	NH	SO_3^-	c-Hexyl					
D-Tryptophan	35	NH_3^+	CO_2^-	Indolyl					
Ethylene glycol	0·6	OH					OH		

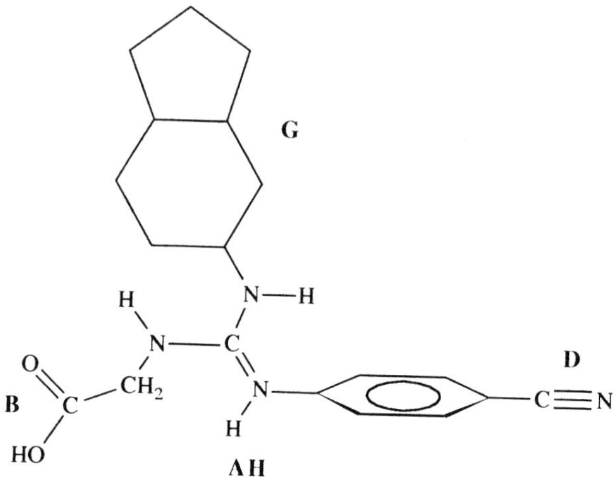

FIG. 9.34. Glycophoric units in sucrononic acid (Tinti & Nofre, 1991*b*).

superaspartame replaced by a thioureido group. That compound had a sweetness potency of 50 000×. By employing the knowledge that an *N*-

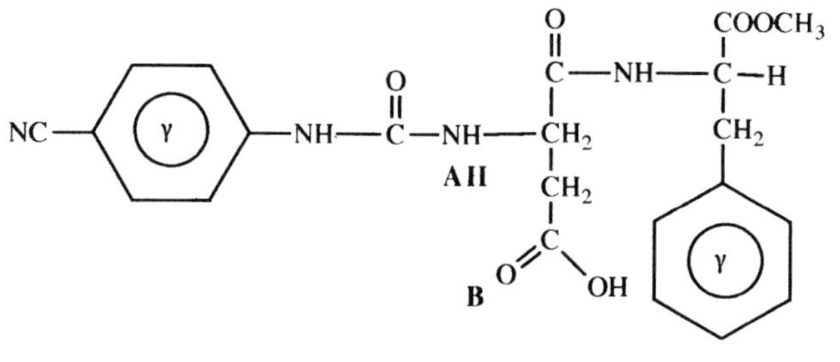

Superaspartame (8,000x)

cyanoguanidine unit is an efficient biosteric substitute for the thioureido group, the N—CN guanidine derivative of superaspartame was prepared. Its sweetness was 40 000×.

Subsequently, sucrononic acid was prepared with a potency of 200 000×. As a result of their obvious success, Tinti & Nofre (1991*b*) postulated that there are eight functional categories that contribute to sweetness, and these were grouped into high-affinity (activity) and secondary sites. The high-activity sites are: (1) AH (NH, NH_3, OH), and are hydrogen bond donors; (2) B, an anionic group with a negative charge (CO_2^-, SO_3^-, CN_4^-); (3) D, a hydrogen bond acceptor group (CN, NO_2, Cl); (4) G, a hydrophobic group. Clearly, site D takes over the traditional role assigned to B in the AH,B couple. If it is not the ionic form of D that interacts with the receptor, the D category becomes, ultimately, further examples of traditional B.

The secondary sites are XH, again a hydrogen bond donor group, in the usual AH sense and Y, E_1 and E_2 are hydrogen bond accepting ligands, again in the usual B sense. This approach to classification of functional groups in sweetness phenomena is shown for selected compounds in Table 9.3 to emphasize the need for the new classification for extraordinary (high-potency) sweetness. Many of the structures for the compounds in the table are found in Chapter 8.

An example of many of the glycophoric sites is given for super-aspartame in Fig. 9.33. The reactive sites for sucrononic acid are given in Fig. 9.34 employing the perspective model of Tinti & Nofre (1991*b*) in which the carbocyclic ring is oriented perpendicular to the AH,B,D-component of the structure. Here again, sucrononic acid can be seen to have multiple glycophores accompanied with special groups with remarkable inductive effect potential. Table 9.3 lists compounds in decreasing order of the intensity with which the glycophoric groups interact with the receptor to initiate sweet taste. Ethylene glycol does not have an H-bond acceptor group in this scheme, unless it is the secondary AH (XH) acting as an acceptor. The demotion of the imino group of saccharine to secondary AH status might very well be justified on the basis of the indirect (hydrated salt) manner with which it (and other sulfamate imino groups) act as a hydrogen bond proton donor (cf.

TABLE 9.4
Glycosteric groups for sweetness

AH	*B*	γ
Hydrated Na ion	Hydrated chlorine ion	—O—Alkane
—OH	—OH	—NO_2
—NH	—C=O	—CN
—NH_2	—HC=CH—	—$(CH_2)_n$
—CH (Special)	—Cl	—Cl
	—SO_2, SO_3	
	—NO_2	

Section 9.3.11.3). It is stated that simultaneous interaction of all sites (including AH) with the receptor is not required in order for a compound to generate a sweet response, but sweetness potency is correlated with the number of sites involved in interaction with the receptor. Potency is also stated to depend on the space-filling properties of the molecule as well as the orientation of the glycophoric groups.

Another interpretation that might be applied to the significance of the various sites proposed by Tinti & Nofre is that they are also examples of conventional AH,B, and γ, but with varying activity for interaction with commensurate sites at the receptor. There seems to be little question that placing an electron-rich single substituent such as Cl, CN, or NO_2 at the proper position for a third (X or γ) site leads to extraordinary sweetness, and that another group then adds the element of lipophilicity necessary to gain rapid access to the receptor. In the favored conformation of sucrononic acid, the CN group occupies the position for X (γ) described by Kier (1972). As sweetness intensity is related exponentially to the interaction energy of the glycophore(s) and receptor, and to the equivalency for the number of glycophores, the notions described by Tinti & Nofre cannot yet be taken to preclude a common AH,B,γ glycophore.

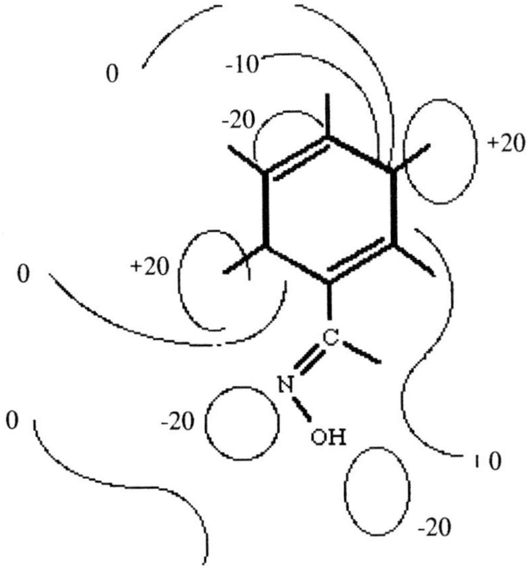

FIG. 9.35. Diagrammatic representation of the electrostatic potential map of (E)-1,4-cyclohexadiene-1-carboxaldehyde oxime in a plane 1·6 Å away from the CNOH plane (Venanzi & Venanzi, 1991).

9.13.3.3 *AH,B,γ Glycosteric Groups*

A list of the various isosteric groups for sweetness, under the general heading of glycosteric groups that function primarily in a proton donating (AH) capacity, a proton accepting B capacity, and as an electron rich or hydrophobic/lipophilic center (γ), with possible strong inductive effect as a γ function, is summarized in Table 9.4. Whether a group serves as B or γ depends in many cases upon its location in a molecule. The NO_2 group in the nitroaniline series of substances serves as a nucleophilic center (B), along with an adjacent ring proton, which is rendered electrophilic by virtue of the nature of other ring substituents, and can act as a hydrogen bond donor. Similarly, protons of halogenated alkanes are electrophilic for the same reason (chloroform). When NO_2, along with CN, are appended to a conjugated system remote from other nucleophilic/electrophilic groups, they can act as γ by virtue of decreasing the electron density of the conjugated system. In this capacity, they have a negative inductive effect and are capable of attracting electrons even more strongly than a hydrogen atom (Korolkovas, 1970). A group such as COOH can be expected to function in opposite capacity.

9.3.13.4 *Electrostatic Potentials*

As an aid to deducing the nature of a sweet compound's glycophore, particularly those with ambiguous or uncertain AH units, Venanzi & Venanzi (1988*a,b*, 1989, 1991) initiated a study of the molecular electrostatic potentials of the nitroanilines, oximes, and acesulfame substances. The procedure focuses on the long range electrostatic energy pattern molecule as a whole, and results obtained were viewed (Venanzi & Venanzi, 1988*a*) as being a recognition electrostatic pattern that guides and dominates the alignment of a sweet-tasting substance with the complementary electrostatic pattern at the receptor. The phenomenon is viewed by the authors as a preliminary step to actual tastant–receptor binding, but the present author believes that the recognition chemistry alone is sufficient to elicit taste.

Significantly, Venanzi & Venanzi (1991) found that the electrostatic patterns of the sweet nitroaniline substances and perillartine analogues were essentially the same. Two negative electrostatic potential patterns were found, one for the oxime group, and one for the NO_2 group. The second was associated with the ring systems. The results appear to confirm the conclusion of Tinti & Nofre (1991*a*) to the effect that AH is not necessary for extraordinary sweetness, but two B functions are necessary. On the other hand, as sweet and bitter perillartine substances yield the same electrostatic recognition patterns, Venanzi & Venanzi (1991) caution that the patterns cannot be used as a criterion for sweet-taste potency. A diagrammatic representation of the map obtained for (E)-1,4-cyclohexadiene-1-carboxaldehyde oxime is shown in Fig. 9.35 illustrating that one of the electropositive areas associated with the hydrocarbon ring protons is in accord with Kier's (1972) assignment of AH to those substances. The present author would also point out that the magnitudes for the areas, as potential AH and B, are of equal, but opposite sign.

9.4 THE PICROPHORE

As with the glycophore, it seems highly unlikely that there is a common element or common group of elements among the compounds of varying classes, structure, and molecular weight that is responsible for the bitterness of diverse substances. Bitterness is elicited by potassium and magnesium salts, quinine, picric acid, peptides, L-amino acids, denatonium chloride, and the sugar β-D-mannose. Among ionic substances, the picrophore is associated with the increasing size of either the anion and/or cation (Section 3.4), and such combinations invariably are described as bitter or unpleasant.

The significant feature of the β-D-configuration of the sugars is that it imparts an unfavorable dipole–dipole feature to a molecule known as the anomeric effect (Section 6.5). In the case of β-D-mannose, it brings three oxygen atoms into a planar arrangement on the same side of the molecule, and as one C—O bond bisects the other two, an arrangement known as the Δ-2 configuration, imparts an element of conformational instability to the molecule, as shown below. It also generates, through the unfavorable dipole–dipole interaction, an element of basicity to the ring-oxygen end of the molecule.

For the amino acid picrophore, Wieser & Belitz (1975) chose the ammonium nitrogen atom, surrounded by a zone 3 Å from the electrophilic nitrogen and extending for a length of 3 Å, as an area for a hydrophobic function. In subsequent studies (Belitz *et al.*, 1981), one component of the picrophore was recognized to be either an electrophilic or nucleophilic function.

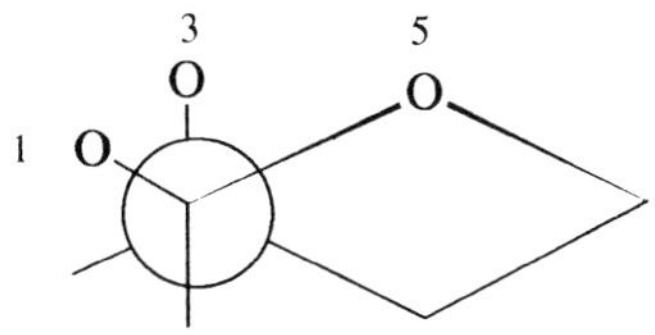

The Delta-two configuration for β-D-mannose

Upon examining a series of bitter and tasteless diterpenes, Kubota & Kubo (1969) concluded that whenever a compound possessed a proton donor system (AH), such as OH, $CHOCOCH_3$ and $CHCOOCH_3$, plus a second group (B), which might be CHO, CO, COOH, $COOCH_3$, $OCOCH_3$, and C=C—C—C—O—, the AH,B unit served as a picrophore when the AH to B distance was about 1·5 Å. This is the case for Isodonal, which is bitter, but the isomer trichodonin, with the opposite configuration at position eleven, and greater AH,B distance, is tasteless. The postulated picrophore seems to apply also to quinine, and the bitter principles of beer (isohumulones). Beets (1978) pointed out there are sterically adjacent polar features in a number of bitter compounds, but none can be reasonably assumed to serve as an AH,B picrophore. Among the examples given was the observation that a single lactone group in a sesquiterpene molecule that is devoid of an AH unit seems sufficient to cause bitter taste, as in germacrolide. Beets (1978) pointed out that a proton is not required for bitterness in other cases also, and that bitterness is associated with the general group X = C – Y in which X and Y are heteroatoms (e.g., N, S), as in anethole trithione (S), and caffeine (N).

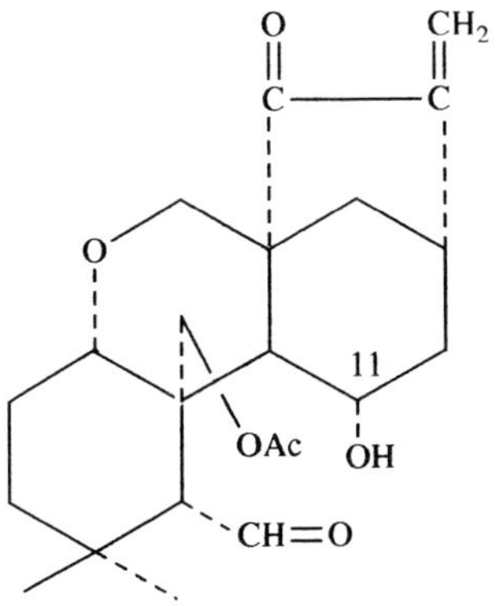

Isodonal (Bitter)

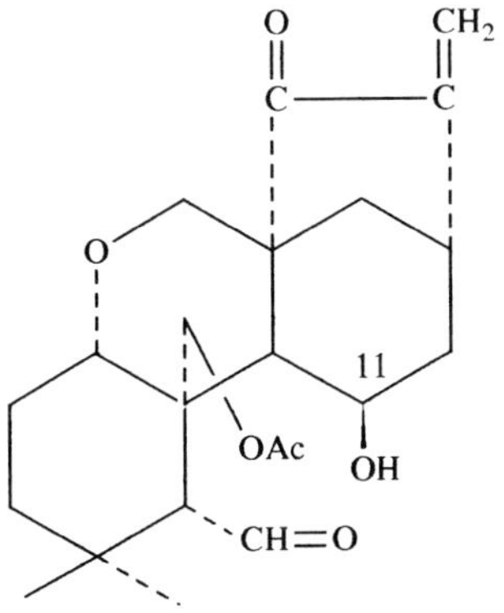

Trichodonin (Tasteless)

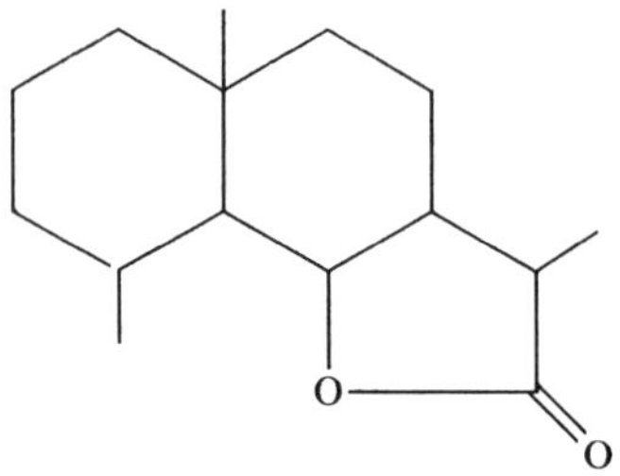

Germacrolide (Bitter)

Shinoda & Okai (1985) designated NH_2 and γ to be the picrophore of various aspartame derivatives. For the amino acids, Belitz *et al.* (1981, 1988) pointed out that the picrophore for the L-amino acids has, in common with the glycophore for the D-amino acids, either an electrophilic group to act in concert with γ to elicit bitterness, or a nucleophilic group to act in concert with γ to elicit bitterness. Perhaps the significant observation with respect to bitterness up to this point is that bitter taste is elicited by organic substances that have a lower number of saporous groupings (two) than is required for sweetness (three). Certain bitter organic molecules, such as the relatively small nitrogen heterocyclic substances, act only as a negatively charged monopole (e.g., pyrrole, pyrazole, and pyrazine).

Saroli (1984, 1985) studied a number of derivatives (Section 8.2.4) of the extraordinarily bitter substance, denatonium chloride. It has a bitter recognition threshold of $1-2 \times 10^{-8}$ M. The picrophore of this substance was assigned to two polar groups, an electrophilic group (the ammonium N atom) and the nucleophilic carbonyl group, with a distance between them of 3·2 Å. The interaction with the receptor is believed to be augmented by two hydrophobic centers, the benzdiethylammonium group and the 2,6 dimethyphenyl group. The latter were 6·8–7·0 Å

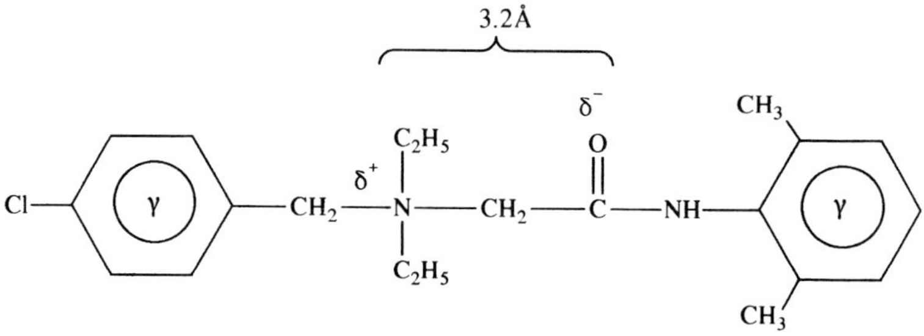

Denatonium chloride

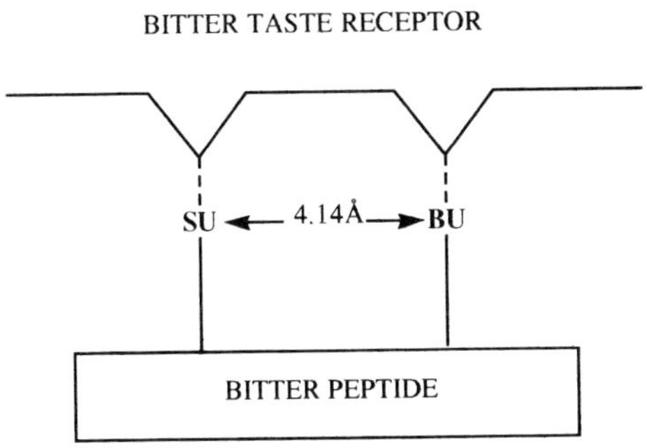

FIG. 9.36. Interaction of a bipartite bitter peptide picrophore with a bitter taste receptor (Ishibashi *et al.*, 1988).

removed from the ammonium nitrogen atom. The picrophore was then viewed to correspond with the Wieser-Belitz amino acid picrophore, and applicable to other bitter substances, such as brucine and morphine. Another possibility for the high-potency bitterness is that the picrophore for denatonium chloride is either one of the monopolar functions, coupled with either one of the benzene groups as γ. In addition to the strong inductive effects of the benzene substituents, the picrophore equivalency of denatonium chloride is greater than one.

According to Belitz & Wieser (1985) the picrophore for quinine is a flat hydrophobic area with three reactive groups (the OMe, OH and quinoline N groups). Alternatively, quinine has two monopolar groups for bitterness acting in concert with two γ sites, much on the order of the two picrophores in denatonium chloride. Protonation of the vinyl group of quinine (Section 1.5.1) diminishes bitterness.

Asao *et al.* (1987) found that bitterness of amino acids and peptides was related to chain length and hydrophobicity, and an interaction with the receptor through the charge on the compound was viewed as necessary for bitterness. Ishibashi *et al.* (1988) proposed that the feature common to all bitter-tasting peptides is a 'stimulating unit' (SU), and a 'binding unit' (BU). The stimulating unit was either a hydrophobic group or a bulky basic group such as the guanidino- or α-amino-group. The binding unit was always a hydrophobic group, and the distance between BU and SU was found to be 4·1 Å (Fig. 9.36).

In the final analysis, and in contrast with sweetness, sourness, and saltiness, there probably is not a common picrophore. There does seem to be a general disparate interaction caused concertedly by ions of different size, or monopoles of different charge. The latter interaction is probably augmented by a lipophilic/hydrophobic function. Finally, the sensation is caused by a large group of substances characterized by their basicity. The nature of bitterness receives further attention in Chapter 10, where the disparity in reactivity among ionic substances, and the disparate monopolar charge on certain bipolar substances is discussed. The fundamental features of the bitter modality are also presented in the application of symmetry principles in Chapter 11.

CHAPTER 10

The Initial Chemistry of Taste

From the discussion set forth in Chapter 9, it is clear that the acidophore, halophore, glycophore, and picrophore for substances that elicit sour, salty, sweet, and bitter tastes have potential for different chemical functions. This chapter explores the extent to which the case for different induction chemistry mechanisms may account, at the outset, for the sour, salty, sweet and bitter taste modalities. It will also explore varying taste potency (Section 1.3.1), and temporal aspects of taste.

Those sections are followed by one that attempts to bring special physical, and other qualitative features of taste, particularly sweetness, into line with the different inductive (recognition) chemistry tenets. A large section is then devoted to empirical procedures (mathematical models) for gleaning structure-taste activity relationships from empirical chemical property parameters that are not easily determined experimentally. While mathematical models have valuable application to the understanding of taste induction chemistry, the parameters employed are also based on heuristic assumptions, and the manner with which the chiral features of taste substances are treated is not always appropriate. The nature of the mathematical models is presented in considerable detail for two reasons. The first is that an understanding of them is required to judge the validity of the conclusions derived from their application. Second, the procedures readily lend themselves to computer analysis, and can be expected to become extensively employed. The chapter concludes with a presentation of the kinetic nature of the recognition chemical phenomena responsible for the differentiation of the four basic tastes, and how taste inhibitors may possibly function at the induction phase of taste-perception chemistry.

10.1 INITIAL CHEMISTRY OF SOURNESS

With an acidophore assigned to the simplest ion possible (Section 9.1), and in view of the simple total-acid chemical equilibria nature of sourness potency (Section 3.6), it might be assumed that sourness is the quintessential chemical structure-activity relation. Nevertheless, several theories have been proposed to explain the initial chemistry of sourness. Organic acids are incompletely dissociated, but just as sour as those acids (inorganic) that do dissociate completely. Beets (1978) attempted to resolve this feature of sour taste by first pointing out that there are different definitions of the term 'threshold' (cf. Pfaffmann, 1959; Marin *et al.*, 1991). Hydrogen bonding was then assigned to be the principal mechanism for the induction of sourness. It was reasoned that the solution concentrations of the three acid species (one molecular and two ionic) are totally interdependent (*sic*). The proposed scheme for the hydrogen bonding of the three species for an organic acid, using three acid species (HA, H^+, and A^-) with a receptor, presumed to be an AH,B unit, is shown in Fig. 10.1. An inorganic acid and its ions may be substituted without any modification of the scheme. As shown in Fig. 10.1, an ionized acid may react stepwise, through hydrogen bonding, with the AH,B receptor postulated by Shallenberger & Acree (1967) for sweetness (Section 10.3.1) to initiate sour taste. Alternatively, the undissociated acid may react concertedly, again through hydrogen bonding, to elicit the sour taste.

The ideas of the concerted and/or stepwise interaction of acids, through hydrogen bonding, does not afford an answer to several critical

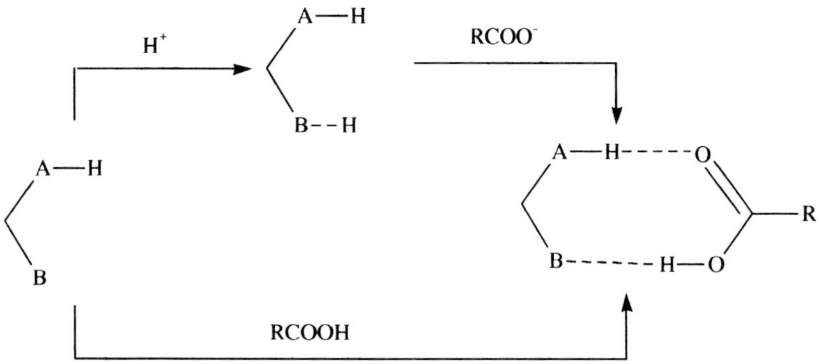

FIG. 10.1. Proposed hydrogen bonding mechanism for the initial chemistry of sourness for a weak acid (Beets, 1976).

questions. The first is why the molar threshold for acetic acid is only 1/100th that for sucrose sweetness, which indicates an activity toward the receptor of high magnitude. The second is that it does not offer a reason for dependence on the definition of the term threshold. While the initial chemistry of sourness and sweetness has some common features, and may even be interrelated, the initial chemistry of those two tastes is probably not the same. An explanation similar to that of Beets for the chemistry of the near-equal sour taste of strong and weak acids is based on the definition of ions (Boudreau *et al.*, 1976). The hydrogen ion and the undissociated acid are viewed as Brönsted–Lowry acids (Section 2.3.1), and each contributes to total sourness.

Such explanations ignore the simple titration-like features of the sour taste sensation. Weak acids are, on an equivalency basis, just about as sour as HCl (Section 8.3.1). Correspondingly, the rank *molar* order for acid sourness correlates with the amount of NaOH needed to titrate a solution to pH 4·4 (Ganzevles & Kroeze, 1987). Therefore, chemical equivalency principles are operative. Even if the environment at the receptor site has reduced polarity, which is probably the case, then strong acids, such as HCl, prevail as incompletely dissociated (weak) acids. Again, there would be little sourness difference among the acids as the hydrogen ion equivalency would be the same. Finally, the sourness of acid foods, particularly fruit (Nybom, 1962), is related to total potential acidity, and not to pH. For sourness intensity, it seems obvious that the receptor serves in some capacity as a base, and in this sense, sour taste recognition chemistry is firmly established as being due to a chemical reaction.

One probable mechanism for the initial chemistry of sourness is a strong ionic interaction by the hydrogen ion with a nucleophilic receptor component, accompanied concertedly by a weaker, nearly irrelevant, ionic interaction between the anion and an electrophilic receptor component. A second possibility is a displacement reaction, and the displacement takes place on the hydrogen ion (Section 2.7.7). In either case, and by viewing the receptor as an n/e unit (Belitz *et al.*, 1979), a general equation for the initial chemistry of sourness that emphasizes the titration-like nature of sourness, with the receptor as the 'base,' is:

$$\underbrace{HA \rightleftharpoons H^{+} + A^{-}}_{\text{Acid}} \quad + \quad \underset{\text{Receptor}}{n/e} \quad \rightleftharpoons \quad \underbrace{Hn \quad + \quad Ae}_{\text{Acid/receptor interaction}}$$

In the equation, the role of the hydrogen ion concentration, or potential hydrogen ion concentration, is far more significant than the role of the anion. It is more significant not only concentration-wise, but also in terms of chemical activity.

Compared with inorganic acids, acetic and other organic acids have a sourness potency higher than might be expected from their equivalent concentrations (Pfaffmann *et al.*, 1971). That feature has been traced to the degree of lipophilicity of the anion (Beidler, 1967; Gardner, 1980*b*), and is a manifestation of the sourness temporal profile. In a homologous series of organic acids, sourness potency parallels the length of the hydrocarbon chain, and butyric acid > propionic acid > acetic acid > formic acid. Presumably, a higher degree of lipophilicity affords either easier, or more rapid, access to the receptor site.

There are cases where a compound is described as tasting sour when there is no immediately recognizable proton to account for the sourness. Water tastes sour after tasting sodium chloride (Section 3.1.4), and the sourness of lithium chloride (Section 3.4.3) and lithium sulfate (Section 3.4.4) is an example. Among the answers that may be given for these cases is the psychological confusion between the sour and bitter tastes. A reason based on chemical principles (displacement and symmetry considerations), is given in Section 11.1.1.

10.2 INITIAL CHEMISTRY OF SALTINESS

The hydrogen ion is one of the most chemically active substances that elicits taste, and the induction of sourness, and its intensity are related to that activity. From recognition threshold values, however, saltiness, appears to be a characteristic taste attribute of a substance (NaCl) that has, at best, only 1/30th of the taste activity of the hydrogen ion. Correspondingly, it might be expected that the effort to derive the nature of induction chemistry of saltiness might be proportionally difficult.

To explain the probable initial chemistry of saltiness, Beets (1978) used the idea of a hydrated metal ion, $M^+(H_2O)_n$, associated with a bare (nonhydrated) anion, X^- to be the halophore. It was suggested that the complex interacts, through hydrogen bonding, with an AH,B hydrogen bonding receptor either through a stepwise interaction, or through concerted interaction. The proposal in diagram form is shown in Fig. 10.2. Beets did not consider the sweetness of dilute sodium and

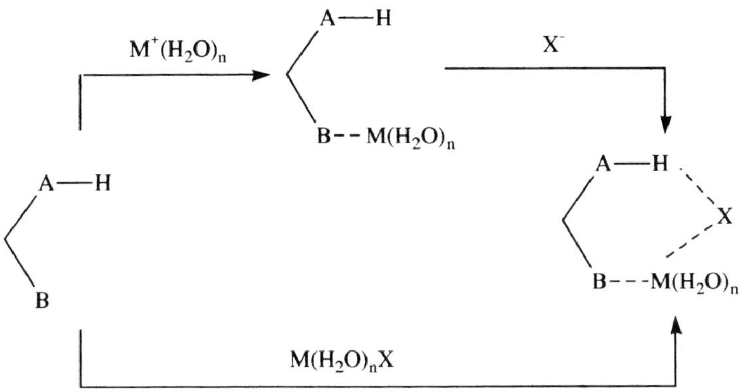

FIG. 10.2. A concerted and stepwise mechanism for interaction of a hydrated metal cation and a bare cation with an AH,B hydrogen bonding receptor for initiation of saltiness (Beets, 1978).

potassium chloride solutions, but the glycophore assigned to those substances to explain their sweetness (Section 9.3.9.1), used the hydrogen bonding scheme to differentiate the sweetness glycophore from the saltiness halophore. However, while there are problems encountered with the assignment of different chemical mechanisms to different tastes, the ambiguities and contradictions that arise when a common set of initial chemical phenomena is assigned to different tastes are even greater.

The initial chemistry of sourness is best explained by an acidophore and either ionic or displacement on hydrogen mechanisms at the base-like (nucleophilic) receptor component. It follows that saltiness is best explained by a special halophore whose taste recognition chemistry involves the *joint* ionic interactions particularly characteristic of the ion pair Na^+ and Cl^-. Among the ionic substances with an anion that has widest ability to *attract* electrons are lithium, sodium, and potassium chlorides. The bond between the anions and chlorine cation in the crystalline lattice is purely ionic (Sections 2.2.7 and 3.2.1). On the other hand, in solution they are among the only strictly ionized substances that elicit saltiness with minimal additional tastes. Thus saltiness recognition chemistry is elicited by ionic interactions, and in addition to Section 9.2, special features of the Na and Cl ions receive further attention in Chapter 11.

With respect to equivalency, the majority of salts are mono-basic, and an appropriate equation for saltiness recognition chemistry is

$$\underbrace{\mathrm{MX} \Longrightarrow \mathrm{M}^+ + \mathrm{X}^-}_{\text{Salt}} + \underbrace{\mathrm{n/e}}_{\text{Receptor}} \rightleftharpoons \underbrace{(\mathrm{M}^+)(\mathrm{n}^-) + (\mathrm{X}^-)(\mathrm{e}^+)}_{\text{Salt/receptor interaction}}$$

and the case for true saltiness occurs only with NaCl. While also ionic in form, the primary distinction from the sour-taste equation is the *near-equal* importance of both the anion and the cation. Apparently, then, the initial chemistry for saltiness is a concerted bi-ionic interaction with the ionic interaction for the cation and the anion approximately equal in intensity. The umami taste is basically similar to saltiness. It has the required sodium cation features, plus additional taste attributes that can be assigned to the chiral glutamate ion.

While both the sour and saltiness tastes are elicited by ionic recognition interactions, where the former is associated with the hydrogen ion, and the latter with sodium and chlorine ions, their recognition chemistry is clearly distinguishable from sweetness and bitterness recognition chemistry at the very outset. In addition, there are no subtle changes in structure or facile structure interconversions, as there are for sweetness/bitterness (Section 1.5.2), that will transform a sour-tasting compound to salty and vice versa. One can only neutralize an acid with a base, or add an acid to a salt to interconvert the tastes. Finally, there are no compounds that can be said to possess high-potency sourness or high-potency saltiness.

10.3 INITIAL CHEMISTRY OF SWEETNESS

From the foregoing discussion, it was found that it was relatively easy to establish the probable recognition chemistry for sourness, but somewhat more difficult for saltiness, and it would logically seem that chemical reactions are also responsible for sweetness recognition chemistry. Now, however, the recognition identity problem is made all the more difficult because only minor changes in the structure of a compound are required to transform bitterness/sweetness attributes (Section 1.5.2). In addition, only subtle changes in the structure of a compound are required to impart sweetness to an otherwise tasteless compound, or to render a sweet compound tasteless. An example of the facility of the latter phenomenon is the dimer–monomer, sweetness–tastelessness transformations of the simple, but sugar-like, compounds glycolaldehyde, glyceraldehyde and dihydroxyacetone (Sections 4.4.1–4.4.3). The crystalline hemiacetal ring dimers are essential for sweetness, but dissociate, with loss of sweetness, merely by dissolving them in water. On the other hand, the formation of hydration shells leads to sweetness in dilute solutions of NaCl (Section 3.4.1). As water of hydration *per se* is not particularly known for ionic interaction capability, it can only be concluded that the sweetness recognition chemistry for hydrated salts, and for sugars, is a weak recognition chemistry allied with the very processes that occur during solvation with water, i.e. weak dipole interactions.

The initial chemistry of sweetness is also a structure-activity phenomenon, and is related not only to the structure and composition of the sweetener, but also to the structure of the receptor, and the nature of the receptor–sweetener interaction. There is also a symmetry attribute to the latter phenomenon, but discussion is reserved for the next chapter.

In sweetener research, the occupancy theory (Section 2.10.1), through reference to pharmacological chemistry seems, at first, to be applicable. It is then often said that the *intrinsic activity* for sweetening agents is about the same. The justification for this notion maintains that all sweetening agents can produce the maximum sensory response if the concentration is sufficient (about 70% w/w for sugars). As sweeteners differ in their relative sweetness, and sweetness potency, it is then said that the chemical affinity (Section 2.10.1) of the sweeteners toward the receptor must be different. Alternatively, it is possible that it is the activity that is different, and there is no binding affinity in sweetness recognition chemistry *per se*.

Blocking the sweetness of sugars and high-potency sweeteners with gymnemic acid and/or lactisol suggests that those sweetness-blocking agents also have great binding affinity for the sweetness receptor, but no intrinsic sweetness activity. However, they may also simply disrupt the recognition chemistry. In view of the latter alternative presented above, the present author prefers to view the different sweetness potency of substances to be due to the different recognition activity demonstrated toward the receptor to be in compliance with the chemical equilibria definitions offered in Section 2.4.2.1. As for the recognition chemistry for sweetness, it appears to be a manifestation of the dipolar interactions that sugars display toward water to alter its very properties and 'reactivity' (Section 6.6.1). Detailed elaboration on that thought now follows, and serves to establish the fundamental nature of sweetness recognition chemistry.

10.3.1 The Concerted Bipartite AH,B Interaction

As sugar sweetness varied with the propensity of sugar glycol groups to intramolecularly hydrogen bond (Shallenberger, 1963), the corollary to

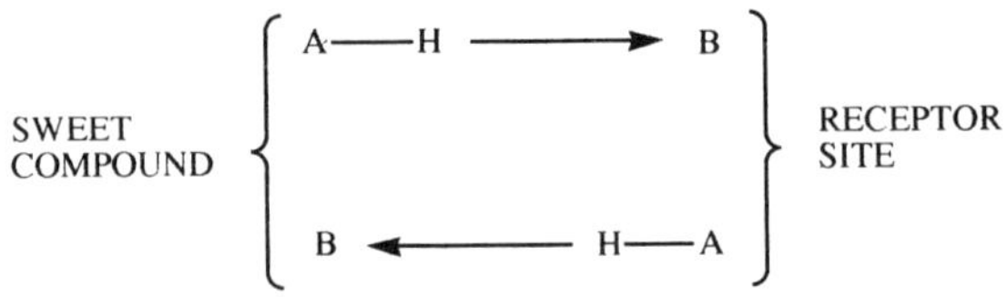

FIG. 10.3. Antiparallel concerted intermolecular hydrogen bonding between a sweet-tasting substance and a chemically and sterically commensurate system at the receptor site for sweet taste (Shallenberger & Acree, 1967).

the bipartite AH,B glycophore hypothesis was developed. It states that the initial chemistry of the sweet taste response must be analogous to concerted and antiparallel intermolecular hydrogen bonding with a chemically and sterically commensurate receptor (Shallenberger & Acree, 1967). The interaction was viewed as neither a proton transfer nor an electrostatic interaction, but probably involved London dispersion forces, a principal element of hydrogen bonds. Furthermore, it needed to be of antiparallel nature (Fig. 10.3). An expanded interpretation for the initial chemistry of sweetness soon followed employing the notion that an intramolecularly hydrogen-bonded AH,B glycophore may serve as a hydrophobic probe (Section 9.3.8.3). The expanded notion also recognized that the receptor is probably also intramolecularly hydrogen-bonded, giving it an element of hydrophobicity. Then, as the hydrogen-bonded glycophore approaches the receptor, water molecules associated with the receptor become delocalized, and intermolecular hydrogen bonding occurs between the glycophore and the receptor. The energy for the event is derived from the disruption of organized water in the immediate vicinity of both the glycophore and the receptor leading to an increase in entropy. The explanation is shown in diagrammatic form in Fig. 10.4. As the structure of hydration water is seldom tetrahedral (Section 3.1.3), such departure from tetrahedral angles allows water to align the saporous unit with the receptor, and a more complete reason for the varying sweetness of the sugars, as related to intramolecular hydrogen bonding, results. Apparently, whenever a potential glycol component of a sugar is hydrogen bonded to a nonglycophore atom, that glycophore has lowered potential to function either as a hydrophobic probe, or as a glycophore.

The above explanation for the initial chemistry of sweetness is supported by the taste equation derived (Section 10.7.1) to explain the magnitude of the neural response to sweet tasting substances. The aforementioned description of the initial chemistry of sweetness, in terms

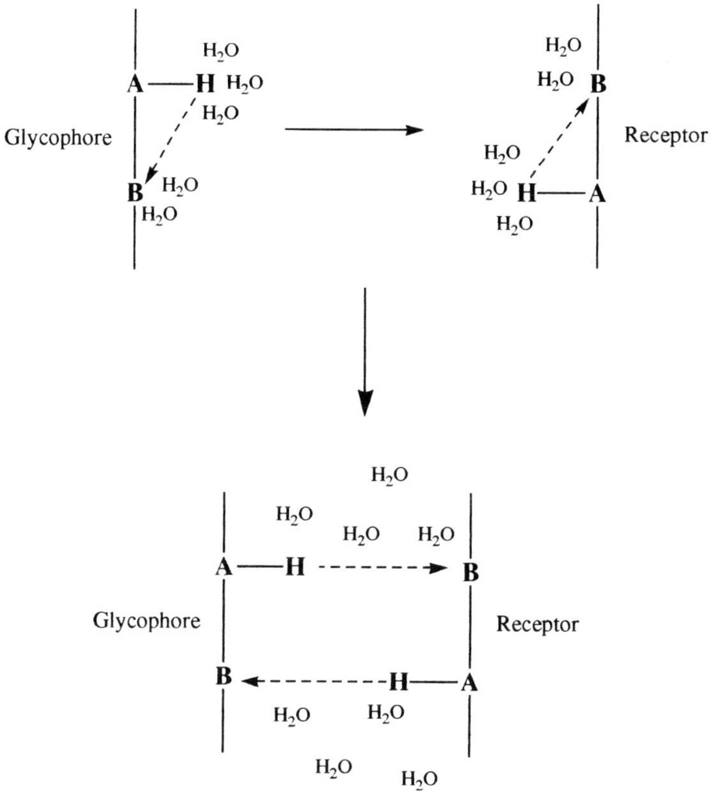

FIG. 10.4. The intramolecularly hydrogen-bonded glycophore as a hydrophobic probe interacting with an intramolecularly hydrogen-bonded hydrophobic receptor.

of concerted intermolecular hydrogen bonds, is also in accord with the speed with which the reaction occurs. It would also seem that the interaction energy for each of the antiparallel bonding components of the bipartite interaction would need to be about equal.

The notion of intermolecular hydrogen bonding (or hydrogen bonding recognition) seems favored as a feature common to all sweet substances, regardless of their structural diversity. The common bonding has profound effect on the structure and properties of water, and serves to distinguish the chemistry of sweetness from the chemistry for other tastes. Presumably, it is the strong directional influence of the hydrogen bond that positions the glycophore at the receptor site to initiate formation of noncovalent interactions (Lee, 1987). In addition, a dynamic representation of the concerted AH,B interaction neatly fits in with much more that is known about the chemistry of sweetness.

10.3.2 The Dynamic (Equilibrium) AH,B Model

A dynamic version of the initial AH,B recognition chemistry of sweetness that takes into account sweetness onset time and activity of the interaction (sweetness intensity) is outlined in Fig. 10.5. In the dynamic model for the initial chemistry of sweet taste, the AH,B glycophore is in equilibrium with a hydrogen-bonded form, and a number of heuristic chemical equilibrium interpretations become possible. Viewing the intramolecularly bonded glycophore form as the active species, when k_1 is slightly larger than k_2, the hydrophobic probe should lead to more rapid response. When k_1 for the equilibrium is very much greater than k_2, then the intramolecular hydrogen bonding potential is much larger than that for intermolecular hydrogen bonding. The interaction with the receptor would then be delayed and the onset time for sweetness longer.

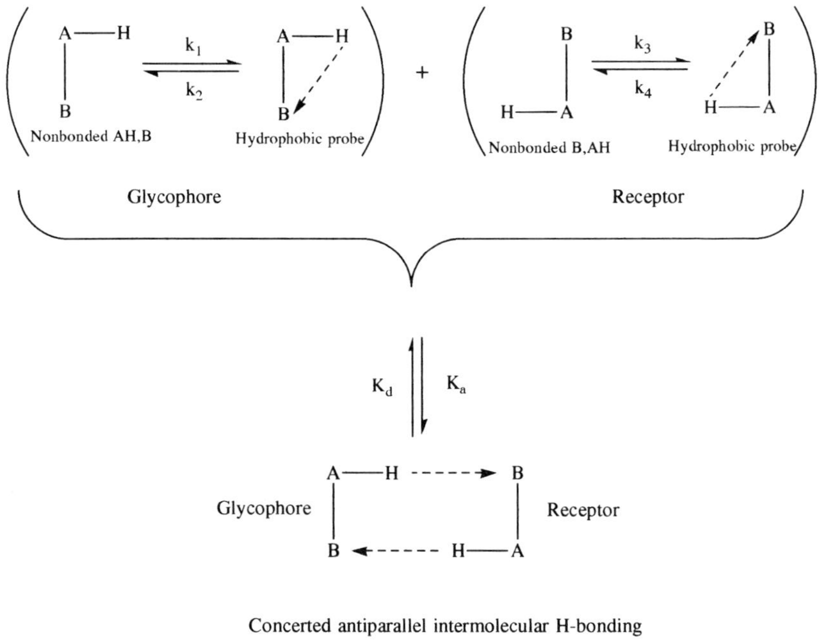

FIG. 10.5. Dynamics of the bipartite AH,B glycophore–receptor interaction.

All AH,B isosteric glycophore–receptor pairings would also not be energetically equal, and persistence would vary. Finally, taste intensity may be quite high even though k_1 is not intrinsically large. This leads to a probable inductive role for γ as a component of a tripartite glycophore (Section 10.3.4.1). While the variations possible for the scheme, and the themes, are numerous, the initial bipartite chemistry of sweetness recognition in equation form is,

$$(AH_g^{\delta+}, B_g^{\delta-}) + (B_r^{\delta-}, AH_r^{\delta+}) \rightleftharpoons (AH_g^{\delta+}, B_r^{\delta-})(B_g^{\delta-}, AH_r^{\delta+})$$

Glycophore Receptor Glycophore-receptor interaction

One possibility for the common mode of sweetness action is that cooperatively hydrogen-bonded water at the receptor is polarized by a distant sweetener (Section 10.4.3.2). The equation is therefore distinguished from the equations for the initial chemistry of sourness and saltiness by virtue of being a concerted antiparallel dipole–dipole interaction.

The equation for sweetness is also to be distinguished from those for sourness and saltiness in a second, and highly significant, manner. There are no isosteric saporous groups for H^+ and NaCl, i.e. the saporous groups have constant activity, and taste intensity varies only with concentration and saporous group equivalency. This, however, is not true for a sweet substance. In the latter case, taste response (T_r) is proportionate to some function of both glycophore activity (a) and the concentration (c) of the sweetener. One possibility is that activity is related to the product of the activity and concentration ($T_r = a + c$), but most probably, there is an exponential relation, i.e. $T_r = c^a$.

Even a simple activity-concentration product serves as a foundation to explain why there are compounds with high glycophore activity (e.g. $\Delta E = -10\,\text{kcal}\,\text{mol}^{-1}$), but with low concentration (e.g. 0·1%) that are just as sweet as compounds with an activity of $-0{\cdot}1\,\text{kcal}\,\text{mol}^{-1}$, but only when the concentration of the latter is 10%. If changes in activity should lead to exponential changes in taste intensity, which will soon be shown to be the case, reasons for the wide range in sweetness potency begin to emerge. Moreover, it is at this point that the notions of recognition chemistry and affinity (binding) chemistry, particularly for sweet and bitter substances, also begin to diverge.

The model for sweetness described now also bears some resemblance to the model for the sourness of weak acids. There, the phenomenon is equated with an acid/base titration (Section 10.1), and sourness intensity is related to total *potential* hydrogen ion concentration. In the case of sweetness, the intermolecular hydrogen bonds for the concerted interaction effectively 'titrate' the intramolecular hydrogen bonds of both the glycophore and the receptor. Treatment of the tongue with the protein Miraculin causes acids to taste sweet (Section 7.5.1). The sourness, and the sweetness potency of the acids after treatment of the tongue with the protein, are directly related (Kurihara, 1971). The observation supports the postulate (Henning *et al.*, 1969) that there is a chemical relation between the sour and sweet tastes, as the sweetness rendered unto the acids is also a stoichiometric titration-like process. Therefore, the glycophore equivalency of sweeteners can be greater than one, with parallel molar or weight-basis sweetness potency consequences.

A final case for the recognition mechanism for the sweet taste modality is the general inhibition of the sweetness of all sweet compounds by several substances, but especially 2-(4-methoxyphenoxyl)propionic acid (Section 10.8). If glycophores with different chemical mechanisms are responsible for the sweetness of diverse substances (Section 9.3.13.2), it would seem very unlikely that equally sweet solutions of those substances would have their sweetness equally suppressed, which is the case (cf. Lindley, 1991).

10.3.3 Multiple AH,B Equivalency

In many cases, the presence of more than one saporous unit, or even one component of a saporous unit for a given taste in the structure of a single compound, is capable of exponentially enhancing taste potency. Using the dihydrochalcones (Fig. 9.25) as the example, Crosby *et al.* (1979) first introduced this possibility. When the equivalency for the AH,B units is two, greater sweetness potency may be derived from two bipartite noncovalent bonding steps. As shown below, the first leads to K_1 between one AH,B unit (A) and a commensurate receptor site (X). The second interaction leads to a double interaction with a second receptor site, and a second equilibrium constant, K_2. It is proposed that the second equilibrium constant (K_2) is much larger than the first. The notion is somewhat in accord with the effects of cooperative hydrogen bonding (Section 2.7.3.4), and might even be considered to be a form of cooperative hydrogen bonding. The formation of the first noncovalent bond makes the formation of the second noncovalent bond all the more favorable.

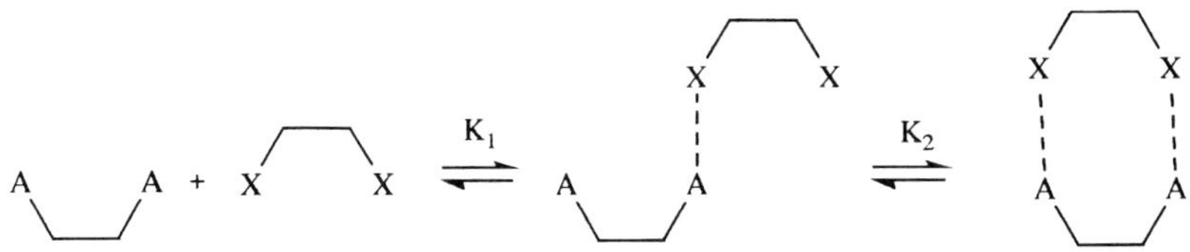

Cooperativity between two AH,B glycophores

Another interpretation of the significance of the multiple glycophores is that the probability of alignment of the glycophore with the receptor is doubled. The intensity then is related to an unknown exponential function of the number of hits as well as to an exponential function of the activity (Section 10.6.3.5). Other compounds with a glycophore equivalency greater than one are sucralose (Section 9.3.11.4), and the high-potency peptide sweeteners (Section 7.4.1.3). In some cases, the complete glycophore is not multiply represented, but the compound may possess multiple AH units for joint interaction with a single B unit. This may be the reason for the higher degree of sucrose relative sweetness, on a molar basis, as compared to other sugars (Section 9.3.8.3). A single AH,B glycophore might also be accompanied by multiple γ functions, as in the extraordinarily sweet substances.

10.3.4 Function of γ

The apparent primary function of γ is to induce, through reaction with a commensurate site, glycophore and picrophore activity. As a result, there is a good correlation between sweetness and bitterness potency and hydrophobicity within a homologous series of substances. The notion is also in accord with the fact that there are no high-potency substances for sourness and saltiness.

Upon recognizing the probable role of a third component in a compound's glycophore, Kier (1972) calculated the group polarizability (Σ_α) for nitroaniline side chains. He used the Blanksma & Hoegen (1946) nitroanaline data, and also the Verkade *et al.* (1946) data (Section 8.1.1). It was found that these values correlated well ($r > 0{\cdot}95$) with the logarithm of the relative sucrose sweetness scores (w/v basis). The group polarizability function was then interpreted to be an example of a biological *effect* correlating in a significant way with a single chemical parameter (dispersion bonding). Expanding the view of its function to include lipophilicity as well as hydrophobicity led to the designation of the third glycophoric site as γ (Section 9.3.10.1). The functions of γ also include facile access to the receptor by imparting an element of general lipophilicity to a compound. It might also act as an additional recognition chemistry hydrophobic probe for positioning of AH,B, over B,AH of the receptor. Finally, as discussed below, it may also serve, through γ, γ recognition chemistry, to convert a potential glycophore and receptor from an inactive sweetness-recognition chemical form to one more appropriate for concerted AH,B unit interactions.

10.3.4.1 *Activation of Glycophore and Receptor AH,B*

The equilibrium constant discussion applied to Fig. 10.5 can be made even broader by inclusion of a γ function for both the glycophore and the receptor. Interactions involving γ can affect (enhance) sweetener and receptor AH,B activity, which can be related to tastant–receptor occupancy theory (Section 2.10.1), but only in recognition chemistry terms. In Fig. 10.6, the sweetener T (Tastant) is in equilibrium between nonhydrogen-bonded and a bonded form. The same is true for the receptor (R). When the γ, γ hydrophobic recognition interaction between the tastant and receptor occurs, both are made active (T^* and R^*). The position of the k_1 and k_2 equilibria are then shifted to the right, toward the intramolecularly bonded forms. This is one sense with which it is implied that γ, as a glycophore component, has an inductive effect on AH,B. As a consequence, eqn (2.20) (Section 2.10) is modified for sweet taste to include an activated tastant, also. This possibility may be unique to taste chemistry. The subsequent tripartite interaction between T^*R^* and R^*T^* would then yield the sweet taste response.

To elaborate on the notion that the γ site, and the γ, γ interaction serves to transform a potential AH,B unit to the proper form for action as a glycophore, the following argument is presented. It is based on the fact that an amine AH, and a carbonyl oxygen B glycophore does not need to possess an ionic charge to initiate sweetness. Amides, such as methylformamide (CH_3CONH_2) and glycollic amide ($CH_2OHCONH_2$), are sweet, and are nonionic substances. For amino acids, Wieser *et al.* (1977) demonstrated that the threshold for amino acid sweetness is a function of the length of the side chain R. Their data are plotted in Fig. 10.7 to illustrate this relation. The intercept for the amino acid threshold in the plot shown in Fig. 10.7 is log 2·1138. Therefore, with 'no' side chain, amino acids are nearly tasteless as the intercept threshold for

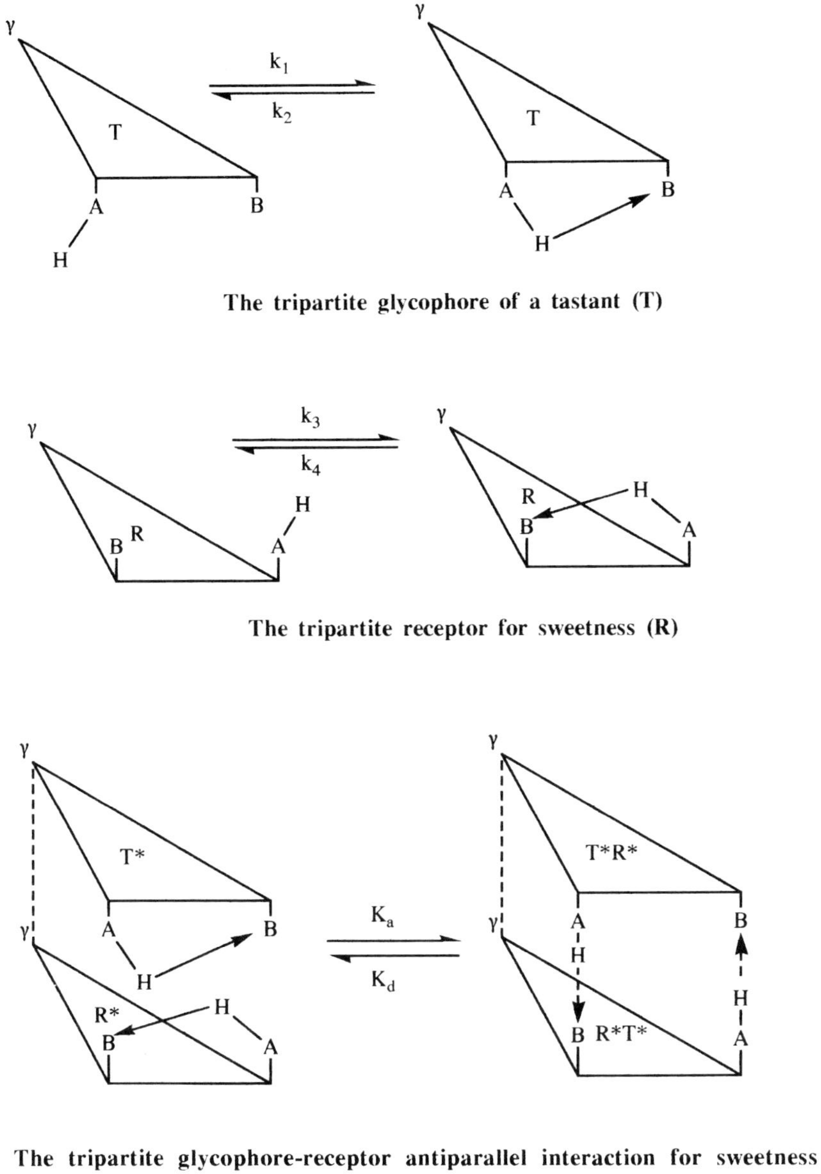

FIG. 10.6. The tripartite glycophore–receptor interaction for the recognition chemistry of sweet taste (cf. Fig. 10.5).

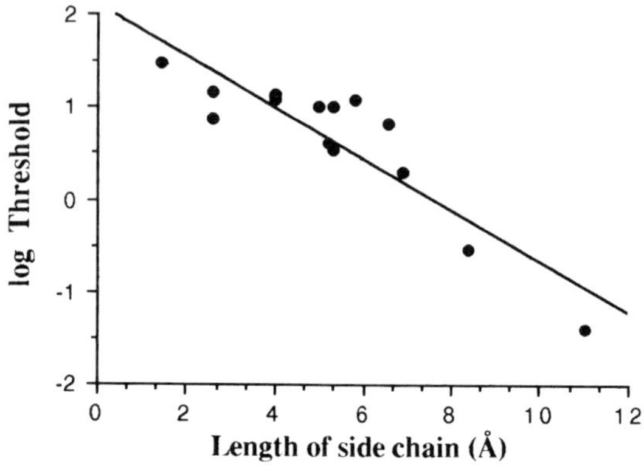

FIG. 10.7. Relation between the length of the D-amino acid side chain and threshold values (μmol ml^{-1}) for sweet taste (Wieser *et al.*, 1977).

amino acid sweetness is 130μmol ml^{-1}, or about 10 times *greater* than the threshold for sucrose relative sweetness (11μmol ml^{-1}). For the amino acids, then, the presence of R is an absolute requirement for sweet taste, even for glycine. It unquestionably serves, through a hydrophobic interaction, to convert a zwitterionic form of AH,B to an active form.

It can also be argued that the zwitterionic character of peptides is weaker than for amino acids (Section 7.2.1), and yet they have higher sweetness potency. Crosby *et al.* (1979) pointed out that for sweet amino acids and dipeptides, in the pH range 4–9, the compounds exist in a dipolar or zwitterionic form in which the amino acid bears little net charge. In essence then, the two oppositely charged ionic groups presumably generate an intramolecular bond with an average bond energy in aqueous solution of about 5 kcal mol^{-1}. An inductive, or activation role for the function of γ in aspartame is clearly evident in the synthesis of the aspartame glycophore using intermolecular aspartame glycophore components (Section 9.3.11.9). Unless contained in the same molecule, the intermolecular aspartame glycophore components have only a sweetness potency of 3 times in one case, and are devoid of sweetness in the other.

Feinstein *et al.* (1991) and Douglas & Goodman (1991) also view the aspartyl dipeptide AH,B unit as an intramolecularly bonded chelate ring. While they admit that the ring cannot be proved conclusively, it is held, *a priori*, on the basis of n.m.r. studies. The reasoning to support the bond is that the C_α—C_β bond of the aspartyl residue has a staggered conformation with the carboxyl moiety and the amino group in the gauche minus position. Then the sp^2 plane of the terminal aspartyl carboxylate carbon atom and the C_α—C_β bond are coplanar. The structure would then favor a hydrogen bond that leads to a six-membered

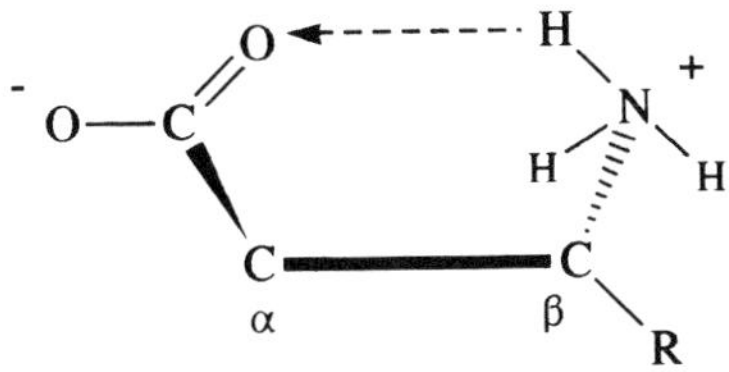

chelate ring. Whereas the intramolecular bond is a hydrogen bond, but the group is charged, it has ionic *character*. The reasoning is in accord with the assessment of the function of the hydrophobic and electron-releasing ability of substituents (Hansch, 1970). In any event, the nature of the bond requires a distinction between an intramolecular ionic bond, and an intramolecular hydrogen bond leading to a group with largely ionic character. The latter seems to be, ultimately, a relatively moot point, and, whenever the zwitterionic function of dipeptides is quenched, sweetness is increased. An example is the trifluoroacetyl aniline analogues of aspartame (Goodman *et al.*, 1985). Similarly, when the zwitterionic character of a bitter peptide is quenched, bitterness is enhanced (Section 10.5). The proposed role of the γ,γ interaction in converting an amino acid or peptide potential AH,B glycophore to the active form for eliciting sweetness is diagrammed below.

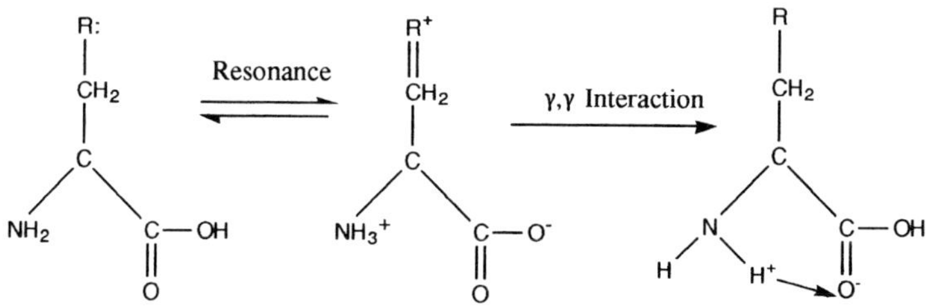

The possible inductive effects of AH,B group 'side-chains' also apply to other compounds, such as oximes (Section 9.3.11.2) and isocoumarins (Section 9.3.11.5), and seem related to the function of γ. In addition, whenever the C-6 hydroxyl group of sucrose is esterified, esterification substituents larger than the methyl ester (acetate, benzoate, phosphate) do not taste sweet (Hough, 1989).

10.3.4.2 *Groups with Strong Inductive Effect on AH,B*

The notion is that γ acts primarily in a lipophilic/hydrophobic capacity, with inductive effects on AH,B and the antiparallel interaction receives support from other studies. Appending a group with strong inductive effect to the γ function further enhances the activity of AH,B. Examples are Cl, CN, and NO_2, or the groups associated with high-potency sweetness (Sections 9.3.11.4 and 9.3.13.2). It is through the inductive effect of NO_2 groups on the dipole moment that influences the sweetness of the nitroaniline compounds (Hansch, 1970). As a consequence, the Hammett σ^+ value is superior to σ when relating hydrophobic/electrostatic parameters to the relative sweetness of these substances (Section 10.6.1.1).

At times a 'second B' seems to be needed, as Crosby *et al.* (1979) pointed out, to explain the unusual (lingering) taste properties of the dihydrochalcones, and as Tinti *et al.* (1982) indicated for an NO_2 group in suosan (Section 9.3.11.7). Such groups are also a requirement for the extraordinary sweetness of substances such as superaspartame and sucrononic acid (Tinti & Nofre, 1991*a*, *b*; Nofre & Tinti, 1993) which have a sweetness potency up to four logarithm units higher than other compounds with the same glycophore (Section 9.3.13.2). The 'second B' is not necessarily a component of a dipolar AH,B function. It is strictly associated with an area of high electron density, usually in an approximate position for γ, or as a substituent of γ. In this sense, such groups can be viewed as highly inductive adjuncts of γ. Other examples are the chlorine atom of 6-chlorotryptophan, chlorine in the mono- or polychlorinated sugars (van der Heijden *et al.*, 1985*b*), the sulfur atoms in 1,6-di-*S*-methyl-1,6-dithio-D-fructofuranose (Section 6.4.5.3) and a 'second' AH,B system of the dihydrochalcones (Lee, 1987). It is to be expected that the sweetness of a compound will rise if a chlorine atom is introduced at a suitable γ position. Sweetness will decline if A or B from an AH,B,γ system is removed upon substitution (Hooft *et al.*, 1993). Thus, to account for high degrees of sweetness potency on a chemical basis alone, it can be stated that there are substituents and groups that: (1) have a high degree of lipophilic/hydrophobic character; (2) have high inductive (amplification) effect on AH,B; and (3) lead to an AH,B,γ equivalency greater than one.

A final comment on inductive groups is that such groups seem to have positive effects on certain glycophores, e.g. NO_2 at a γ position in suosan. Here, the glycophore is an amino acid group. An NO_2 at the γ position in saccharine (Table 8.2) has negative effect on the activity of AH,B (see also Section 11.1.3). Here, the glycophore is an imine and sulfoxide oxygen atom.

10.3.5 Alternatives to the Concerted AH,B Chemical Interaction

The concerted hydrogen bonding interaction, with an antiparallel bilateral symmetry element serves to characterize the initial sweet taste mechanism. It also serves to distinguish sweetness from other tastes. Nevertheless, chemical alternatives for the initial chemistry of sweetness have been proposed. The chemical alternatives are based on the corollary to the tenet that there is not likely to be a common glycophore among the various compounds that elicit sweetness. *Ergo*, there is not likely to be a common chemical interaction for that basic taste quality, either.

10.3.5.1 *Ionic Versus Hydrogen Bonding Mechanisms*

Sweet amino acids and dipeptides exist in a dipolar or zwitterionic form in which the amino acid bears little net charge. The electrostatic attraction between the two oppositely charged ionic groups generates an *ionic* bond with an average intramolecular bond energy in aqueous solution of about 5 $kcal\,mol^{-1}$. Crosby *et al.* (1979) believe that it is unlikely that the strong intramolecular ionic bond present in amino acids and dipeptides would disrupt to form weaker noncovalent bonds with polarized groups of the receptor. Instead, an ion–dipole interaction of the rigid hydrogen-bonded zwitterion (*sic*) with the receptor is favored. The reasoning holds provided that the premise is true. The hydrogen bonding notion for sweetness induction chemistry seems more probable if the zwitterion is quenched (Section 10.3.4.1) in the apolar environment of the receptor, or through other means (the concerted hydrophobic interaction).

As applied to a dihydrochalcone, Crosby *et al.* (1979) state that an intramolecular (hydrogen bonding) reaction is energetically more favorable than the corresponding intermolecular reaction. The statement is somewhat in contradiction to that which is known about inter- and intramolecular hydrogen bonding (Section 2.7.3.2). Using the dihydrochalcone molecule shown in Fig. 10.8 as a reference model, it is stated that the strength of the intramolecular hydrogen bond between C_2—OH and the adjacent carbonyl may be as large as 7 $kcal\,mol^{-1}$ (cf. Dietrich

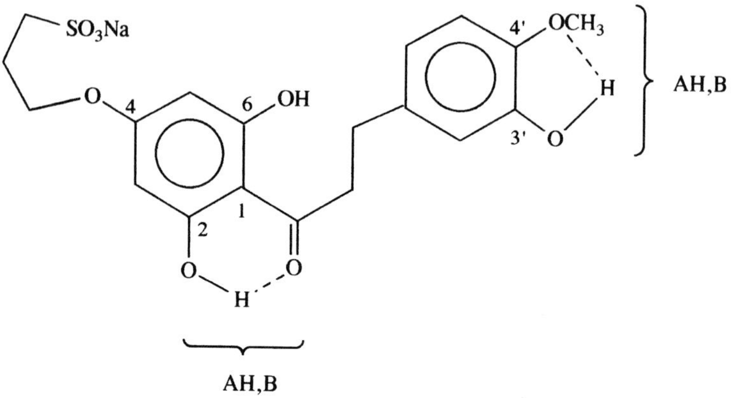

FIG. 10.8. The two unequal intramolecular hydrogen bonds in a selected dihydrochalcone structure where C_2—OH to O ≫ than $C_{3'}$—OH to O (Crosby *et al.*, 1979).

et al., 1976). The strength of the hydrogen bond between the C_3'—OH and C_4'—OCH_3 is worth not more than 2 kcal mol^{-1}.

The argument then goes on to state that rather than break the stronger intramolecular hydrogen bond of the hydroxyketone group, the entire unit may participate in a binding interaction with a polarized region of the receptor through dipole–dipole or ion–dipole forces. The argument can only be countermanded by stating that the idea that an intramolecular hydrogen bond precludes the formation of an intermolecular hydrogen bond seems contrary to a wealth of other data. Dietrich *et al.* (1976) believe that a wide variety of biologically active compounds possess, as *necessary for activity*, functional groups capable of forming intramolecular hydrogen bonds. The thought is in accord with the tenet that the bond acts as a hydrophobic probe, and the proposed recognition chemistry of sweetness, and not with the aforementioned interpretation of the significance of dihydrochalcone intramolecular hydrogen bonding.

10.3.5.2 *Concerted AH,B as an Electrophilic/Nucleophilic System*

Belitz *et al.* (1981) expanded the AH/B system for sweetness into a more general electrophilic/nucleophilic (e/n) system to better jointly account for the sweetness/bitterness of compounds. In this treatment, the glycophore is an e_s/n_s stimulatory unit and the receptor is an n_r/e_r unit. Presumably, the initial chemistry of sweetness can be treated as a form of a weak displacement reaction. Although Belitz *et al.* did not explain their ideas of an e/n system using displacement reaction terms, displacement

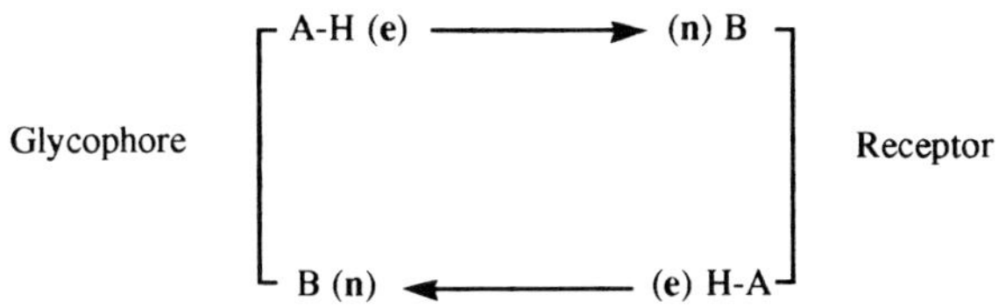

FIG. 10.9. Concerted antiparallel nucleophilic displacement reaction mechanism for sweet taste (e = an electrophilic group, n = a nucleophilic group).

(Section 2.7.7) seems to be the most probable mechanism for e/n interactions that elicit sweet and bitter taste. In this sense, the e/n proposal has considerable merit, not only as an alternative to AH,B, but also to abide by the symmetry prerequisites for a glycophore (Chapter 11). However, displacement implies collision, an event not particularly commensurate with the notion of recognition chemistry.

To correspond with displacement mechanisms, the Lewis base (B:) needs to replace B in an AH,B unit. The neutral molecule required is then R-AH. Thus the initial chemistry of sweetness for the e_s/n_s and e_r/n_r interaction is a concerted antiparallel nucleophilic displacement reaction. The Lewis base is supplied by the receptor in one case, and by the sweet molecule in the second case (Fig. 10.9). Although solvation of dipolar groups tends to make them less reactive with respect to participation in displacement reactions, they may possibly yet retain sufficient nucleophilicity to partake in taste displacement recognition phenomena. The mechanism is in accord with the notion that the interaction chemistry for the sugars is weaker than that of more highly potent sweeteners. On the other hand, not all hydrogen bond interactions are equivalent for different classes of sweeteners with different AH,B activity. Indeed, intermolecular hydrogen bonding itself can be viewed as weak nucleophilic displacement reactions to partly explain why the sugars are not as sweet as the intense sweeteners. Thus, while peptide NH_2 and C=O groups generally partake in displacement reactions more readily than hydroxyl groups, they also partake in hydrogen bonding phenomena more readily. Nevertheless, the fact that AH and B for the amino acids does not need to be the zwitterionic form, but a neutral form instead, seems to fit better with an e/n scheme. Kier's (1972) concept of the third dispersion site was also expanded by Belitz *et al.* (1981) to encompass a more general hydrophilic/lipophilic function, i.e. an apolar system (a), somewhat along the lines of the γ site proposed by Shallenberger & Lindley (1977).

Based upon the expanded notion e/n/*a*, and with special reference to the taste of amino acids and amines, Belitz *et al.* (1981) proposed that there are six types of sweet/bitter stimulants. A general description of their structure with respect to a hypothetical e/n receptor, and its relation to the apolar feature(s) is shown in Fig. 10.10. The types of sweet and bitter structural parameters are: (1) compounds with e/n only that are sweet (glycine, D- and L-alanine); (2) compounds with e/n and a third, apolar function that have enhanced sweetness (D-norleucine); (3) compounds that taste both sweet and bitter (cycloleucine, i.e. 1-amino-cyclohexane-1-carbocyclic acid); (4) compounds that taste bitter (L-norleucine); (5) compounds that taste very

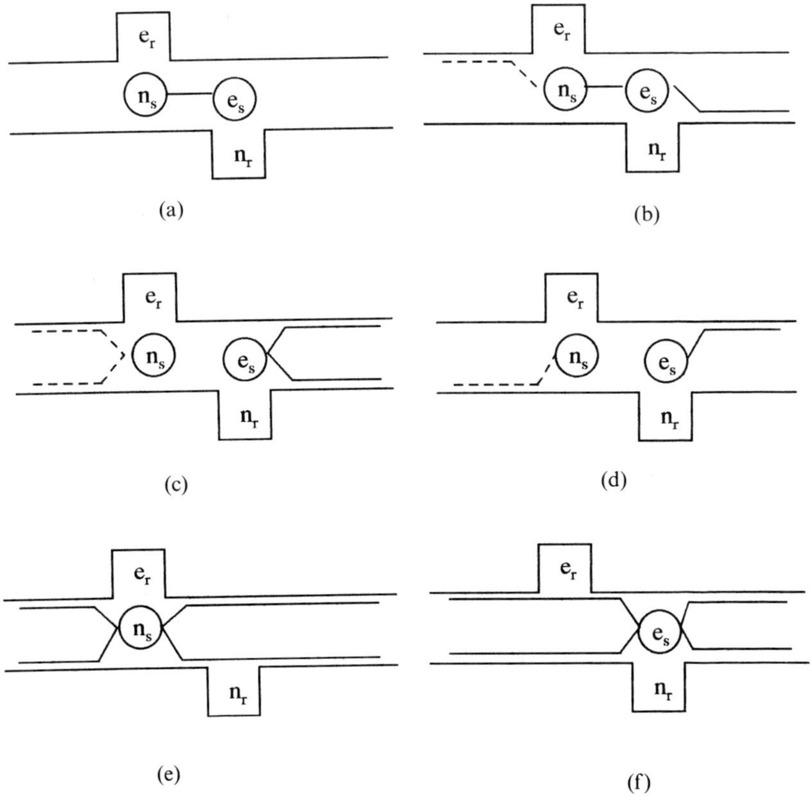

FIG. 10.10. Structural parameters for sweet and bitter taste (Belitz *et al.*, 1981). (a) Sweet, (b) enhanced sweet, (c) sweet and bitter, (d) bitter, (e) and (f) very bitter.

bitter (octane sulfonic acid); and (6) compounds that again taste very bitter such as *i*-pentylamine.

The steric arrangement of the e_s/n_s system and the apolar groups determine whether compounds taste sweet (type 2), bitter (type 4), or bitter-sweet (type 3). Among the organic class of substances, bitter taste was found to require suitable apolar and polar groups. The latter may be either p^+ alone, along with an apolar function (type 5), p^- alone along with an apolar function (type 6), or both p^+ and p^-, along with an apolar function (types 3 and 4). It was further observed that a distinction between the sweet and bitter taste of the D- and L-amino acids only becomes apparent when R (the side chain) is larger than the propyl group. Perhaps equally important, it was observed that taste threshold values for both sweetness and bitterness depend on the length of R, and pass through a minimum for both acyclic and cyclic amino acids. For acyclic compounds, when R = the hexyl group (six-carbon chain), both D- and L- unbranched amino acids are tasteless. The pentyl group (5-carbon chain), while not imparting sweetness, does impart bitterness. This finding suggested that there is a spatial barrier about 7 Å removed from the receptor's n_r function that quenches potential for both sweet and bitter tastes.

For 2-amino acids with a cyclic side chain containing four carbon atoms, the substance is sweet, and with 5–8 carbon atoms, the substances are sweet/bitter. But when the cyclic side chain contains more than eight carbon atoms, sweetness is lost, bitterness progressively diminishes, and the compounds ultimately become tasteless. Further examination of other substances showed that the barrier prevails, not only for acyclic and cyclic amino acids, but for aminosulphonic acids and alkyl halides as well. This relation is shown for the acyclic amino acids. It also applies to

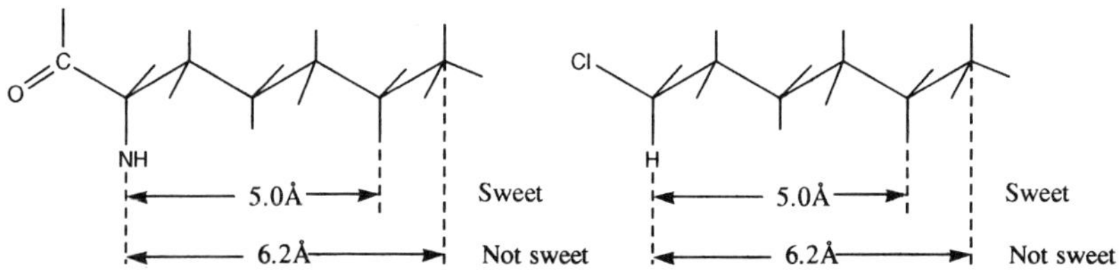

an alkyl chloride. It can be noted that in the latter compound, the electron withdrawing power of the chlorine atom renders a methane carbon atom capable of acting as an electrophile in the same manner as

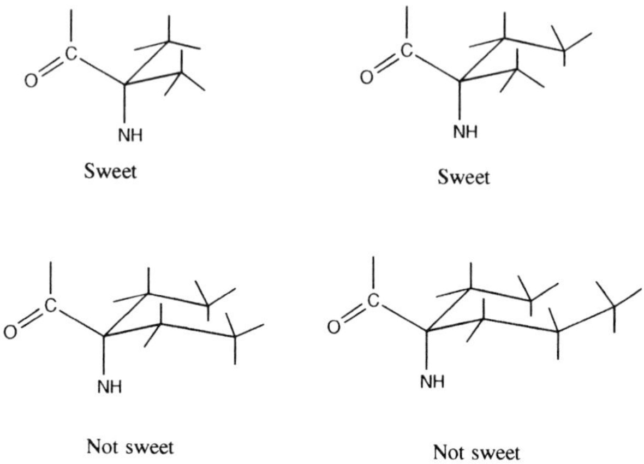

FIG. 10.11. Taste of amino acids with branched chains (Belitz *et al.*, 1981).

assigned by Shallenberger & Acree (1967) to an AH,B unit of chloroform. Interestingly, and in terms of AH,B,γ theory, one reason the compounds with γ removed by 6·2 Å compounds are not sweet is that it is too far removed from AH,B to have inductive effect on the potential glycophore. Alternatively, for steric reasons, there is simply no γ,γ interaction at the receptor. The taste attributes of the branched chain amino substances are particularly significant in this latter respect, as when both groups are larger than the ethyl group (Fig. 10.11), sweetness is lost.

The distance between the side chains of the branched chain amino acids in Fig. 10.11 is 2·5–3·4 Å. The case for the 2-amino acids is even more intriguing. D- and L-Valine are sweet and bitter, respectively, but D- and L-tertiary leucine are neither sweet nor bitter (Fig. 10.12). The nature of the taste of the leucines versus tertiary leucines suggested that, in addition to steric hindrance, charge distribution changes invoked by the nature of the side chains affects the taste of these substances. Another example of differences in charge distribution affecting the taste is found in the taste of the halogenated aryl (PhX) compounds. Here, only the fluorine substituted compound is sweet while Cl, Br, and I substitutions lead to tasteless compounds (Belitz *et al.*, 1981). The case is opposite to that for halogenated nitroaniline compounds and saccharine (Sections 8.1.1 and 8.1.2). As with the acyclic amino acids, the results seem to reflect the notion that steric and inductive effects need to be jointly considered to assess the effect on taste. Belitz *et al.* (1981) concluded that

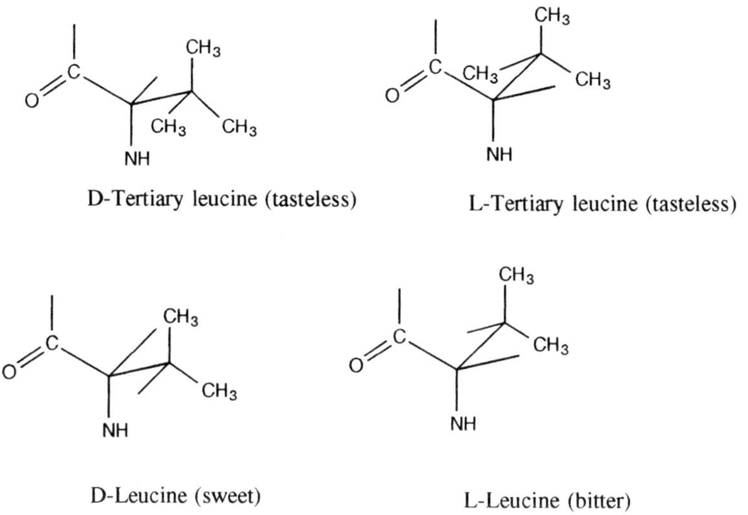

FIG. 10.12. Taste of various leucine isomers (Belitz *et al.*, 1981).

for taste quality (bitterness and sweetness), and intensity, three factors are important. They are : (a) the kind, number, and steric arrangement of polar groups; (b) the charge distribution within the molecule; and (c) the size and shape of the non-polar groups.

As an example of steric hindrance of taste, the amino acids (Belitz *et al.*, 1981) were employed. It was noted that when the side chain (R) of amino acids with unbranched side chains is larger than the propyl group (7 Å), there is taste differential between D- and L-amino acids (cf. Shallenberger *et al.*, 1969). It was also observed that when the distance from the most distant carbon atom of the apolar group to the e_s group was less than 0·5 Å, the taste of compounds is either sweet or bitter. This was observed for amino acids with either cyclic or acyclic side chains, aminosulphonic acids, and alkyl halides. When the distance is greater than 0·54 Å, the molecule has no taste.

Many of the compounds studied had optical antipodes with opposite taste, namely the D- and L-amino acids. Others had mixed (sweet and bitter) tastes. Belitz *et al.* (1987, 1988) attempted to resolve the varying taste properties by reference to a probable three-dimensional rectangular coordinate receptor for both the sweet/bitter taste attributes. The results obtained led to the view that the receptor for the sweet and bitter taste attributes is a hydrophobic pocket containing an n_r/e_r system, as shown

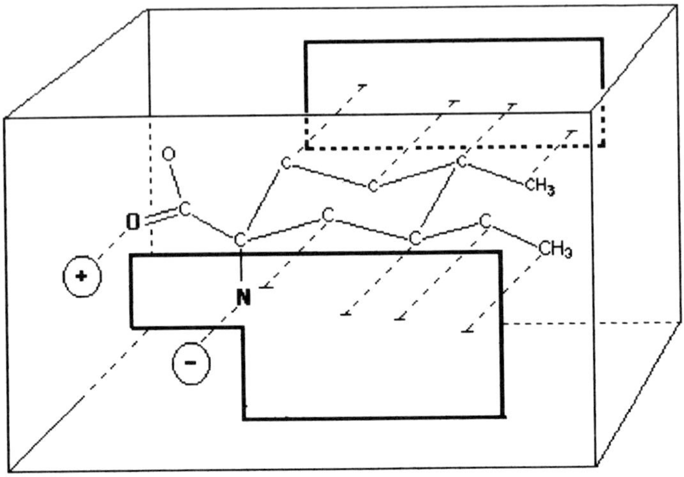

FIG. 10.13. Rectangular coordinates for the receptor for sweet and bitter taste (Belitz *et al.*, 1988).

in Fig. 10.13. In Fig. 10.13, Belitz *et al.* (1988) designated the circled plus and minus sites shown as the e_r and n_r sites, respectively. The parallel rectangular areas were designated as apolar areas for the receptor. In the rectangular coordinate structure, two hydrophobic sites are used to accommodate simultaneously the tripartite, and relatively intense sweetness of D-norleucine, the proportionately intense bitterness of L-norleucine, and the combined bitter-sweet attributes of 1-amino-cyclohexane-1-carbocyclic acid. The respective structures of those three compounds, for interpretation of the significance of Fig. 10.13 are shown in Fig. 10.14. To assess the dimensions of the pocket, Belitz *et al.* (1988) superimposed a number of substances with sweet and bitter taste to describe the allowable and forbidden positions in the pocket. The forbidden positions apparently either disturb the compound's bipolar e/n system, or are located in sterically forbidden positions (Belitz *et al.*, 1979, 1981). Continued studies by Belitz (cf. Rohse & Belitz, 1991, 1993) ultimately led to topological inferences as to the nature of the sweetness receptor.

10.3.5.3 *Nonchemical Mechanisms for Sweetness*

The chemical explanations given for the sweetness of compounds can be countermanded by employing a host of nonchemical generalities. The nonchemical mechanisms are usually employed to differentiate

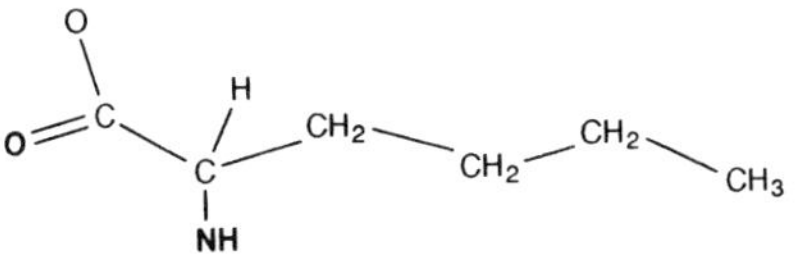

D-Norleucine
(Sweet)

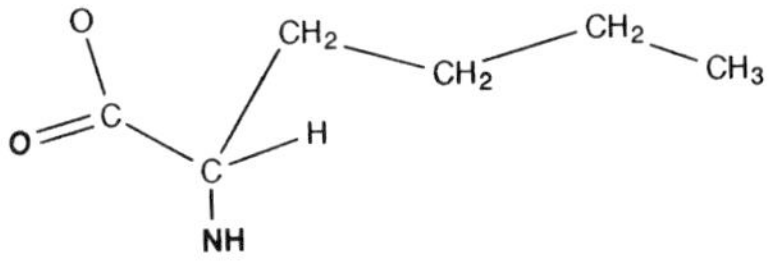

L-Norleucine
(Bitter)

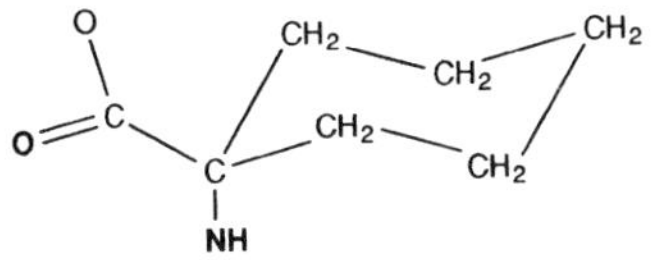

1-amino-cyclohexane-1-carbocyclic acid
(Bitter/sweet)

FIG. 10.14. Structures of amino acids used to formulate the composite structure shown in Fig. 10.13 (Belitz *et al.*, 1988).

high-potency sweetness from the sweetness potency of the sugars. The major premise for these generalities is a reversal of Ockham's razor principle, i.e., chemical mechanisms, such as AH,B,γ, are not sufficiently complex to explain the enigmatic observations associated with sweetness induction (DuBois *et al.*, 1991, 1993). Parenthetically, Ockham's razor states that the simplest answer is usually the best and the more assumptions needed to explain a theory, the less likely that it is correct. Thus, one of the alternatives to AH,B is that the sugars are full agonists and high-potency sweeteners are partial agonists, but it is difficult to envisage a partial agonist that is 300 times sweeter than a full agonist. Nevertheless, the notion is in accord with the model proposed by Lancet & Ben-Arie (1991) to account for the number of elementary interactions required for the binding of a tastant to a receptor. A second is that sugars have alternate modes of receptor activation (including an osmotic effect), and bind as agonists with low receptor affinity, which is also taken to be evidence of a poor 'fit.' The evidence offered for these conclusions is based largely on concentration/response data. Other explanations for high-sweetness potency are also based on receptor 'fit' (e.g., Temussi, 1991, pers. comm.; Tinti & Nofre, 1991*b*). Further discussion of this alternative to AH,B,γ glycophore activity is reserved for a more general discussion of topology (Chapter 11). To explain the sweetness of enantiomeric sugars, Simon (1991) proposed that small saccharides diffuse across tight junctions between taste cells and interact with putative receptors at the basolateral membrane, i.e. there is no receptor for sugar sweetness in the structure of the taste microvilli.

10.4 SPECIAL QUALITATIVE ASPECTS OF THE INITIAL CHEMISTRY OF SWEETNESS

There are a number of special qualitative features unique to the chemistry of sweet taste. Foremost among them is that while the sweetness of substances increases with increasing concentration, it does so in varied ways. For substances that are not as sweet as sucrose (molar basis), increasing the concentration increases the relative sweetness of the material, e.g. glucose. With substances that have higher sweetness potency than sucrose, increasing the concentration results in decrease in relative sweetness. Some examples are sulfamates (including the saccharines), cyclamates, acesulfams, peptides, chlorosugars, and the dihydrochalcones.

A number of reasons have been proposed for the phenomenon. Among them is that: (1) There are not as many receptors for the high intensity sweeteners as for the sugars; (2) there are different receptors for different substances; (3) there are different chemical mechanisms for the sweetness of different substances; and (4) in making a tripartite contact, the large size of some high intensity sweeteners (partial agonists) serves to 'block out' other potential AH,B sites. However, as shown below, the phenomenon seems to be simply a manifestation of the psychophysical laws.

10.4.1 High-Potency Sweetness Variation with Concentration

The ability to distinguish whether decrease in the relative sweetness of a high-potency sweetener is chemical, psychological, or otherwise has direct application to the nature of sweetness recognition chemistry. Notions (1)–(4) of the previous section are derived from observations that, while the sensory response to increasing concentrations of sugars over a small range of concentrations is linear (Fig. 6.2), that of the high-potency sweeteners is parabolic. Interestingly, the panel employed in the latter studies (DuBois *et al.*, 1991) had also been trained to 'linearize' their response to the reference substance (sucrose). Usual plots of perceived sugar sweetness over a wide range of concentrations are parabolic (Fig. 6.1).

The data of Guadagni *et al.* (1974) afford joint examination of the phenomenon from the stoichiometric viewpoint, and from the viewpoint

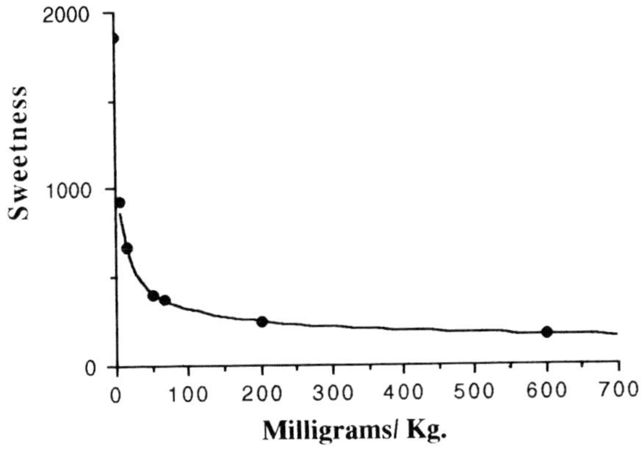

FIG. 10.15. Isosweet concentrations of neohesperidin dihydrochalcone, relative to sucrose. Sweetness $= 1677x^{-0{\cdot}358}$ (Guadagni *et al.*, 1974).

of the psychophysical laws. According to the latter (Chapter 1), geometric increments in concentration are required to elicit arithmetric increments in perceived sweetness intensity. As can be seen from Fig. 10.15, the relative sweetness of neohesperidin dihydrochalcone (and other high-potency substances) decreases markedly with increasing concentration.

When the increasing equal-sweetness concentrations of the dihydrochalcone and sucrose are plotted on a molar basis, the result obtained indicates that the amount of dihydrochalcone required to equal the sweetness of increasing concentrations of sucrose increases by a constant

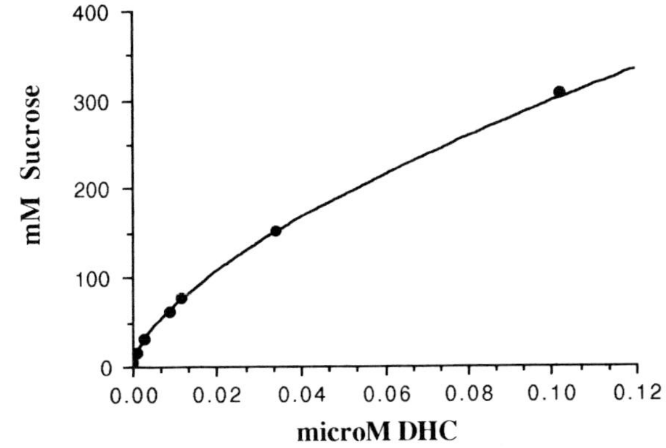

FIG. 10.16. Plot of equi-sweet molar sucrose concentrations (M sucrose $\times 10^{-3}$) and neohesperidine dihydrochalcone concentrations (M DHC $\times 10^{-6}$).

exponential amount. The constant amount is indicated by the curvilinear relation shown in Fig. 10.16. In that figure the mathematical nature of the curve is mM Sucrose $= 1267x^{0 \cdot 626}$ mM dihydrochalcone ($r = 1{\cdot}00$).

The data suggest that, in view of the psychophysical laws (which are also exponential), it merely takes increasingly higher concentrations of high-potency sweetener to match the diminishing perceived sweetness associated with increasing sucrose concentrations (cf. Chapter 1). If that is the case, then a plot of equi-sweet concentrations on log/log coordinates should be linear. As shown in Fig. 10.17, that is precisely the result obtained. Therefore, the decrease in relative sweetness of the high-potency sweeteners with increasing concentration, and the self-synergism of low-potency sweeteners (Section 6.3) is not of chemical origin. It is due instead to the dynamics of sensory perception. Not only are the notions (1)–(4), presented in Section 10.4, highly improbable, the decreasing relative sweetness of the high-potency sweeteners serves to support the contention that there probably is a common chemical recognition mechanism for all sweeteners, and there also may be a common generic receptor for all sweeteners.

In contrast to the nonchemical basis for decreased relative sweetness for high-potency substances, adaptation rate and magnitude do have a chemical basis. DuBois *et al.* (1993) have shown that the rate and magnitude of adaptation is directly proportionate to sweetness potency. Beverages sweetened with high-potency sweeteners may show rapid, and extensive adaptation, while those sweetened with sugars do not.

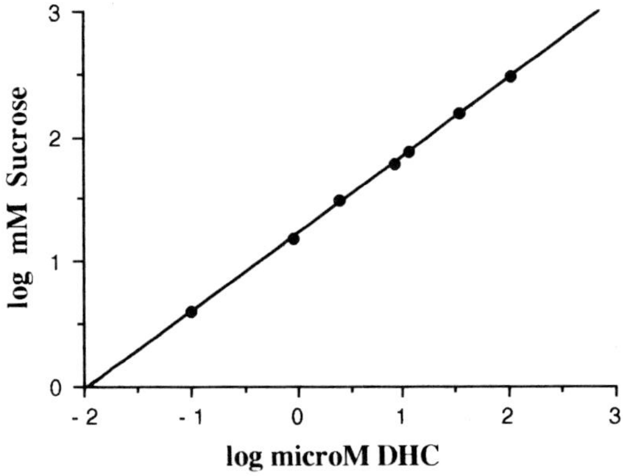

FIG. 10.17. Plot of equi-sweet molar sucrose concentrations (log mM sucrose) and neohesperidine dihydrochalcone concentrations (log M DHC $\times 10^{-6}$).

Presumably, the beverages initially had near-identical sweetness intensity. If so, it appears that mass action (concentration) phenomena do not govern adaptation rate, but chemical activity does. The observation is also strong evidence for a common chemical recognition mechanism for sugars and the high-potency sweeteners.

10.4.2 The 'Orderly Queue'

Birch *et al.* (1980) proposed an 'orderly queue' mechanism to account for the temporal and intensity characteristics shown by some sweeteners. The notion that the temporal sweetness ionophore mechanisms are different is derived from the observation that the sweetness of the protein thaumatin, as compared with sucrose, has a delayed onset time and lingering persistence. Those characteristics are purported not to be explainable on structural grounds. Although the phenomena are somewhat concentration dependent (Birch *et al.*, 1980), when sucrose and thaumatin are tasted, the onset time for the former is only 1/3 that for latter. The sweet taste sensation for thaumatin also persists about four times longer, and persistence and maximum intensity did not seem to be simply related. While the subjective intensity for sucrose sweetness shows a plateau after a given concentration, persistence time continues to increase. As shown in

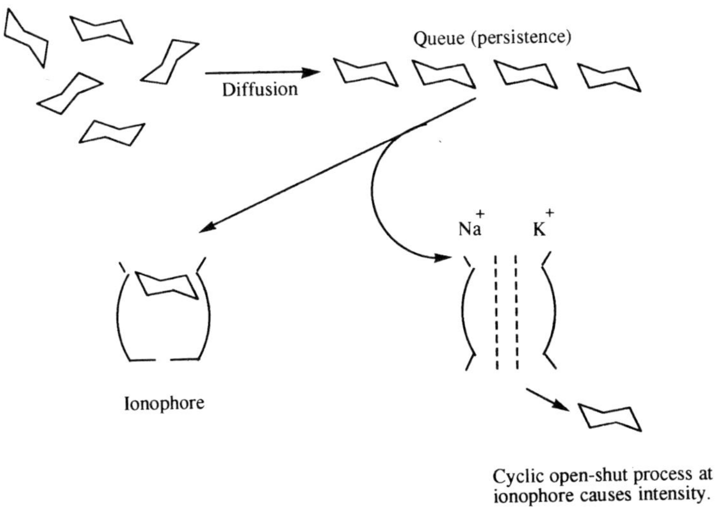

FIG. 10.18. Chemoreception events in the orderly queue hypothesis (Birch *et al.*, 1980).

Fig. 10.18, sugar molecules are thought to pass irreversibly into a queue to account for a localized supply of molecules to the ion-channel mechanism, leading to persistence. Intensity at the ionophore results from triggering of the ion-channel mechanism by sugar molecules, and is a cyclic process. It is governed by the 'fit' of the stimulus with the ionophore mechanism, and the rate with which these sites are occupied and vacated.

The slow onset behavior for thaumatin does not seem to be all that unusual, and can be partly explained by the binding of the substance to a noninteractive site. The compound has affinity for the receptor at an unreactive site, and is held 'in reserve' at some unspecified nonreceptor area of the membrane surface. Following release from such sites, and diffusion from extraneous areas, the molecules then become available for recognition interaction at a sweet–taste receptor site. Persistence and maximum intensity would not be simply related in this case, either. One criticism that can be directed toward the thesis is that if there is indeed an orderly queue mechanism in the chemistry of sweet taste, it is difficult to envisage the force or the source of energy responsible for aligning the molecules. However, the driving force may simply be the recognition orientation mechanism presented in Section 10.3.2 (see also Section 10.4.3.2).

An actual structural basis for the behavior of thaumatin is contained in the analogy of the 'zippering action' of the multiple AH,B units contained in one structure with multiple AH,B receptor sites. In this sense it is analogous to one theory for the initial chemistry of antigen–antibody formation. A structural (size) and property (insolubility) basis is available for the slow onset time for the dihydrochalcones, and for their persistence, i.e. binding of dihydrochalcone phenolic groups with the protein receptor (van der Wel *et al.*, 1987). Finally, persistence can be explained by straightforward application of equilibria phenomena as applied to a reversible system (Sections 10.3.2 and 10.7).

10.4.3 Solution Properties of Substances and Initial Taste Chemistry

The physical characteristics and properties of substances can sometimes be correlated with their taste attributes whenever the physical properties are, in part, a manifestation of chemical properties. Among the sugars, Loginov (1945) reported that there is a relation between solubility and sweetness potency, but it is tenuous, and limited (Tables 5.3 and 6.1). That approximate relation can be traced to the role that intermolecular hydrogen bonding has on the extent of molecular packing in the crystal structure. Other observed relations between taste and other physical parameters of substances follow, but they too seem to be rather tenuous.

10.4.3.1 *Molar Volume and Intrinsic Viscosity of Substances*

Shamil *et al.* (1987), Birch & Shamil (1988), and Shamil *et al.* (1988*a*) initiated a series of studies to assess the probable significance of several solute solution parameters to taste. The premise employed is that various features of taste, such as access efficiency, are governed by the specific solution properties of substances. It was found that the apparent specific volume (ϕV/molecular weight) of substances were as follows:

Salt	0·00–0·33
Sour	0.33–0·52
Sweet	0·52–0·71
Bitter	0·71–0·93

The results were interpreted to indicate that the receptors for the four taste modalities reside at different layers of the taste epithelium, and indicate compatibility with the structure of water. However, no relation was found between apparent specific volume and sweetness potency, and no relation was found between an acid's anion and sourness potency. The latter is to be expected if the common acidophore can be assigned only to the hydrogen ion (Section 9.1). That acidophore assignment also makes the above specific molar volume limits assigned to various sour-tasting substances open to serious question.

Subsequently, molecules with mixed tastes, such as L-proline (sweet and bitter), and with a specific volume of 0·725, were found to be borderline between the specific volume ranges associated with sweet and bitter tastes (Birch & Kemp, 1989). On the other hand, D-gluconic acid has potential sweet and sour saporophores, but its apparent specific volume (0·515) places it in the sour region and, although sourness is 'indeed its taste' the point made seems to stretch the thesis a bit.

It was also found (Shamil *et al.*, 1988*a*) that there was no relation between published sweetness values and the intrinsic viscosity and apparent specific volume of the sugars. Birch & Kemp (1989) observed that two of sixteen L-amino acids are both sweet and bitter, as might be expected from their molecular volume. In addition, a low specific volume value for aspartic acid (sour), and a high value for L-leucine, is believed to be in accord with their respective sourness and bitterness. More recent thoughts along these lines have been published (Birch *et al.*, 1993).

10.4.3.2 *Role of Water in Molecular Recognition*

A comprehensive theory governing the role of water in molecular recognition remains to be formulated. Galema *et al.* (1990) proposed that molecular recognition in aqueous carbohydrate solutions is probably governed by an interplay between the rigid polyol structure, dictated by its stereochemistry, and the flexible three-dimensional structure of water.

Jeffrey (1993) has proposed that the string of dipoles that arises from the electronic structure of the O—H···O or N—H···O cooperative hydrogen bonding system (Section 2.7.3.4) is clearly very polarizable. It will respond to the electrostatic potential of a sweet molecule and transmit this information along the chain to a receptor molecule. The chains are probably dynamic, as in water or in aqueous solution, and the energetic advantage will ensure that there are always sequential chains even if they are part of dynamically changing networks. Jeffrey then envisaged two mechanisms for the transfer of electrostatic information. In the first, the hydrogen-bonded water structure at the receptor is polarized by a distant sweetener, or *vice versa*. In the second, the water structure operates as a 'radar' system to guide the sweetener molecule on to the receptor site. To comply with occupancy theory, the latter would seem favored. It could also serve as the mechanism for an 'orderly queue.' To comply with recognition chemistry theory, the former would seem to be favored (Section 10.3.2).

10.5 INITIAL CHEMISTRY OF BITTERNESS

Unlike the sweet, salt, and sour tastes, the initial chemistry of bitterness appears to be associated with a number of *different*, weak or strong recognition interactions. For the bitterness/unpleasantness of salts, it appears certain that the initial chemical interaction is ionic, but either the anion (NaI) or the cation (KCl) may be more important, i.e.

$$\underset{\text{Bitter salt}}{\mathbf{M}\mathrm{X} \text{ or } \mathbf{MX}} \quad + \quad \underset{\text{Receptor}}{\mathrm{n/e}} \rightleftharpoons \underset{\text{Bitter salt/receptor interactions}}{(\mathbf{M}^+)(\mathrm{n}) + (\mathrm{X}^-)(\mathrm{e}^+) \text{ or } (\mathbf{M}^+)(\mathrm{n}^-) + (\mathbf{X}^-)(\mathrm{e}^+)}$$

The sensation may also be assigned to substances with a hard or soft acid or basic nature (Pearson, 1963) such as picric acid and alkaloids. High-potency bitterness is also associated with the presence of a plus or minus monopolar function and a hydrophobic group within a molecule (Belitz *et al.*, 1981), and *N*-acetyl ester derivatives of amino acids, which are devoid of a zwitterionic attribute, taste bitter, often very strongly (Asao *et al.*, 1987). An equation for monopolar and high-potency bitterness is

$$\underset{\text{Bitter monopole}}{(AH^{\delta+}, \gamma \text{ or } B^{\delta-} \gamma)} + \underset{\text{Receptor}}{(n/e, \gamma)} \rightleftharpoons \underset{\text{Bitter monopole/receptor interaction}}{(AH^{\delta+},\gamma)(n, \gamma) \text{ or } (B^{\delta-},\gamma)(e, \gamma)}$$

Should dipolar nonbonded interactions be unbalanced, principally by virtue of composition and/or configuration, then an assessment of bitter taste also results. Put another way (Section 9.4), an unbalanced hydrogen bonding mechanism can be associated with the bitterness of amines (Shinoda & Okai, 1985) and diterpenes (Kubota & Kubo, 1969). This is probably one of the reasons why saccharine is intrinsically sweet and bitter. One potential set of AH,B,γ glycophores is unbalanced (dissymmetric), but the other set is not. A third equation for bitterness then would be identical to that given in Section 10.3.2 for sweetness, except that a dissymmetry element is introduced (Chapter 11). The same

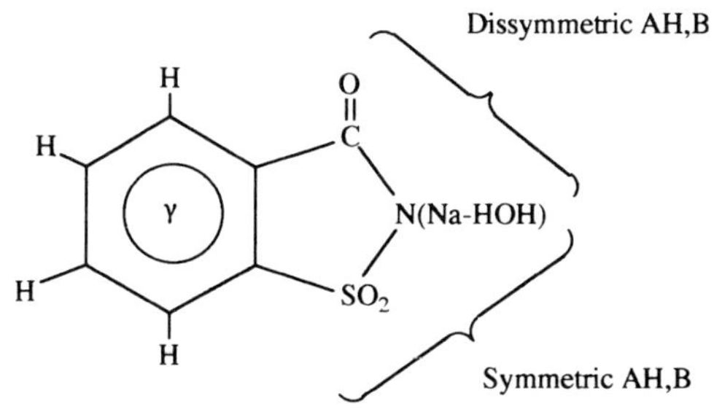

Saccharine

rationale can be applied to bitterness of the amino acids by application of the e/n thesis. Belitz *et al.* (1981) assigned the bitterness of amino acids to either of the polar amino acid groups, in addition to the apolar (γ) group

(Section 10.3.5.2). Several schemes for the initial chemistry of unbalanced and concerted hydrogen bonding interactions leading to bitterness are then possible. In one, *pseudo*-glycophore AH is more significant than B or in the other the *pseudo*-glycophore B is more significant than AH. Then, the AH,B unit is a picrophore, and not a glycophore. Herein lies the basis for one of the criticisms to the initial AH,B glycophore tenet. Compounds that do not taste sweet, but possess an AH,B unit, also possess an AH,B unit that is not, for various reasons, electronically disposed to elicit sweetness.

In another scheme, either AH or B is totally absent or inactive in the structure of a compound, and the lone polar group, coupled with the apolar function, serves as the picrophore for varied dissymmetric interactions with the receptor. If there is an element of truth in the latter surmise, then it is clear why there is an intrinsic relation between the sweetness and bitterness of compounds (Sections 1.5.1 and 1.5.2). Finally, as with sweetness (Section 10.3.2), the magnitude of the bitter taste response for organic substances would seem to be proportionate to the product of the activity of the picrophore times the concentration.

10.6 STRUCTURE-ACTIVITY RELATIONSHIPS

The rationale behind structure-activity relationships in taste had its origins in pharmaceutical chemistry with its attendant interaction theories (Section 2.10). It is also based on the large effects that substituents can have upon chemical recognition rate and equilibrium constants. Consequently, taste potency is believed to be a function of several properties operating simultaneously to produce the biological effect (McFarland, 1971). Those properties are, of course, related to structure, and a number of structure-activity relations have been shown to be related to the sweetness of the variously substituted 2-amino-4-nitrobenzene compounds. Foremost among them is the Hansch equation (cf. Hansch & Fujita, 1964), or

$$\log 1/C = a\pi^2 + b\pi + c\sigma + f$$

In the equation, $\log(1/C)$ is a measure of biological activity, C is the concentration of the tastant that yields a standard response (e.g., sweetness recognition threshold, and relative sweetness). The term π is an experimentally determined partition coefficient ($k/1$) for the compound determined between water and octanol, or

$$\log \frac{(k/1)_X}{(k/1)_H} = \pi = k\Delta\Delta G_{\text{hydrophobic}}$$

In essence then, π is a property related to a compound's lipophilicity/hydrophobicity, and there is an optimum degree of lipophilicity in a series of tastants. A compound can not be so hydrophilic as to limit adsorption by the receptor membrane, and it can not be so lipophilic as to quench sweetness (cf. Hansch & Fujita, 1964). There is also a parabolic relation between π and sweetness potency in the Hansch equation.

The term σ in the Hansch equation is the Hammett constant, or

$$\log \frac{K_{(X)}}{K_{(H)}} = \sigma = k\Delta\Delta G_{\text{ionization}}$$

where $K_{(X)}$ is the ionization constant of a benzoic acid substituted by a group X and $K_{(H)}$ is that of benzoic acid (McFarland, 1971). The Hammett σ constant is also more of a property-activity term than a structure activity relation, and is associated with the electronic properties of a molecule, including a compound's dipole moment (Charton, 1965).

The free energy associated with the Hammett σ constant is thought to be an index to the intrinsic activity of a compound. It relates chemical structure both to equilibrium constants and to reaction rate constants, and represents the ability of a group to attract or repel electrons through a combination of inductive and resonance effects. If a substituent has a positive σ-value, it will attract electrons. If it has a negative σ-value, it will donate electrons (Korolkovas, 1970).

For application of the structure-activity terms, a notion of intrinsic activity is employed. Through reference to the occupancy theory of drug action, intrinsic activity (efficacy) is defined (Korolkovas, 1970) as being composed of two stages: (1) complexation (recognition interaction) of the compound with the receptor; and (2) production of an effect. In the application of the structure-activity models, receptors are viewed as being in a heterogeneous phase of a hydrophilic/lipophilic matrix, i.e. at an interface (McFarland, 1971), and taste then becomes an exchange phenomenon. Here, for complexation *per se*, the tastant, initially solvated by water, is thought to become solvated by the receptor. The principal forces for tastant receptor complexing are hydrophobic bonding, ion–dipole interactions, hydrogen bonding, and van der Waals interactions. Unfortunately, hydrogen bonding *per se* cannot be quantified, and its effects seem to be buried somewhere in the property constants that are employed.

10.6.1 Hansch π and Hammett σ Constants

The first attempt to relate substituent effect to relative sweetness was the demonstration by Deutsch & Hansch (1966) that the relative sweetness values for a series of substituted 2-amino-4-nitrobenzene compounds (Section 8.1.1) depend on hydrophobic bonding. The authors demonstrated that the logarithm of the relative sweetness of these compounds correlated satisfactorily with the logarithm of an octanol–water partition coefficient. Thus, log RS $= 1{\cdot}214\pi + 1{\cdot}970$, $r = 0{\cdot}766$. However, inclusion of the calculated value for the Hammett σ constant gave log RS $= 1{\cdot}610\pi - 1{\cdot}831\sigma + 1{\cdot}79$, and the relation was markedly improved ($r = 0{\cdot}936$).

Including a π^2 parameter in the computation did not improve the correlation, and the linear dependency on hydrophobicity led Crosby *et al.* (1979) to suggest that the function of π in sweetness phenomena probably does not involve passage across a membrane. Instead, it was believed to be associated, for the most part, with binding to the receptor. Interestingly, the findings of Deutsch & Hansch are only slightly improved when molar sweetness values rather than w/v relative sweetness values are employed. In theory, molar values seem to be needed.

While the overall relation between π and relative sweetness alone is satisfactory ($r = 0{\cdot}776$), reexamination of the Deutsch & Hansch data yielded a plot (Fig. 10.19) that shows two distinctly different patterns to the data. One is due to the OMe, OEt and OPr substances ($r = 0{\cdot}999$), and the other is due to the remaining compounds ($r = 0{\cdot}953$). It is the intermediate plot (not shown) that uses all of the data that yields the correlation coefficient of 0·776. Part of the reason for the separation of the plots

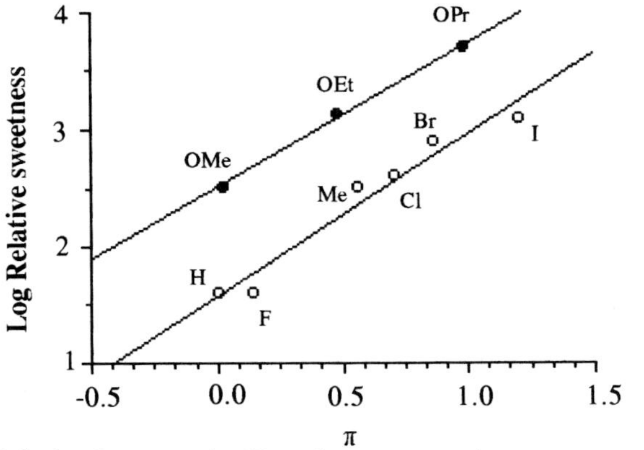

FIG. 10.19. Relation between the Hansch π constant (Deutsch & Hansch, 1966) and relative sweetness of the 2-amino-4-nitrobenzene series of compounds.

therefore seems to be due to the different type of inductive effect that the two groups of substituents have on AH and B. The net effect of multiple correlation (multiple linear regression) using both π and σ is to minimize variance in the π data. The latter was obviously imparted by the influence of σ on the methoxy, ethoxy and propoxy substituents, or *vice versa.* While Deutsch & Hansch did not report the relation of the Hammett σ value alone versus sweetness for the compounds, it is found to be, on the whole, minimal. The reason it correlates poorly is that σ alone versus relative sweetness yields an inverse parabolic plot (Fig. 10.20), with minimum sweetness when the σ constant is zero. The constant increases in positive sign with decreasing ionization potential for the halogens (Table 3.6), and increases in negative sign with increasing alkane chain length.

Subsequently, Hansch (1970) also reexamined the above data, and substituted σ^+ for σ. This was done to account for the possibility of a direct resonance interaction between the substituent X and the NO_2 groups. This tenet, in particular, strongly supports the notion that one function of the γ, γ interaction is to render the AH,B glycophore active (see Section 10.3.4). Substitution of σ^+ for σ in the initial multiple regression equation yielded $\log RS = 1{\cdot}434\pi - 1{\cdot}026\sigma^+ + 1{\cdot}584$. The coefficient for this correlation is 0·972. In addition to the improved correlation, π was brought more in line with its importance in other biological systems (values beyond $\pi = 1{\cdot}2$ are rare).

On the thesis that chromatographic procedures on inert substrates, such as polyamide layers, provide separations of substances that involve

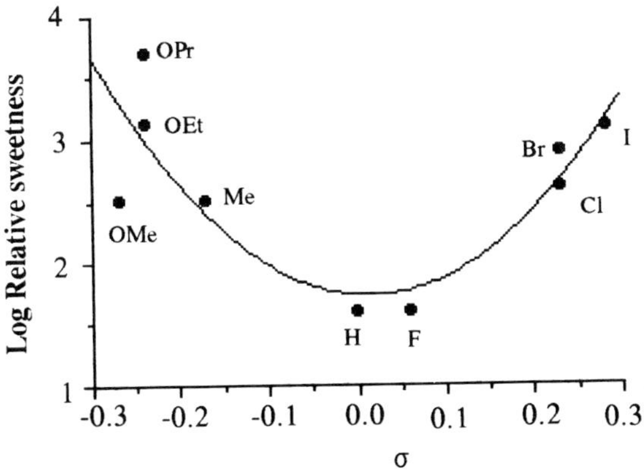

FIG. 10.20. Plot of Hammett sigma constants (Deutsch & Hansch, 1966) and relative sweetness of 2-amino-4-nitrobenzene series of compounds.

both partitioning and hydrogen bonding, Polański & Ratajczak (1991) determined chromatographic R_M values, where $R_M = \log(1/R_F - 1)$. These were related to hydrophobic constants. The latter, in turn, served as an index to sulfone sweetness (Ratajczak & Polański, 1991*b*). The calculation of chromatographic R_F values (the migration distance of a substance divided by the migration distance of the irrigating solvent), or the use of chromatographic retention times, is a simple method of determining a parameter related to partition coefficients.

Several other structure-activity models have application to sweetness chemistry, and introduce additional concepts and terms. Some (McFarland, 1971) are the parameter μ, which is a measure of the free energy associated with a substituent or compound's dipole moment. The parameter α, is a polarizability parameter for a given substituent, or compound. Presumably, the chemistry of hydrogen bonding is buried in the σ, μ and α parameters. Still other models employ calculated energy interactions, and an empirical bond connectivity parameter.

10.6.2 The McFarland Probability Model

The McFarland (1971) model for structure-activity relations converts the linear free energy model of Hansch to a relation based on probability and free energy. It begins with consideration of the steps necessary for interaction and response to occur. These are based upon the probability of the events and the free energy changes that occur. As applied to sweeteners, the steps envisaged are: (a) The sweetener gains access to the receptor by passive means, i.e. through a 'random walk' process; (b) a sweetener–receptor complex occurs, i.e. the sweetener binds to the receptor; and (c) the sweetener–receptor combination initiates a chemical event that leads to a biological (sensory) response. The probability that a sweetener will proceed through the three steps is the product of the probability that it will go through the individual steps, or:

$$P_{o,c} = P_a \times P_b \times P_c$$

If there is no impedance for any particular step, the probability that it will complete that step is 1·0. That step has no influence on the overall probability, as the nonlimiting steps have been removed from consideration. As a sweetener is introduced near the receptor, it is proposed (Crosby *et al.*, 1979) that P_a is about 1·0. If the compound–receptor complex alone (P_b) is sufficient for a substance to exert its effect, then the probability of step c occurring is 1·0. Here, however, the compound acts as a competitive inhibitor because the *complex formation* is tantamount to the tastant achieving its final state. The example seems to neatly explain the effects of sweetness inhibition displayed by substances such as lactisol.

If the equilibrium constant of complex formation is very large, then the probability that complex formation will occur is about 1·0. It would then be nearly impossible to detect any limiting effect by a biological test, such as a sweetness response. The number of molecules undergoing all three steps within the time limits of a test is N^*, and

$$N^* = N_o \times P_{o,c} = k \times C_o \times P_a \times P_b \times P_c$$

N_o is the number of sweetener molecules given for tasting, C_o is the corresponding molar concentration, and k is a proportionality constant. P_o is the overall probability that a tastant molecule from some initial state will achieve the transition step c; P_a, P_b, and P_c are the individual probabilities of the separate steps occurring in a particular system.

The status of the problem as presented by McFarland (1971) has now become the relation of the number of molecules arriving at step c to the sensory response. This is best obtained by comparing the different tastants for a given quality to the amount of each compound needed to elicit a standard biological response. In this case, it is the recognition threshold. Apart from the criticism leveled toward the significance of threshold data, the advantage of the procedure is that the range of concentrations are theoretically without limit. Consequently, there is less ambiguity as to whether one compound is more potent than another. Finally, a standard biological response (sweetness) can be chosen to give optimum precision of measurement.

On the assumption that the *number of molecules* needed to give a standard, less than maximum, response is the same for each sweetener, N^* in the above equation is that number, and it is constant for each sweetener. This assumption seems to be invalid. Nevertheless, to continue on, the sweetener-related variables are then C_o (the concentration of sweetener in the external phase giving the standard response), P_a, P_b, and P_c. The constant k remains unaltered for different sweeteners. Thus, since k and N^* are constants, the above equation becomes

$$1/C_o = k/N^* \times P_a \times P_b \times P_c$$

or

$$\log(1/C_o) = \log P_a + \log P_b + \log P_c$$

The three probability terms of the logarithmic equation above then become for step a:

$$\log P_a = a\pi^2 + b\pi + f$$

where π is the Hansch partition coefficient, and the equation represents transport and/or access to the receptor. For step b:

$$\log P_b = d\mu_z^2 + e(a_z - a_H) + f$$

where μ^2 is a dipole–dipole interaction term that can be either positive or negative, and $(a_z - a_H)$ is a polarizability term that can also be positive or negative. Z is a particular sweetener substituent. For step c:

$$\log P_c = c\sigma + f$$

where σ is the Hammett constant.

When assembled, the original $(\log(1/C_o) = \log P_a + \log P_b + \log P_c)$ equation becomes:

$$\log(1/C_z) = a\pi^2{}_z + b\pi_z + cs^2{}_z + d\mu^2{}_z + e(\alpha_z - \alpha_H) + f$$

McFarland (1971) then took the Hansch approach to sweetness one step farther. He used a σ constant calculated from benzoic acid substituted at the *ortho* position, rather than at the *para* substitution, as Hansch had done. He also added two of the aforementioned constants to the model, as these are related to the free energy changes that occur in the elicitation of sweet taste. They were the μ^2 parameter (a group dipole moment index), and the polarizability constant, α. Tabulation of all constants for the 2-substituted-5-nitroanilines is given in Table 10.1. From the data shown in Table 10.1, McFarland derived the expression

TABLE 10.1
Free energy terms employed by McFarland (1971) to determine the relation to the relative sweetness (RS) of 2-substituted 5-nitroaniline compounds

R	π	σ	μ^2	$(\alpha_R - \alpha_H)$	*Log RS*
H	0·00	0·00	0·00	0·00	1·602
OMe	−0·02	0·12	1·56	6·261	2·519
OEt	0·48	−0·01	1·00	6·261	3·146
OPr	0·98	−0·04	1·00	6·261	3·699
F	0·14	0·93	1·82	−0·15	1·602
Cl	0·71	1·28	2·50	4·867	2·602
Br	0·86	1·35	2·31	7·765	2·903
I	1·20	1·34	1·66	12·800	3·097
Me	0·56	0·29	0·185	4·618	2·519

$$\log \mathrm{RS} = 1{\cdot}3\pi - 1{\cdot}08\sigma_\mathrm{o} + 0{\cdot}45\mu^2 + 0{\cdot}052\,(\alpha_\mathrm{x} - \alpha_\mathrm{H}) + 1{\cdot}66$$

In the equation, the terms μ^2 and α are associated with the binding to the receptor, and the term σ is an index of the intrinsic activity of the sweetener. The π, σ, and μ^2 coefficients were found to be highly significant. The expression $(\alpha_\mathrm{x} - \alpha_\mathrm{H})$ was also found to be significant, but only at the 5% probability level. The correlation coefficient for the multiple regression equation was $0{\cdot}988^{**}$, and the dependence of the relation on π is taken to indicate that the binding energy is associated with a favorable entropy change.

Upon further inspection of the McFarland (1971) data, clearly only two parameters are directly related linearly, and about equally, to relative sweetness. These are π ($\mathrm{RS} = 0{\cdot}1214p + 1{\cdot}970$, $r = 0{\cdot}766^*$) and α ($\mathrm{RS} = 0{\cdot}136a + 1{\cdot}899$, $r = 0{\cdot}768^*$). On a 1 : 1 basis with relative sweetness, the other constants are not significant. When π^2 is included in the computation, the equation becomes

$$\log \mathrm{RS} = -\,0{\cdot}572p^2 + 1{\cdot}795p - 1{\cdot}019s + 0{\cdot}412m^2 + 0{\cdot}070(\alpha_\mathrm{x} - \alpha_\mathrm{H}) + 1{\cdot}66$$

with a coefficient of correlation of $r = 0{\cdot}999$. In the latter equation, all terms are highly significant. Here, the significance of π^2 would again seem to relate to facile access of the substances to the receptor, as opposed to transport through a membrane. In other words, for sweetness, step a in the probability sequence is not 1·0. Finally, aside from the fact that molar sweetness scores were not used, an additional note of caution with respect to the significance of hydrophobicity estimates and correlation with sweetness of the nitroaniline substances is in order. Iwamura (1980) found that when the number of compounds is increased from the original nine employed by Deutsch & Hansch (1966) to twenty-one, the correlation between taste intensity and hydrophobicity is not nearly as good.

10.6.3 Interaction Energy

Using the same series of compounds that were used to derive the relation between relative sweetness and the Hansch π constant, Höltje & Kier (1974) calculated the interaction energy of those substances, and related it to relative sweetness. They employed a monopole-bond polarizability method and tabulated the interaction energy as the sum of the electrostatic (E_o), polarization (E_p), dispersion (E_d) and repulsion (E_r) energy.

The description of the nature of the electrostatic (E_e), polarization (E_p), dispersion (E_d), and repulsion (E_r) energy by Höltje & Kier (1974) is particularly significant to this treatise, and their description is presented in detail.

10.6.3.1 *Electrostatic* (E_e) *Energy*

The electrostatic energy is the work required or the energy gained when two charged points approach each other. The charge associated with each atom is assumed to be centered on the nucleus. The total electrostatic energy for two interacting molecules is then the sum of all pairs of atoms in the two molecules. Charges of like sign are repulsive while opposite signed atoms are attractive. This energy is related to the first power of the distance between two atoms, and it has significant value even at long distances.

10.6.3.2 *Polarization* (E_p) *Energy*

The polarization energy is a measure of the distortion produced in the electron cloud of a bond by the presence of a charged atom in its vicinity. The ease with which a cloud of electrons can be perturbed is the polarizability. The magnitude depends on the relative freedom of the peripheral electrons from the influence of the atom nucleus. The distortion or polarization of the electrons in a bond by a nearby charged atom leads to an attraction that is related to the third power of the distance between the atom and the bond. This distance is a vector from the atom pointing to the midpoint of the bond. The energy of interaction of this type is a summation between all atom-bond pairs in the two molecules.

10.6.3.3 *Dispersion* (E_d) *Energy*

The dispersion energy is the attractive force derived from the charge fluctuation in an orbital inducing a charge fluctuation in a nearby orbital. The two dipoles thus formed are complimentary and are attractive. The influence is related to the sixth power of the distance and is highly dependent upon proximity and molecular similarity. The phenomenon is prominent in neutral nonpolar molecules. The ability of the peripheral electrons in the molecule to oscillate or be perturbed by induction is dependent on their polarizability in the molecule. The molecular polarizability is approximated by the ionization potential of the molecule, i.e. the energy required to remove an electron. The energy is derived from a summation over all atoms in both molecules.

10.6.3.4 *Repulsion* (E_r) *Energy*

According to Höltje & Kier (1974), all the aforementioned interactions yield, upon calculation, an increasing energy of binding as atoms or molecules approached. Even coalescence of nuclei would calculate to be a favored process. In reality, a repulsion occurs as the atoms approach. The repulsive energy quickly rises as atoms penetrate each other's electron domains or van der Waals radii. It is necessary then to correct the attractive energies with an expression simulating the repulsive energy E_r. This equation is empirically derived, using the van der Waals radii in its formulation. These repulsions are summed over all interacting atoms in both molecules.

Results are tabulated, with minor corrections to the original data, in Table 10.2.

10.6.3.5 *Relation Between Intensity and Binding Energy*

The relative sweetness of the compounds in Table 10.2 varies over two log

TABLE 10.2

Contributions of partial interaction energies (kcal mol^{-1}) from electrostatic (E_e), polarization (E_p), dispersion (E_d), and repulsion (E_r) energies to the total energy (E), and the relative sweetness score for a 2-amino-4-nitrobenzene series of compounds

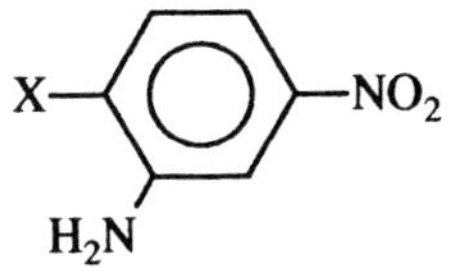

X	E_e	E_p	E_d	E_r	E	*Log RS*
OC_3H_7	−0·129	−0·041	−6·545	+1·318	−5·397	3·61
$OCH_2CH{=}CH_2$	−0·122	−0·040	−6·136	+1·051	−5·247	3·30
I	+0·234	−0·032	−6·588	+1·425	−4·931	3·10
OCH_2CH_3	−0·115	−0·046	−5·921	+1·168	−4·914	2·98
Br	+0·255	−0·026	−5·056	+1·063	−3·766	2·90
$OCH(CH_3)_2$	−0·115	−0·046	−5·921	+1·168	−4·914	2·78
Cl	−0·200	−0·025	−4·609	+0·841	−3·993	2·60
OCH_3	−0·099	−0·051	−4·969	+0·902	−4·217	2·34
CH_3	−0·220	−0·045	−4·645	+0·948	−3·962	2·34
F	−0·191	−0·100	−3·607	+0·657	−3·241	1·50
H	−0·190	−0·036	−3·514	+0·632	−3·108	1·50

units (i.e., by a factor of 100), while the difference in total energy varies by only 2·3 kcal/mol. In the realm of weak interaction energy phenomena, e.g. taste perception, a change in interaction energy of about 0·1 kcal/mol is capable of causing a fivefold change in sweetness intensity. Thus, while AH,B, as an integral component of a compound's structure is prerequisite to sweetness, the effect of γ on the intensity of the sweetness is pronounced.

The plot of the total energy data versus the logarithm of the relative sweetness is shown in Fig. 10.21. The coefficient of correlation (0·904) is quite good, even though the relative sweetness scores for the halogen derivatives are taken from the data of Blanksma & Hoegen (1946) while the remainder of the data are taken from Verkade *et al.* (1946). The latter relative sweetness scores of the compounds for which comparisons can be made (Table 8.1) are generally 20–30% lower than those offered by Blanksma and coworkers. In any event, the salient feature contained in the data of Fig. 10.21 is that the relative sweetness (RS) of substances increases exponentially with binding energy (E), i.e.

$$\mathrm{RS} = E^{0\cdot771}$$

Therefore, only slight increases in the activity of the taste recognition reaction yield large and accelerating increase in taste intensity. Similarly, in a study directed toward the effect of the polarity of the substituents of sugar rings on sweetness, Hooft *et al.* (1993) found a direct relation

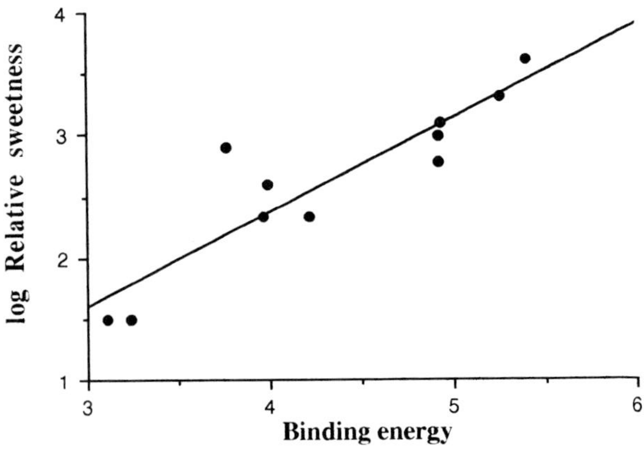

FIG. 10.21. Relation between total binding energy of 2-amino-4-nitrobenzene compounds and their relative sweetness (Höltje & Kier, 1974). The equation for the plot is $y = -0{\cdot}711 + 0{\cdot}771x$.

between sweetness and the Bolzmann-population of a specific sweet conformation, or

$$\log_{10}(\text{Sweetness}) = B + A\,\Delta E$$

In the equation, A equaled $-0{\cdot}73\,\text{kcal mol}^{-1}$. The energy difference between the favored conformation and the absolute minimum energy found in molecular dynamics simulation are about $5\,\text{kcal mol}^{-1}$ for sucrose and $1\,\text{kcal mol}^{-1}$ for sucralose. Using threshold-relative sweetness scores it is found that an energy difference of $1\,\text{kcal mol}^{-1}$ will cause a sweetness change of 0·75 log units, which agrees remarkably well with the value derived for the nitroanilines.

10.6.3.6 *Correlation of Binding Energy Parameters with Taste Intensity*
The one factor that all structure-activity relations point to is that sweetness potency is invariably tied into hydrophobic, lipophilic, electrostatic, and inductive effects. The nature of the chemical recognition interaction that leads to the sweetness quality, as distinguished from saltiness, sourness, and bitterness, is probably buried somewhere in those terms. Höltje & Kier directed their attention only to sweetness potency. The electrostatic and polarization energy for the different compounds varies only slightly (Table 10.2), and contributes only about 10% of the total energy in the first case, and about 1% of the total energy in the second case. On the other hand, the dispersion and repulsion energy together account for about 90% of the total.

TABLE 10.3
Correlation coefficients and standard error of estimate for the correlation between relative sweetness (RS) and the binding energy parameters E_e, E_P, E_d and E_r of Höltje & Kier (1974)

Regression parameters	*Coefficient of correlation*	*Stnd. error of estimate*
RS × E_e	0·387	0·657
RS × E_p	0·519	0·609
RS × E_d	0·937**	0·249
RS × $E_e + E_p$	0·681	0·628
RS × $E_e + E_d$	0·937**	0·263
RS × $E_p + E_d$	0·952**	0·232
RS × $E_e + E_p + E_d$	0·952**	0·247
RS × $E_e + E_p + E_d + E_r$	0·955**	0·336

Continued examination of the data in Table 10.2 for correlations between the various parameters and perceived sweetness yield additional information on the sweetness chemistry for those substances. Correlations found for the data in Table 10.2 are given in Table 10.3. Preliminary examination of all possible simple and multiple relations available from Table 10.2 indicate that the E_d and E_r parameters were highly correlated, but also redundant. E_r is merely a weaker property index of E_d. While removal of E_r from the relations yielded about the same correlation coefficient, the standard error of estimate improved vastly, as shown in the lone three-way regression and the lone four-way regression in Table 10.3. The apparent role of E_e is negligible, and serves only to confound the straightforward relation between sweetness and E_d. It also increased the standard error of estimate (Compare RS $\times$ E_d with RS $\times$ $E_e + E_d$). Of utmost significance to the relation between chemical mechanism and taste modality, the relation between sweetness and the ionic parameter is minimal.

The only significant simple relation in Table 10.3 is between relative sweetness and the dispersion parameter, E_d. Yet, while E_p alone is not significant, the interaction between E_p and E_d is the strongest relation observed, and the one with the lowest standard error of estimate. Therefore, the interaction between dispersion and polarization parameters is most important, and the chemistry of sweetness for the nitroaniline compounds is largely related to those two parameters. The relation probably reflects the inductive effect that dispersion interactions have on polarizability. Presumably, that relation is an application of the significance of γ to AH,B (or of X to e/n). The greater the energy of the dispersion interaction, the greater the energy of the interaction associated with a compound's polar attributes.

10.6.4 Molecular Connectivity

An empirical structure-activity procedure, that obviates experimental determination of chemical properties, was developed by Kier & Hall (1976). It is known as *molecular connectivity*, and has been found to correlate with a number of physical features of a compound such as solubility, boiling point, heat of atomization, and molecular polarizability. It can also be correlated with molecular features that have as their basis, more fundamental properties such as hydrogen bonding activity, and molecular volume (Crosby *et al.*, 1979). Therefore, it is not surprising that connectivity can be correlated with chemical attributes such as sweetness, and even with bitter thresholds for amino acids (Gardner, 1980*a*). Because the procedure is readily adapted to computer analysis of data, its application is becoming widespread.

The rationale behind the connectivity method is that there is a relation between the properties of molecules and compounds and the *pathway* among atoms of either the same or of different type. For example, as a series of alkanes is ascended, the molecular weight, heat of atomization, molar refraction, molecular volume, refractive index, boiling point, and specific gravity increase with regularity. Many of the properties are purely additive, such as molecular weight and molar volume. Others, such as the increment in boiling point for successive members of a series, are not quite arithmetrically additive, but nearly so.

While there is a linear relation between many properties and the number of carbon atoms for a homologous series of hydrocarbons, such relations are not straightforward when branched molecules are included. The boiling point of branched alkanes decreases with an empirical order of increased branching. Other properties also decrease, but still others, such as molar volume and specific gravity, show no correlation whatsoever with the degree of branching. The concept of molecular connectivity sets out, in place of a complete quantum mechanical treatment for a series of molecules, to quantify the *topology* of molecules. The word topology is used here in the sense that there can be a description of the connections between atoms, i.e. the chemical bonds between them. The topological features are then related to molecular properties. Not all molecular properties have strong dependence on this type of molecular topology, but others, e.g. hydrophobicity, do. One example of the former is the ionization potential, which arises from a single Schroedinger equation eigen value, and is probably dependent on a singular structural feature (functional group). Dissociation constants are another example.

The procedural basis of the connectivity method is application of graph theory, and identification of topological indices. A graph is defined as a diagram of a molecular structure, and is not a plot of data points on coordinates. It is also the fact of connection that is emphasized, and not the type of connection. A graph (G) then, is a set of points (vertices) connected by lines (edges), and a structural formula can be written as a matrix from which algebraic expressions can be derived. The structural formula for methane itself is only a short step from being a graph (Kier & Hall, 1976). The *valence* (v_i) or *degree* of the vertex is defined as the number of edges that have the valence as an endpoint. The valence is

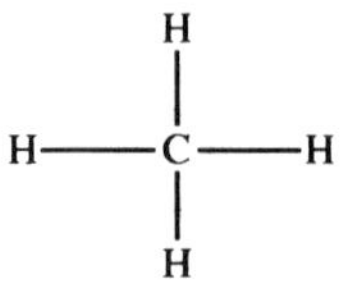

Methane

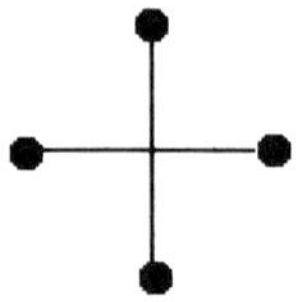

Graph for the methane molecule

designated as $\delta(v_i)$, or simply δ_i. An edge sequence in a graph is defined as a set of edges connected in G and is symbolized as $(v_i, v_j), (v_j, v_k), (v_{m-1}, v_m)$. Each edge terminates on two vertices that are the endpoints of other edges in the sequence, with the exception of terminal vertices. The *length* of a sequence is the number of edges in the sequence. A *path* is an edge sequence in which all edges and vertices occur only once. An edge or vertex that occurs only once in a sequence is said to be *distinct*. Thus in the *hydrogen-suppressed* (hydrogen atoms not shown, or used for computations) graph of isopentane, there are *two* paths of length 3. They are $(v_1, v_2), (v_2, v_4), (v_4, v_5)$ and $(v_3, v_2), (v_2, v_4), (v_4, v_5)$.

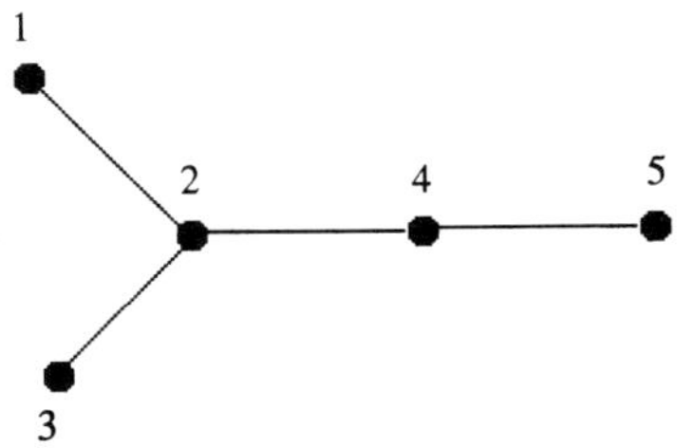

A *chain* is an edge sequence in which all edges are distinct, and a *circuit*, in which the first and last vertices are distinct, is a special case of a chain. For a subgraph, which is a designated portion of a graph, Kier & Hall (1976) offer the following examples: number I below is a cyclic graph, while the bolder lines in II describe a subgraph, III is a star subgraph, and IV is a circuit. The next step in development of the connectivity

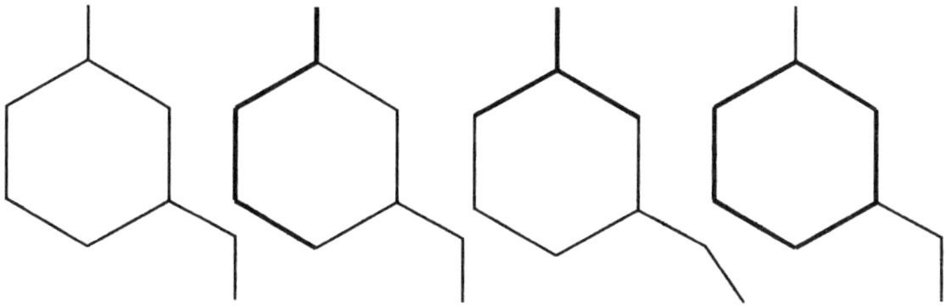

I II III IV

procedure was the derivation of a topological index from the topological matrices, as a chemical graph can be written as a matrix. These lend themselves to manipulations that yield a single number, or small sets of numbers, that are characteristic of the graph. They also lend themselves to quantitative comparison with physical or chemical properties, as in a structure-activity relation.

To derive a topological index, the authors modified the Randić (1975) abstract notion of the *branching index*. What follows next is development of the notion of a valence-weighted graph G_v, and the assumption that a satisfactory quantitative relation is contained in the relation between structure and many of the properties of a compound. A second assumption is that there is a functional relation between a compound's properties and the connectivity characteristics of the graph. The connectivity function $C(\chi)$ for the property p for the relation is a sum of the terms that depend in linear manner on the graph characteristics, or

$$C(\chi) = b_o + \sum_{m,t} b_t(m)\,{}^m\chi_t$$

The quantities $b_t(m)$ depend on the property, and may be determined by multiple regression with experimental data, or calculated by a theory or model of the property. In a multiple regression study, experimental values of a property are regressed against $C(\chi)$. The expression ${}^m\chi_t$ are terms defined for a subgraph of type t containing m edges connected in the valence graph G_v. The order of the subgraph s, designated as m, and disconnected subgraphs are not considered.

There are four types of subgraphs. Those of type 1 ($t = \mathrm{P}$) are called *path* terms and consist of subgraphs whose subgraph valencies are no greather than 2. Examples are:

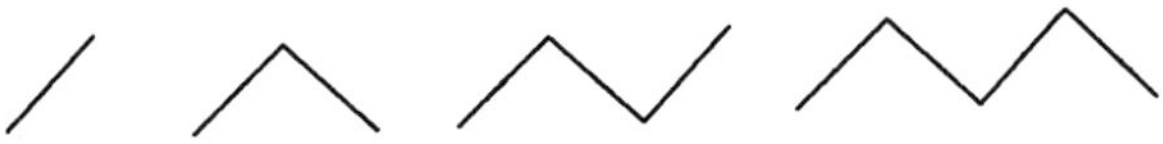

Type 1. subgraphs

Type 2 subgraphs ($t = \mathrm{C}$) are called *cluster* terms and consist of subgraphs whose subgraph valencies include at least 1, 3 or 4, but do not include 2. A star subgraph K_{1n} is a special case of the second (central) diagram below. Examples are:

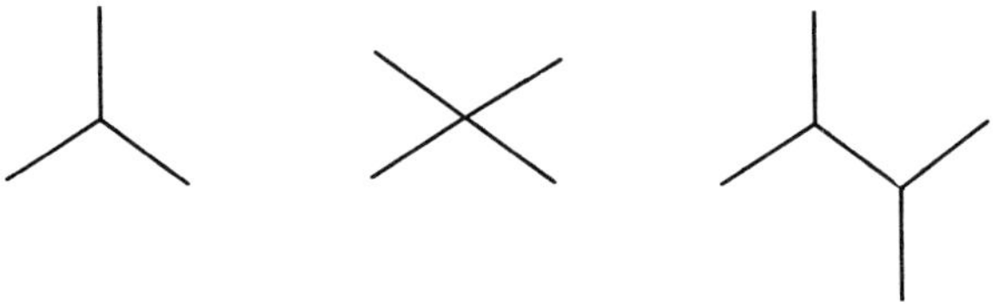

Type 2. subgraphs

Type 3 ($t = \mathrm{PC}$) subgraphs are called path/cluster terms whose subgraph valencies must include 2 in addition to 3 and/or 4. Examples are:

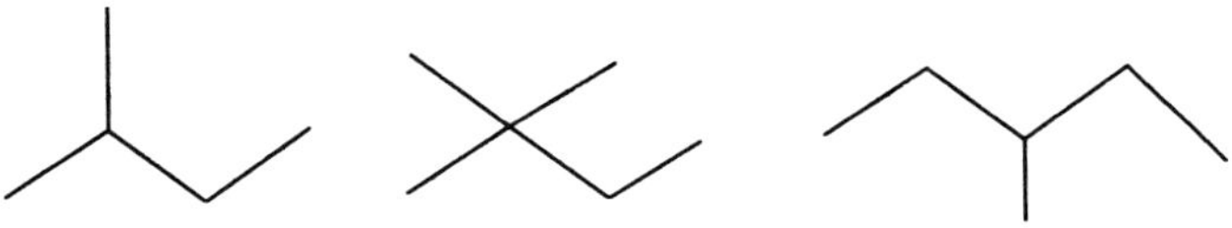

Type 3. subgraphs

The type 4 subgraphs ($t = \mathrm{CH}$) are called *chain* terms, and have edge sequences that contain at least one *circuit*, which is a special case of a chain. Examples are:

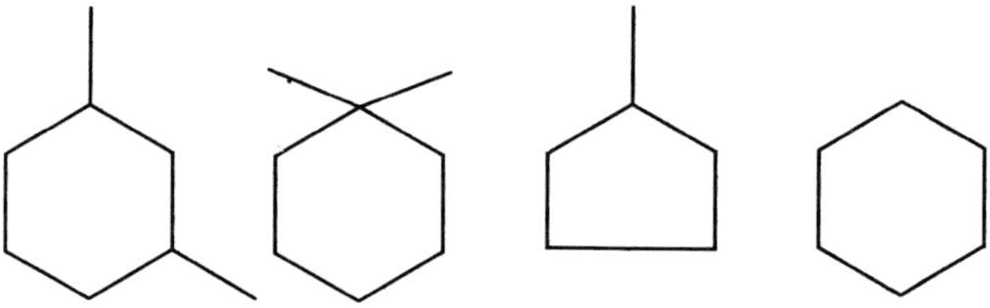

Type 4. subgraphs

The connectivity indexes ${}^{m}\chi_{t}$ are then evaluated as a sum of terms over all distinct connected subgraphs:

$$ {}^{m}\chi_{t} = \sum_{j=1}^{n_m} {}^{m}S_{j} $$

where n_m is the number of type t subgraphs of order m. The subgraph term ${}^{m}S_{j}$ is a quantity calculated for each subgraph and is defined as follows:

In molecular connectivity theory, the most critical aspect of the relation between molecular topology and properties is the dependence on vertex valence. This leads to a third basic assumption. *The subgraph terms depend on the reciprocal square root of the vertex valence in a multiplicative manner*:

$$ {}^{m}S_{j} = \prod_{i=1}^{m+1} (\delta_{i})_{j}^{-0{\cdot}5} $$

where j denotes the particular set of edges that constitute the subgraph. The number of subgraph valencies multiplied together depends on the subgraph type. Chain subgraphs are defined by m vertices, and all others by $m + 1$ vertices, where m is the subgraph order.

Thus the connectivity indices ${}^{m}\chi_{t}$ are valence-weighted counts of connected subgraphs. The vertex weighting scheme introduces the vertex valence into the relation between significant topological features and properties of the molecules. *Weighting* is a key feature in the molecular connectivity method. To enumerate and evaluate the ${}^{m}\chi_{t}$ indices, and to reveal the nature of the chi terms, a number of various terms are employed. They are:

(1) A zero order term (${}^{0}\chi$) consisting of a single vertex, i.e. it has no edges, and ${}^{0}\chi$ is the sum of the reciprocal square root valencies for the graph.

(2) A first-order term (${}^{1}\chi$) that is a sum of all edges in the graph, weighted by the reciprocal square root valencies. There is only one type of graph edge, and only one type of subgraph of order one: The first order term is the Randić (1975) branching index. In essence, ${}^{1}\chi$ describes the size of a molecule in terms of the number of bonds. In a discriminate analysis application (Kier, 1980) to the sweetness/bitterness of oximes, where $y = c_{1\times 1} + c_{2\times 2} + \cdots + c_{n}x_{n}$, compounds were judged to be sweet when $y > -3{\cdot}27$ and bitter when $y < -3{\cdot}27$.

(3) A second order term ($^2\chi$) for subgraphs that contain two pairs of connected, adjacent edges that are of one type only, the path type ($t = \mathrm{P}$).
(4) A third order term ($^3\chi$) wherein path, cluster, and chain terms are encountered.
(5) A fourth order subgraph term ($^4\chi$) in which all types of subgraph terms are considered (path, cluster, path/cluster, and chain). Higher order terms are calculated in a manner analogous to the $^4\chi_t$ terms, including m edges in each case. As applied to the sweetness/bitterness of oximes (Kier, 1980), $^4\chi_p$ was employed as a term that is influenced by the size of the molecule as well as by the substituents.

10.6.4.1 *Application of the Connectivity Terms to Alkanes*
As an example for application of these terms, the first order term, $^1\chi$, which has subsequent application to sweet substances, is herein applied to hexane using the expression:

$$^1\chi = \sum_{k=1}^{N} C_k = \sum_{k} = \frac{1}{\sqrt{\delta_i \delta_j)_k}}$$

The structure below shows the hydrogen suppressed graph of hexane. The numbers above and below the carbon atoms are the vertex valences, or the number of single covalent bonds for each carbon minus the number of hydrogen atoms. Deriving the reciprocal of the square root for each connectivity, and summing them, i.e. $(0{\cdot}707 \times 2) + (0{\cdot}500 \times 3) = 2{\cdot}914$, the $^1\chi$ value for hexane is obtained. The zero-order term is 4·828, the second order term is 1·707, and the third order term is 0·957.

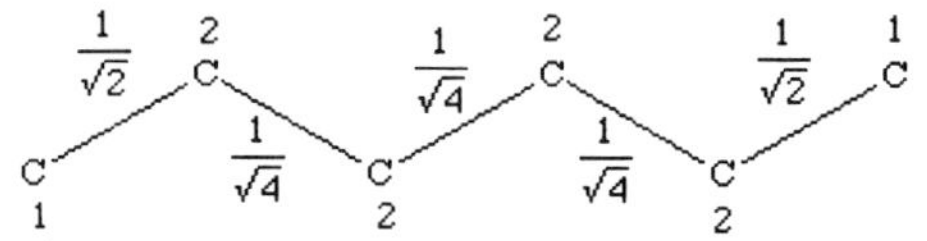

Vertex valences and connectivities for the bonds in hexane

For compounds that contain heteroatoms, the connectivity index requires treating heteroatoms differently, and one method for oxygen is as follows. Whereas all δ values can be considered to be the difference between the number of valence electrons Z^v and the number of hydrogen atoms h_i, which are suppressed in a graph, then

$$\delta^{v}_{i} = Z^{v} - h_i$$

The δ for a methyl group then becomes 4 minus 3 = 1, a methylene group δ = 4 minus 2 = 2, and a methine group δ = 4 minus 1 = 3. As applied to an oxygen atom in alcohol, $\delta^{v} = Z^{v} - h_i = 6 - 1 = 5$. As applied to pentanol, a breakdown of the graph leads to a hydrocarbon chain component of $^{1}\chi^{HC} = 2{\cdot}207$, and a hydroxyl group connectivity (χ^{OH}) of $1 \div$ the square root of $2 = 0{\cdot}707$. Thus, $^{1}\chi$ for this substance is $2{\cdot}207 + 0{\cdot}707 = 2{\cdot}914$, for No. I in Fig. 10.22. The value $^{1}\chi^{v}$ for graph II in Fig. 10.22 is $2{\cdot}207 +$ the reciprocal of the square root of 10, or $2{\cdot}207 + 0{\cdot}316 = 2{\cdot}523$. Both $^{1}\chi^{v}$ and $^{1}\chi^{v}$ have application to sweet substances.

10.6.4.2 *Significance of Connectivity Indices*

To relate molecular connectivity to a compound's chemical and biological characteristics, the connectivity procedure sets out to determine two types of molecular connectivity: (1) a straightforward *bond connectivity* index, $^{1}\chi$; and (2) a *valence bond connectivity* index, $^{1}\chi^{v}$. The reason given for inclusion of the second connectivity parameter δ^{v} into the computations is to minimize redundancies. It adjusts $^{1}\chi$ to obtain better correlations between $^{1}\chi$ (or $^{1}\chi$ times $^{1}\chi^{v}$) and various physical, chemical, and biological properties.

From the above discussion, it is apparent that connectivity does not consider the chiral nature of either individual carbon atoms, or the molecular conformation, which are of paramount importance to the

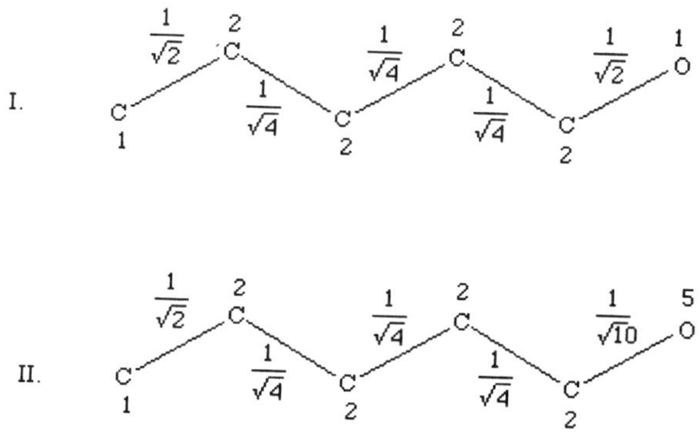

Fig. 10.22. Vertex valencies and connectivities for pentanol.

sweetness of amino acids and sugars. The connectivity values for either a D- or an L-amino acid are the same, and the constants are not capable of indicating that amino acid enantiomers may be either sweet or bitter. The constants are also the same for the enantiomeric sugars, which may or may not differ in their sweetness attributes. The concept of connectivity is designed to account for molecular shape, such as branching, in a manner more sophisticated than such gross properties as partition coefficients. Unfortunately, it neither takes into account the three-dimensional array of atoms in space, nor what the significance of that array of atoms might have on structure-activity relationships.

On the other hand, when the significance of the various parameters is understood, the computations are easy, and yield information about molecules that might otherwise be derived only by laborious laboratory procedures. Crosby *et al.* (1979) pointed out that the arithmetic simplicity can also be a disadvantage. The single parameter gives no indication of the mechanism of the compound's action, i.e. the one fundamental parameter has no intrinsic functional significance. Finally, the use of an unreasonably high number of predictor variables in multiple regression analysis tends to minimize the significance of the relations.

10.6.4.3 *Application of Connectivity to Nitroaniline and Oxime Taste*
Application of the connectivity constants to sweetness (Kier & Hall, 1976) used the forms ${}^1\chi$ and ${}^1\chi^v$, and multiple correlation with the relative sweetness of the 2-substituted 5-nitroanilines, on a *molar* basis is quite good. The data used are shown in Table 10.4.

By using two forms for χ (${}^1\chi$ and ${}^1\chi^v$), Kier & Hall adjusted the connectivity constants to simulate a compound's hydrophobicity and polarizability, i.e. the two most important parameters in the energy relations of Section 10.6.3. The equation for the relation of ${}^1\chi$ and ${}^1\chi^v$ to relative sweetness (RS) is $\log RS = 0{\cdot}350\,{}^1\chi + 0{\cdot}694\,{}^1\chi^v - 3{\cdot}856$, where $r = 0{\cdot}953$. Although the range for the constants ${}^1\chi$ and ${}^1\chi^v$ is about the same (1·9 and 1·8, respectively), the latter is about twice as important as the former with respect to its effect on the coefficient of correlation. This is because of the reciprocal relation of connectivity properties. The term ${}^1\chi^v$ compensates for the significance of polarization properties alone (which are somewhat constant for this closely related series of compounds). Emphasis is placed on the significance of the apolar hydrophobic properties. In other words, ${}^1\chi$ in the regression equation is forcibly associated with the Hammett sigma constant, and ${}^1\chi^v$ is forced to relate to the π function in the Deutsch & Hansch (1966) equation. The

TABLE 10.4
Data used for correlation of relative sweetness (molar basis) of 2-substituted 5-nitroaniline compounds with $^1\chi$ and $^1\chi^v$ (Kier & Hall, 1976)

X	$^1\chi$	$^1\chi^v$	*Log RS*
O-*n*-Pr	6·647	4·264	1·406
I	5·109	4·279	0·675
OEt	6·147	3·764	0·885
Br	5.109	3·556	0·566
Cl	5·109	3·166	0·365
OMe	5·647	3.176	0·293
Me	5·109	3·064	0·337
F	5.109	2·452	−0·591
H	4·716	2·653	−0·538

correlation between $^1\chi$ only with the sweetness potency is 0·765, while that for the relation between $^1\chi^v$ and sweetness potency is 0·915. The relation of each alone to the relative sweetness *score* (RS) on a weight/volume basis leads to coefficients of correlation of 0·729 and 0·943, respectively. The latter single correlation is nearly as good as the multiple correlation coefficient obtained when both $^1\chi$ and $^1\chi^v$ are used to relate $^1\chi$ and $^1\chi^v$ to sweetness potency. In a second application of the connectivity procedure, Kier (1980) applied the first and fourth-order χ constants to a series of cyclohexylaldoximes. Seventeen out of twenty molecules were correctly correlated with respect to their sweet or bitter tastes, and nine out of ten sweet substances were correctly correlated (see also van der Wel *et al.*, 1987).

10.6.4.4 *Application to the Varying Sweetness of the Sugars*

In a modification of Kier & Hall's (1976) approach to qualification of sweetness phenomena, Daniel & Whistler (1982) calculated a molecular connectivity index ($^3\chi_m$) for various sugars and sugar derivatives. Their modification includes abandonment of the $^1\chi^v$ parameter, and subsequently (Daniel, 1989) accounting for hydrophobicity-lipophilicity with an experimentally determined parameter. The $^3\chi_m$ constant was then related to literature values for the relative sweetness scores of the sugars. It is stated that this connectivity method allows conversion of a three-dimensional structure into a number ($^3\chi_m$) that represents that three-dimensional structure. It is construed to be a measure of a sweet compound's intrinsic activity after it is bound to the receptor (Daniel, 1989).

For calculation of sugar molecular connectivity parameters, nonhydrogen, noncarbon atoms are assigned the specific vertex valence number δ_i. The vertex valence is defined as the atom's number of valence-shell electrons minus the number of hydrogen atoms normally attached. For carbon atoms, δ_i is the number of atoms to which the carbon atom is bound minus the number of hydrogen atoms attached to it. A number χ for an entire molecule can be calculated from the sums of ${}^0\chi, {}^1\chi, {}^2\chi, \ldots {}^n\chi$, which are 'clusters' of vertex values. For the ethylene glycol cluster, O—C—C—O, the vertex values are 5,2,2,5, respectively. The ${}^3\chi$ path term for the molecule is *the reciprocal of the square root of the multiplied vertex values*, namely $(5 \times 2 \times 2 \times 5)^{-0{\cdot}5}$ or 0·1. The multiplication of these values, before deriving the reciprocal of the square root, seems to be the major modification of the Kier–Hall method. The answers, or cluster terms, are then said to characterize the chiral centers of carbohydrates. In fact, the characterization is an adjustment for known chiral effects. Different values are assigned to sugar ring hydroxyl substituents, depending upon whether they are axially or equatorially disposed on a sugar ring. However, axial and equatorial dispositions are not related to the *rectus* or *sinister* nature of an individual carbon atom's substituents, and the method does not use different signs for clusters about carbon atoms with opposite handedness. When the chirality of a cluster unit is deemed to be significantly with respect to its effect on sweetness, it is given an appropriately lower or higher empirical δ value. The anomeric cluster unit for β-D-glucopyranose is assigned a δ_i value for O-1 that is smaller for the α-D-anomer because it is presumed that α-D-glucose is sweeter than β-D-glucose, which is probably a moot point (see Section 6.2.1).

To obtain a definitive molecular value, ${}^n\chi$, values are assigned to vertices, δ, that attempt to take into account the consequences of chiral configurations. They are then used to construct subgraph values for known partial molecular contributions to ${}^n\chi$. As the sugar glycol groupings are glycophoric units, all O—C—C—O terms were construed to be integral parts of the sweetness contribution to the molecular ${}^n\chi$ term. To incorporate Kier's (1972) dispersion sites into the computation of ${}^n\chi$, all O—C—C—O′ (where C′ is a methyl or methylene unit) paths, but with no common C—C path, were included. In addition to the aforementioned treatment of the anomeric glucoses, and because both D-mannose and D-galactose are not as sweet as D-glucose, the value assigned to axial secondary hydroxyl groups is 6. The value assigned to equatorial secondary hydroxyl groups is 4. The reason for the arbitrary assignment of the higher and lower values is because of the inverse square root nature of the adopted $^3\chi$ path term. The term $^3\chi$ needed to be either correspondingly increased or decreased, based on the known relative sweetness of the sugars.

Ring oxygen atoms are given the δ_i value 8 as they are associated (Birch, 1976) with sweetness depression. Glycosidic disaccharide oxygen atoms are given values of 7 when axial and 9 when equatorial, based on the reported sweetness of the nonreducing disaccharide sugar alcohols maltitol and cellobiitol (Lee, 1979). Ether linkages at secondary positions are assigned values of 9 (axial) and 7 (equatorial). As glycosides are not as sweet as the parent sugars, the glycosidic oxygens are assigned values greater than 6. Oxygen atoms at primary positions, for example the C-6 hydroxyl group of the hexose sugars, seem to contribute little to the sweetness of the sugars, and are assigned the molecular connectivity value of 5 when free, and 6 if etherified.

To use the connectivity parameters to determine the χ value of a molecule, all OCCO paths, all distinct OCCC$'$ paths with no common C—C paths, and all chiral cluster terms are counted to provide an equation $^3\chi = \text{OCCO} + \text{OCCC}' + \text{chiral cluster}$. A chiral cluster is simply designated as

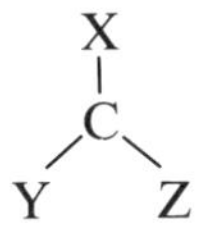

As will subsequently be demonstrated, the chiral cluster parameter is essentially a tetrahedral carbon atom path with the hydrogen atom eliminated. It is not included in the computation, but the carbon atom itself is included. It will be recalled that the feature that renders a carbon atom chiral is the four substituents of different size attached to it (Section 2.11.3.3). The carbon atom *per se* is not chiral, but resides in a chiral environment. To indicate the chiral nature of the atom's appended substituents, an antipodal specification needs to be made, such as R or (+) for *rectus*, and S or (−) for *sinister*.

The chiral terms are individually related to the configuration of each ring carbon atom, and in total to the unique configuration of the carbohydrate. For α-D-glucopyranose, the calculation of $^3\chi_m = \text{OCCO} + \text{OCCC}' + \text{chiral cluster terms}$ for the compound, it is found that there are five OCCO combinations, as shown below. There are three OCCO clusters (C-1,O-1,O-2,C-2 through C-3,O-3,O-4,C-4), all of

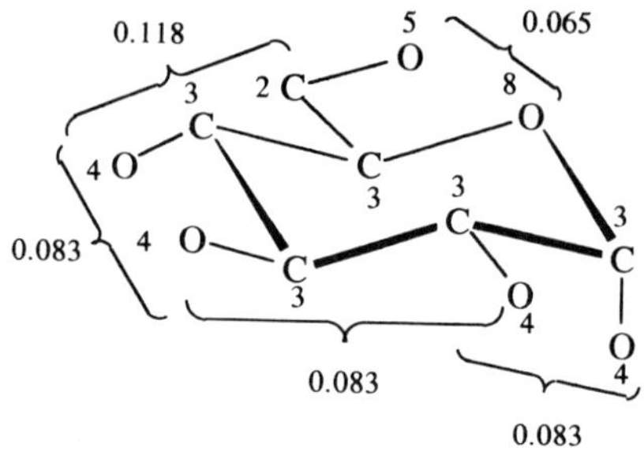

which have a vertex sequence of $4 \times 3 \times 3 \times 4 = 144$. This leads to an inverse square root value of 0·083. Apparently C-4, O-4, C-5, C-6 are included in the total OCCO groupings, and it has an inverse square root value of 0·118. If so, the COOC group is used again for the OCCC′ group. The final COOC grouping seems to be C-5, O-5, C-6, O-6, with an inverse square root value of 0·065. The sum of these values $(3 \times 0{\cdot}83) + 0{\cdot}118 + 0{\cdot}065$ is in good agreement with the value indicated by Daniel. This value is not obtainable without using the C-4, O-4, C-5, C-6 unit as OCCO also.

For α-D-glucopyranose, OCCC′ is O-4, C-4, C-5, C-6, and the inverse square root for the sequence 4,3,3,2 is 0·118, again in agreement with the value given by Daniel. With the carbon atoms of α-D-glucopyranose specified as being either right (+) or left (−) handed, the so-called chiral terms for the asymmetric carbon atoms are as follows.

Carbon atom	*δ values*	*Square root*	*Inverse square root*
C-1(+)	$3 \times 4 \times 8 \times 3$	16·97	0·059
C-2(+)	$3 \times 4 \times 3 \times 3$	10·39	0·096
C-3(−)	$3 \times 3 \times 3 \times 4$	10·39	0·096
C-4(+)	$4 \times 3 \times 3 \times 3$	10·39	0·096
C-5(+)	$3 \times 3 \times 2 \times 8$	12·0	0·083
			0·430

The total for the chiral cluster values is 0·430. The chiral parameters yield the same vertex product and inverse square root values for asymmetric carbon atoms that have opposite configuration (e.g., C-2 and C-3). The tetrahedral carbon atom connectivity paths include asymmetric carbon atoms in the calculation, but does not distinguish between them. Finally, for α-D-glucopyranose, ${}^3\chi_m$ = OCCO + OCCC′ + chiral cluster =

0·980. To compare sugars of different molecular weight, the $^3\chi$ values are multiplied by 1000 and then divided by the molecular weight to yield $^3\chi_m$. For the sugar in question, it is 5·44.

The first application made for $^3\chi_m$ values was to correlate them (Daniel & Whistler, 1982; Daniel, 1989) with relative sweetness values for sugars and sugar alcohols. A graph of the approximate relation found, using thc sugars only, is shown in Fig. 10.23. The plot of the relative sweetness of the sugars only on a weight/volume basis is excellent ($r = 0{\cdot}97$). The plot of the relative sweetness of the sugars on a molar basis, which is favored because of the molecular nature of the connectivity calculation, is somewhat less ($r = 0{\cdot}86$). If the connectivity value for β-D-mannose is included in the calculation, the relation is strained. Although β-D-mannose has a connectivity value of about 5·0, it is not sweet, but bitter. With respect to the taste of the L-sugars, it follows that the calculated $^3\chi_m$ values would be identical with respect to magnitude and sign for both sugar enantiomers. Neither asymmetric carbon atom nor ring conformational chirality specifications are included in the calculation of $^3\chi_m$. That feature then led Daniel & Whistler (1982) to state that because the rules for calculating molecular connectivity values for D- and L-sugars are the same, the method predicts that D- and L-sugars should be isosweet. By the same reasoning, molecular connectivity calculations would predict that the enantiomeric amino acids should be isosweet, but they are not.

In continuation of the effort to quantify the sweet taste response, Daniel (1989) emphasized the importance of a compound's hydrophilic-

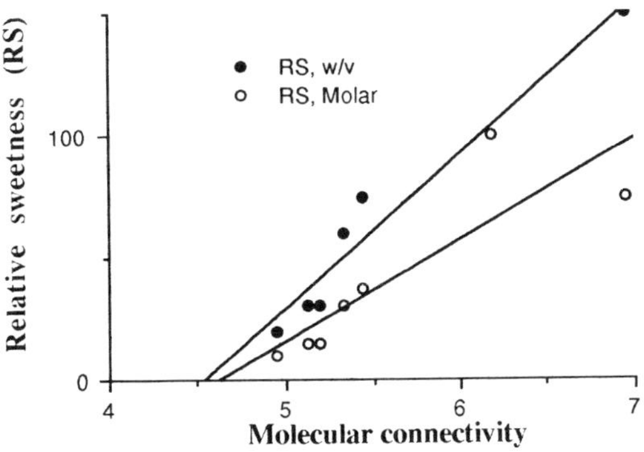

FIG. 10.23. Relation between relative sweetness of sugars and molecular connectivity. Connectivity values versus w/v sweetness scores from Daniel (1989).

lipophilic balance (HLB). The rationale employed is that a relative sweetness score should be a function of both HLB and the intrinsic activity ($^3\chi_m$) of the compound when bound to the receptor. Considering the concentration of sugar found in the isobutanol component of an equilibrated water–isobutanol partition to be the partition coefficient *P* of the system (Collander, 1950; Table 5.4), Daniel calculated the hydrophile-lipophile balance of selected sweeteners from the expression

$$(\text{HLB-7}) = 0{\cdot}36 \ln (1/P)$$

It was then demonstrated that relative sweetness of the sugars is an inverse parabolic function of HLB (maltose excluded), and a sigmoid function of the intrinsic activity ($^3\chi_m$). With fructose assigned a sweetness score of 110, instead of 150, and again eliminating maltose from the plot, the relative sweetness of glucose, mannitol, fructose and sucrose was plotted against $[(\text{HLB})^2 \times {}^3\chi_m]$. The resulting plot is a near-ideal linear relation. It, along with the aforementioned parabolic HLB and sigmoidal $^3\chi_m$ plots, is intended to convey (*sic*) that extremely hydrophilic compounds (carbohydrates) have a low level of sweetness. Increasing the hydrophobic character of the compound, while maintaining the integrity of the AH,B,γ glycophore, will increase the sweetness up to a maximum value (RS_{max}). Beyond this HLB value, increases in hydrophobicity will decrease sweetness, and eventually lead to a maximum level of bitterness (RB_{max}) (cf. Hansch & Fujita, 1964). Although the HLB values used for developing the relationship are in error (except for maltose, all are too high by 1·0 HLB units), the constant error should not appreciably alter the significance of the results obtained.

With respect to the molecular significance of the aforementioned relations, it is found that none exists when the molecular parameters are plotted against sweetness scores recalculated to a molecular basis. For example, RS fructose, w/v × 180/342. In addition, even for practical application, the significance of Collander's partition coefficients for the sugars is open to question. Collander (1950) cautioned that the error term for the coefficients given is as much as ±20%. The partition coefficients among the sugars vary by ±40%. Put another way, the difference in lipophilicity and/or hydrophobicity among the sugars is quite small, and has questionable application to their varying relative sweetness. Daniel (1989) concludes the discussion with a mathematical model for sweetness, but because of errors contained in the derivation of molar volumes, and HLB constants, it is of questionable validity.

10.6.4.5 *Sulfamate Side-Chain/Connectivity Relations*

In an extension of earlier studies, Spillane & Sheahan (1989) attempted to relate a spatial parameter for the side chain R in the structure of a series of the carbosulfamate ($RNHSO_3^-Na^+$) compounds to sweetness. They also studied the heterosulphamates (where the N-substituent possesses a carbon skeleton that includes one or more heteroatoms). In the earlier study (Spillane & McGlinchey, 1981), the authors calculated side-chain parameters x, y, z, and V (xyz) for a series of carbamosulfamate substances using Corey–Pauling–Koltun space-filling models. They then plotted x (the length of the side chain) against V. It was found that 90% of the sweet-tasting compounds fell within a triangular area described as being between 5·2 Å and 7·2 Å on the y axis and at 250 Å on the x axis. In the most recent report, Spillane & Sheahan (1989) found that of the 17 new compounds examined, the above procedure predicted that three of the four that were sweet ought to be, and that one that was not sweet ought to be sweet. The latter compound did not possess an AH function. Three of the others that did not possess an AH function were correctly predicted to be not sweet. As a result of these findings, the method is claimed to confirm the lack of sweetness in those three and other cases.

For the heterosulfamate substances, linear discriminate analysis using five parameters (x, y, z, V, and the 1st order connectivity index $^1\chi^v$) was computed. While x, y, z, and V were applied to only the side chain, the connectivity index was calculated for the entire molecule. Using seven combinations (subsets) selected from the five parameters, including one that included all five, the authors determined the ability of those seven subsets to predict sweetness. Linear discriminate analysis was employed for the three substances, among a total of 33, that were known to be sweet. All parameters failed to assign sweetness to at least two compounds observed to be sweet, and the subset containing all five parameters failed for all three. The method also predicted that many substances that were not sweet ought to be.

Based upon the dimensional parameters obtained for the compounds, Spillane & Sheahan (1989) then calculated xy (area) and z for the sweet carbosulphamates to be $31{\cdot}2 \pm 3{\cdot}3$ Å and $6{\cdot}2 \pm 0{\cdot}63$ Å, respectively. For the sweet heterosulphamates xy and z calculated to be $46{\cdot}1 \pm 14{\cdot}7$ Å and $6{\cdot}3 \pm 0{\cdot}59$ Å, respectively. It was concluded, perhaps unjustifiably, that the analysis reinforces the idea of different receptor sites being required by different types of sweeteners, even within a class of compounds (the sulfamates).

10.7 KINETIC THEORY

Taste is perhaps the weakest known 'chemical reception' mechanism. The sweet taste dissociation constants for the sugars are so high (half-maximal responses are obtained at 0·1 to 1·0 M concentrations) that a pharmacologist would consider them to be non-specific or artifactual (Lancet & Ben-Arie, 1991). Nevertheless, special kinetic patterns can be associated with taste, and the significant kinetic theories for the recognition chemistry of taste are as follows.

10.7.1 The Beidler Equation

Renqvist (1918) proposed that the intensity of taste response is proportionate to the amount of stimulus adsorbed to the receptor per unit time, or $\mathrm{d}A/\mathrm{d}t = k$. Beidler (1954) then stated the mass action phenomena in terms of *concentration*, and assumed that the magnitude of the response is proportionate to the total amount of stimulus adsorbed. Either way, the interaction complies with first-order kinetics, and Beidler's treatment established the means for deriving constants for the equilibria phenomena. Traditional treatment, i.e. derivation of constants using a time sequence, is not practicable due to the rapid nature of taste interactions, which are usually about 50 μs. In its initial form, the Beidler taste equation is

$$\frac{C}{R} = \frac{C}{R_s} + \frac{1}{kR_s} \tag{9.1}$$

where R is the magnitude of the response to the taste stimulus at concentration C. $R_{\max}$ is the maximum response at saturating stimulus concentration. The equation is of the same order as those used to assess the binding of ions with proteins (Scatchard, 1949), and for determining the affinity of an enzyme toward its substrate.

10.7.2 Taste Association Constants

The Beidler equation has several applications. The value of K can be calculated for any taste quality once response magnitudes have been determined for several concentrations of stimulus. The logarithm of K is directly proportionate to the binding energy (recognition interaction), and ΔG, the change in free energy, $= -RT \ln K$. Because the values for ΔG are small (several kilocalories/mole), the initial interaction, whether chemical or physical, is very weak.

When R in the Beidler equation is plotted against the logarithm of a wide range of concentrations (approx. 4 log units), a sigmoid curve is obtained with R approaching R_{max}. In addition, a plot of C/R versus C is linear. The slope of the curve is the reciprocal of R_{max}, and the intercept is $1/KR_{max}$. K for the neurophysiological response to NaCl by rats is about 9·8 (7–15), depending upon the methodology used. As the taste stimulation process can be considered to be a thermodynamic equilibrium, the free energy (ΔE) can be calculated from the expression

$$\Delta E = -RT \ln K$$

In the equation, R is the universal gas constant, and T the absolute temperature. For the binding of NaCl to the taste receptor, $\Delta E = -1{\cdot}37\,\mathrm{kJ\,mol^{-1}}$, which is a reasonable value for an ionic interaction. As the magnitude of the response to salt stimulation varies little with change in temperature, Beidler (1954) concluded that most of the free energy for the salt taste can be attributed to change in entropy.

It is not always possible to derive the complete sigmoid curve for different tastants. Such studies are also complicated by the onset of adaptation and other problems. Therefore, Maes (1985) developed an alternative procedure for deriving taste equilibrium parameters. For this purpose, the Beidler equation was expressed as

$$R/R_{max} = 1/1 + KC$$

where the sensory response R is to the maximum sensory response R_{max} as 1 is to $1 + KC$. This generalized sigmoid form of the Beidler equation is shown in Fig. 10.24 where K is set to 1. To apply the equation, a short-cut was used to determine R_{max} for R obtained for different values of C (Maes, 1985). As the plot shown in Fig. 10.24 is a hyperbolic tangent function, it possesses several salient features. First among them is that the slope b (first derivative) is maximum at its inflection point ($C = K$; $R = 0{\cdot}5\,R_{max}$), and has the value $0{\cdot}25\,R_{max} \ln 10$, or $0{\cdot}58\,R_{max}$ per log concentration unit. As $R_{max} = 1{\cdot}74b$, and b is directly proportional to R_{max}, it can be employed in its estimation.

As the curve in Fig. 10.24 is linear about the inflection point, a regression line through responses obtained to at least two concentrations in this range will contain slope b. If b is determined using two concentrations one log-unit apart, i.e. differing by a factor of 10, then R_{max} will be 1·92 times b when K is at the mid-point of the interval, and 2·44 b when K coincides with one of the concentrations. b will be even larger when K lies beyond the test interval. As K is essentially unknown in advance, tests need to be performed to reduce the intrinsic variability. If K is to be restricted in this manner, i.e. to the above test interval, R_{max}, on the average, turns out to be about 2·09b.

The Beidler equation, and the method of Maes, applies to taste neurological responses obtained with animals. McBride (1987) has shown that the Beidler equation applies equally to human sensory response to the four basic tastes, either using taste intensity scores obtained by just-noticeable-difference (JND) accumulation, or by category rating. McBride then found that the Maes procedure can be applied to JND scores without alteration. Modification is needed to use rating data because a rating scale is an equal-interval (linear) scale with an arbitrary zero. As a consequence, an empirical rating response is equal to its true subjective value except for an additive constant C_0. For example, a rating of 10 on a 13-point category scale is not necessarily twice as strong as one rated 5. As a consequence, an empirical rating response r is not equal to $R_{max}/2$, but $r + C_0$ does equal $R_{max}/2$.

Applying the procedure to taste data, McBride (1987) demonstrated an ideal fit of human sensory data to plots of the Beidler equation for substances representing sweetness, saltiness, sourness, and bitterness. For

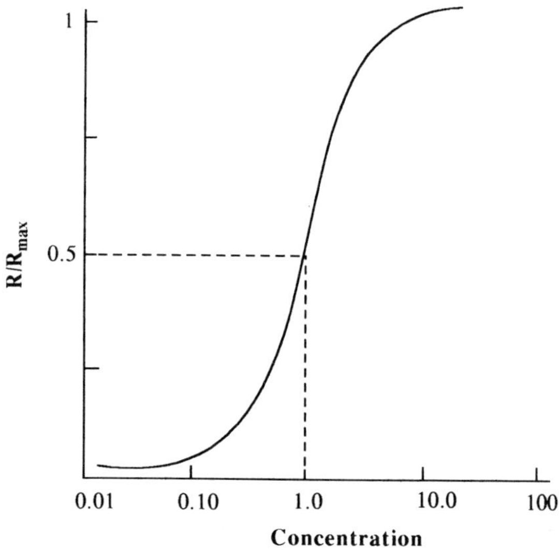

FIG. 10.24. Sigmoidal relation between taste response and stimulus concentration specified by the Beidler equation. The inflection point (maximum slope) occurs at $R = 0{\cdot}5R_{max}$ when $C = 1/K$ (Maes, 1985; McBride, 1987).

different sugars (Fig. 10.25) R_{max} and $R_{max}/2$ is constant, and the three Beidler-type curves are identical except for lateral displacement due to different relative sweetness.

McBride then proceeded to accumulate data for other tastes, and tabulated a series of significant constants derived from the data. Perhaps most significant was a set of K values for the sweetness of glucose, fructose, and glucose, along with NaCl saltiness, caffeine bitterness, and citric acid sourness. When plotted against threshold concentration, which corresponds to 2 JNDs on the Beidler curve, the result is linear and highly significant (Fig. 10.26). For the McBride association constants, which are equal to $1/C$ in moles/liter, the ΔG associated with sweetness and saltiness ($-1{\cdot}98 \times 298 \ln 5{\cdot}3 = 984\,\text{cal mol}^{-1}$) is $-0{\cdot}984\,\text{kcal mol}^{-1}$. For the sourness of citric acid, $-3{\cdot}369\,\text{kcal mol}^{-1}$, and for the bitterness of caffeine, $-3{\cdot}019\,\text{kcal mol}^{-1}$. For substances with varying sweet and bitter taste potency (i.e., different isosteric glycophores and picrophores, the importance of the chemical recognition activity parameter (Section 10.3.2) would seem to be buried in the response/concentration terms.

10.7.3 Taste Dissociation Constants

To those familiar with the kinetics of drug–receptor interactions, the kinetics of taste is an enigma (Lancet & Ben-Arie, 1991). The half-maximal effect of sugars is obtained at concentrations of 0·1–1·0 M, and ΔG is quite small. It is also difficult to ascertain if the initial chemistry of

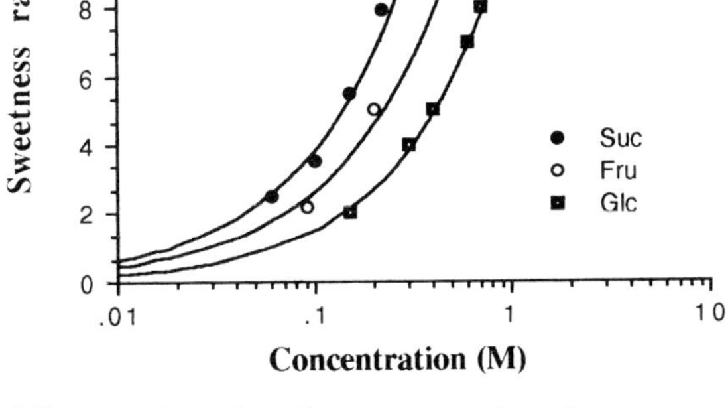

FIG. 10.25. Beidler equation plot of sweetness ratings for sucrose, fructose, and glucose (McBride, 1987).

taste, particularly the sweetness of the sugars, is chemical or physical (cf. Crosby *et al.*, 1979; DuBois *et al.*, 1991). This, on occasion, leads to the suspicion that the action of the sugars at the receptor may be physical (DuBois *et al.*, 1993), but there are significant problems with that suspicion also. Sugars with the same osmotic properties (glucose and fructose) have different sweetness intensity, and the anomeric D-mannoses have different tastes.

Taste receptors are presumed to respond to weak multiple elementary interactions such as hydrogen bonds, ionic pair interactions, and hydrophobic interactions. Each is believed to contribute -1 to -2 kcal mol^{-1} to the free energy change. To generate a typical affinity for a dissociation constant of 1 nM to 1 mM, Lancet & Ben-Arie (1991) state that it is necessary to have 6–9 elementary interactions. They pointed out that a K_d of 100 mM is equivalent to only *one* elementary interaction, which is difficult to reconcile with the known stereospecificity of sweet taste, let alone with subsequent transduction requirements. It was suggested that the problem can be partly resolved if sugars have a multiplicity of functional groups, and that the receptor is large and complex, but the stereospecificity requirements yet remain difficult to reconcile. The problem also seems to begin to resolve if it is granted that K_a and K_d are not affinity constants, but activity (recognition) constants, instead (Sections 2.4.2.1 and 2.12).

Simon (1991) pointed out that some typical mM values for K_d are 0·1 for yeast hexokinase, and 0·0002 for galactose binding enzyme. The K_d

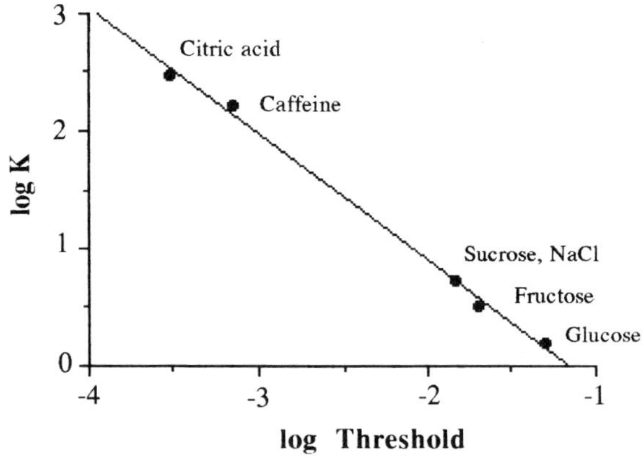

FIG. 10.26. Plot of threshold values and activity coefficients of taste for sweet, sour, salt, and bitter substances (McBride, 1987).

for sweet taste in the human is 200–300. Accordingly, the K_d calculated for sucrose from the data of McBride in Section 10.7.2 is 188. It is also pointed out that the K_d of 0·2 mM for the galactose binding enzyme corresponds to a free energy change of $-9{\cdot}2\,\text{kcal mol}^{-1}$. A single hydrogen bond between the sugar and the receptor would reduce the ΔG to about $-5\,\text{kcal mol}^{-1}$, and K_d would increase to about 0·2 mM. As only a minor change in the amino acid composition of the receptor (enzyme in this case) is needed to accomplish this, Simon concluded that the low affinity of the sugars cannot be used as an argument against receptors for sweet taste.

10.8 MECHANISM(S) OF TASTE INHIBITION

Preliminary discussion of the mechanism of sweetness inhibition was presented in Section 1.1.6.1, and is continued here for application to the mechanism for the initial chemistry of sweetness recognition. Essentially, the action of sweetness inhibitors varies (Jakinovich & Sugarman, 1988). Gymnemic acid solutions, which first must be applied to the tongue, are believed to act noncompetitively (Bartoshuk *et al.*, 1969). The mechanism of inhibition for gymnemic acids and ziziphins is believed to be selective inactivation of the sweet taste transduction mechanism (Kennedy & Halpern, 1980). Kennedy & Halpern also proposed that there is nonselective general disruption of the taste cell membrane.

Ray & Birch (1981) also found that the inhibition of sucrose sweetness by gymnemic acid over a large range (0–50%) of sucrose concentrations is not characteristic of either competitive or noncompetitive inhibition. They suggested that the two mechanisms may operate simultaneously. However, a plot of the lower sucrose concentrations (to 20%) versus the reciprocal of taste intensity (Fig. 10.27) strongly suggests a competitive type of inhibition (cf. Section 2.10.3.1). Adams (1985) pointed out that the gymnemic acids, and sodium dodecyl sulfate (and probably α-naphthyl sulfate) are 'surface active' taste modifiers. The ziziphins also probably fit that category, and the observation possibly serves to explain the mechanism for their inhibitory action. In all probability, an inhibition mechanism that disrupts a recognition process, such as hydrogen bonding, would also yield a plot suggestive of competitive inhibition.

Among the known competitive inhibitors (Section 2.10.4.1) of sweetness, or those inhibitors whose effect is surmounted by increased concentrations of sweetener, and are effective when added to sucrose solutions, are methyl

4,6-dichloro-4,6-dideoxy-α-D-galactopyranoside (Jakinovich, 1983) and the phenoxyalkanoic acid substances (Lindley, 1991). Like the aforementioned substances, these too suppress the sweet tastes of all sweet substances, while the effect on salt, sour, and bitter is minimal. A plot of the action of lactisol, or Na-(+)-2-(4-methoxyphenoxy)propionate (PMP) on sucrose at 8% and 10% concentrations, and on aspartame at a concentration (0·07%) equal to the sweetness of 12% sucrose, is shown in Fig. 10.28. It was pointed out that a key observation to link sweetness induction with inhibition, from a structural point of view, is that the enantiomer of PMP is quite inert. It was also pointed out that many inhibitors of sweetness are component parts of the structure of otherwise sweet compounds, but they lack an AH unit. PMP (Lactisol, Fig. 1.6) itself is a component part of the dihydrochalcone structure (Fig. 8.9). The inhibitor 4,6-dichloro-4,6-dideoxy-α-D-galactopyranoside is a component part of a sweet 'galactosucrose' structure. Cyclohexyl acetic acid (Fig. 1.6) is a component part of the cyclohexylsulfamic acid structure (Fig. 8.3), and indole acetic acid (Fig. 1.6) is a component part of the D-tryptophan structure (See Fig. 7.1c for L-). Free PMP tastes slightly bitter, and the monopole/γ interaction with the receptor suffices to explain that attribute. Finally, the ability of all inhibitors to inhibit the sweetness of all sweeteners is strong evidence that the chemical mode of interaction for sweetness is the same, but the mechanisms of interaction of substances that cause other tastes are different because those are relatively unaffected by sweetness inhibitors.

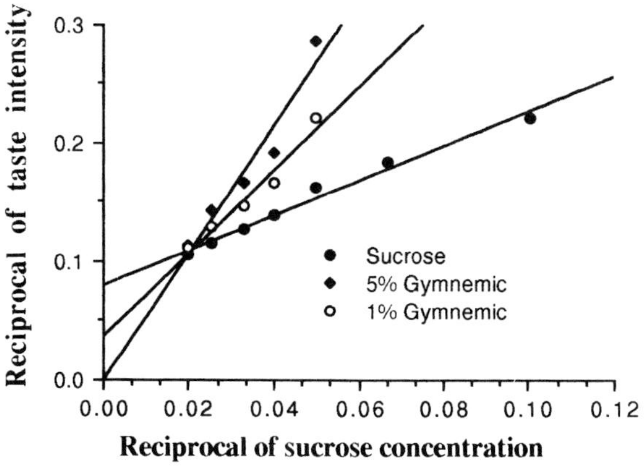

FIG. 10.27. Inhibition of sweetness with gymnemic acids. Data of Ray & Birch (1981).

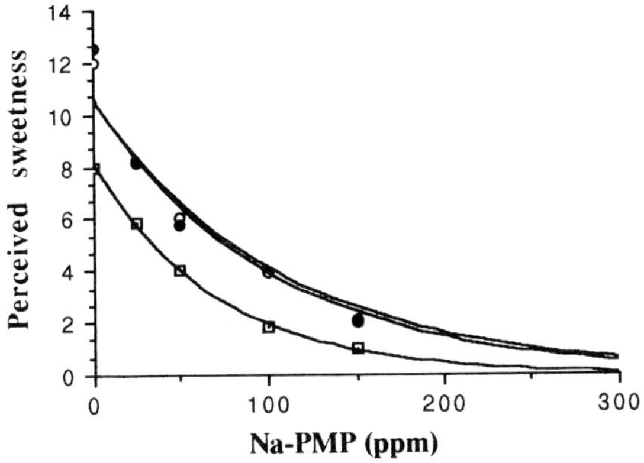

FIG. 10.28. Inhibition of sucrose and aspartame sweetness at different concentrations of (+)-2-(4-methoxyphenoxy)propionate. Solid circles, 0·07% aspartame; open circles, 12% sucrose; open squares, 8% sucrose (Lindley, 1991).

It has also been reported (Schiffman *et al.*, 1983) that the diuretic amiloride suppresses the salt taste response to NaCl and LiCl, but not to KCl and NH_4Cl. This finding has been confirmed by many others (cf. Hellekant *et al.*, 1988). In this case, the action is believed to be on ion channels in the transduction cascade. While the structure of the compound (Fig. 1.6) indicates that it might have taste, it also has other biological activity (it is a potassium-sparing diuretic and generally blocks nerve responses). It is also reported to suppress the sweetness of substances, but that contention has been challenged (Hellekant *et al.*, 1988; Lindley, 1991), and the inability to block the taste of KCl and NH_4Cl might very well be due to an inability to block the 'bitter' taste that those compounds possess. Hellekant *et al.* (1988) proposed two mechanisms for the action of amiloride. The first is blocking of apical taste cell sodium channels, and the second is an effect on perimembrane sodium balance due to decreased entry by sodium ions into the taste cells.

The fact that the initial chemistry of taste is never inhibited by mechanisms that can be strictly classified as noncompetitive and uncompetitive (Sections 2.10.4.2 and 2.10.4.3) casts doubt on analogies drawn between taste chemistry and enzyme-substrate or even drug chemistry. It would also cast doubt on analogies that invoke lock and key topological criteria to account for taste chemistry phenomena, a major topic in the following chapter.

10.9 RECOGNITION CHEMISTRY VERSUS BINDING AFFINITY

The significance of the role of the dynamic receptor for sweetness, and the significance of the exponential product–activity interaction to sweet taste responses cannot be overemphasized (Section 10.3.2). The distinction between recognition and binding chemistry in taste phenomena began to show at that point. Subsequent sections served to fully augment the notion, and the suspected exponential relation between activity and response was verified (Section 10.6.3.5), albeit by employing a binding precept. A recognition reaction that occurs in solution, as opposed to binding at a receptor site, would also serve to increase exponentially, through an orientation process, the effective concentration of a sweetener. There would then follow a sequence of factors, *each* of which leads to an exponential increase in response. Among them are the inductive effects of a hydrophobic glycophore component (Sections 10.3.4 and 10.3.4.1), and the enhancement of that inductive effect by appending additional electron-withdrawing groups to the molecule (Section 10.3.4.2). As a concentration-response curve is sigmoidal (Marin *et al.*, 1991), an increase in one or more glycophore components, or the total glycophore equivalency, also serves to exponentially increase the response (Section 10.3.3). The net result is that, even for sweeteners with a common glycophore, e.g. an L-aspartyl group, sweetness potency might be expected to vary by as much as five log units (Section 9.3.9.4).

CHAPTER 11

Symmetry, Chirality and Topology in Taste

Three intuitive principles have application to taste chemistry. They are symmetry, chirality, and topology. Symmetry, and in particular point-group theory and chirality, have discrete definitions, although conventions need to be invoked (Sections 2.11.1 and 2.11.3). Topology has only general definition with no applicable conventions other than those that are derived from symmetry and chirality considerations.

As a tool, point group symmetry theory does not include the concept of bilateral symmetry. The reason is because two- and three-dimensional point-group theory is formulated for application to crystal forms of compounds, and/or for application to the three-dimensional molecular structure and properties of compounds. That bilateral symmetry might have application to biological structure-activity relations was not among the objectives of point-group theory development. Nevertheless, the notion of bilateral symmetry has significant application to taste recognition chemistry. This was realized when the case for the probable mechanisms for the induction chemistry of the sweet, salt, sour, and bitter tastes was examined (Chapter 10).

In a similar vein, development of the intuitive notion of chirality is usually directed toward three-dimensional basic structures with four ligands. Consequently, the intuitive application of chirality in taste chemistry is the usual three-dimensional form. On the other hand, the role of chirality in lesser geometric dimensions has long been known to have application to the comprehension of biological chemical phenomena, for example prochirality (Section 2.11.3.6). It is usually associated with an apparent chiral attribute of otherwise symmetrical substrates in enzyme recognition reactions. Prochiral considerations, and the consequences of those notions, have application to taste chemistry and are also further developed in this chapter.

Lastly, the notion of topology is usually and intuitively applied to chemistry as a variation on the theme of three-dimensional topological symmetries, asymmetry, and/or dissymmetry (chirality). The principal manifestation of topology is in enzyme, immunological, and pharmacological chemistry. One general form is the lock and key theory of Emil Fischer. It was proposed to account for stereoisomerism phenomena that are configurational (diastereoisomeric) and enantiomeric. Another topological analogy, that encompasses three-dimensional chirality, is hand and glove congruence. Both stochastic models are widely used. The forms of topology that seem to have application to taste chemistry are also presented in this chapter.

11.1 SYMMETRY IN TASTE

A preliminary case was made in Section 9.3.8 for the role of symmetry/dissymmetry in determining if a potential AH,B glycophore in a molecule was functional or not. In this chapter, the primary emphasis on the role of symmetry in taste is directed toward the nature of the taste modality (sweet, salt, sour, bitter). Indeed, when possible chemical mechanisms for the induction chemistry of sweet, salty, bitter and sour tastes were examined, evidence accrued to indicate that principles of symmetry and dissymmetry played a role in distinguishing among them. The sensory response described as sourness was associated with an excess of hydrogen ions (the acidophore). The true salt taste of sodium chloride was associated with a metal characterized by the difficulty with which it yields an electron, and a nonmetal characterized by the ease with which it accepts an electron. That observation suggested that a variation on the notion of bilateral symmetry (i.e., proper combination of antipodal opposites) might be characteristic of the saltiness attribute.

Sweetness was viewed as an antiparallel hydrogen bonding phenomenon, and a case for a bilaterally symmetrical interaction arose with the case for the sweetness of dilute solutions of sodium chloride. Furthermore, substances that belong to the C_{2v} symmetry point group have a dipole moment in common, and they also have characteristic taste, including sweetness (Section 2.11.1.6). Lastly, ethylene glycol, the general model for the nature of the glycophore, belongs to the C_{2v} symmetry group. That observation suggested that chemical bilateral symmetry may be a general characteristic of sweetness induction. In this case, however, the bilateral symmetry element is a dipole–dipole interaction, wherein the charge on the dipoles are about equal, but of opposite sign ($\delta^+ = \delta^-$). The observation (Section 9.3.8.1) that cyclitols that are sweet have a dissymmetric distribution of axial/equatorial ring hydroxyl substituents was particularly intriguing, and applicable to the latter notion. Finally, it seemed apparent that bitterness was in the main, associated with elements of bilaterally dissymmetric bipolar or dissymmetric dipolar properties. The following symmetry sections examine the symmetry attributes of substances, and the possible relation to differences among taste modalities.

11.1.1 The Sour and Alkaline Tastes

The sour and alkaline tastes are dimensionless phenomena in the sense that one is elicited only by hydrogen ions, and the other only by hydroxyl ions. With respect to charge, a solution of anions and cations has an element of bilateral symmetry. However, with respect to numbers, they do not. Thus, when pH is neutral, the number of oppositely charged ions is the same (bilaterally symmetric), and the solution is devoid of sourness. The sour and alkaline tastes can therefore be said to be initiated by dissymmetric concentrations of hydrogen and hydroxyl ions, respectively. In this sense, dissymmetry and chirality are not equated on a 1 : 1 basis, as was proposed by O'Loane for molecular structures (Section 2.11.3). Instead, the degraded symmetry is simply degraded bilateral symmetry (bilateral dissymmetry). Intriguingly, this type of dissymmetry can be detected experimentally, e.g. by pH measurement, ion migration rate, etc., and it can now be proposed that it is also detected by taste.

The initial chemistry of sourness was attributed (Section 10.1) to either a strong dissymmetric cationic interaction with the receptor, or to a displacement on hydrogen, where the receptor serves as the Lewis base. The alkaline taste is also either a dissymmetric anionic interaction, or displacement on hydrogen, but in this case the tastant provides the Lewis base, and the receptor provides the Lewis acid. For those cases where a compound (e.g., lithium sulfate) is described as tasting sour when there is no immediately recognizable proton to account for it, symmetry may again play a role. One possible answer is that there is a disproportionately higher degree of solvation of the anion by an OH^- group of water than there is for solvation of the cation by H^+. Consequently, a sour sensation results.

11.1.2 Saltiness/Bitterness

In addition to the antipodal difficulty/ease with which sodium and chlorine respectively donate and accept an electron (Sections 2.7.2, 3.2.1, 3.2.2, and 10.2), the two ions have other symmetry features. Among them are the equal coordination number, the ability to enter into a purely ionic bond (Section 2.7.2), and the neutral pH of NaCl solutions. Moreover, at high salt concentrations, the bipolar identities of the halophore are truly transposed, but there is little apparent effect on the perception of the saltiness modality (Section 9.2). Such transposition would only be possible for antipodal ions with bipolar symmetry attributes.

There is even an ionic size-electrode potential symmetry for NaCl. The product of the ionic radius and electrode potential yields a symmetry coefficient ($\chi_{\text{saltiness}}$) close to one. The coefficient is determined from the expression

$$\chi_{\text{saltiness}} = \frac{r_{\text{m}} e_{\text{m}}}{r_{\text{x}} e_{\text{x}}}$$

where r_{m} is the ionic radius of the metal ion (Å), e_{m} is the electrode potential of the metal ion (Ev), r_{x} is the radius of the nonmetallic anion, and e_{x} is the electrode potential of the nonmetallic anion. Presumably, the nearer $\chi_{\text{saltiness}}$ is to unity, the more 'pure' the saltiness attribute. Symmetry coefficients for selected ions, calculated from ionic radii and electrode potential data (Chapter 2) are shown in Table 11.1. On the basis of the bipolar symmetry coefficients it would be expected that LiCl, LiBr and NaCl have predominately salt taste, which is the case (Kionka & Strätz, 1922). Those three compounds, along with KF have a symmetry coefficient average of $0{\cdot}975 \pm 0{\cdot}007$. It might also be expected that KCl, with $\chi_{\text{saltiness}} = 1{\cdot}58$, would have a taste something other than salty. That additional taste is usually described as bitter (Skramlik, 1921) or unpleasant. Lithium chloride might also be a candidate for near 'true'

TABLE 11.1
Bipolar symmetry coefficients for selected salt anions and cations

Anion	*Lithium*	*Sodium*	*Potassium*	*Caesium*
Fluorine	0·54	0·69	**1·02**	1·28
Chlorine	**0·84**	**1·07**	1·58	1·99
Bromine	**0·97**	1·23	1·83	2·29
Iodine	1·75	2·22	3·30	4·14

salt taste, and Hellekant *et al.* (1991) describe the saltiness of LiCl as 'clean.' However, the saltiness symmetry coefficient of LiCl is somewhat low. As LiCl salts have an alkaline pH, the mixed taste is probably an admixture of the salt and the alkaline tastes rather than the bitterness assigned to it by Dzendolet & Meiselman (1967). That other factors are involved is suggested by the fact that KF is a candidate for pure saltiness taste ($\chi = 1{\cdot}02$) when in fact KF is poisonous, and corrodes glass and porcelain. Its taste is not known.

Based upon their deviation from bipolar symmetry, it might be expected that caesium salts are invariably bitter, which is also the case (Moncrieff, 1967). The same is true for iodide salts. This indicates that there is evidence to support the tenet that true saltiness only occurs when the anion and cation properties have an element of bilateral symmetry. The symmetry is related to several parameters, and includes reactivity, ion size, and electrode potential. The symmetry expression for the relation is that they are bilaterally symmetrical, not only with respect to number, but with respect to other properties. From the discussion of symmetry associated with saltiness it is apparent that an ionic mismatch with respect to reactivity, size, numbers, and electrode potential leads to varying degrees of bitterness, and admixture with other tastes. In this case the ions are essentially bilaterally dissymmetric.

11.1.3 Sweetness/Bitterness

From the previous bipolar symmetry discussion, the sweetness quality might be expected to vary when the symmetry coefficient χ sweetness, or

$$\chi_{\text{Sweetness}} = \frac{\mu\delta^{+}}{\mu\delta^{-}}$$

deviates from unity. In this taste quality coefficient expression, μ is the magnitude of the monopolar attribute. Unfortunately, there is no practical way to measure μ so that the magnitudes of two antipodal dipole components might be compared. The symmetry coefficient, along with the hydrogen bonding parameter, is buried in property terms such as the σ and σ^{+} and μ^{2} constants (Section 10.6.2). However, it is manifested in other empirical ways. It can be presumed that in ethylene glycol, and in glycine, the sweetness bipolar symmetry coefficient is about 1·0. The AH,B of ethylene glycol is probably an equilibrium between two symmetrical dipolar forms. Glycine is a neutral amino acid with about equal charge on the mono-ionic components. The same is true for the

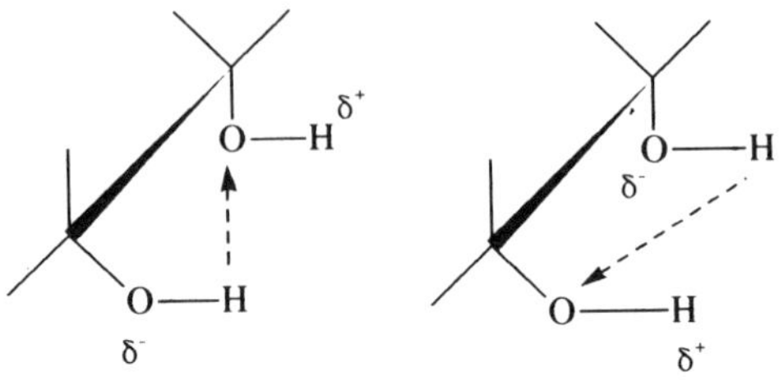
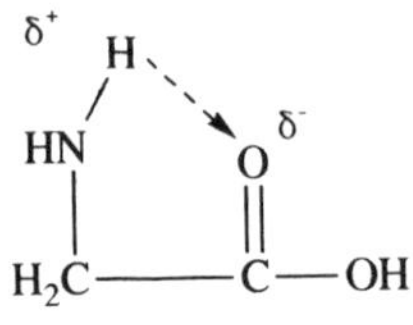

$\mu\delta^+ = \mu\delta^-$ $\mu\delta^+ = \mu\delta^-$ $\mu\delta^+ = \mu\delta^-$

Ethylene glycol Glycine

dipeptide aspartame. When the coefficient $\mu\delta^+/\mu\delta^-$ is greater than unity, or less than unity, other tastes, particularly bitterness, can be expected. Ethylene glycol and glycine can also be expected to be relatively free from additional and intrinsic sourness, saltiness, and bitterness.

The intrinsic bitterness of saccharine (Section 10.5) in symmetry terms might be assigned to two B functions. In one (SO_2), the symmetry coefficient for it and AH is about one, but for the other (C=O), it is not. When a sulfur atom is substituted for a carbonyl oxygen atom in the transformation of dulcin to ethoxyphenylthiourea, with transposition of sweetness to bitterness (Fig. 1.16), it would seem that the taste transformation is due to degradation of the glycophore's dipolar symmetry. When a sulfur atom is substituted for oxygen in a heterocyclic ring system, and the function of the oxygen atom is not as a glycophore, then the effect of the substitution is purely inductive, as in the thiosugars (e.g., 1,6-di-*S*-methyl-1,6-dithio-D-fructofuranose), isovanillyl compounds, and thiophenesaccharine.

For dilute solutions of sodium chloride, the energy of hydration for the two ions is about the same. The source of sweetness for this ionic substance is probably due to concerted hydrogen bonding between the fully hydrated ions and the appropriate receptor site. If so, it might be expected from the enthalpy and free energy of hydration (Na $\Delta H = -397$ and $\Delta S = -87{\cdot}4$; Cl $\Delta H = -377$ and $\Delta s = -96{\cdot}7$) that the dipolar interaction with a dipolar receptor might be about equal. The respective coefficients for ΔH and ΔS for sodium and chlorine is 0·95 and 1·1, and the sweetness dipolar symmetry coefficient $\mu\delta^+/\mu\delta^-$ is probably about 1·0.

An example of a negative induction effect on the activity of AH,B, obtained through substitution, is apparent with saccharine (Table 8.2).

When an NO_2 group is placed at the position on the benzene ring usually assigned to the third glycophoric site (Kier, 1972), saccharine is only slightly sweet. When the NO_2 group is positioned elsewhere on the benzene ring, the scalene distance parameters for the Kier triangle are altered. The predominate taste then becomes bitter. Apparently, the optimum position of γ in saccharine imparts symmetry to AH,B. With substitution, that symmetry is destroyed. The effects of such substitutions on the taste of saccharine, and the alterations of the Kier triangle are shown in Fig. 11.1. Introducing an inductive group (Cl) into the structure of sucrose, and other substances can be construed to have a positive inductive effect on the activity of AH,B. Whether a group with strong

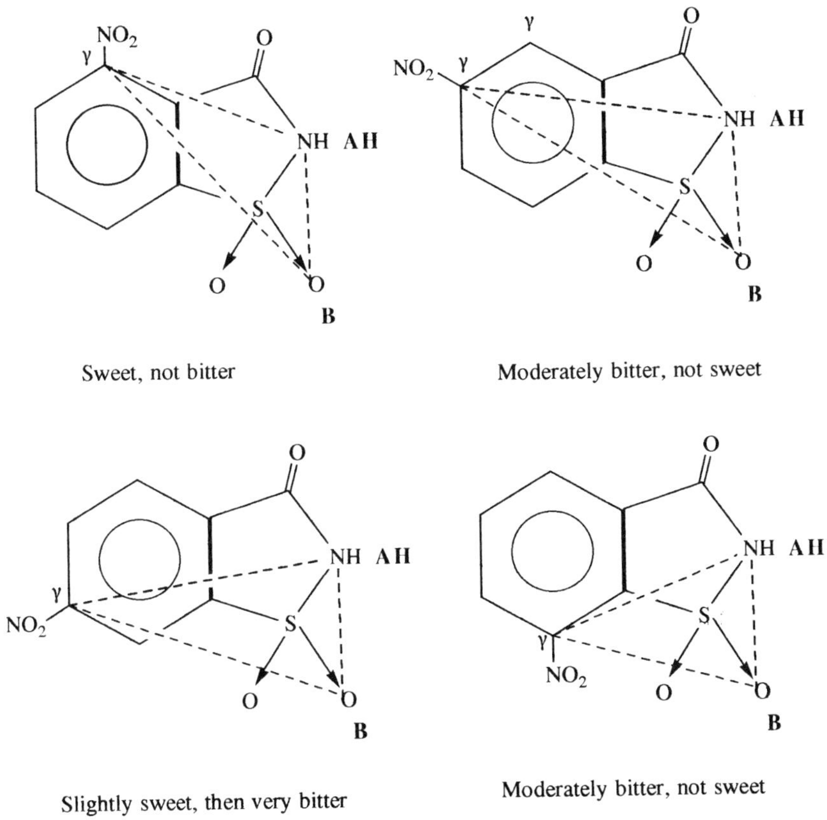

FIG. 11.1. Effect of NO_2 substitution on the position of γ and on the taste of saccharine.

inductive capacity has positive or negative effect on AH,B activity then varies with the nature of the glycophore (cf. Section 10.3.4.2).

The effect of substituents and side chains on a sweet compound would also seem to influence, through inductive effects, both the symmetry and the intensity of a concerted interaction. If so, the need for size and topological considerations to account for many observations on the nature of sweetness induction chemistry are minimized. The effects would also serve as the basis for the observed interrelations and transformations between the sweetness and bitterness of substances when structure is subtly altered. Symmetry considerations also obviate the need for separate sites for the two taste interactions, and even the need for different receptors for them.

A diagrammatic symmetry representation of the concerted, and initial interrelated inductive chemistry of sweet and bitter taste perception is shown below. There, interaction $a = b$, leads to sweetness perception. Concerted interactions where $a > b$, or $a < b$, leads to bitterness. In either of the latter two cases, a δ^{+} or δ^{-} monopole assumes precedence

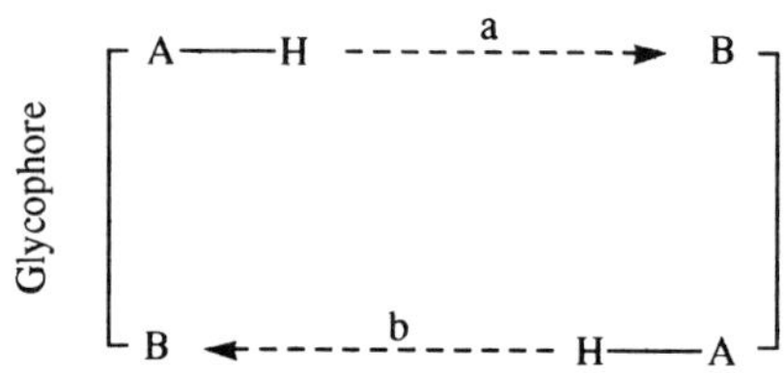

Interaction a = interaction b, sweet
Interaction a > interaction b, bitter
Interaction a < interaction b, bitter

(cf. Belitz *et al.*, 1981). Organic compounds elicit bitterness when a dipole unit is disproportionately charged, and the symmetry is again a case of dissymmetry. When these compounds have only a single monopole function, a condition for bitterness (Belitz *et al.*, 1981), the bitterness is truly due to dipolar asymmetry. Some examples are quinine and phenylthiocarbamide.

11.1.4 Symmetry Classification of Tastes

The application of symmetry to taste chemistry leads to a symmetry and mechanism classification of the primary tastes (Table 11.2), and serves to supplement the earlier ionic–nonionic (Chapter 1) classification of tastes.

The earlier classification merely listed compounds that generally ionize in solution (saltiness and sourness), and those that do not generally ionize (sweetness and bitterness). The salient feature of the symmetry classification is the implication that all taste recognition chemical phenomena are at the most, two-dimensional.

TABLE 11.2
Symmetry and inductive chemistry classification of tastes

Symmetric saporophores	*General chemical character*
Salty	A concerted ionic interaction, pleasant.
Sweet	A concerted hydrogen bonding (e/n) interaction, pleasant.
Dissymmetric saporophores	*General chemical character*
Sour	Dissymmetric displacement on H^+, or strong unbalanced ionic interaction, unpleasant.
Bitter	Any disproportionate interaction, e.g. ionic, H-bonding, etc. Unpleasant.

11.2 CHIRALITY IN TASTE

In 1886, Piutti presented a paper before the Paris Academy of Science entitled 'Asparagine.' In it he described the synthesis of the D-form of an amino acid. Surprisingly, in contrast to the tasteless L-form, D-asparagine tasted sweet. In a note appended to Piutti's paper, Pasteur pointed out that the receptor for sweet taste must, therefore, be dissymmetric (Section 2.11.2).

The significance of Pasteur's dissymmetric receptor site proposal has far-reaching conceptual consequences. It forms the very basis for understanding the chiral nature of taste induction chemistry, even though application of chiral principles initially lead to an enigma. For example, one interpretation of Pasteur's conclusion is that enantiomeric substances must, of necessity, have different tastes, but there is an inconsistency contained in that conclusion. Glycine, a symmetrical amino acid, and L-alanine and other L-amino acids (proline, hydroxyproline, threonine), taste sweet. Therefore, the specificity shown by the dissymmetric receptor toward D- and L-amino acids is not absolute. In addition, the L-sugars are now known to taste sweet.

11.2.1 The Bipartite Glycophore and Receptor

11.2.1.1 *D- and L-Amino Acids*

To adapt the AH,B tenet to the varying sweetness of amino acid enantiomers, Shallenberger *et al.* (1969) proposed that a spatial barrier, erected 3 Å removed, but perpendicular to receptor AH,B, can account for the D,L-amino acid tastes. The distance was chosen so that L-amino acids with a side chain larger than the ethyl group (L-alanine), would generally be tasteless. The steric barrier would then prevent the fixed amino acid AH,B glycophore from interacting with receptor AH,B, as shown in Fig. 11.2. The model also covered the case for the other sweet-tasting L-amino acids. Those substances that possess a hydroxyl group that can serve as an AH group, and, along with the carboxyl carbonyl oxygen atom, have the proper configuration for sweet taste.

The notion of a spatial barrier at the receptor has been employed by Castiglione-Morelli *et al.* (1990) to account for the ability of favored conformations of aspartame to fit over the general receptor, while the nonfavored conformations do not fit (Fig. 11.3). The barrier has also been employed by Douglas & Goodman (1991) to explain the sweetness of alitame and other isomers. However, discussion of that report is reserved for Section 11.2.2.4, where the full significance of the purported barrier is developed.

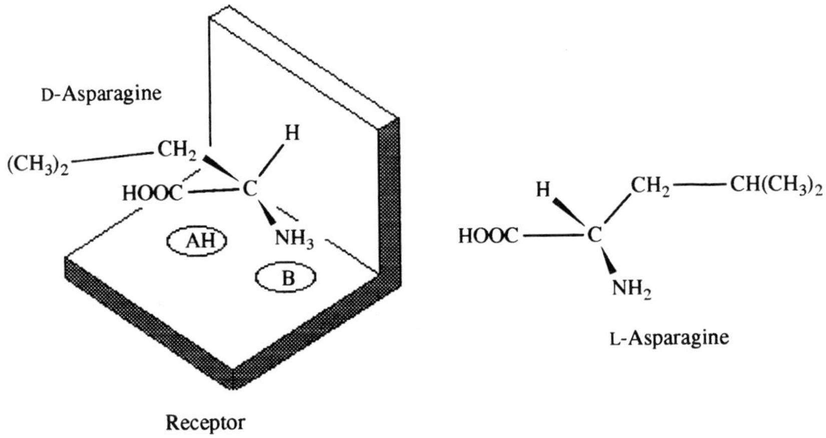

FIG. 11.2. The spatial barrier at the sweetness receptor site to account for the sweetness of the D-amino acids and tastelessness/bitterness of L-amino acids (Shallenberger *et al.*, 1969).

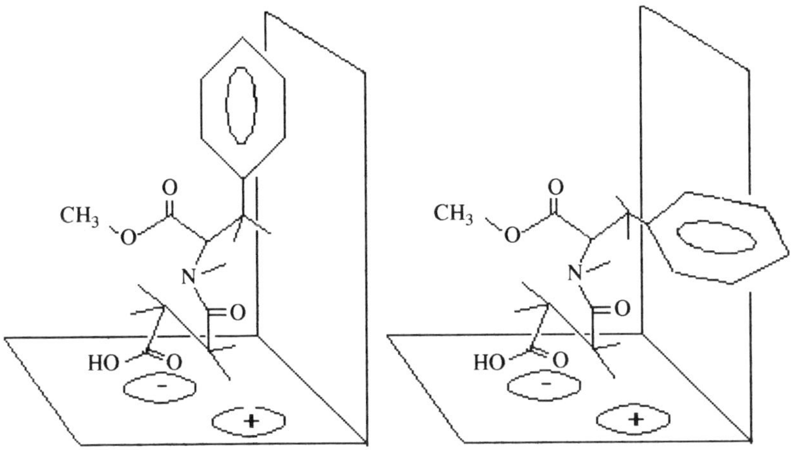

FIG. 11.3. The superposing of the glycophore for the proposed reactive conformation for aspartame over the receptor for sweet taste, and the inability, through interference with the spatial barrier, to fit a second 'favored' conformation onto the receptor (Castiglione-Morelli *et al.*, 1990).

11.2.1.2 *D- and L-Sugars*

At the time the AH,B tenet was proposed, it was widely believed that as the D-forms of the sugars were sweet, the L-forms must be bitter or tasteless. The belief not only logically follows from Pasteur's rationalization of the case for amino acids, but is also in accord with the report (Boyd & Matsubara, 1962) that the L-sugars, particularly L-glucose, do not taste sweet. For the original AH,B tenet, it was also thought that any non-hydrogen bonded α-glycol pair of sugar ring hydroxyl groups would be equally capable of serving as an AH,B sugar glycophore. For example, if AH,B for β-D-glucopyranose is arbitrarily assigned to OH-1 and OH-2 respectively, it can neatly be positioned over an AH,B receptor. Dismayingly, however, it was also observed that those very same hydroxyl groups for β-L-glucopyranose could be equally positioned over the purported AH,B receptor (Fig. 11.4). For the AH,B tenet to have substance, the L-sugars, and in particular, L-glucose, would need to taste sweet. Otherwise, it seemed that either the Pasteur taste dissymmetry principle was flawed, or the AH,B thesis was flawed. When a series of L-sugars was assembled (courtesy of N. Richtmeyer), it was found that a taste panel could not distinguish between the sweetness of D- and L-sugar pairs, including D- and L-glucose. Subsequently, it was learned that Wolfrom & Thompson (1946) had reported that L-fructose tastes 'very

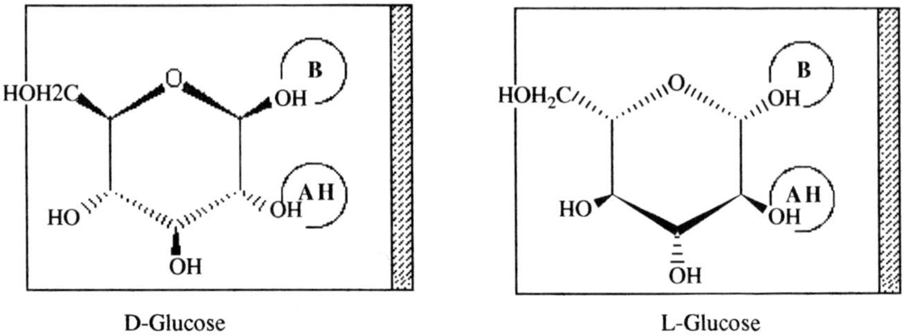

FIG. 11.4. Positioning of OH-1 and OH-2 of D- and L-glucose over receptor AH,B 3 Å removed from a hypothetical spatial barrier.

sweet.' That assessment, along with taste scores for the other L-sugars (Table 11.3) was then included in the report (Shallenberger *et al.*, 1969) that added the notion of a spatial barrier as a complement to the AH,B receptor site.

According to the proposal, the barrier not only explained why D- and L-amino acids have different tastes, but also required that the L-sugars possess sweetness. Eventually, it was discovered that every chemist who had first synthesized an L-sugar, beginning with Emile Fischer's (1890) preparation of L-glucose, had duly reported that they taste sweet. Others are L-glucose (Fischer, 1890), L-allose and L-altrose (Austin & Humoller, 1934), and L-fructose (Wolfrom & Thompson, 1946). More recently, it has been reported that L-sucrose (Szarek & Jones, 1980) tastes sweet. Ironically, it appeared that earlier chemists, along with contemporary chemists (Richtmeyer, 1968, pers. comm.; Wolfrom, 1968, pers. comm.),

TABLE 11.3
Sweetness scores of D- versus L-sugars (10% w/v) (Shallenberger *et al.*, 1969)

Sugar	D-*Enantiomer*	L-*Enantiomer*
Arabinose	5·2	5·6
Xylose	4·6	4·4
Glucose	5·4	6·0
Rhamnose	4·6	6·5
Mannose	4·9	5·0
Galactose	5·6	6·0
Glucoheptulose	5·2	6·6
Fructose	Very sweet	Very sweet

did not consider the sweetness of substances to be a particularly significant attribute.

Other reasons given for the sweetness of the L-sugars is that there is a chemical mechanism for sweetness unique to the sugars, or that sugars and polyols do not elicit sweetness by a chemical mechanism, but by a physical (osmotic or volume-response) process instead (Dubois *et al.*, 1993). An even more unusual interpretation of the sweetness of both D- and L-sugars is the suggestion (Simon, 1991) that receptors for the sugars do not exist. A final explanation is that, with a common receptor for the amino acids and sugars, the sweetness of the D- and L-sugars can be attributed to OH groups with transposed AH and B functions (James *et al.*, 1989). That thought had also occurred to Shallenberger & Acree, and while it was true that an OH group can act in either acceptor or donor hydrogen bonding capacity, it was rejected because the overall electronic character of the hydroxyl groups would also need to be transposed.

Using the premise that a chiral anomaly requires a chiral explanation, it will subsequently be shown that neither the Pasteur principle nor the AH,B thesis is in error, although the initial positioning of the spatial barrier at the receptor may have been premature. On the contrary, it is the conventional (three-dimensional) interpretation of the significance of the Pasteur chiral principle that is inadequate, and application of the three-dimensional implication leads to even greater enigmas. For example, one recourse to an explanation for the tastes of enantiomeric sugars and amino acids is to invoke a highly improbable chiral operation (transformation) at the receptor site. The improbable operation would be tantamount to everting a right-handed glove so that it would be congruent with a left hand. In other words, the chiral receptor would need to be everted for L-sugars, but not for L-amino acids. Conversely, the L-sugars might be totally everted, but that operation requires severance and reformation of all single covalent bonds.

11.2.2 The Tripartite Glycophore and Receptor

In two dimensions, the AH,B,γ tripartite glycophore proposed by Kier (1972) is doubly chiral (Section 2.11.3.4). Not only do the three points describe a triangle that may be labeled in either clockwise or counter-clockwise sequence, but the triangular arrangement is also skewed to the left (Fig. 9.17). The net result is that configurational diastereoisomers (isomers that are not enantiomers), conformational diastereoisomers, and an enantiomer, are possible. These, along with their chiral specification, are described below.

11.2.2.1 *Tripartite Glycophore Isomers*

The established tripartite glycophore of Kier, or (I) in Fig. 11.5, has a two-dimensional configurational diastereoisomer. It is (II), and the label sequence proceeds in the opposite direction. There is also a two-dimensional conformational diastereoisomer (III), where the configuration is the same, but the conformation is skewed in the opposite direction. Finally, there is a two-dimensional enantiomer (IV), where both configuration and conformation 'go the other way.'

As depicted by the planar drawings in Fig. 11.5, if (I) is the planar glycophore, then (II) must be the planar receptor for concerted antiparallel AH,B hydrogen bonding and tripartite hydrophobic interaction. Template I can be superposed on II for the required interactions by a slide (translation) operation. As a planar glycophore, II is incapable of concerted hydrogen bonding interaction with the receptor. Template (III) can interact with the receptor, but in bipartite AH,B fashion only, but (IV) cannot interact at all, as a slide operation fails to superpose either the groups required of concerted interaction, or the two hydrophobic groups. In stereochemical parlance, the glycophore and the receptor are two-dimensional configurational diastereoisomers.

As applied to sweetness chemistry, only the receptor structure (II in Fig. 11.5) can be construed to be truly one-sided, as it lies on the surface of, perhaps, a membrane or a globular protein. Because of the one-sidedness, the receptor site is a prochiral feature or component of a receptor protein, even though the protein itself is essentially chiral (Section 2.11.3.6). The glycophores must then be construed to describe a plane on a surface of a three-dimensional object that can be oriented over the prochiral receptor component of the protein in many ways. Put another way, the glycophores have a three-dimensional topological feature, and unless the third dimension is affixed to structures I–IV, application of their significance to the chiral nature of sweetness chemistry, as potential glycophores, is ambiguous.

Glycophore structures I–IV are rendered three-dimensional in Fig. 11.5 by adding (for simplicity) a single point below the tripartite planes of I–IV to form V–VIII. It might have been placed elsewhere, but the same arguments and conclusions would result. The fourth point, and the irregular tetrahedron that results, is assigned to the remainder of the molecules containing the tripartite AH,B,γ glycophore. The chiral nature of the tripartite glycophore and the ability/inability to interact with the one-sided prochiral receptor is now analogous to the interaction of prochiral compounds, such as ethanol, with the planar alcohol dehydrogenase receptor (Section 2.11.3.6).

The singular difference is that the substrate used to explain prochirality in Section 2.11.3.6 is three-dimensionally symmetrical, while in this exercise, a three-dimensionally chiral substance may, at times, be recognized as a 'symmetrical' object. It is just the opposite of prochirality, and herein defined as *prosymmetry*, as anticipated in

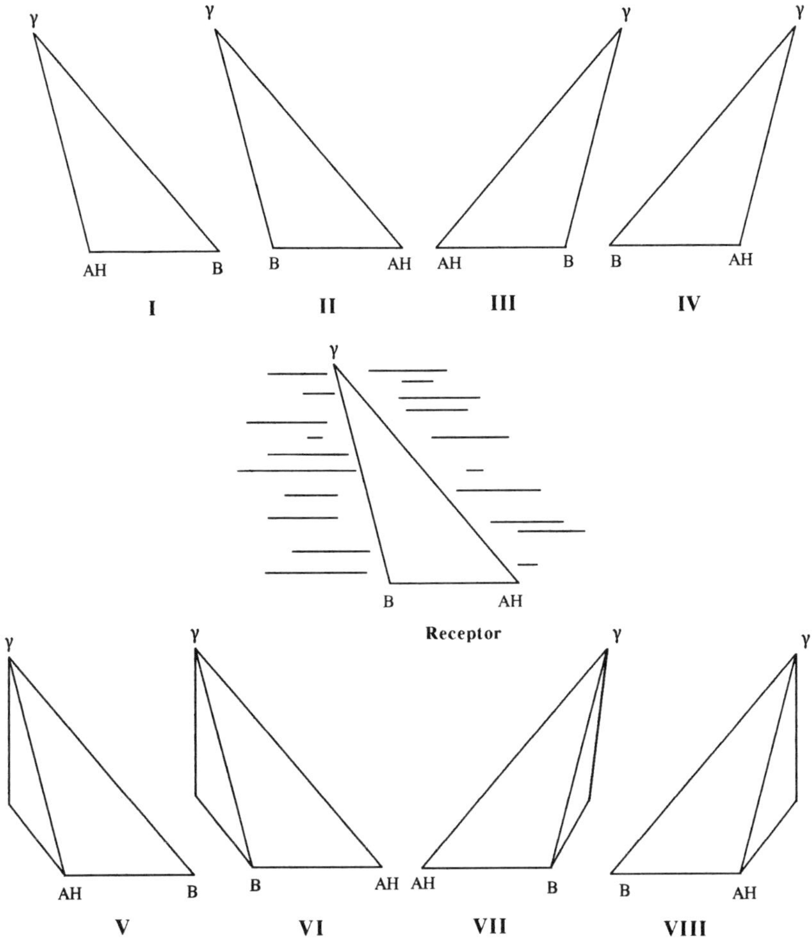

FIG. 11.5. Various possible chiral arrangements of the planar and three-dimensional tripartite AH,B,γ glycophore with different AH,B configuration for comparison with the probable chiral nature of the receptor.

Chapter 2. The substance containing the glycophore can then be chiral, but with a face commensurate with that of the receptor for appropriate tripartite antiparallel interaction with the prochiral receptor. If, in certain enantiomers, the face is not commensurate with that of the receptor, there will be no interaction. In certain cases the enantiomers of a chiral compound have prosymmetrical glycophoric faces, and although the three-dimensional enantiomers are not superposable, their prosymmetric glycophoric faces are identical, and equally capable of interaction with the receptor. Finally, even achiral (symmetrical) compounds may have a prochiral tripartite glycophoric face that is commensurate for interaction with the receptor. In any event, the receptor site is prochiral. The fact that the receptor site is probably contained on the surface of a chiral substance (protein), or on a membrane surface is moot.

Among the three-dimensional glycophores, structure V, or the three-dimensional counterpart of I, can not be superposed for tripartite interaction by a slide operation because of steric hindrance. Rotating V for superpositioning generates planar face IV, which also can not interact in tripartite antiparallel fashion with the receptor. Similarly, neither VI nor VII corresponds to the tripartite glycophore, after rotation, but VI can at least partake in concerted bipartite hydrogen bonding interaction. The only three-dimensional glycophore that can be positioned over the receptor in commensurate form for tripartite interaction is VIII, with the enantiomeric planar face of IV. A simple symmetry (C_2) operation brings VIII into antiparallel hydrogen bonding, and hydrophobic congruence with the receptor. The aforementioned and somewhat intricate interrelation between planar glycophore faces, the receptor face, and the general topology of sweet compounds is a continuation of the attempt outlined by Shallenberger (1983). That original article failed to make clear the distinction between results obtained from conceptual chiral transformations, and the actual case for chemical structures (Buitenhuis & Kanters, 1991). It also did not recognize the potential for prosymmetry. However, it now seems clear that resolution of the varied chiral phenomena associated with taste can be effected by employing a receptor that is multiply chiral, but the chirality is manifested in only two-dimensions. As the receptor resides on a surface, and is one-sided, the multiple chirality simply becomes multiple prochirality. Therefore, it is capable of interacting with a chiral substance as though it were symmetrical, i.e. the receptor recognizes the prosymmetrical surface attributes of an otherwise chiral substrate. By the same token it remains capable of distinguishing certain types of three-dimensional substrate chirality. Put another way, the receptor is a one-sided prochiral key, and the tastant is a three-dimensional object (lock). If the latter is chiral, it may nevertheless have a prosymmetrical one-sided face that is also congruent with that of its enantiomer.

The principal chiral, prochiral, topological, and prosymmetrical feature of sweetness chemistry is that the initial chemistry of sweetness obviously occurs in Edwin A. Abbot's (1884) 'Flatland.' Sweet substances are three-dimensional objects, and only a planar one-sided receptor is needed to establish the probable nature of subsequent chemical events. This is particularly true when planar AH,B,γ occurs either on a planar molecule, or on a surface (face) of a molecule that is not planar. For that purpose, a glycophore chiral-nature calculation-specification procedure, analogous to that applied to the sugars in Section 2.11.3.5, may be applied.

11.2.2.2 *Specification of the Chiral Nature of the Glycophore and Receptor*
As the receptor is one-sided (Fig. 11.5), it can only be viewed from above. From that view, the AH to B to γ sequence is clockwise, and the scalene AH,B,γ arrangement is skewed to the left. The configurational chirality is *rectus* (r) and the conformational chirality is *sinister* (s). The multiple chirality specification is r, s. If a (+) operand is assigned to the (r) configuration, and a (−) operand to the (s) conformation, the product is $(+)(-) = [-]$, or S. The only glycophore, contained on either a planar or topologically three-dimensional sweetener, that is capable of antiparallel and tripartite interaction with the receptor is I and VIII in Fig. 11.5, respectively. The multiple chirality of the glycophore is $(+)(+) = [+]$, or R. For a given glycophoric face of a molecule to interact with the one-sided receptor, it must have a combined configurational and conformational element of R chirality. Otherwise, sweetness is diminished (bipartite antiparallel interaction only), something other than sweet (e.g., bitter), or the compound is tasteless.

To summarize Section 11.2.2.2, in order for a concerted antiparallel and hydrophobic interaction between the glycophore and the receptor to occur, only an overall right-handed (R) glycophore can interact with a planar receptor. Furthermore, the receptor must have an overall left-handed (S) element of chirality. The case is contrary to the chiral nature requirements for the interaction of three-dimensionally chiral objects. As a result, the conclusion tends to offend sensibility. For example, a glove made for the right hand (R) cannot be fitted over the left hand (S). It can only be fitted over a homochiral hand. Similarly, only a right hand can shake a right hand, and only a right-handed shoe will fit a right-handed foot. However, in a prochiral or prosymmetrical system, only an overall three-dimensionally left-handed object can be made to superimpose with an overall three-dimensional right-handed object. The analogy here is that only the palm of the right hand can be superimposed on the palm of the left hand. Although the right and left hands are chiral, they are prosymmetrical with respect to the two-dimensional geometry of their palms.

The salient feature of sweetness chemistry chirality is that, not only must the planar geometry of the glycophore and the receptor be opposite, the labeling (functional property) sequence must also be opposite to comply with an antiparallel interaction prerequisite. Hence, the so-called chiral anomaly resolves by application of the required interaction between two-dimensional prochirality and three-dimensional chirality. In this case, however, and in contrast with the traditional thinking of prochirality, it is the receptor that is prochiral, while the tastant (substrate) may be either three-dimensionally chiral (prosymmetrical) or achiral (symmetrical). There seems to be little need to execute a chiral transformation (analogous to everting a glove) on the tastant to account for the chiral features of taste perception. There also seems to be little need for both chiral and achiral receptors. Finally, whereas the known chiral nature of sugar and amino acid glycophores is known to be absolute (Section 5.1.1.2), the proposed prochiral nature of the receptor is also absolute. The topology of taste recognition chemistry is therefore manifested in 'silhouette' chemistry, i.e., the interaction between planar glycophore and receptor faces.

11.2.2.3 *Application of Chiral Glycophore Calculations to Planar Compounds*

The structure of saccharine and glucose is essentially planar. They are two-sided molecules, and can be said to have a top and a bottom face (Fig. 11.6). In either case, the glycophore chirality is *R* or $(-) \times (-) = [+]$ or *R* for one side, and $(+) \times (+) = [+]$ or *R* for the other side. Either side of saccharine or glucose can interact with the receptor in Fig. 11.5 by a slide operation, and/or a rotation (C_2) operation. L-Glucose (Fig. 11.6) also has a tripartite glycophore, but in this case the $(+)(+) = [+]$ form lies on the top of the molecule, while the other lies on the bottom. The glycophore chiral calculation-specification procedure is in accord with the known sweetness of L-glucose.

To specify the chiral nature of a sugar glycophore, James *et al.* (1989) proposed that when AH,B,γ of fructose is viewed along the C-2 to C-3

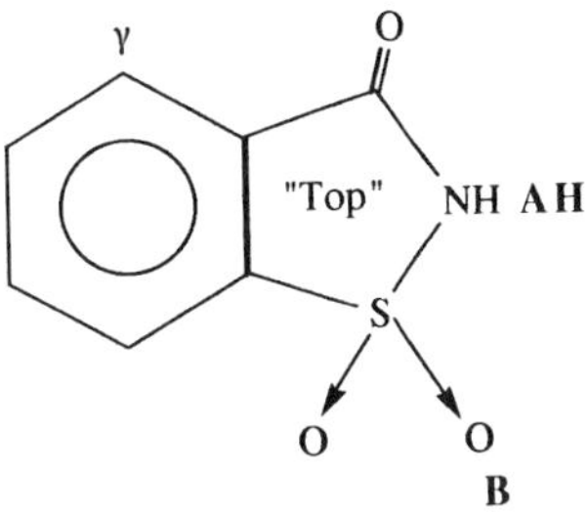

Saccharine, AH,B,γ [+]

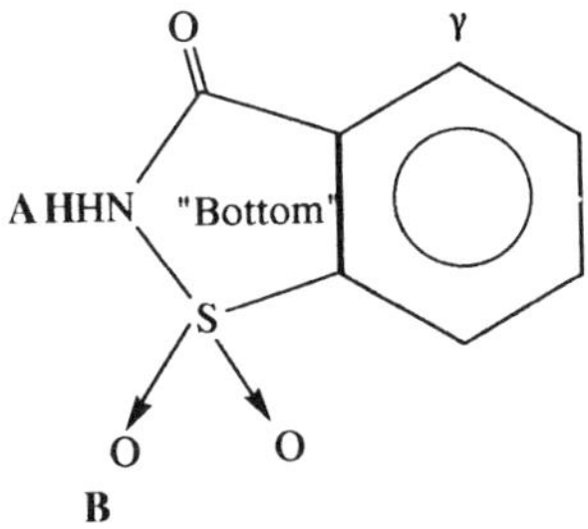

Saccharine, AH,B,γ [+]

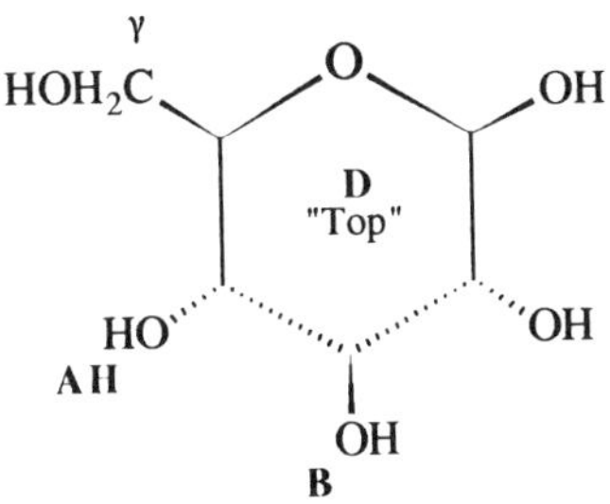

D-Glucose, AH,B,γ [+]

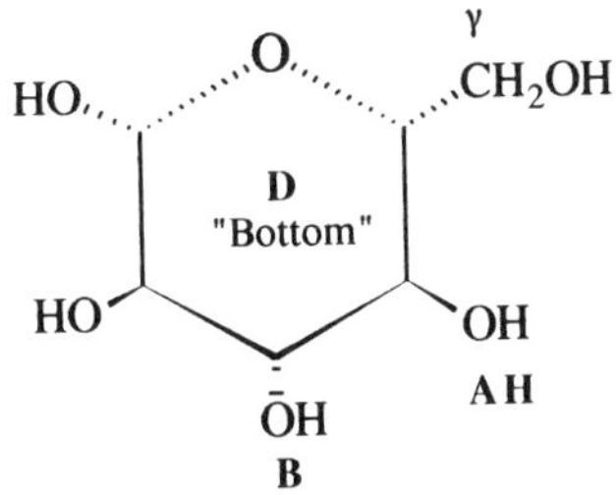

D-Glucose,AH,B,γ [+]

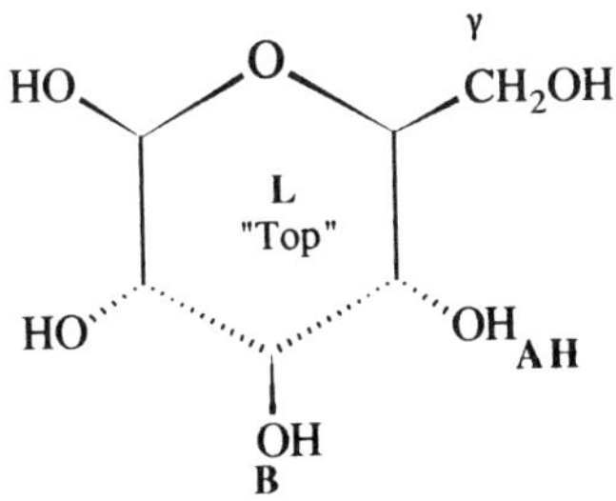

L-Glucose, AH,B,γ [+]

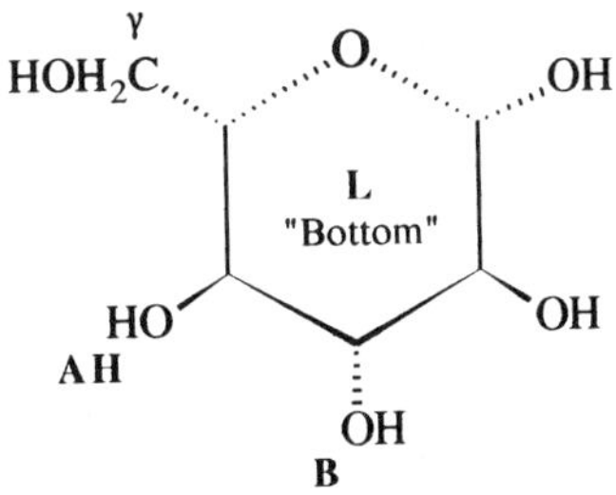

L-Glucose, AH,B,γ [+]

FIG. 11.6. Specification of the chiral nature of the saccharine and glucose tripartite AH,B,γ glycophores for either side of the molecule.

fructose bond, the sequence AH to B to γ proceeds clockwise for the D-form. It proceeds counterclockwise for the L-form. This thought is shown in the diagram below. However, it needs to be emphasized that those

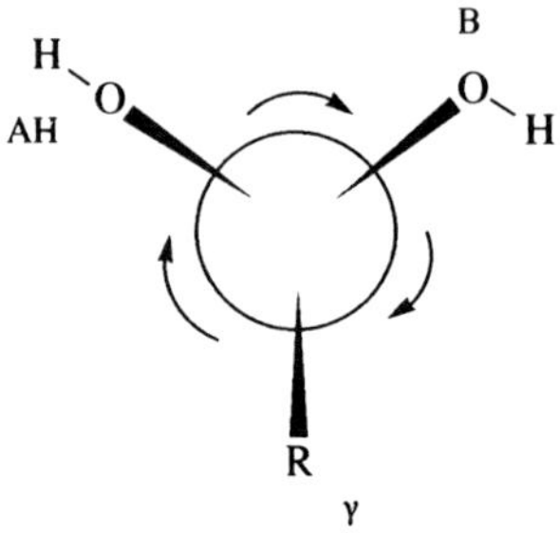

Priority sequence for specifying the chiral nature of a glycophore.

authors consider AH of fructose to be OH-1 and B to be OH-2. That the configurational sequence specification for the two different fructose glycophores is the same as above is due to the different viewpoint procedures. The procedure of James *et al.* (1989) requires that γ must always be oriented below AH,B. However, it is not always possible to determine the relative position of γ, or the direction of skewing. The latter is required for complete glycophore conformational specification.

11.2.2.4 *Application of Glycophore Calculations to Nonplanar Compounds*
Because of their general topology, many enantiomeric molecules do not have a planar AH,B,γ face that can be made congruent by a simple C_2 symmetry operation. The operation may introduce a barrier between AH,B and γ. Thus, the AH,B,γ functions do not describe an uninterrupted configurationally and conformationally identical plane for both enantiomers. Fructose is an example among the sugars, and asparagine is an example among the amino acids.

Fructose. For fructopyranose, the usual groups assigned to AH,B,γ (OH-2, O-1, and C-6) do not lie in the general plane of the molecule, as indicated by Shallenberger & Lindley (1977), and others. They are perpendicular to it instead. Conformational drawings of fructopyranose, unfortunately, give the illusion that the glycophores reside in the plane of the molecule. Unlike the glucoses, which has a planar and two-sided glycophore, fructopyranose is an example of a compound with only a one-sided glycophore.

Due to the uncertainty of the identity of AH,B,γ for fructose (Section 9.3.8.2), the chiral nature of potential fructose glycophores was examined using the chiral calculation/specification procedure. The results obtained are summarized in Fig. 11.7. The specification of AH,B (Shallenberger &

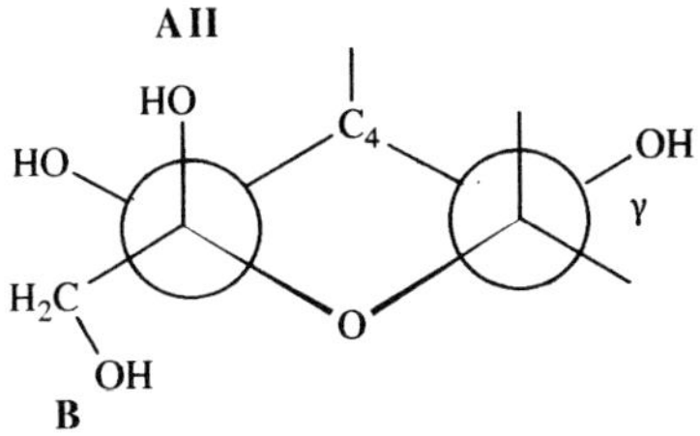

D-Fructopyranose, AH,B,γ = V

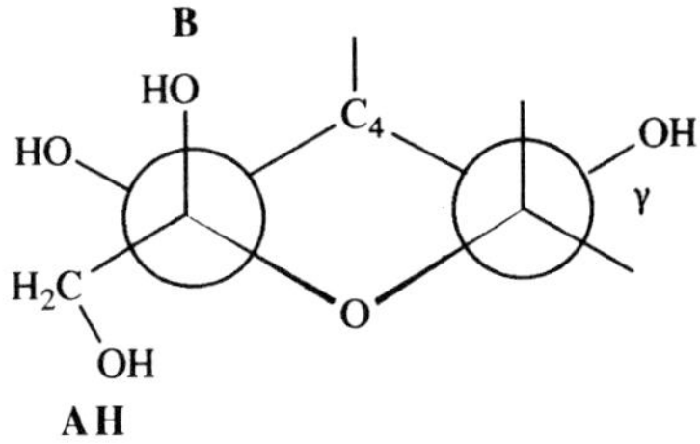

D-Fructopyranose, AH,B,γ = VIII

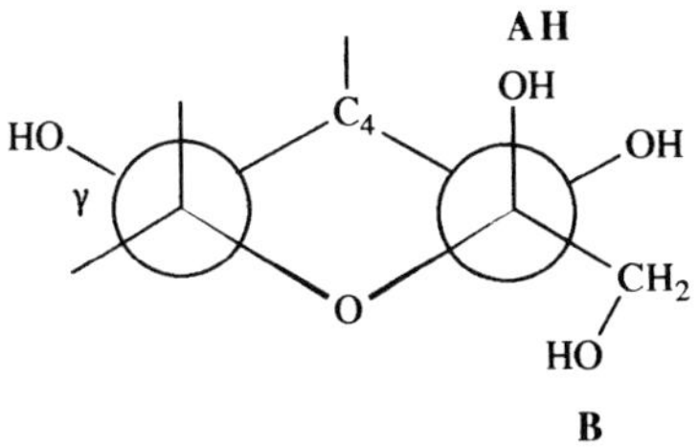

L-Fructopyranose, AH,B,γ = VIII

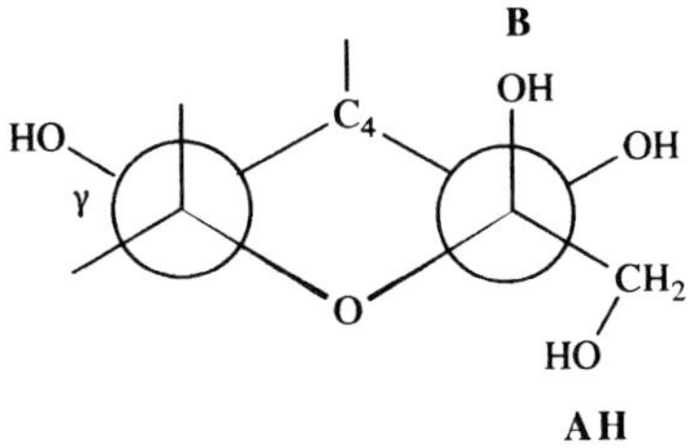

L-Fructopyranose, AH,B,γ = V

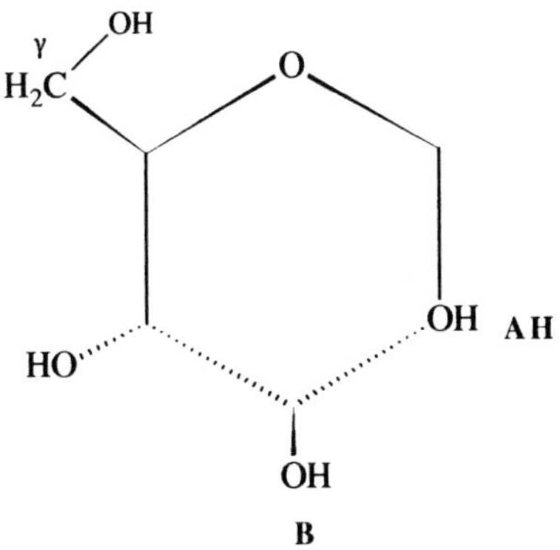

D-Fructopyranose, AH,B,γ = VIII

FIG. 11.7. Chiral nature of various tripartite glycophores proposed for fructopyranose.

Acree, 1967), and of AH,B,γ by Shallenberger & Lindley (1977) for D- and L-fructopyranose, where OH-2 is AH, O-1 is B, and C-6 is γ, yielded glycophore V in Fig. 11.5 for the D-form. A glycophore hybrid between VI and VIII is obtained for the L-form. The L-glycophore can be made to interact with the receptor in antiparallel tripartite fashion, but the D-glycophore cannot. The result is not consistent with their sweetness potency. Attention was then directed toward the identity of AH,B,γ proposed by Hough (1989) and Mathlouthi & Portman (1990), where OH-1 is AH, O-2 is B, and C-6 is γ. In this case, a hybrid between VI and VIII is found for the D-form. However, that glycophore for the L-form is unacceptable, as L-fructose would not then have any potential to elicit sweetness.

Only one case for an acceptable AH,B,γ glycophore, using the calculation method, is found for β-D-fructopyranose. It is on the bottom of the average plane of the fructopyranose molecule, where OH-5 is AH, O-4 is B, and C-1 is γ. That glycophore is unimpeded by axial hydroxyl groups, and corresponds to VIII in Fig. 11.5. The calculated assessment of the fructopyranose tripartite glycophore can be compared with the D-fructopyranose AH,B proposal of Birch & Shamil (1988). In the latter, OH-3 is AH and O-4 is B. However, the OH-5, O-4, C-1 glycophore is not allowed for the L-form, as it is VI for the enantiomer. To account for the sweetness of L-fructose then, a probable (hybrid) glycophore is that mentioned at the outset of this section, or OH-2 as AH, O-1 as B and C-6 as γ. As two different glycophores now need to be invoked to account for the sweetness of the antipodal fructoses, it is quite likely that they are not equally sweet. The form with the non-ideal glycophore (L-fructopyranose) would probably not be as sweet as the other. The proposal is in accord with the observations of others. Ogawa *et al.* (1987) reported that *pseudo*-L-fructopyranose is not quite as sweet as *pseudo*-D-fructopyranose.

Asparagine. With respect to the sweetness of the D-amino acids and tastelessness/bitterness of the L-forms, asparagine is used as the example. In this case, AH is assigned to the NH_2 group, B to the carbonyl group of –COOH, and γ to the methane side-chain atom. One significant difference between the amino acids and the sugars is that the amino acid AH and B unit is associated with a single chiral center. The four substituents project toward the apexes of the tetrahedral carbon atom with an angle of 109°. For sugars, AH and B are single entities contained on contiguous chiral centers. Inspection of the structure of D-asparagine shows that AH,B,γ, form a common face of the D-asparagine molecule. It corresponds configurationally and conformationally to template VIII in Fig. 11.5. That for L-asparagine corresponds to template V. In terms of the chiral calculation/specification procedure, D-asparagine has an *R*, or $(+)(+) = [+]$ glycophore, which is commensurate for interaction with the receptor by rotation. L-Asparagine also has an *R* glycophore or $(-)(-) = [+]$ glycophore that would be commensurate with the receptor through a slide operation, but the slide is prohibited by steric hindrance. As shown in Fig. 11.8, using the irregular tetrahedron models, model V cannot be made to interact in tripartite fashion with the receptor, but there is a possibility of a bipartite-only interaction. This feature of amino acid structure–sweetness relations led to the idea that the γ,γ interaction is necessary for activation of an amino acid's AH,B unit (Section 10.3.4.1).

With an L-asparagine bipartite AH,B interaction with the receptor completed, the γ function of L-asparagine can be positioned, through rotation, over the receptor γ site, but can not reach it. The result of rotation is shown in Fig. 11.9 using framework models of the two asparagines. In this case, the lack of taste for the L-enantiomer is commensurate with the plus-z chiral coordinate proposed for the tastelessness of certain dipeptide substances (Section 9.3.12.2).

For the L-amino acids that are reported to taste sweet, serine has a side chain not too different in size from that of alanine. This accounts for its weak sweetness. However, as both D- and L-α-aminobutyric acid taste

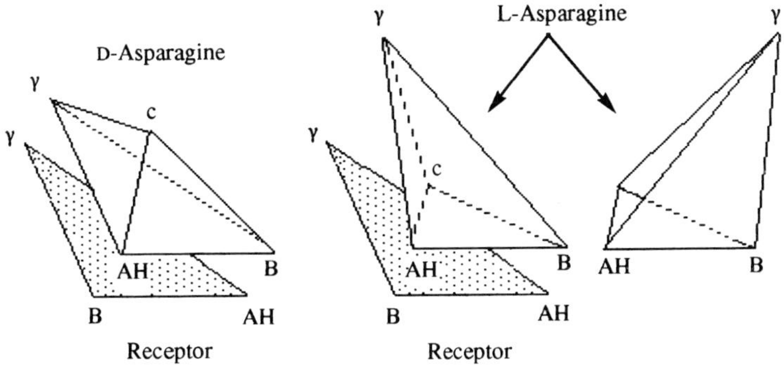

FIG. 11.8. Depiction of the topography of D- and L-asparagine as an irregular tetrahedron with apex c serving as the remainder of the molecule that makes up the chiral tripartite AH,B,γ glycophore.

sweet, rationalization of their taste on the basis of the size of the side chain casts doubt on that side-chain size 'principle.' D- and L-Proline (Fig. 11.10) and hydroxy proline, however, have planar template (I) for the D-isomer, and VIII for the L-isomer, and both should taste sweet. L-Threonine also presents little problem as the AH,B,γ glycophore is also template I.

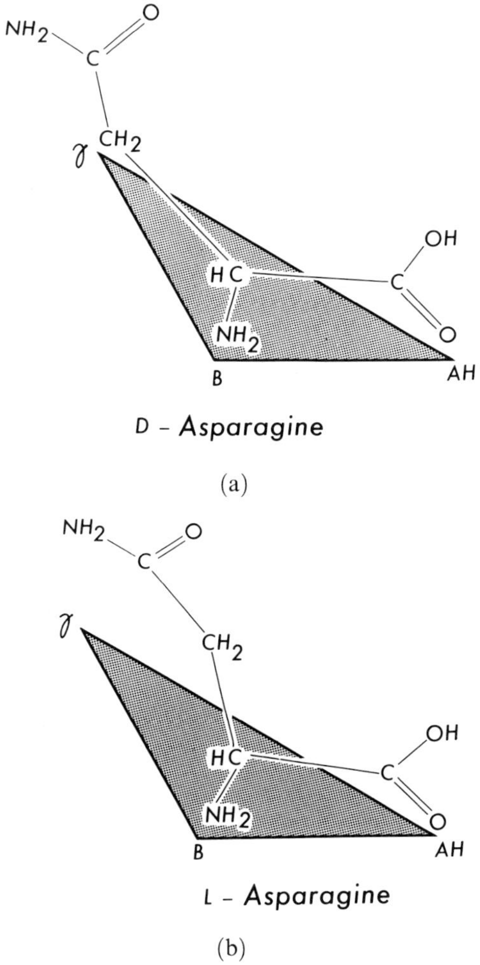

FIG. 11.9. Positioning of (a) D- and (b) L-asparagine over a common tripartite planar receptor for sweet taste.

D-Threonine has no sweet template, unless through rotation of the side chain methyl group, a weak VIII form is possible. Inui & Kaneko (1961) report that D-threonine is sweeter than the L-form, but Wieser *et al.* (1977) list the L-form to be slightly sweeter than the D-form.

It can now be stated that inclusion of Kier's third site, and evaluation of the significance of the chiral tripartite AH,B,γ arrangement for both the glycophore and the receptor, leads to better insight for the role of chirality and topology in taste. There is no need for a spatial barrier at the receptor. There is, however, a spatial barrier involved in the chemistry of sweetness. It is contained in the intrinsic three-dimensional structure of the potential tastant itself.

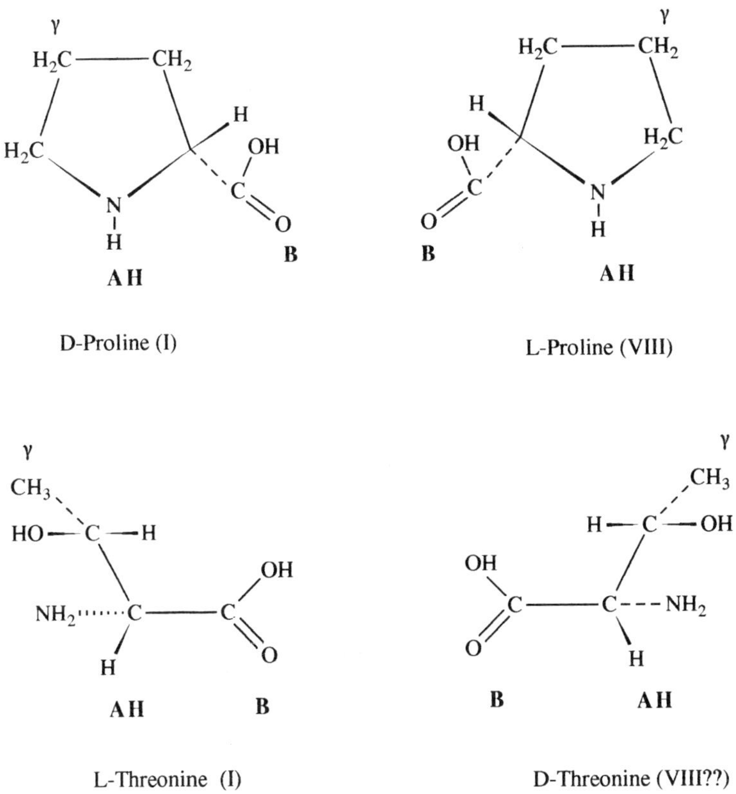

FIG. 11.10. AH,B,γ units in sweet-tasting L-amino acids.

11.2.2.5 *Application to the Sweetness/Bitterness of Amino Acids and Peptides*

It has been proposed that the γ unit of the amino acids serves to make an amino acid active by interaction with a hydrophobic unit of the receptor. The interaction serves to convert the zwitterionic form to a hydrogen-bonded AH,B unit (Section 10.3.4.1). The thought receives considerable support from the observation that it is the L-form of the tertiary butyl ester of alanine that is sweet (75×) while the D-form is bitter (Nofre *et al.*, 1974). In this case the more lipophilic, and presumably hydrophobic tertiary

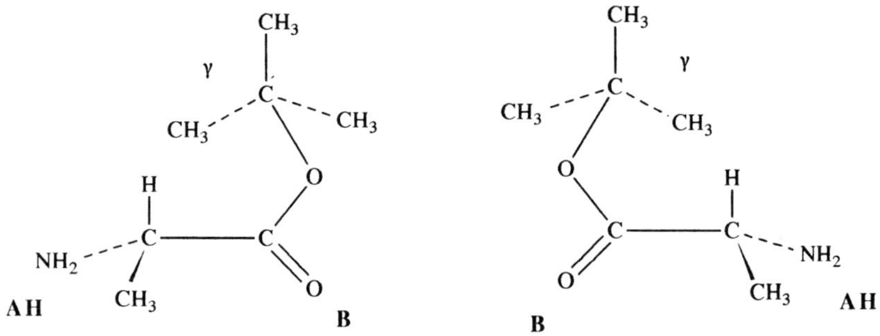

L-Alanine tertiary butyl ester (I) D-Alanine tertiary butyl ester (II)

butyl group, and not the methyl group side chain of the ester serves as γ. Accordingly, it is the L-enantiomer that has the proper AH,B,γ chirality (planar template I) for interaction with the receptor. The bitter D-form has planar template II. In the latter case, only AH,γ, or B,γ can interact with the receptor. This observation is also in accord with the proposal (Belitz *et al.*, 1981) for bitterness among the amino acids (Section 10.5).

The transposition of γ by substitution also seems applicable to the taste of aspartame derivatives, where it is the multiply equivalent identity of γ that leads to an appropriate AH,B,γ configuration, in spite of the L-aspartyl configuration (Section 9.3.11.9). Application of the observations of the Goodman school on alanine retro-inverso dipeptide amides (Section 7.4.1.2) led Goodman *et al.* (1987, 1988) to observe that the favored conformation for sweet tasting isomers had the 'L' conformation characteristic of aspartame. The bitter-tasting structures had the minus-z conformation. Examination of four stereoisomeric cyclopropane analogues of aspartame (Mapelli *et al.*, 1988) led to the conclusion that those isomers were not sweet. They are not sweet because the phenyl group is rigidly positioned to preclude the 'L' planar conformational relation to the zwitterionic (AH,B) groups.

Subsequently it was observed (Feinstein *et al.*, 1991; Douglas & Goodman, 1991) that sweet tasting alitame stereoisomers also had the 'L' conformation, while the minus-z conformations were bitter and the plus-z conformations were tasteless. Compounds with the reverse L-conformation were predicted to be tasteless. The L- and minus-z conformations of a 'glycine' dipeptide stereoisomer, *N*-(aspartyl)-*N*′-[(2,2,5,5-tetramethylcyclopentanyl)carbonyl]diaminomethane are shown in Fig. 11.11. For the L-conformation, the relation of AH,B, and γ with respect to the rest of the molecule is commensurate with template VIII in Fig. 11.2, and commensurate for reaction with the planar tripartite sweet receptor. The minus-z conformation, associated with bitter-tasting substances (Section 7.4.1.2) has the tripartite AH,B,γ group configuration similar to that for L-asparagine (V in Fig. 11.5), but γ 'goes the other way' (VII in Fig. 11.5).

11.2.2.6 *Chirality in the γ Function*

The effect of chirality in the hydrophobic unit of the tripartite glycophore seems to be that there is none. Among the aspartyl dipeptides the configuration of the L-aspartyl group is critical. However, the configuration of the second amino acid is important only in so far as it directs the relative position of the γ function. While only the L-,L-form of aspartame is sweet, L-aspartyl derivatives of D-amino acid esters are also sweet (Mazur *et al.*, 1970). In the L-aspartyl-α, α-dialkylglycine esters, Tsang *et al.* (1984) found that whether the terminal carbon atom is chiral or not

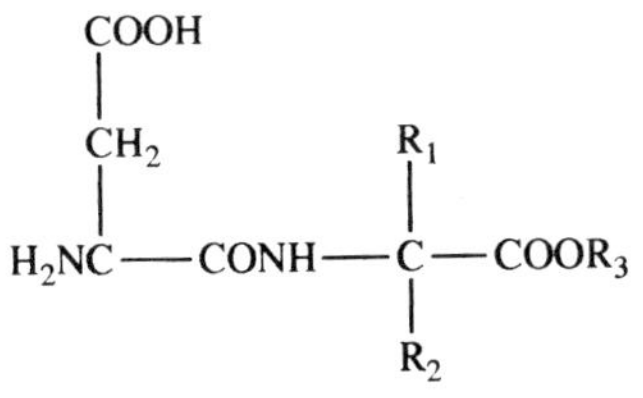

L-Aspartyl-α,α-dialkylglycine esters

is immaterial. When $R_3 = CH_2Ph$, and R_1 and R_2 are H and CH_3, or *vice versa*, the compounds are sweet. When both R_1 and R_2 are methyl groups, the compound is tasteless. The authors explored the possibility that the lack of chirality at that center was responsible for the lack of taste by preparing L-aspartyl-α-aminoisobutyric acid benzyl ester (both

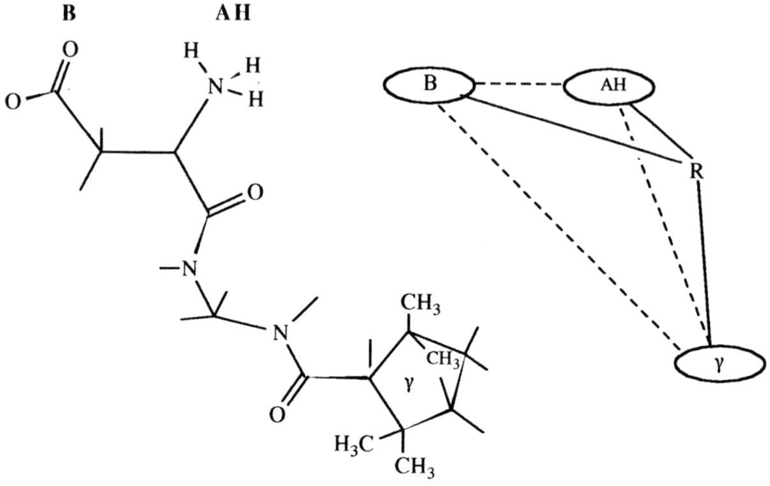

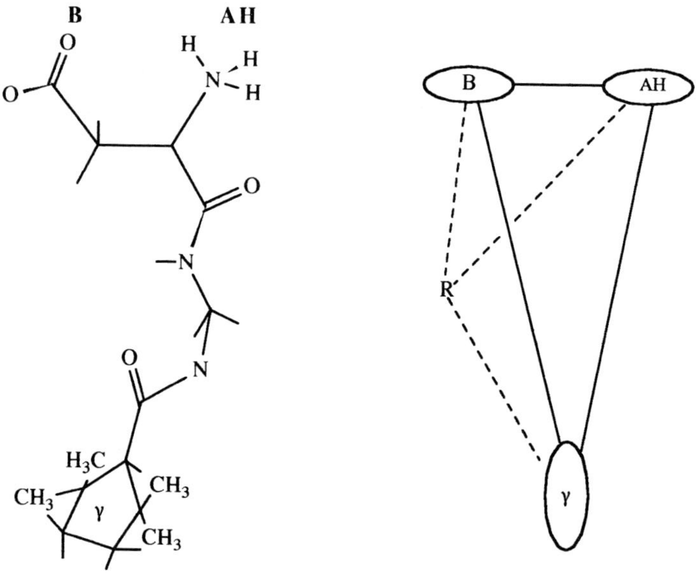

FIG. 11.11. The 'L' and minus-z conformations for *N*-(aspartyl)-*N'*-[(2,2,5,5-tetramethylcyclopentanyl)carbonyl]diaminomethane (the 'glycine' retro-inverso dipeptide amide stereoisomers) (Goodman *et al.*, 1988).

R_1 and R_2 are $-CH_2CH_3$ and $R_3 = CH_3$). That compound too did not taste sweet, and it can be concluded achirality at the C-terminal amino acid residue is not a factor in sweetness. The results obtained can then be attributed to a conformational effect. Presumably, that thought can be construed to mean that the conformational position of γ, in relation to AH,B, is the important factor. To test the idea, several nonsweet α, α dialkylglycine peptides were prepared (L-aspartyl-DL-α, α-methyl-*n*-propylglycine methyl ester, L-aspartyl-DL-α, α-methylphenylglycine methyl ester, and L-aspartyl-DL-α, α-methylbenzylglycine methyl ester). When both alkyl groups in L-aspartyl-α, α-dialkylglycine ester substances were ethyl groups, the compound was again tasteless. Thus, in addition to conformational effects, it was concluded that the overall size of the groups at the C-terminal amino acid residues were important. Extension of those side chains cannot apparently be accommodated by the taste receptor. Alternatively, it would seem that the size-shape parameter has inductive effect on AH,B. The same arguments apply to the retro-inverso compounds and their dipeptide analogues (Section 7.4.1.2).

The sweetness difference between 2S,3S neoastilbin and its 2R,3R isomer (Section 8.1.5.2) can also be attributed to chirality. It affects the position of γ in such a manner that, through lack of induction for AH,B (no hydrophobic interaction), AH,B is inactive. The argument applies to the 1′R and 1′S isomers of hernandulcin as well (Section 9.3.11.6).

The seemingly unimportance for the chirality of the glycophore hydrophobic group is not restricted to the peptides. Spillane & Sheahan (1989) observed that when R in the sulphamate sweeteners is adamantane, each of the resulting two possible enantiomers is described

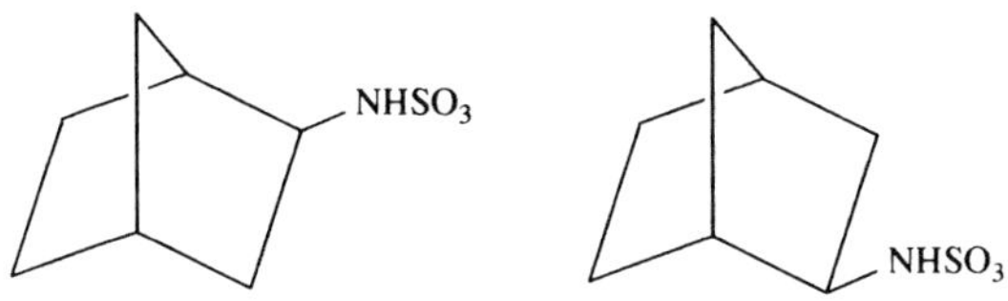

Chirality in the hydrophobic side-chain of sulfamate sweeteners

as 'very sweet.' Unlike the peptides and amino acids, where the AH,B units are at a center of chirality, the sulfamate AH,B grouping is achiral (but the planar AH,B,γ triad may show planar chirality). On the whole, results obtained that have application to the nature of the glycophore indicate that it is approximately planar, and the chiral attributes of side chains are essentially those that might be expected of a 'flatland' receptor.

11.3 CHIRALITY PLUS TOPOLOGY IN TASTE

Without question, the importance of chirality plus topology in taste has enigmatic features. Amino acids are differentiated by the chiral nature of substituents (AH and B) about a single chiral center, and there is an obvious difference in the taste of D- and L-amino acids. For sugars, AH and B reside on different chiral centers. (+)-Hernandulcin (Section 8.1.6) is a case where only AH is attached to a chiral center, and only the (+) enantiomer is sweet. The same is true of the sulfones (Section 8.1.8), but both enantiomers are sweet. The hexoses have five chiral centers and chiral ring forms, and are more complex in terms of chiral topological structure than the amino acids or the sesquiterpenes. Eversion operations are possible for the sugars. A right-handed ring form can be converted to a left-handed ring form by everting it (turning it inside-out). Yet, there is little or no difference in the taste of their enantiomeric forms that have the configuration at each chiral center transposed and also the conformation of the ring transposed. As stated above, but reemphasized here, the significance of the overall chirality of the molecule is also not important for peptides, as there is no significance for chirality in the peptide side chains. Only L,L aspartame, among the four possible configurational isomers, is sweet (Section 7.4.1.1). However, only the L,L dipeptide amide, among the four possible amides and retro-inverso analogues (Section 7.4.1.2), is bitter!

It is difficult to envisage a topologically-defined receptor with only partial elements of handedness, or one that can respond to chirality in one case, and indiscriminately not respond to it in another case. Finally, all substances that taste sweet have their sweetness depressed by a single compound, 2-(4-methoxyphenoxy)propionic acid (PMP), that is active in only one enantiomeric form. That observation (Lindley, 1991) indicates that different compounds elicit sweetness by the same general chemical mechanism. It also indicates that their two-dimensional AH,B,γ chirality is the common topological chemical parameter, and not their size and shape.

11.3.1 Mirror Image Models

An expedient, and obvious way to account for the different taste of enantiomers, is to propose that there are receptors for those tastes that are alike in all respects, except that one receptor is the mirror-image of the other. Thus, Lehmann (1978) stated that the stereoselectivity of the response can only be explained (*sic*) in terms of a three-point association. To account for the near-equal magnitude of the sweet and bitter tastes of the enantiomeric amino acids (Section 7.3), Lehmann envisaged a case whereby AH and B of a bipartite AH,B/barrier sweetness receptor are transposed (Fig. 11.12). The thought needed immediate alteration to account for the sweet taste of L-alanine. That amino acid's taste is explained by an additional isoteric sweetness receptor that is the mirror image of the initial sweetness receptor. To explain the sweetness of L-aspartic acid amides, the receptor pocket for sweetness shown in Fig. 11.12 for the *rectus* amino acid is lengthened. Whether a chiral site is required for bitterness seems to depend again on the definition of the dimensional kingdom for the chirality. Like sweetness, there is a need to invoke chirality in some cases, but not in others. Among dipeptides, L-D- and D-L-diastereoisomers, and the D-D- and L-L-enantiomers may all be bitter (cf. Belitz & Wieser, 1985).

To account for the sweet taste of L,L-aspartame, and the bitterness of the D,D-enantiomer, Castiglione-Morelli *et al.* (1990) also proposed sweet and bitter AH,B receptor sites. The functions are sterically transposed, but they have identical topological boundaries. The bitterness receptor is then the three-dimensional configurational *diastereoisomer* of the sweetness receptor, as shown in Fig. 11.13. In the figure, the contours for the receptor boundaries have been adjusted to conform with the most recent sweetness receptor proposed by Temussi and coworkers (Temussi, 1991). In an earlier study (Temussi *et al.*, 1984) the sweetness of *para*-tolylurea, and the bitterness of *meta*-tolylurea was explained with

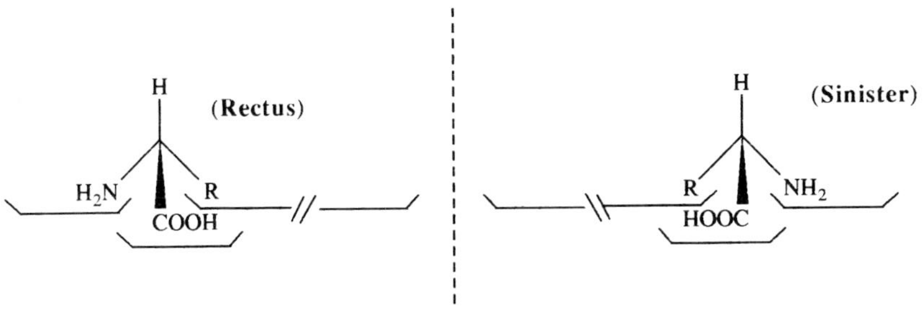

FIG. 11.12. Mirror-image receptors for sweet and bitter taste of amino acids (Lehmann, 1978).

sweetness and bitterness receptors with definite boundaries. However, in this case, the bitter receptor was not the configurational diastereoisomer of the sweet receptor, but was the conformational isomer instead.

11.3.2 Topological Receptor Models

It has long been known that dipeptide ester sweetness varies with the size and length of side chains (Brussel *et al.*, 1975). Accordingly, Ariyoshi (1976) proposed that the receptor for sweetness was a pocket. The proposal is based on the assumption that sweetness activity does not depend on the L- or D-configuration of the second amino acid in peptides, but on the size and shape of side-chain substituents. The pocket contained spatial barriers to the back (for amino acids), to the right (for sulfamates), and to the left of an AH,B site. The latter was employed to explain the lack of sweetness for a phyllodulcin derivative.

To explain the taste of dipeptides, the receptor was described (Ariyoshi *et al.*, 1991) as a pocket with a barrier at the site facing the C-terminus of sweet tasting peptides (cf. Iwamura, 1981). The model served to explain why peptide bitterness potency diminished sharply with elongation of the C-terminus, and terminates with the pentapeptides. In final form, the topological receptor for sweet and bitter peptide substances took on a tubular shape with numerous AH,B and hydrophobic interactive sites, as shown in Fig. 11.14. Earlier, Wieser *et al.* (1977) proposed that the binding site for amino acid bitterness/sweetness probably is a

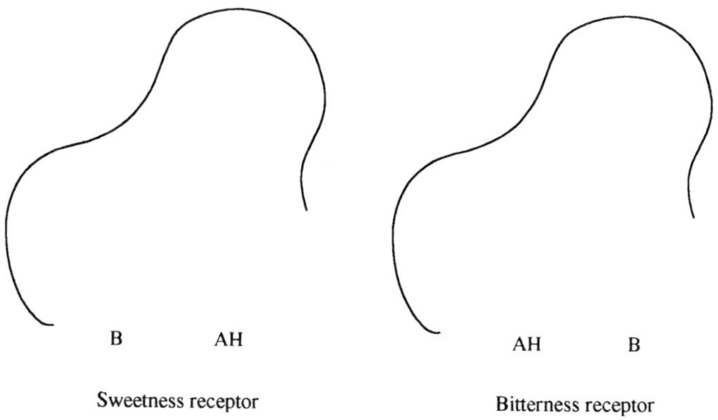

FIG. 11.13. The topologically identical but configurational diastereoisomeric sweetness and bitterness receptor sites proposed by Castiglione-Morelli *et al.* (1990) to account for the taste of aspartame isomers.

hydrophobic tube with a bipolar contact point at one end. Such topology would seem to be supported by application of the orderly queue hypothesis for tastant accession to the receptor (Section 10.4.2), but the notion seems to have been abandoned.

Because the magnitude of amino acid bitter/sweet attributes is about equal, and the bitterness of amino acids and other compounds is essentially a monopole/γ bipartite interaction, Belitz *et al.* (1981) proposed a composite receptor for those tastes. The reason for the equal magnitude of amino acid bitter/sweet attributes would then be more in concert with a monopole/γ interaction with the same monopole that acts in concert with an antipodal monopole to elicit sweetness. The common receptor for sweetness and bitterness contained two areas for hydrophobic interaction, as shown in Fig. 10.13. Liberty has been taken to change the Belitz *et al.* (1988) specification of the electrophilic (e) receptor site to AH and the nucleophilic (n) site to B, for facile comparison with the AH,B,γ tripartite receptor in Fig. 11.5. The side of the receptor in Fig. 10.13 that contains a common AH,B,γ unit is the backside of the Fig. 11.5 tripartite receptor. By the same token, there is a bitter tripartite AH,B,γ site with the same geometry as IV in Fig. 11.5. It spans the schematic receptor shown in Fig. 10.13. The net effect of doubling the number of hydrophobic sites is to place the second γ site above the common receptor AH,B,γ. This yields two planar tripartite sites, one for both sweetness and bitterness, and the other for bitterness only. The two sites are therefore two-dimensional conformational diastereoisomers.

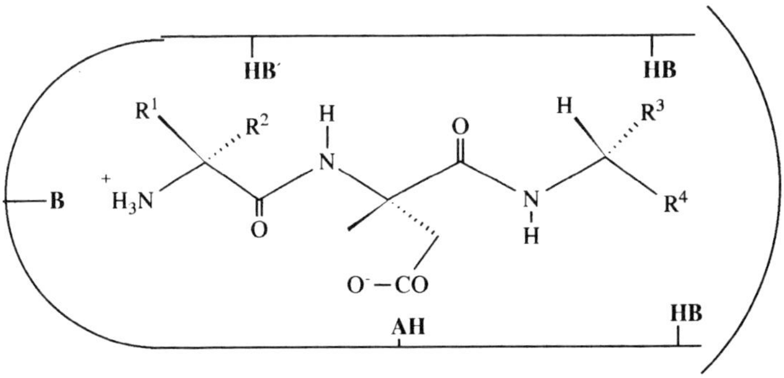

FIG. 11.14. Schematic drawing of the interaction between sweet peptides and the receptor where $R^1 = H$, R^2 and R^3 are small hydrophobic groups, but $R^3 \leq R^4$, AH is a proton donor, and B is a proton acceptor. HB is a hydrophobic binding site (Ariyoshi *et al.*, 1991).

Through application of the double hydrophobic receptor site, it might be expected that D-norleucine would taste sweet (see Fig. 10.12) because of interaction with an AH,B,γ-1 receptor. L-Norleucine would taste bitter by virtue of reaction with an AH,B,γ-2 receptor. Alternatively, if bitterness among the amino acids is due to a bipartite monopole/hydrophobic interaction with the receptor (Belitz *et al.*, 1981), then the enantiomeric amino acids are also capable of reacting at a common sweetness/bitterness receptor that is tripartite only (Fig. 11.5). In the case of L-asparagine, even a monopole/hydrophobic interaction can be forbidden, not by a receptor spatial barrier, but by the barrier imposed by the relation of AH to B to γ in the structure of L-asparagine itself. The latter assessment also applies to dipeptides where a reverse L-conformation and minus-z conformations were associated with bitterness, while a plus-z conformation was associated with tastelessness (Section 9.3.12.2).

The net effect of two hydrophobic sites is tantamount to extinguishing the significance of two-dimensional chirality by transposition to a higher dimension. In essence, the consequences of any apparent stereospecific chiral phenomenon can be rationalized by generating or subtracting spatial parameters (see Section 2.11.3). Whether it should be attempted or not seems to depend on the degree with which one desires to resolve chiral problems by application of chiral principles. The latter seems to the present author to be the most rewarding road to ultimate comprehension of the role of chirality in taste.

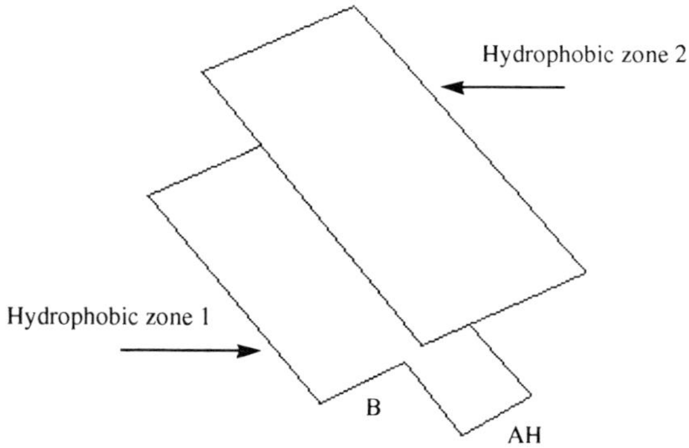

FIG. 11.15. Schematic sweet–bitter receptor with an AH,B system and two hydrophobic zones for sweet *or* bitter amino acid taste (Belitz *et al.*, 1987*a*). See also Fig. 10.13.

The initial chiral anomaly for sugar and amino acid sweetness might have been resolved by adding spatial parameters. For example, if D-amino acid sweetness is due to reaction with a three-dimensionally appropriate sweetness receptor, then the L-forms would not be expected to taste sweet. In the fourth dimensional spatial continuum, however, the three-dimensional chirality of the receptor is extinguished. Therefore, as both chiral families for the sugars taste sweet, the recognition chemistry for the sugar–receptor taste interaction would then seem to occur in some fourth-dimensional geometric continuum. This form of rationalization is capable of generating complex receptors *ad infinitum*, but offers little new information on what the stereochemical features of taste might ultimately be.

As demonstrated in the summary to Section 11.2.2.2, there are much simpler answers for such sweetness/bitterness observations, i.e. a prochiral receptor. In order for the receptor to be prochiral, it need only be a one-sided skewed planar arrangement of three groups with different functionality. The receptor topological requirements would then seem to be minimal, at best. Nevertheless, many studies have been directed toward sweetness receptor topology, and are presented below.

11.4 TOPOLOGY IN TASTE

The final case for the structural relations between a tastant and receptor for sweetness and bitterness emphasizes the significance of topology and/or receptor site numbers. The emphasis on topological features in taste is not without precedent. The relation between size and tastelessness is well documented in the earlier literature (cf. Moncrieff, 1967). The abruptness with which taste is altered or disappears upon ascending a homologous series of substances can be striking. This is especially true for narrow ranges of size for the side chains of homologous series of substances, such as peptides, sulfamates, ureas, etc. (cf. van den Heijden *et al.*, 1985*a*, *b*). The effect, however, could either be conformational (Iwamura, 1980; Ariyoshi, 1984) and/or inductive.

The application of strict topological considerations alone becomes enigmatic also. Homologous series of substances aside, sweetness and bitterness are caused by compounds that range in size, on an atomic/molecular level, from minute to preponderous. Sweetness is found among salts (mol. wt ≈ 60), sugars, amino acids, and peptides (mol. wt ≈ 150–

300), sesquiterpenes and dihydrochalcones (mol. wt $\approx$ 400–600), and proteins (mol. wt $\approx$ 10 000–22 000). Moreover, while topological receptors are predicted on lock and key analogy, with tastants exhibiting varying affinity for the receptor, there is little evidence to support the notion of affinity, *per se*.

11.4.1 The Temussi Model for Bitterness/Sweetness

One of the earliest suggestions for the significance of receptor topology for sweetness was advanced by Lelj *et al.* (1976), with subsequent modifications (Temussi *et al.*, 1978, 1991). The receptor incorporated AH,B functional sites, and the spatial barrier of Shallenberger *et al.* (1969). The rationale for the topological study was the conclusion (Temussi *et al.*, 1978) that the AH,B tenet alone was probable justification for a common chemical site for sweetness, and purported exceptions to it might be resolved by consideration of the receptor's topological attributes. As sweetness (within a homologous series) is usually lost beyond a certain critical molecular size, the authors sought to identify the essential geometric features of many sweet and bulky compounds, and to construct a template for their shape.

α-L-Aspartyl-L-phenylalanine methylester (aspartame) was elected as a starting compound as it combined the properties of high sweetness and large molecular volume. The conformation selected for study, based upon conformational energy minima, is described as $F_I D_{II}$ (Lelj *et al.*, 1976; Temussi *et al.*, 1978). It corresponds to the component projection structures in Fig. 11.16. The $F_I D_{II}$ conformation is characterized by flatness (planarity), and when the amine (AH) and the carboxyl carbonyl (B) functions are placed on a plane surface to simulate interaction with the receptor B and AH functions respectively, the phenyl group of the molecule is skewed to the right, in the yz plane. This form of the molecule was circumscribed by a contour map in the yz plane resulting in a description of the yz boundaries for the receptor.

The bottom of the receptor was viewed as being equivalent to the spatial barrier used to describe L- and D-amino acid taste (Shallenberger *et al.*, 1969), and the receptor described a shallow pocket with a depth of about 3 Å. As viewed in the xz plane, the compounds showed near-perpendicular and vertical progression along the z axis with minimal tendency to approach the barrier. That initial model was modified slightly, through study of many compounds in addition to those of the 1978 article, and a description of it (Castiglione-Morelli *et al.*, 1990; Temussi *et al.*, 1991) is shown in *pseudo*-perspective in Fig. 11.17. In the hands of Temussi and coworkers, the receptor site was seen to accommodate the sweetness of many other substances, including derivatives of saccharine, and 5-ethyl-6-methyl-oxathiazinone dioxide that, although sweeter than aspartame, does not fully occupy the receptor cavity. If the $F_I D_{II}$ conformation of aspartame is the sweet conformation, and γ is the benzene ring, the tripartite AH,B,γ glycophore template is glycophore VIII in Fig. 11.5. It is capable, after rotation, of reaction with the receptor template in Fig. 11.5.

Among the largest of known sweet compounds are the sulfonaphtho-imidazole substances, of which 3-anilino-2-styryl-3*H*-naphthol[1, 2-*d*] imidazole-5-sulfonate (SSN) is an example. A related compound, 3-anilino-2-phenyl-3*H*-naphtho-[1, 2-*d*]imidazol-5-sulfonate (TSN) is tasteless (Castiglione-Morelli *et al.*, 1990). The compounds are shown in Fig. 11.18. From conformational studies (n.m.r measurements and molecular dynamics calculations) it was determined that the compounds are, with the exception of the aryl group attached to the imino nitrogen atom,

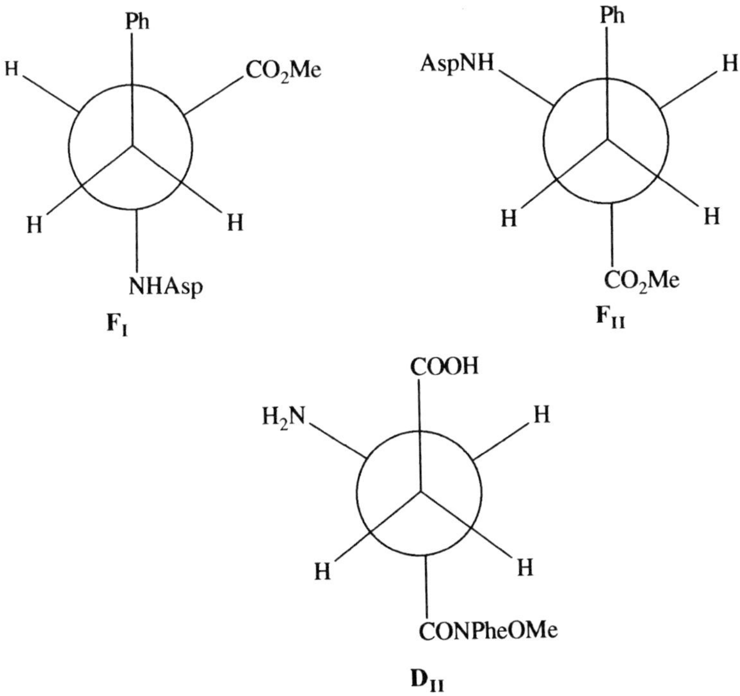

FIG. 11.16. Conformations of component parts of aspartame (cf. Castiglione-Morelli *et al.*, 1990).

essentially planar. The aryl group is perpendicular to the remainder of the molecule. Castiglione-Morelli *et al.* (1990) tentatively identified AH,B in those substances as the SO_3H group, but rejected the notion, as the extremely low pK_a values make the occurrence of the acidic form very unlikely. Therefore, they choose SO_3^- to represent an incomplete AH,B entity that can be tolerated by the site, and elicit sweetness, provided that the steric fit for the remainder of the molecule is sufficient. That principle was also employed (Temussi *et al.*, 1984) to circumvent the uncertainty as to the identity of AH in 2-propoxy-5-nitroaniline.

The fit with the proposed topological receptor was quite good for the SSN substance, but steric impedance was found for the lone benzene ring in the planar portion of the TSN molecule. That interference was offered as the reason the TSN substance was tasteless. Alternatively, AH and B can be assigned to SSN as shown below (AH might also be assigned to an aromatic ring proton). Then γ lies at the ten o'clock position, and corresponds to planar glycophore template I (Fig. 11.5). Although the aryl group projects toward the observer, the entire AH,B,γ unit is planar. By this reasoning, the TSN compound would then be tasteless because it does not possess an appropriate B function. A second explanation, although multiple choices are available for AH,B, is that the TSN

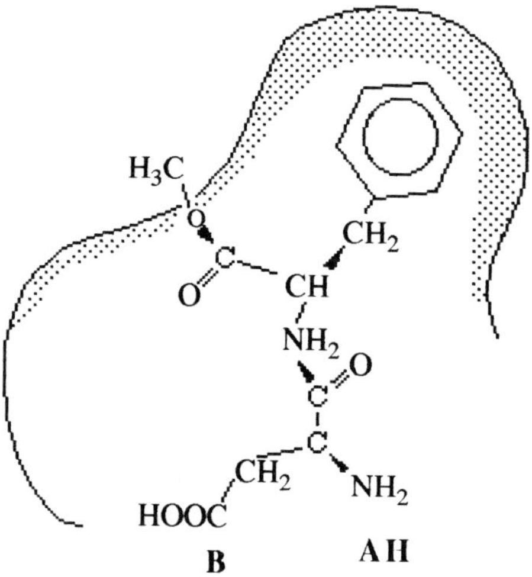

FIG. 11.17. Topological depiction of the sweetness receptor of Temussi *et al.* (1978) drawn to contain aspartame in the F_ID_{II} conformation (Castiglione-Morelli *et al.*, 1990).

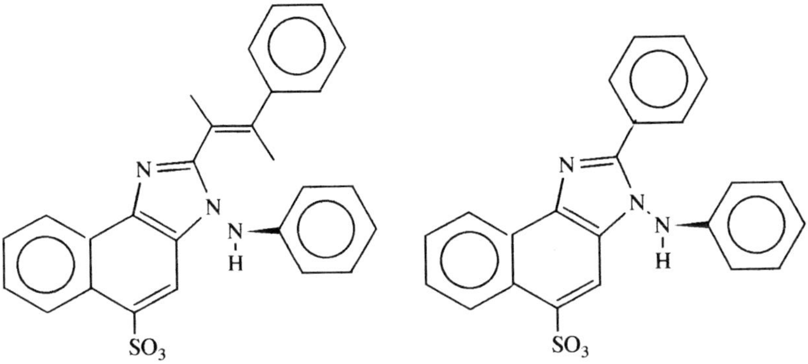

SSN TSN

FIG. 11.18. Structure of sulfonaphthoimidazole substances.

substance simply does not have a complete tripartite glycophore. The overall receptor topology, as viewed by the Temussi group, is relatively shallow and open-ended. The model explains why some bulky substituents, out of the main molecular plane of the compound, have no topological influence on the taste of some sweet (or bitter) molecules (van der Wel *et al.*, 1987). On the other hand, the protein sweeteners, with a globular diameter of 25 Å, and high-potency sweetness, cannot be accommodated by such a receptor.

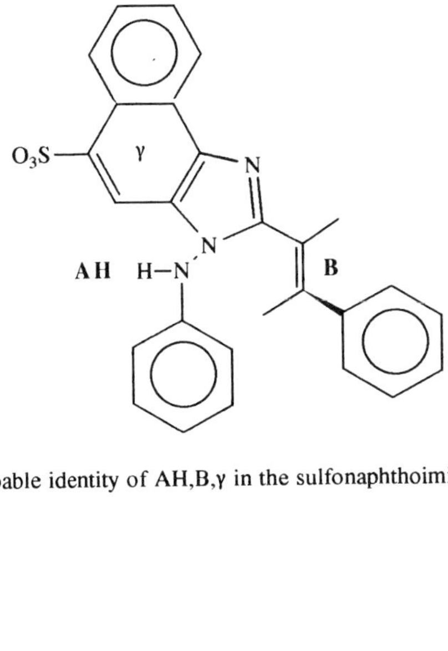

Probable identity of AH,B,γ in the sulfonaphthoimidazole, SSN

11.4.2 The Culbertson-Walters Model for the Sweetness Receptor

Culbertson & Walters (1991) constructed a general model for high-potency sweetness based on electrostatic patterns and computer assisted conformational analysis. The compounds employed were the guanidine acetic acid compounds of Tinti & Nofre (cf. 1991*a*, *b*), an aryl (aromatic) ring urea derivative of aspartame, an aspartic fenchyl ester derivative, and aspartame. A three-dimensional construction of the combination receptor reveals the presence of a major and a secondary NH responsive area, a carboxylate responsive area, a depressed methylester site, a hydrophobic region, and an aryl region. The receptor, as seen from above, is shown in Fig. 11.19. A special feature of the Culbertson & Walters receptor is the inclusion of a second hydrophobic (aryl) site to account for the extraordinary sweetness of substances, i.e. compounds with four primary reactive sites. On a statistical basis, these sweeteners have a structure (two γ components) that allows for interaction with the planar AH,B,γ receptor in either of several orientations. The second hydrophobic receptor site may not be necessary to account for extraordinary sweetness, *per se*, but may act in a complementary inductive capacity. Nevertheless, application of the model showed that it was consistent with known structure-activity relations in aspartic dipeptides, aspartic amides, and guanidine compounds. The finding is not particularly surprising, as those substances were the models used to construct the receptor.

To the Culbertson & Walters way of thinking, compounds that effectively fill the hydrophobic region tend to have highest sweetness

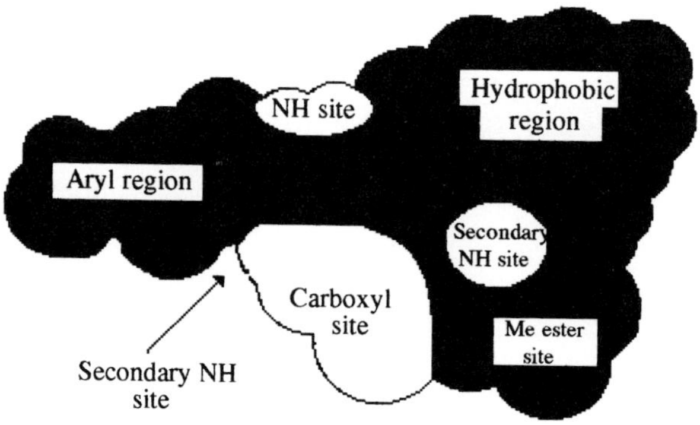

FIG. 11.19. Topology of a sweetness receptor constructed from the topology of extraordinary sweet substances (Culbertson & Walters, 1991).

potency. Those that only partially occupy the hydrophobic region, or extend beyond its boundaries, either have low sweetness potency or are nonsweet. From the discussion and examples presented, the potential of the model to predict sweetness, tastelessness, and even the ability for sweetness inhibition for certain substances (e.g. succinyl-L-phenylalanine methyl ester, or aspartame minus the amino group) appears to be quite good. With respect to the need for two hydrophobic sites in the receptor itself, it can be noted that extraordinarily sweet compounds also neatly fit the receptor proposed by Temussi and coworkers (Temussi *et al.*, 1991, pers. comm.). In the final analysis then, the receptor proposed by Culbertson & Walters (1991) is a multiple example of the proposed planar AH,B,γ template (II) indicated in Section 11.2.2.1. As a consequence, it is incapable of explaining the most fundamental stereochemical requirement of a sweetness receptor, i.e. the different tastes of amino acid enantiomers. In addition, chirality in side chains would now need to be recognized, but apparently it is not (see Section 11.2.2.6).

11.4.3 The Tinti/Nofre Topological Model for Sweetness

Upon recognizing a general family of glycophores for sweet taste (Section 9.3.13.2), Tinti & Nofre (1991*b*) proposed a multifunctional sweetness receptor that contains several hydrogen donors and hydrogen acceptor groups separated by 3·0 Å. Through reference to Table 9.3 they are AH–Y, AH–B, XH–Y, XH–E_1 and XH–E_2. No reason is given for omitting other possible combinations, such as XH–B. There is also a site (D) for the nucleus of a nitrogen atom in a CN group, and a large hydrophobic component (G). The authors also assigned x, y, z coordinates to the positions of the functional groups, but the point of origin for determining those coordinates is not clear. The proposed receptor-site is shown in Fig. 11.20. The complementary groups for the glycophore sites are shown in I. They are again shown in II, where the functionality of those glycophore sites is specified in terms of the antiparallel functionality of the composite glycophore, i.e. as the functionality of the receptor groups, in AH,B,γ terms. The receptor contains at least five AH,B units, a secondary electrophilic site, and a primary hydrophobic site.

The distinguishing feature of the multiple receptor is that AH,B and D are taken to be high-affinity sites. The XH and the secondary B sites are taken to be hydrogen bonding acceptor and donor groups, eliciting a sweetness potency about that for amino acids and sugars. However, AH for saccharine is an XH group. The suggestion implies that the chemistry of sweetness for sugars and amino acids is elicited by a mechanism different from that for intense sweetness. If that is indeed the case, then, through application of symmetry principles, it would seem that mechanism must nevertheless be bilaterally symmetrical.

The model is attractive for the ease with which it resolves the significance of a strong inductive group (CN, NO_2) when that group is viewed as a 'second B' equidistant (9 Å) from a primary hydrophobic (G) group. However, the multicomponent receptor cannot now account for the taste of enantiomeric amino acids and peptides. When the second 'B' ('D' group) is viewed as an electron rich (X or γ) glycophore function, with strong inductive effect on AH,B, the need for such topological and functional complexity disappears.

All told, if there is a single topological receptor for high-potency sweet compounds, a topological consideration seems to mandate that the topology of the receptor is pliant and flexible. A parallel topological endeavor was the attempt (Amoore, 1963*a*, *b*) to fit the compounds that elicit 'primary' odors into a topological classification. At the outset, many examples were found to fit seven prerequisite topological shapes, but in the end more exceptions were found for each shape than there were molecular shapes with the same odorant quality that fit the receptor. Similar problems are also encountered with the notion of lock and key explanations for enzyme specificity. Ultimately, 'induced-fit' supplements to the lock and key tenet need to be invoked (cf. Whitaker, 1972).

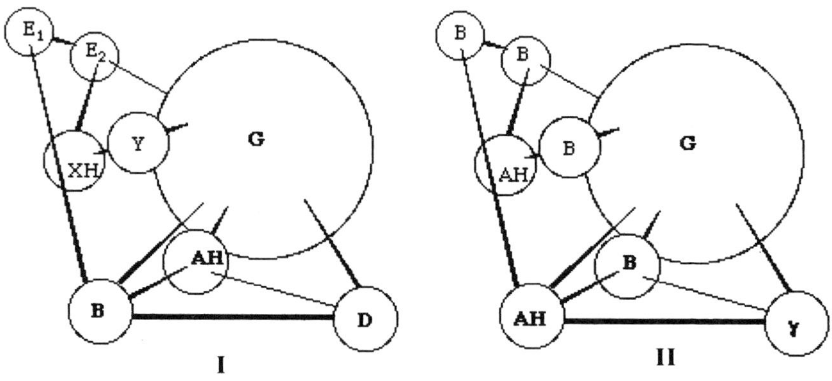

FIG. 11.20. The Tinti & Nofre (1991*b*) receptor site (I) for the sweetness, and the functionality assigned to those sites (II).

Topological receptors are also predicated, of course, on the tenet that there is no common glycophore nor common chemical mechanism for sweet and high-potency sweeteners, i.e. there is something *structurally* unique about high-potency tastants. However, the need for a four-point contact requirement for high-sweetness potency seems to be self-contradictory and high-potency recognition activity alone (Section 10.3.4.2) is probably the source of the 'uniqueness.' Moreover, a common glycophore for *all* sweeteners can be simply described as an uninterrupted planar arrangement of three different functional groups, and the receptor for such a glycophore also need only be an uninterrupted antiparallel planar arrangement (silhouette) of only three different functional groups.

11.5 SYMMETRY, CHIRALITY AND TOPOLOGY IN RETROSPECT

11.5.1 Role of Bilateral Symmetry

The notion of bilateral symmetry seems to bring the subject of taste chemistry to a higher level of comprehension, and the classification scheme of Section 11.1.4 seems to be a significant improvement on the standard ionic-nonionic chemical classification of tastes. The symmetry scheme also neatly organizes the exceptions contained in the standard scheme. For example, the notion of dissymmetric bitter reactions applies to both nonionizable and ionizable compounds.

11.5.2 Role of Prochirality and Prosymmetry

The ultimate role of prochirality and prosymmetry in taste chemistry phenomena was not anticipated, but in retrospect, it probably should have been. The need to render the receptor prochiral, while the tastant might be either chiral (prosymmetrical) or symmetrical, serves to resolve the aforementioned chiral anomaly/enigma. Enantiomeric amino acids can usually vary markedly in their ability to elicit sweet taste, while the enantiomeric sugars usually vary only slightly. A significant feature is the application of the notion of prochirality to the receptor, and not to the substrate (tastant). The substrate can then interact with the receptor as if it (the substrate) were either chiral, chiral but prosymmetrical, or achiral. The tenet also serves to minimize the need for complex topological forms for receptors, or for multiple receptors, in order to account for anomalous chiral and symmetrical taste phenomena.

11.5.3 Role of Topology

The general lack of any particular topological significance, other than inductive effects in taste chemistry interactions, might have been anticipated from the general recognition features of taste chemistry. One of the reasons for the need for topology (lock, key) analogy in enzyme processes is that there is always a chemical transformation of the substrate. In pharmacological and immunochemical interactions, there is a physiological consequence (biochemical transformation). If there is a common denominator among taste induction, enzyme, and drug chemistry, it is the *prochiral nature* (Section 2.11.3.6) of the *recognition chemistry* for each phenomenon. However, even here the phenomena are different. For taste chemistry, the prochirality of the receptor is significant. For enzyme and drug chemistry, it is the prochirality of the substrate that is significant. In the latter structure activity relations (SAR) and *subsequent* to the recognition chemistry, a chemical transformation occurs (hydrolysis, reduction, esterification), or there is a physiological response. It is at this point that the three-dimensional attributes of a molecule become important. Diastereoisomers and enantiomers are then differentiated in all cases, either absolutely, or relatively. If it is granted that taste chemistry is essentially a form of recognition chemistry, then taste chemistry phenomena are processed before there is any need for absolute discrimination of three-dimensional topological features. Stated another way, the spatial requirements for binding are much more rigid than they are for recognition. Parenthetically, many pharmacologically active substances taste bitter, but at least one (nitroglycerin) is distinctly sweet. If a drug is pharmacologically active, that activity occurs *subsequent* to its ability to elicit taste. Without doubt, the notion of prosymmetry has its counterpart in enzyme and drug recognition chemistry, and such observations have previously been characterized as 'nonspecific' or 'nonallosteric' phenomena.

CHAPTER 12

The Taste Receptor(s)

12.1 THE TASTE CELL MEMBRANE

The interaction of a tastant with the receptor has long been believed to occur at the surface of the taste cell membrane (Beidler, 1954, 1962; Kennedy & Halpern, 1980). Among the evidence cited by Beidler (1962) for that conclusion is: (1) the threshold concentration of sodium chloride is reached within a millisecond after 0·1 M sodium chloride has been applied to the tongue; (2) sucrose does not penetrate the membrane of any type of cell that has been investigated, and lead acetate and beryllium chloride, which also taste sweet, are very toxic when they reach the interior of a cell; (3) a taste receptor rapidly elicits a steady electrophysiological response, and the magnitude is proportionate to the stimulus concentration; and (4) the taste response declines rapidly following a water rinse. It also stands to reason (albeit, teleologically) that the taste response is a chemical property index to chemical activity before ingestion. Many investigators subscribe to the Beidler tenet. Others believe ions must pass through the cell membrane to elicit saltiness (cf. Teeter & Cagan, 1989; Lancet & Ben-Arie, 1991; Schiffman, 1990). On the whole, it seems to be generally agreed that the initial chemistry of all tastes occurs at a membrane surface and good account of the probable chemical events responsible for different tastes can be formulated, based on that premise alone.

Because lipid soluble substances penetrate membrane surfaces rapidly, it was first believed that the membrane surface of a cell was lipoid in nature, with small holes to allow passage of water soluble substances. Subsequent surface tension measurements, however, indicated that it also had proteinaceous nature. As indicated by optical polarization studies, Schmitt & Bear (1939) proposed that the lipids were oriented radially, and the protein molecules were oriented tangentially. Those findings led Davson & Danielli (1952) to propose that the membrane was a phospholipid bilayer, with the hydrocarbon ends of the phospholipids oriented inward to form a continuous hydrocarbon phase, and the polar ends oriented outward. Each surface of the membrane was then composed of polar ends of a bilaterally symmetrical lipid bilayer, and each surface was loosely bound to a thin layer of protein.

At the present time, the generally accepted membrane model is the fluid-mosaic structure initially proposed by Singer & Nicholson (1972). It consists of a fluid phospholipid bilayer with globular protein molecules embedded in the membrane. Other proteins penetrate either side of the membrane, and extend through it (transmembrane proteins) to form pores or channels. The proteins are thought to be either lipoproteins, glycoproteins, or both. The model also postulates that the proteins are globular, as their α-helix content is quite high ($> 40\%$). The diagram shown in Fig. 12.1 is a cross-section of the fluid mosaic membrane. The Singer-Nicholson membrane model provides for membranes of widely different protein content and protein constitution, although a single protein is present in the case of some sensory receptor cells, e.g. retinal rods. It can also account for the electrical properties, and for membrane semipermeability. Electrophilic and nucleophilic taste receptor units are probably components of transmembrane glycoproteins analogous to the

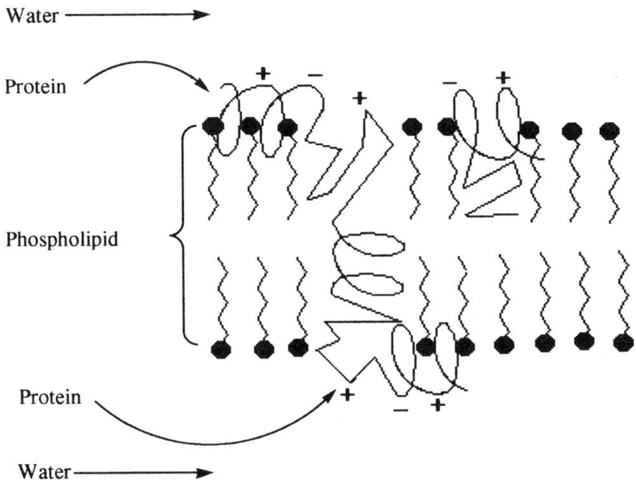

FIG. 12.1. The Singer & Nicholson (1972) fluid-mosaic membrane model.

sialoglycoproteins found in human erythrocytes (Dill, 1987). Glycophorin A, the predominant protein of red blood cells, extends from the cytoplasm and through the lipid membrane, where the repeating amino acids are largely hydrophobic, and into the exterior. Because it extends through the membrane, glycophorin A has been associated with a number of receptor functions, including binding of calcium and magnesium ions.

12.2 PROTEIN NATURE OF THE TASTE RECEPTOR(S)

The general receptor chemical functions, and the stereochemistry of those functions, has been covered in previous chapters. Little was said about the chemical nature of the purported receptor, molecule(s), but it has also long been suspected that the receptor for taste is a protein with a combination of the polar and nonpolar properties characteristic of globulins (Section 7.2.2.3). It is presumed to be embedded in a phospholipoid matrix. The matrix, in turn, is exposed to an aqueous environment.

12.2.1 Isolation and Enzyme Studies

Purported sweet taste protein receptors (Dastoli & Price, 1966), and a bitter-taste protein receptor (Dastoli *et al.*, 1968) have been isolated from animal taste buds. The isolation methods employed parallel standard enzyme extraction and purification procedures, and include solubilization of the protein with buffer, and precipitation with ammonium sulfate. Using spectroscopic techniques, the isolated proteins were then shown to complex with sweet and bitter tasting substances. One of the sweet-taste protein receptor preparations was active toward D-amino acids and inactive toward L-amino acids (Price, 1972). Among the criticisms of the taste-protein isolation reports are that other substances also cause change in optical properties when added to the preparations, such as dimethylsulfide (Nofre & Sabadie, 1972). The physical procedures employed to assay taste activities are also nonspecific. In addition, the amount of protein isolated is excessively high (Price & DeSimone, 1977).

Hiji & Sato (1973) used ^{14}C-labeled glucose and fructose to measure direct binding to pertinent protein fractions from rats, but the results obtained were not particularly convincing. Hiji (1975) found that proteolytic enzymes (pronase and semi-alkaline protease, but not trypsin and papain) suppress the response to sweeteners, but not other tastes. However, upon application of pronase to the tongue, Giroux & Henkin (1971) report nonselective decrease in thresholds of each of the four basic tastes in man.

12.2.2 Immunological Studies

That the receptor is probably a protein has been indicated in a unique way. The immunological reaction for the formation of antibodies is usually a protein–protein interaction. Antibodies raised in the rabbit to sweet tasting proteins are capable of cross reaction with a variety of nonprotein sweeteners (Hough & Edwardson, 1978). On comparing immunological reactivity with the relative sweetness of substances on a molar basis, and with thaumatin as the sweetness reference standard, an excellent correlation (0·9606) was obtained by Hough & Edwardson (1978), and is shown in Fig. 12.2.

van der Wel & Bel (1978) also found that rabbit antibodies are produced against thaumatin I and monellin. In addition, it was found that antibodies raised against thaumatin cross react with, and compete for, monellin. As the proteins are closely related immunologically, the latter authors proposed that the proteins have conformationally identical antigenic determinants that probably coincide with the active site responsible for sweetness. This occurs even though the amino acid sequence for monellin and thaumatin is different. Kim *et al.* (1991) also found that antibodies raised against the two proteins have reciprocal reaction, and report that antibody–sweet protein complexes do not taste

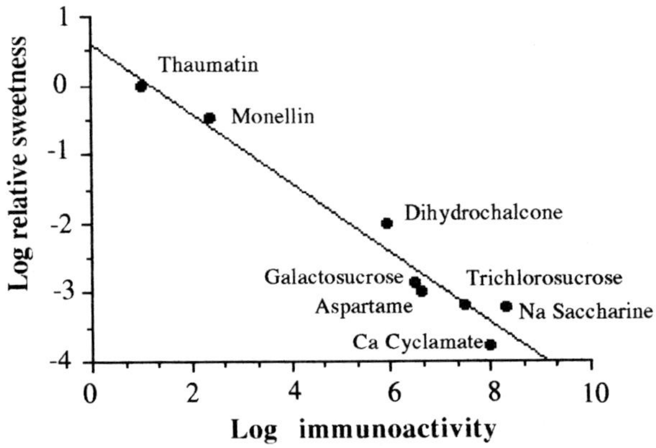

FIG. 12.2. Relation between immunological activity and the relative sweetness of selected compounds (Hough & Edwardson, 1978).

sweet. The relation between the initial chemistry of sweetness and immunology is also implied in the deliberations of Boch & Lemieux (1982).

12.2.3 Nature of the Protein–Tastant Interaction

While a protein receptor at the taste cell surface is the most likely candidate for a general taste receptor, it does not act in an enzymic manner (Beidler, 1961). Unlike the action of enzymes, taste phenomena are independent of both temperature and pH, and many enzyme poisons initiate a taste sensation that disappears rapidly with a water rinse. A more general reason why a protein is a good receptor candidate for sweet taste is that the amino acids in a protein are capable of significant perturbations when reacting with small ligands (Hölte & Kier, 1974). This can lead to alteration of function, as the fine structure of proteins is dependent upon influences across space in a *folded* molecule (as globular membrane proteins indeed are). Hölte & Kier believe that the phospholipid component of a membrane would be a poorer choice for a receptor because the phosphate group, being highly charged, presents an insensitive, nonselective feature to an approaching ligand. On the other hand, very few of the polar groups of phospholipids are phosphate alone. The polar groups can be composed of phosphoglycerides, phosphotidylethanolamine, phosphatidylserine, phosphatidylinositol, etc. Nevertheless, the side chain of tryptophan (3-methylindole) appeared to be an ideal choice for the dispersion (hydrophobic) component of the receptor.

The three chemically different, and chirally arranged receptor functions at a membrane surface might be ligands on three entirely different chemical substances. However, the above data indicate that the principal component of the taste receptor is indeed proteinaceous, at least for the electrophilic and nucleophilic (AH,B) functions. Their sweetness action can be destroyed by extracting the protein (denaturation), or by degrading it with proteases. Although denatured, the protein may still be capable of taste reactions of lesser order than the concerted tripartite AH,B,γ chemical interactions required for sweetness. For sweetness, the receptor can be viewed as having an action entirely dependent upon its form and association with other structural components of the taste receptor cell membrane. Unlike enzymes, which impart structural transformations to a substrate, the function of the taste receptor protein is to transmit ligand interaction energy information without necessarily transforming the substrate. The chemical structure transformation ability of an enzyme may remain intact upon isolation (and even improved by recrystallization), provided appropriate cofactors are included. The total taste response ability of a taste receptor protein is partly lost by disruption of the cell surface membrane, whether by action of detergents, or by protein extraction procedures.

The more general interaction of a taste receptor protein with a tastant would seem to be the various chemical recognition reactions outlined in Chapter 10 for an acidophore, etc., along with concomitant symmetry attributes (Section 11.1.4). The general dynamic membrane electrophilic/nucleophilic structure of Section 12.6 is capable of differentiating among the postulated chemistry for the four tastes. It can even be construed to represent a single generic receptor *type* for each of the four tastes.

12.2.4 Hydration of a Dynamic Membrane Protein Receptor Model

To envisage the role of water in a tastant–receptor mechanism, a dynamic electrophilic/nucleophilic protein receptor can be presumed to be in equilibrium between relatively nonpolar (hydrogen bonded) and polar forms (Section 10.3.2). In the first case the water is relatively unstructured about the hydrogen-bonded form (receptor hydrophobic probe), and somewhat structured about the polar form, as shown in Fig. 12.3. Immediately above the phospholipid dipolar structures, the same condition probably prevails. Further removed from the membrane surface, the water is even more structured. The most important difference in the water structure surrounding the receptor, and that surrounding the polar ends of lipids, is that the receptors are probably component parts of a protein.

In the diagram, the electrophilic and nucleophilic receptor units are envisaged as protruding above the lipoid surface, and are hydrated with

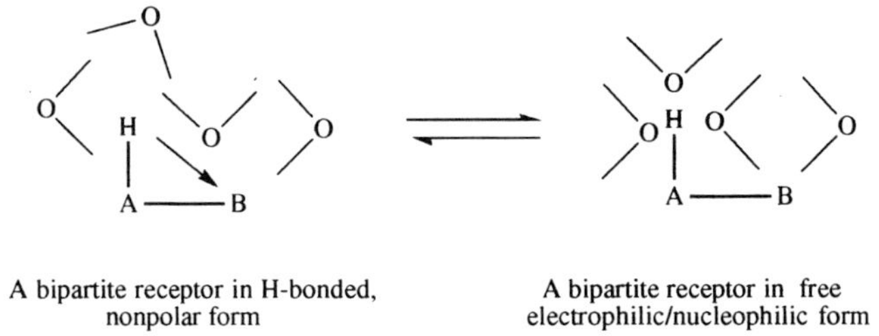

FIG. 12.3. Diagrammatic representation of the occurrence of a hydrated AH,B receptor in hydrogen-bonded form in equilibrium with a free, but hydrated, electrophilic/nucleophilic form.

molecules of water that have lower entropy than the water in the immediate vicinity. In hydrogen-bonded form, which are in dynamic equilibrium with the polar form, the isosteric receptor serves to distinguish among the different chemical mechanisms proposed for the different tastes (Chapter 10). The location of γ in the model shown in Fig. 12.3 might be general (the lipoid matrix itself), or specific (the hydrophobic side chain of an amino acid belonging to the protein helix). In either case, it is embedded in the hydrophobic component of the membrane bilayer. The significance of the receptor models of van der Heijden *et al.* (1985*a*, *b*), as well as those of others (the Belitz and Temussi groups) are particularly pertinent to the assignment for the location of the receptor γ function(s). Their results indicate that γ lies below the receptor polar functions. When a sweet substance interacts in tripartite fashion with the receptor (Section 10.3.4), the energy for the taste sensation is obtained from the increase in the entropy of hydration water about the hydrogen-bonded electrophilic-nucleophilic (AH,B) receptor.

12.2.5 Protein Receptor Models for Taste

As a protein receptor for any taste has never been isolated and characterized, the amino acid sequence of the protein receptor is not known. It is possible that the protein sweeteners and the receptor are similar (Higginbotham, 1979). Indeed, the receptor may be the antiparallel mirror-image functional structure of the protein sweeteners. A proposal by Suami & Hough (1991) for the nature of the sweetness receptor is based on the likelihood that L-asparagine prevails at the N-termination of an α-helix, and the penultimate amino acid is L-proline (Richardson & Richardson, 1988). The receptor AH and B functions are assigned to the NH_3^+ and the C=O (CH_2CONH_2) groups of the asparagine residue, respectively. The γ function for the receptor is assigned to a hydrophobic side chain of an amino acid. It occurs at the fifth position from the N-terminus of the helix. As an overview to its function, the authors propose that the receptor has two interconvertible conformations. The first corresponds to a recessed B receptor function in a resting state, wherein receptor B is bound to an unpaired amide group at the fourth amino acid residue. The other corresponds to the receptor in an active state, with B exposed, as shown in Fig. 12.4. Alternatively, to comply better with Section 10.3.4.1, the intramolecularly hydrogen-bonded form may be the active form, and corresponds to TR* (Section 2.10). Nevertheless, the intramolecularly-bonded receptor AH in the

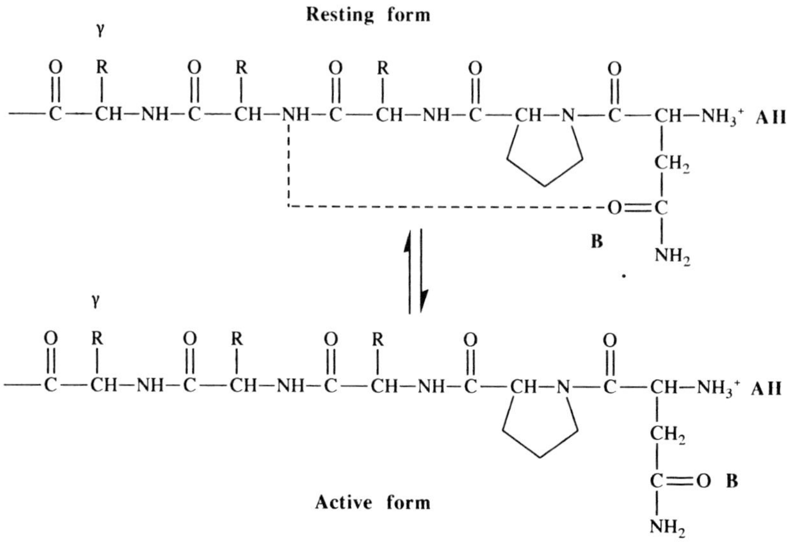

FIG. 12.4. The active and resting form of the sweet–taste receptor AH,B and γ in the N-terminus of a sweet-receptor protein's α-helical structure (Suami & Hough, 1991).

resting form is not necessary for the sweet taste interaction. In the active conformation, a stimulus molecule is adsorbed to (or, perhaps, recognized by) the receptor, the intramolecular hydrogen bond is cleaved, and a new hydrogen bond is formed between AH of the sweetener and B of the receptor. This results in movement of the AH,B,γ groups on the receptor into the correct geometry for recognition by the antiparallel sites on the sweet molecule. The proposal is in compliance with the postulated recognition chemistry (Section 2.12) of carbohydrate interactions with antibodies and lectins (Lemieux, 1982).

Conversely, the initial interaction may be between γ of the sweet substance and γ of the nonhydrogen-bonded (resting) protein form. That interaction would serve to lead to a protein intramolecular hydrogen bond and to TR*, where receptor AH,B is intramolecularly hydrogen-bonded. Then, there is no need for receptor B to be bonded elsewhere in the helix. The latter process would also transform a three-dimensional topological configuration for the receptor into an appropriate, and active two-dimensional AH,B,γ arrangement. Subsequently, antiparallel intermolecular hydrogen bonding occurs. This interpretation considers the

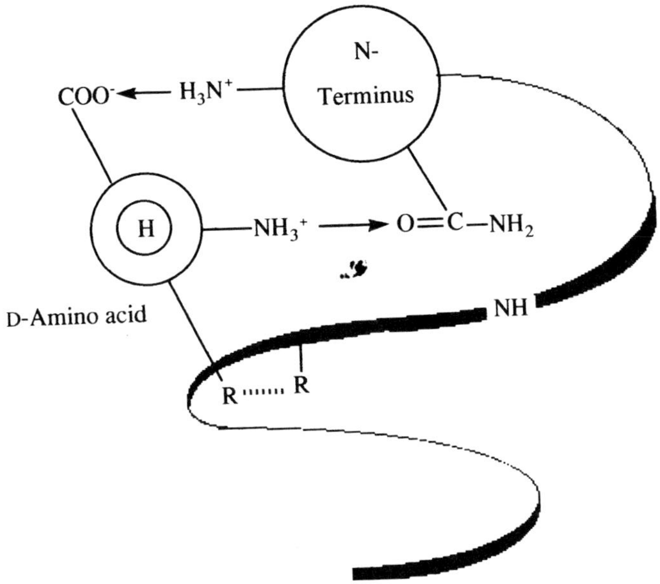

FIG. 12.5a. Interaction of a D-amino acid with the protein N-terminal AH,B unit of an amino acid and with a hydrophobic component of an amino acid in the helical structure (Suami & Hough, 1991).

functions assigned to γ in Section 10.3.4.1. The interaction also leads to activation of the glycophore and the receptor, and quenches the amino acid zwitterionic function to promote hydrogen bonding.

Suami & Hough also demonstrated how D-leucine (sweet) can interact with the receptor to elicit sweetness, but L-leucine (not sweet) cannot. In this case, as shown in Fig. 12.5a, the initial interaction of D-leucine with the helical protein receptor is an attraction between the isobutyl side chain (γ) of D-leucine with the alkyl side chain of the fifth amino acid residue.

For an L-amino acid (Fig. 12.5b), a monopolar interaction with AH of the receptor is possible, accompanied with a γ, γ interaction. Thus, while there is no sweet taste, the model is in compliance with the requirement for bitterness as proposed by Belitz *et al.* (1981). The intensity of the bitterness would then be of about the same magnitude as the sweetness for the D-amino acid.

In an earlier publication, Suami (1987) proposed a similar receptor model. In the latter case, either L-serine or L-threonine was chosen to be

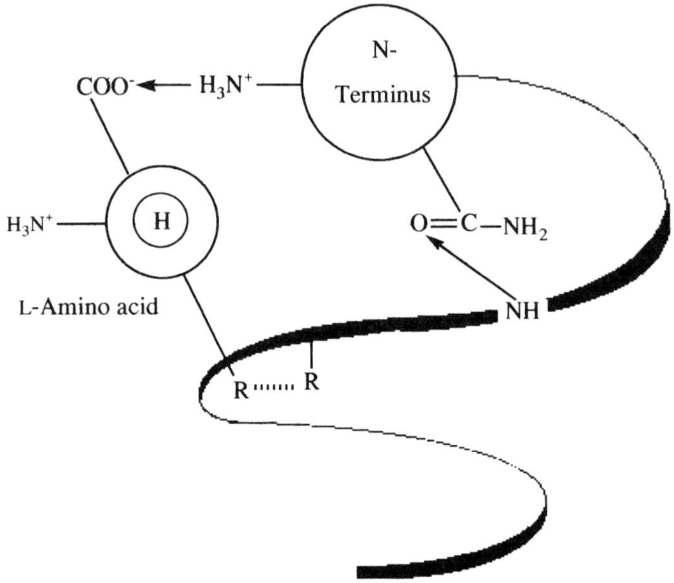

FIG. 12.5b. Interaction of an L-amino acid with a protein N-terminal AH unit only of an amino acid and with a hydrophobic component of an amino acid in the helical structure (Suami & Hough, 1991).

the N-terminal amino acid for the protein α-helix. In addition to accounting for the taste of D- and L-leucine, it was proposed that the poorer tripartite fit of *pseudo*-L-fructose with the receptor accounted for the finding that it was not as sweet as the D-enantiomer. Presumably, as the more recent results for D- and L-leucine were the same, the most recent receptor is also capable of accounting for the sweetness intensity difference between the enantiomeric *pseudo*-sugars.

12.3 MEANING OF THE TERM 'DIFFERENT RECEPTORS'

In presenting the case for single or multiple receptors for the four basic tastes, and to elaborate on the scope of the expression 'different receptor,' there is need to discuss how protein receptors can differ. At the outset, two major class distinctions can be made. (I) It is possible that there is a single taste receptor for all tastes, and the different way that different chemical substances interact with it leads to different transduction mechanisms. (II) There are different isosteric receptors contained on different proteins for each of the basic tastes that lead to different transduction mechanisms. A class II receptor may also be a protein composed of amino acids with varying sequence, a glycoprotein (with carbohydrates incorporated through single covalent bond), a lipoprotein with the amino acid back-bone bonded to lipoid materials, a mixed glyco-, lipo-protein, or even a metallo-protein.

12.4 SINGLE AND/OR MULTIPLE RECEPTORS FOR EACH BASIC TASTE

Different tastes occur with greater or lesser intensity on different sections of the tongue, and at different parts of the oral cavity. Nevertheless, a longstanding problem of taste structure-activity studies is whether there are specific or nonspecific receptors for all four tastes (Bryant *et al.*, 1989). If the receptors for the different tastes are nonspecific, then the discriminating feature for the four tastes probably lies in the special chemistry involved at the outset of interaction with the receptor (Chapter 10). They would then also be differentiated in the chemistry of transduction. If, on the other hand, receptors are specific and include specific binding proteins, then the characteristic and different events that discriminate among the different tastes during the transduction mechanism are simply due to that fact, and several have been suggested. Sodium, potassium, and calcium ions have been implicated in the transduction mechanism (Section 1.4) in all tastes (Teeter & Cagan, 1989), but salt taste is believed to be mediated by an epithelial sodium channel. Sour taste is believed to be mediated by a proton block of an outward potassium channel. Sweetness is probably mediated both by a decrease in conductance to K^+, and activation of Na^+ movement. Much of the evidence for single or multiple receptors for each of the four basic tastes is derived from neurophysiological and psychological studies, with special emphasis on evidence from adaptation studies, and the perceived taste of mixtures.

12.4.1 Neurophysiological Studies

The primary neurophysiological evidence for single receptors is that there are neurons that respond best to salt, or to sugar (Frank, 1985). It is also known that taste cells may respond to two or to all four substances associated with the basic tastes. A single taste fiber may also respond to a variety of taste stimuli (cf. Pfaffmann *et al.*, 1971), and it may do so preferentially in certain cases. From the neurophysiological data presented in Section 1.1.4.3, it now seems clear that there must be receptors that respond best to a given taste modality, but there are also receptors that respond to all taste modalities.

12.4.2 Psychophysical Studies

Modern psychophysical evidence also points to different receptor mechanisms for the four tastes (Bartoshuk, 1977), and trained subjects can discern all four primary taste attributes to greater or lesser extent, in the taste of a given chemical substance (Section 1.1.3.2). The observation strongly suggests that a given tastant can react with a given chemical taste receptor in more than one capacity. For a substance that tastes sweet, there can also be an element of saltiness, sourness, and bitterness intrinsic to the very same reactive groups that elicit the sweetness sensation. Thus, even though there may be different chemical mechanisms for different tastes, there is little need for different receptors for them. Nucleophilic or electrophilic receptor sites can interact with different chemical substances, or different but isosteric groups, in many different ways.

12.4.2.1 *Threshold Studies*

Evidence for a single isosteric and multifunctional receptor for all tastes is obtained upon examining the thresholds of substances in mixtures, and from suprathreshold mixture studies themselves. It was shown in Chapter 1 that part of the reason for the different thresholds for sweeteners is due to different activity toward the receptor for sweetness (Shamil *et al.*, 1988*b*). Presumably then, the activity is shown for a common sweetness receptor. It has also long been known (Heymans, 1899) that the threshold concentrations for substances that elicit the sweet, salt, sour, and bitter taste are increased whenever the concentration of a second substance with a different taste quality is increased. Linear regression analysis of that data (Table 1.3) yields an important result. Any proportionate change in the concentration of the second taste substance results in a proportionately equal change in the threshold concentration of the first substance, i.e. $C_1/C_2 = T_1 - T_0/T_2 - T_0$. As an example, a plot of the data for the sour threshold of HCl in the presence of varying amounts of sodium chloride is shown in Fig. 12.6. The plot in Fig. 12.6 is linear, and the HCl sourness threshold = 0·0030% HCl + 0·0008% NaCl. The relation suggests that a straightforward interaction exists between the threshold concentration for HCl sourness and the concentration of NaCl.

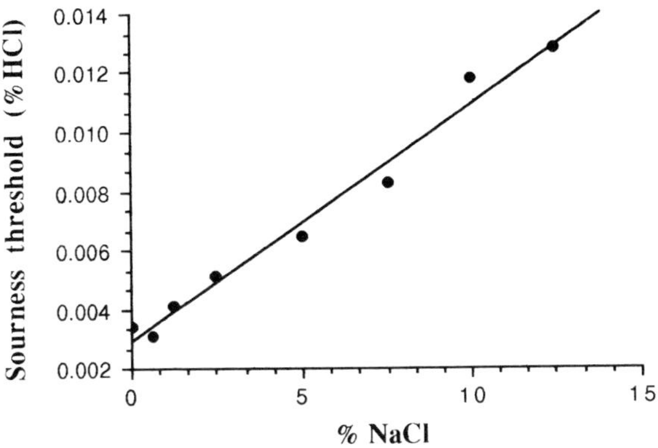

FIG. 12.6. Effect of increasing concentrations of sodium chloride on the sourness threshold of hydrochloric acid (Heymans, 1899).

When the threshold for HCl alone is taken into account, the data show that doubling the concentration of the second substance (salt) precisely doubles the sour recognition threshold of HCl. Some examples are:

% NaCl	*HCl threshold (%)*	*HCl threshold minus intercept*
1	0·0038	0·008
2	0·0046	0·016
4	0·0062	0·032
8	0·0094	0·064

Examination of all possible combinations of substances in Table 1.3, such as the effect of increasing concentrations of sucrose on the threshold for quinine HCl bitterness, yields identical relationships. Chemical competition among substances with different taste qualities for a single multifunctional receptor therefore seems to be more the rule than separate receptors for each taste. The mixed tastant threshold data strongly indicate that there is at least a common generic receptor for the four basic taste qualities.

12.4.2.2 *Taste of Mixtures*

There is also evidence to be found for a single multifunctional isosteric receptor for taste in the assessment of suprathreshold mixtures of substances. Ultimately, the perception and assessment of the taste of mixtures of substances is complicated and difficult, as taste profiles, spectra, synergism and suppression need to be considered. The assessment also again brings up the argument as to whether mixed taste phenomena are synthetic (creation of 'new' tastes) or analytic (Section 1.1.1). Taste perception may be psychologically synthetic, analytic, or some combination of both (Erickson, 1985; Bartoshuk & Gent, 1985; Schifferstein & Frijters, 1990). However, the result of Haefeli & Glaser (1990) on the taste of racemic amino acid mixtures is particularly applicable as evidence for competition for a single receptor, as the taste of amino acid racemic mixtures (Table 7.2) is clearly a mixed taste assignable to the attributes of both the D- and L-forms.

One of the earliest efforts to evaluate the taste of mixtures is that of von Skramlik (1926). von Skramlik attempted to synthesize the varying taste of salts that departed from the basic saltiness of NaCl by using mixtures of NaCl, quinine sulfate, fructose, and potassium tartrate. He obtained some, but obviously varied, success. Along similar lines, Shinoda & Okai (1985) attempted to synthesize sweetness by mixing AH with B, plus a hydrophobic component (γ), from different sources. The present author's interpretation of the results obtained are diagrammed in Fig. 12.7, but the assignment of AH and B receptor components as indicated by Shinoda and Okai (1985) have been altered to give cross-reference continuity to this treatise. A tripartite interaction of AH,B,γ is shown in I using aspartame as the example of an intensely sweet substance. For the case where AH,B alone prevails in a single substance, ethylene glycol was chosen as the example, but other compounds that taste sweet, but lack a site suitable for hydrophobic interaction, could have been chosen (chloroform, glycine). Example III is taken directly from the data of Shinoda & Okai (1985). In example III, a bitter-tasting B-deficient aspartame analogue, plus acetic acid led to sweetness. A two-component mixture exhibiting sweetness was also obtained in III with cyclohexylamine (AH,γ) and acetic acid (B). In addition to obtaining sweetness with mixtures of bitter-tasting aspartame analogues, as shown in III, sweetness was obtained by an admixture of either triethylamine or cyclohexylamine as AH and various benzoylamino acids acting as B,γ (cf. Section 9.3.11.9).

As caffeine and strychnine are alkaloids, and clearly do not have an AH system, those substances were used as an example of compounds with only a B,γ system. As shown in IV, addition of acetic acid, now presumably acting as AH, led to sweetness in the admixture. Finally, solutions of brucine, phenylthiourea and other bitter compounds, representing B,γ in single substances (the authors specified AH,γ), were

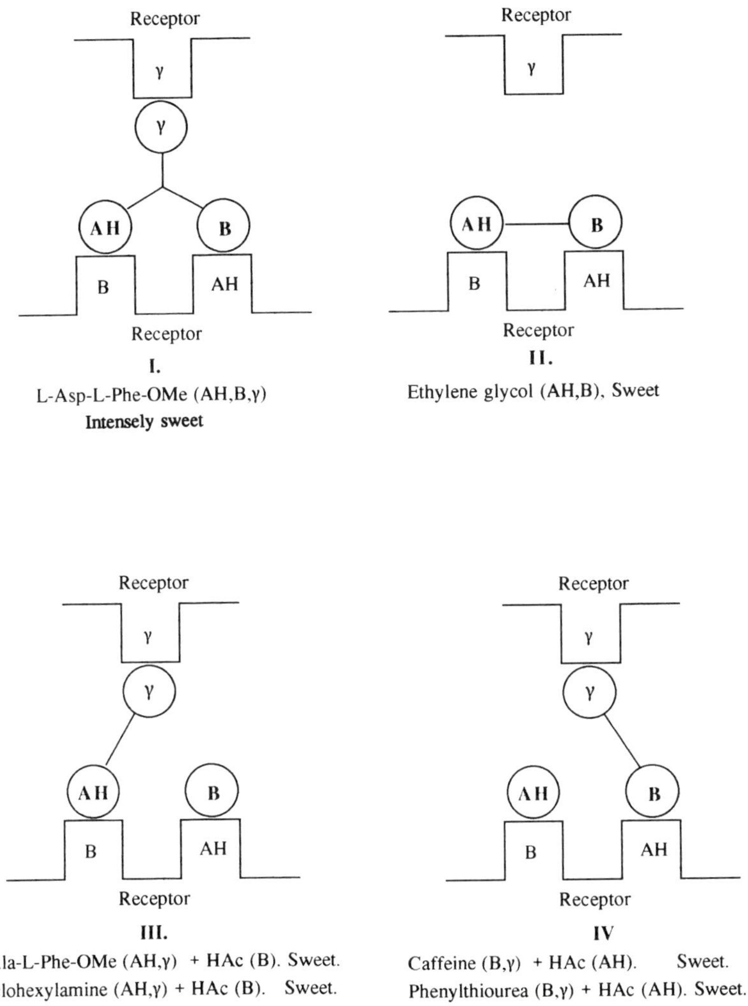

FIG. 12.7. Result of mixing various sources of AH,B,γ on the taste of the mixture (Shinoda & Okai, 1985).

transformed to sweet-tasting solutions with the addition of acetic acid. Presumably, in this case, acetic acid was again a source of AH. These mixtures undoubtedly contained other tastes (e.g., sweetness/sourness; sweetness/bitterness). The findings of Shinoda & Okai lend credence to the Belitz *et al.* (1981) postulate (Section 9.4) that bitterness is elicited by either a positive or negative monopole in the company of a hydrophobic function. The findings also serve to strengthen the notion that the bitterness/sweetness attributes of substances are chemically related. On the whole, it would seem that the unreported sweetness potency of the intermolecular AH,B,γ mixtures was not particularly large.

A more recent attempt to synthesize sweetness using glycophore components from different sources was described by Seki *et al.* (1990). It is also analogous to the initiation of sweetness by the interionic action of hydrated sodium and chlorine ions (Section 9.3.9.1). A relatively nonsweet (0·7×) peptide (δ-acetyl-L-ornithyl-β-alanine methyl ester hydrochloride), with relatively ineffective AH,B,γ components, was rendered more sweet by addition of HCl as 'B' to the solution. The addition presumably enhanced the function of the original AH,γ as sweetness potency of the original compound was more than doubled (1·8×). It is also possible that a more effective conformation was obtained.

Sweetness versus saltiness. For saltiness and sweetness, at low concentrations of NaCl and sucrose, the sweetness of sucrose is increased, presumably because dilute solutions of salt taste sweet (Fabian & Blum, 1943; Beebe-Center *et al.*, 1959; Pangborn, 1960, 1962; Kroeze, 1982). In any event, there is little detectable saltiness in the control solution (approx. 0·05 M; 0·3% w/v). The principal effect of higher concentrations of both substances is to diminish (compress) the intensity

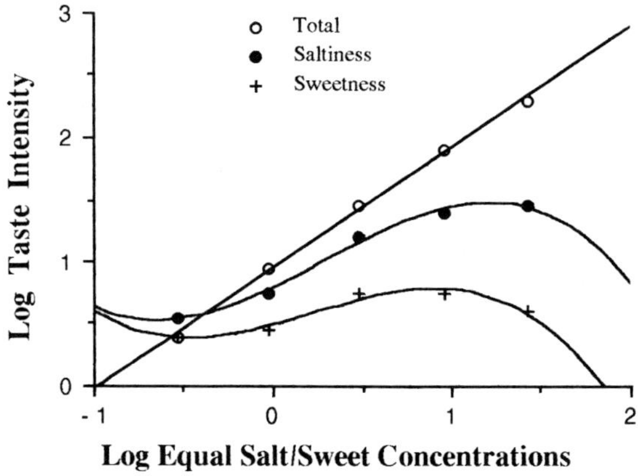

FIG. 12.8. Compression of the sweetness of sucrose and the saltiness of sodium chloride with increase of each substance (Beebe-Center *et al.*, 1959).

of the intrinsic taste quality of each (Beebe-Center *et al.*, 1959; Pangborn, 1960, 1961, 1962; Moskowitz, 1971*b*, 1972). The data of Beebe-Center & Waddell (1948) and Beebe-Center *et al.* (1959), are shown in Fig. 12.8 to demonstrate the taste intensity of NaCl and sucrose in solutions containing concentrations that are about equal in the intensity of their individual tastes.

For each concentration position along the abscissa in Fig. 12.8 the values for NaCl and sucrose respectively, as percentage w/v, are 0·30 and 1·03, 0·94 and 3·08, 2·98 and 9·57, 9·18 and 26·3, 26·5 and 61·2. While the log–log plot of the sum of the taste intensity for saltiness and sweetness, determined individually, are linear, the plots of apparent saltiness in the presence of perceived sweetness, and *vice versa*, are high-order polynomial functions. The complex function indicates that there are multiple factors affecting the perceived saltiness and sweetness of the substances when present in a mixture. Possible reasons for the complex nature of the saltiness–sweetness plots in admixtures of salt and sucrose seem to filter through, provided the following steps are taken. (1) Reports of suppression and enhancement of tastes after adaptation to one or more substances are eliminated (adaptation is not considered to be a factor). (2) Results of studies directed toward the varying sensitivity of different parts of the tongue are eliminated.

Among the possible reasons for the results obtained is the possibility of a chemical interaction between salt and sugar that affects the intrinsic taste quality of each. However, there is little evidence for such an interaction. If salt lowers the sweetness of sugar and *vice versa*, other reasons for the observed phenomena become apparent. The data in Fig. 12.8 show that sucrose diminishes the saltiness threshold of NaCl by a factor of about two, and NaCl diminishes the sweetness threshold of sucrose by a factor of two (cf. Moskowitz, 1972). The two tastes are also readily differentiated in admixture (Beebe-Center *et al.*, 1959; Moskowitz, 1972). DeGraaf & Frijters (1989) found that the perceived taste intensity of a sucrose/NaCl mixture can be predicted using the square root of the sum of the squared intensities for the unmixed components of the mixture. DeGraaf & Frijters also found no evidence to indicate that the sum of the two tastes is less than the individual totals (e.g., Beebe-Center *et al.*, 1959). The perceived total taste intensity of a sucrose/NaCl mixture is equal to the sum of the perceived sweetness intensity and the perceived saltiness of the mixture.

These findings suggest that if there are separate receptors for saltiness and sweetness, the electrophysiological action must feed into the same transduction conduit somewhere along the line. On the other hand, if there is competition for the same receptor, and the solution has equal taste quality/intensity for each tastant, then probability alone would yield an approximate 50/50 response for each taste quality. The notion of 'competition' versus 'compensation' (taste cancellation) is not new to taste theory. To some authors who have considered the complexity of taste mixtures, 'competition rather than compensation seems to be the rule' (Trout & Sharp, 1937). The competition proposal seems to be more in tune with the chemical nature of taste. Figure 12.8 also demonstrates that there is little saltiness suppression by sucrose at the outset of the plot. Dilute solutions of NaCl have a sweetness attribute. As a consequence, there is 70% more suppression of sweetness by salt at higher concentrations than there is suppression of saltiness by sweetness (30%). The observation suggests a second reason for the polynomial nature of the saltiness and sweetness plots in Fig. 12.8.

At a point where the receptor(s) is nearly saturated with tastant, NaCl seems capable of exchanging or displacing sucrose in the receptor recognition reaction to effectively terminate the sweetness recognition interaction, and to initiate saltiness. The conclusion is in accord with the proposed ionic recognition bond for saltiness, which is probably stronger than the proposed concerted intermolecular hydrogen bond for sweetness. If so, the saltiness of NaCl and the sweetness of sucrose would not be cross-adaptable, which is the case (Sections 3.4.1 and 12.4.2.3).

Sweetness versus sourness. There is general agreement that sucrose suppresses sourness, but the suppression of sweetness by acids seems to be pH dependent. Changing the pH from 5·8 to 4·0 caused little suppression of sucrose sweetness, but lowering the pH to 2·5 afforded a 50% reduction in the sweetness of both glucose and sucrose (Stone *et al.*, 1969). Pangborn (1960, 1961) also reported the suppression of sucrose sweetness by acetic, citric, lactic, and tartaric acids.

As with saltiness and sweetness, halving the perceived intensities is the foremost effect (Moskowitz, 1972), again indicating competition for a common receptor. However, the ability of citric acid to depress the sweetness of 0·5–20% sucrose was more pronounced at the low citric acid concentration (0·007%) than it was at the highest (0·073%) citric acid concentration (Pangborn, 1963). This would seem to suggest that the sucrose interacts more strongly to a common receptor than does the single acid proton. The conclusion is contrary to the known interaction energies for sweet and sour tastes. Instead, the result suggests that there are different mechanisms for suppressing the taste of acids at different concentrations. Suppression of citric acid dissociation at higher concentrations of acids is the more likely cause in one case. If so, the action would be noncompetitive (Section 2.10.4.2). For the other, it was shown by Schifferstein & Frijters (1990) that 0·01 M citric acid and 1·0 M sucrose have about equal taste intensity, but 0·01 M citric acid reduces the sweetness of sucrose solutions (Fig. 12.9). This plot is typical of uncompetitive inhibition (Section 2.10.4.3). Kuznicki & McCutcheon (1979) found that at least some of the suppression of sweetness with acid occurs at the receptor site, and proposed competition between the sucrose and the acid for the receptor proton acceptor (B in this case).

Schifferstein & Frijters also found that sucrose solutions at 0·1 M concentrations reduced the sourness of citric acid solutions, but in this case the effect is much more pronounced (Fig. 12.10). It too is also highly indicative of competitive inhibition. As the suppression of sucrose sweetness intensity by citric acid was essentially independent of sucrose concentration, whereas the suppression of sourness increased with sucrose concentration (data not shown), clearly the effects are psychologically asymmetric. The effect of citric acid sourness and sucrose sweetness on each other may have an underlying chemical basis. In this case it seems to be the greater effect that molar solutions of sucrose have on water activity. Decreased water activity suppresses acid dissociation. The weak acid would then not taste as sour as it might if dissociation was not suppressed. In summary, as might be expected from

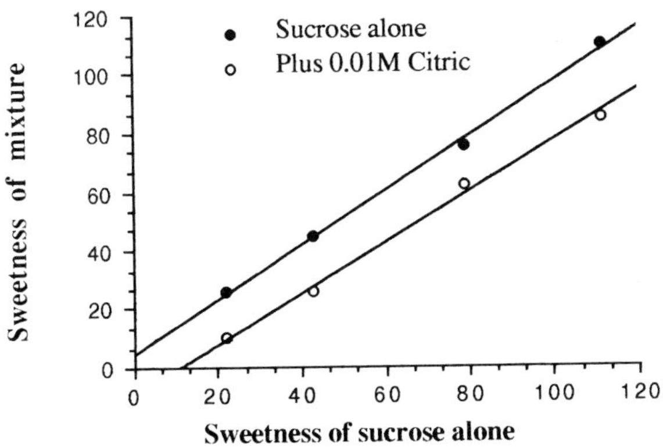

FIG. 12.9. The psychological function for the sweetness of sucrose alone and in the presence of 0·01 M citric acid (Schifferstein & Frijters, 1990).

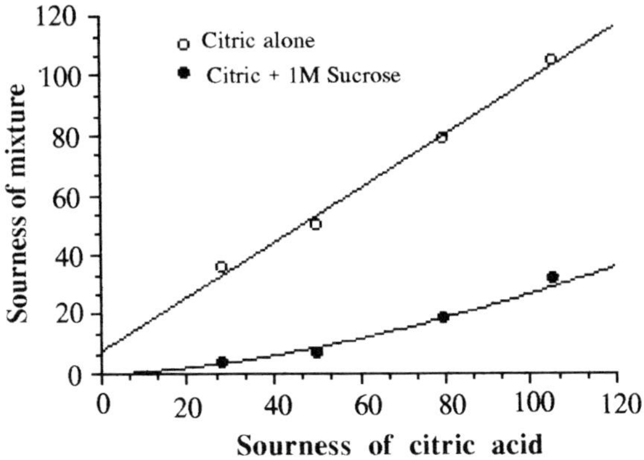

FIG. 12.10. The psychological function for the sourness of citric acid alone and in the presence of 1 M sucrose (Schifferstein & Frijters, 1990).

recognition chemistry phenomena, the nature of mutual taste inhibition for two substances with different tastes varies with their properties at different concentrations.

Sweetness versus bitterness. There is little question that bitterness compresses sweetness. Cameron (1947) found that a solution containing 5% sucrose and 2% urea was equal in sweetness to a 3·1% sucrose solution. The data of Moskowitz (1972) show a progressive decrease (approx. 50%) in the sweetness of 0·1 M glucose and fructose solutions containing 6×10^{-5} to $2{\cdot}5 \times 10^{-4}$ molar quinine sulfate. Generally, data on the effect of sweetness on bitterness is sparse. Bartoshuk (1975) reported that the bitterness of quinine hydrochloride is compressed when in admixture with sucrose, sodium chloride and HCl, but Pangborn (1960) found that sucrose does not mask bitterness in wines. Thus, while there is little question that sucrose can reduce some of the bitterness of substances, and bitter substances can effectively reduce sweetness, the bitter response is obviously the stronger of the two.

12.4.2.3 *Adaptation Studies*

Psychophysiological evidence to indicate that there is more than one receptor seems to follow from adaptation studies (Section 1.3.2). This is particularly true when adaptation to a substance with one predominant taste quality has little or no effect on the taste of a substance with a predominantly different quality. Thus, there is little evidence for any cross-adaptation among taste qualities. However, adaptation study formats do not generally seem to consider the chemical activity that different tastants may have toward a common generic receptor site. Adaptation studies are also conducted under unusual equilibria conditions (Section 2.4.2.2), i.e. using a forced zero-order kinetic system (Section 2.5.1). For these reasons, it is clear that the adaptation mechanism occurs in the transduction sequence of taste events, somewhat removed from the site for the initial recognition chemistry of taste.

12.4.3 Chemical Studies

From the chemical point of view, evidence for the existence of multiple taste receptors for the four basic tastes rests with the highly probable belief that the different sensations are caused by different initial reaction mechanisms (McBurney, 1974). As these forces are most likely to be van der Waals, hydrogen bonding, ionic, and dipole–dipole interactions, Beidler (1963) was the first to find it difficult to reconcile those different mechanisms with a single receptor. On the other hand, a single isosteric receptor with potential for multiple functionality can respond to those different reaction mechanisms. In the final analysis, all chemical reactions involve electrons and protons with different interaction energies. If the initial recognition interaction is *chemical*, with different types of reaction for the different taste modalities, then a relatively simple generic chemical site might be capable of various types of chemical bonding phenomena.

The question that arises next is that, if there are different initial reaction mechanisms for the four tastes, how is this translated into the different taste qualities? One possibility is that the different reactions initially lead to different subsequent transduction reactions. Evidence for different transduction mechanisms such as proton-modulated K^+ and Na^+ channels at the apical taste cell membrane, and opening of Ca^{2+} channels from a cascade of events for sweetness, is mentioned above, and in Section 1.4.

Evidence for different receptors for the different tastes is also found in the observation (Shamil *et al.*, 1987; Birch & Shamil, 1988) to the effect that small molecules have apparent molecular volumes characteristic of the different tastes. The authors concluded that specific receptors for the four tastes reside at different layers of the taste epithelium. However, the fact that substances that elicit different tastes have different molecular volumes might also be caused by a relation between specific volume and the nature of the chemical interaction at the receptor site. Different hydration shell orders among compounds that generally taste sweet, salty, sour, and bitter would then serve as the basis for the tenet that the chemistry for the different taste qualities is different.

Evidence for multiple receptors is also to be found in the action of taste suppressors, such as gymnemic acid and 2-(4-methoxyphenoxyl)propionic acid (Section 1.1.6.1). While gymnemic acid at high concentrations may abolish sweetness, and has some effect on saltiness, sourness, and bitterness, 2-(4-methoxyphenoxyl)propionic acid does not affect those tastes. In the former case, there is a suggestion of uncompetitive inhibition (Section 2.10.4.2), whereby only select sweetness-specific receptors, that can also interact with substances with other tastes, are removed from a general pool of receptors. Evidence for multiple receptors based on sweetness blocking can also be interpreted to be the blocking of only a specific type of taste recognition reaction that is unique to sweetness. While gymnemic acid must be applied to the tongue, the mode of action exhibited by 2-(4-methoxyphenoxyl)propionic acid is intriguing. That substance is effective at such minute concentrations in admixture with the sweetener so as to suggest that its mode of action would seem to be general specific inhibition as opposed to any of the classical inhibition formats. The latter, in turn, appears to be destruction or prevention of intermolecular hydrogen bond formation. The compound 2-(4-methoxyphenoxyl)propionic acid, like gymnemic acid, blocks the sweetness of *all* sweeteners, indicating a common mechanism for sweetness, and may indeed be a general detergent type of action. If that is indeed the mode of action for 2-(4-methoxyphenoxyl)propionic acid, the action strongly supports the contention that the initial chemistry of sweetness, as distinct from the other tastes, is due to hydrogen bonding.

It would also seem that the action of the protein Miraculin, wherein treatment of the tongue with the protein renders the taste of acids sweet, requires at least two different receptors. On the other hand again, one type of site on the receptor might be rendered active by the protein, while the acid renders a second site active. The net response would then equate to sweetness in much the same manner as described for the bimolecular generation of sweetness (Section 9.3.11.9). Thus, it can be concluded from chemical studies that all evidence for a multiple set of receptors for the four basic tastes can be equally interpreted as evidence for a single receptor, albeit one with multiple interactive capabilities.

12.4.4 Mathematical Models

From considerations of mass action phenomena, Lazareff (1922) concluded that all stimuli that show the same taste are mediated by the same receptor (Section 1.3.2). Considerable evidence for a single receptor for each of the basic tastes is also contained in taste psychophysical data and mathematical models (e.g., Beidler, 1954). The thermodynamic data that result from application of mathematical models also support the tenet. As the result of applying the Beidler equation to human sensory perception, McBride (1987) was able to derive equilibrium constants (K) for substances that elicit sweetness, saltiness, sourness, and bitterness. As K can represent the activity with which the stimulus interacts with the receptor, the manner with which it is related to that activity follows.

The thermodynamic significance of an equilibrium constant is that its logarithm is proportional to the maximum work involved in any given reaction, and activity (A) is

$$A = \mathrm{RT} \ln K$$

The equation is essentially the same equation as that for free energy, except that the negative sign for the right side of the equation has been dropped. Thus, the K values given by McBride (Fig. 10.26) for sucrose sweetness (5·3), NaCl saltiness (5·3), citric acid sourness (303·0) and caffeine bitterness (166·7) are, in terms of activity, 1·67 for sucrose sweetness and NaCl saltiness, 5·71 for citric acid sourness, and 5·12 for caffeine bitterness. The sweetness of sucrose and the saltiness of NaCl can then be assigned to symmetrical hydrogen bonding and symmetrical ionic bonding, respectively (cf. Chapter 11). With a probable energy of interaction of about 5 kJ/mol for sweetness and saltiness, the relative binding intensities for sourness and bitterness are $5{\cdot}12/1{\cdot}67 \times 5 - 15\,\mathrm{kJ\,mol^{-1}}$. For sourness, charge transfer would seem to fit the requirement, and for bitterness, a reinforced ionic bond for bipartite interaction, and/or a strong monopole–monopole hydrogen bond, reinforced by a hydrophobic interaction would seem to fit the energetic requirements for bitterness. Evidence from mathematical and thermodynamic models support the notion of different receptors for the different tastes, but, as with cross-adaptation studies, a common receptor with multifunctional capability cannot be strictly excluded.

12.4.5 The Chiral Nature of Taste

Strong evidence for a single receptor *type* for all tastes is found in the observation that the two antipodal tastes, sweetness and bitterness, have representatives to cause them that are simply enantiomers. Alternatively, one site is used by Belitz *et al.* (1988) for both sweetness and bitterness, but it has two γ functions. When one enantiomer elicits sweetness and the other bitterness (amino acids), the intensity of each modality is approximately equal. Saporous groupings for sweetness and bitterness therefore seem to share at least one common chemical function. It might be either an electrophilic or nucleophilic component, and that component probably interacts with a common chemical receptor site to elicit either sweetness or bitterness.

From consideration of the dipeptide substances, it was deduced that all the sweet substances not only possess AH,B, and γ, but they possess a conformation to bring AH,B,γ into an equivalent planar relationship. This indicates a common receptor for peptides and related substances (Douglas & Goodman, 1991).

A single topological site has been proposed by several authors to account for the sweetness of substances. The motivation for topological proposals is the tenet that goodness of fit into the receptor structure is the most significant attribute of a sweetener. A complete bipolar (AH,B) unit is not even required (cf. Temussi, 1991, pers. comm.). Culbertson & Walters (1991) and Tinti & Nofre (1991*b*) proposed multi-site topological sweetness receptors. Those receptors have just one topological structure capable of responding to nearly all chemical variations imposed by sweet substances. Unfortunately (Chapter 11), such versatile topological structures no longer possess ability to interact preferentially with amino acid enantiomers. Irregular topological structures possess potato-like asymmetry (Section 2.11.2), but it is difficult to determine the application to either common or multiple receptor site dissymmetry arguments.

12.5 SINGLE AND/OR MULTIPLE RECEPTORS WITHIN EACH BASIC TASTE

The bulk of arguments for single and multiple receptors within each basic taste is largely directed towards sweeteners. Neurophysiological and psychological evidence for multiple sweet receptor mechanisms appears to be overwhelming, but there are still doubts (cf. Bartoshuk, 1987). The argument can be advanced that if chemical mechanisms for different tastes are different, isosteric receptor groups for the induction chemistry of each taste are possible. Isosteric groups on proteins of different amino acid composition and sequence is also a strong possibility.

12.5.1 Neurophysiological Studies

Evidence for different receptors for sugar sweetness is suggested by data obtained from electrophysiological studies. Single taste fibers in a number of animals give a differential electrical response to different sugars (Beidler *et al.*, 1955; Pfaffmann, 1955). Beidler & Tonosaki (1985) constructed a bar graph summary of data from several sources. The whole nerve tympanic chorda responses to different sugars are the same for rats as for humans. The results seem to indicate that there are multiple receptor sites even for different sugars. On the other hand, steric differences among the sugars is sufficient to alter their ability to interact with a common receptor site. The observations of neurophysiologists, with respect to the fact that a single fiber responds to more than one taste, can also be assigned to a given receptor reacting with saporous groups in several different capacities. Results obtained with threshold diminution in mixtures (Section 12.4.2.1), and results obtained with straightforward mixtures of compounds (Section 12.4.2.2) seem to fit well with the common receptor proposal.

Jakinovich (1976) directed attention to the finding that, among the disaccharides, sucrose (as determined by electrical receptor cell response) is sweeter than all other disaccharides, and it was concluded that there must be a specific sucrose receptor site (Jakinovich, 1985). For monosaccharides, the sweetness order found was fructose > mannose > glucose > galactose (Jakinovich & Goldstein, 1976). The order is not quite the same as for humans (mannose is bitter-sweet), but it is of the same order found for humans by recording the response in the chorda tympanic nerve (Diamant *et al.*, 1963). Jakinovich & Goldstein (1976) also found that methyl α-D-glucopyranoside and methyl β-D-fructofuranoside are the most stimulatory to the gerbil, as measured by electrical response. Therefore, they proposed that the sucrose receptor site is made up of two subsites. One binds the glucopyranosyl moiety of sucrose, and the other binds the fructofuranosyl moiety. However, to humans, methyl α-D-glucopyranoside tastes both sweet and bitter and methyl β-D-fructofuranoside is tasteless. Methylglycoside properties also bear little relation to glucosylfructoside and fructosylglucoside properties. Methylglycosides are lipophilic, sugar–sugar glycosides are lipophobic. More recently, Jakinovich & Sugarman (1988) proposed that while sucrose and other sugars react at a single receptor, a mixture of sucrose and saccharine interact at a different receptor. The electrophysiological results again indicated that there are two major sites for sucrose sweetness.

12.5.2 Psychophysical Studies

Faurion *et al.* (1980) found that human sensitivity to various sweet compounds indicates that sweet taste can be represented in a three-dimensional continuum if twelve different sweet compounds are considered. That observation indicated to the authors that there are several distinct receptor mechanisms for sweeteners. The results serve to also support the tenets that different substances have different activity toward a common receptor, and no two chemical substances have the same taste profile and taste spectrum.

12.5.2.1 *Adaptation Studies*
Further comment on the application of sweetness adaptation studies (Section 1.3.2.4) as an indication of the number of receptors, or even the number of receptor sites at a given receptor, seems now to be in order. In effect, to say that sweetener X cross-adapts sweetener Y is to say that after adaptation to X, Y does not taste sweet either. To say that X does not cross-adapt Z is to say that Z yet tastes sweet. To infer that there is a different receptor for Z than for X and Y ignores the tenet that taste is a chemical sense. Indeed, the notion of asymmetric cross-adaptation, where one compound adapts another, but not *vice versa*, probably is a manifestation of chemical recognition activity toward a common interactive site. In chemical terms, X and Y simply have about the same activity toward the receptor, but Z has even greater activity toward it. In view of the lack of a known mechanism for adaptation (other than, perhaps, 'fatigue'), and lack of knowledge as to where adaptation occurs in the chain of events that leads to it, it can be concluded that the efficacy of adaptation studies as criteria for the taste receptor number is low. The adaptation data that are interpreted to indicate different receptors for given tastes follows, even though McBurney (1972) proposed, on the basis of adaptation data, that there is a common type of receptor for all sweeteners.

The results of cross-adaptation among sugars, sugar alcohols, and high intensity sweeteners are highly variable (Section 1.3.2.4). Nevertheless, Schiffman *et al.* (1981*b*) concluded that there are multiple receptors for sweetness. Even here results of the adaptation studies are mixed. Adaptation to sodium saccharine increased the sweetness of glucose by a highly significant 17%, and adaptation to acesulfame K also enhanced the sweetness of the sugars. Lawless & Stevens (1983) found that several high-potency substances (dihydrochalcone, aspartame, and saccharine) were cross-adapted by themselves, and sucrose was also adapted by the high-potency sweeteners. The finding that either partial or no cross-adaptation occurs when the intensive sweeteners are used as the preadaptation rinse suggested that there are 'multiple peripheral mechanisms for the encoding of sweet taste.' Lawless & Stevens (1983) also found that adaptation to saccharine diminished the sweetness of aspartame, but adaptation to aspartame did not affect saccharine sweetness. Again, such results confirm that different sweeteners have different activity toward receptors with common isosteric features, and the multiple peripheral mechanisms are probably due to a compound's taste profile and taste spectrum. In view of the different rates with which diffcrent sweeteners exhibit adaptation (DuBois *et al.*, 1993), it is clear why Bartoshuk (1987) was led to conclude that the ability of adaptation studies to reveal the number and types of receptors is, in itself, nonconclusive.

Cross-adaptation studies with glucose, fructose, galactose, sucrose, and maltose led Saito & Miyamoto (1988) to conclude that there are pyranose and furanose receptor sites in the human. In this case, it is well known that the different sugars have different activity toward the receptor, as their relative sweetness varies. Whether it is a matter of the concentration of pyranose or furanose tautomers present in their structure, or at mutarotational equilibrium is a moot point.

Clearly, interpretation of results obtained from adaptation studies is difficult, and application of adaptation findings to the problem of different receptor types and numbers for sweetness is probably not valid. This is particularly true if adaptation is caused by a biochemical event somewhat removed from the chemistry for the initial recognition of sweetness, i.e. in the transduction process. The observations do strongly indicate, however, that chemical activity phenomena are operative at the induction of taste. Sweet-tasting substances that interact more strongly with the sweet receptor show either partial or no adaptation with substances that interact more weakly.

12.5.2.2 *Taste of Mixtures*

Bartoshuk & Cleveland (1977), in a study of two-component mixtures of glucose, fructose, sucrose, and maltose, found that the sweetness was about that to be expected from the contribution by each sugar alone. However, sweetness of the three component mixtures of the sugars was significantly higher. The finding suggested that if the receptors stimulated by each of the sugars were not completely identical, the synergism might be explained on that basis (Bartoshuk, 1987). As the synergism observed was only slight, it can also be explained simply on the basis of superimposing different sweetness profiles.

Adaptation studies also suggest that there may be different receptor processes for the sour taste of acids. Those findings are based on the

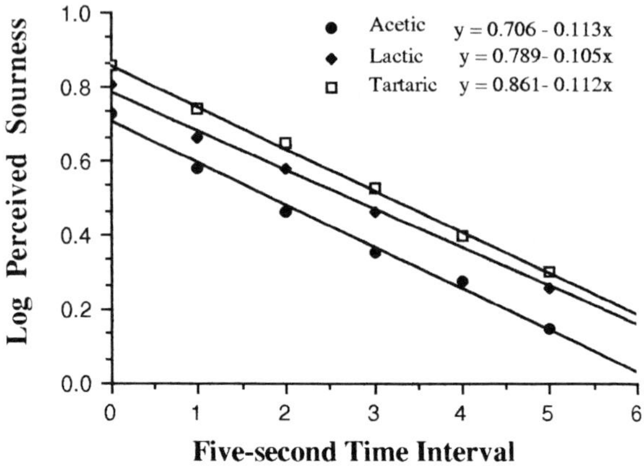

FIG. 12.11. First order reaction kinetics for the perceived sourness of organic acids plotted against elapsed time after tasting (data taken from Ganzevles & Kroeze, 1987).

observation that there is no self- or cross-adaptation with HCl (see, however, Section 1.3.2.2), whereas the carboxylic acids self- and cross-adapt to different degrees (Ganzevles & Kroeze, 1987). The authors also pointed out that the degree of adaptation (acetic > lactic > tartaric) is inversely related to the dissociation constant for these acids. As degree of acid dissociation seems to bear little relation to acid sourness (Section 10.1), there seems to be little reason for it to be now related to the degree of adaptation. In fact, the degree of adaptation is directly related to molecular weight, and to the diffusion coefficient of the acids. It can also be related to the antilog of the extrapolated initial sourness of the acids (5·08, 6·15, and 7·26, respectively) obtained by plotting the data of Ganzevles & Kroeze (1987) according to first order reaction kinetics. This plot of their data is shown in Fig. 12.11, and correspondence to unimolecular reactions is ideal for all the acids. As the acids all have H^+ in common, but differ with respect to the nature of the anion, the varied responses observed are probably due to the activity of the anion at the nucleophilic component of a generic receptor.

12.5.3 Chemical Studies, Including Synergism

There is considerable evidence for multiple receptors for sweetness based on chemical considerations alone. There is great diversity among the substances that cause sweet taste. This led Crosby *et al.* (1979) to conclude that substances having structures of such great divergence must exert their effects at different receptors. However, an equally valid argument is that it seems quite implausible to expect that such diverse substances with the same taste quality do not have some common chemical denominator, although activity toward the receptor might differ. Such compounds might even possess various abilities to elicit other taste attributes, but they have a common feature analogous to the proton common to all substances that taste sour.

Even here, though, it is the attempts to derive the nature and function of a common sweetness denominator, such as AH,B, and γ that also lead chemists to the position that there are multiple receptors for sweetness. Within a homologous series of sweet substances, such as amino acids, while the length of the side chain is important, the size and shape of the side chain also seems to have a role. The effect of size and shape of side chains is usually interpreted to be of topological significance, but it might also be an inductive effect on electron distribution. From assessments of the identity and position of γ, van der Heijden *et al.* (1985*a*) pointed out that there are at least three and possibly four different geometric receptors for sweet taste. However, no matter how receptor geometry is envisaged, the general geometry of three reactive sites (AH,B,γ) appears to be an uninterrupted plane. That fact is also a feature common to the ten series of homologous sweeteners examined by van der Heijden *et al.* (1985*a*), and is commensurate with the notion of a tripartite prochiral receptor.

For sweetness, Rohse & Belitz (1991) concluded that there are at least two different receptors required that are based on e/n distance parameters alone. However, in at least one of those cases (Section 9.3.13.1) it was pointed out that there are other e/n (AH,B) possibilities. The significance of molecular topology has also led a number of authors to conclude that there are different receptors for sweet taste. Indeed, Spillane & McGlinchey (1981) and Spillane & Sheahan (1989) concluded that there is need, based on area and volume parameters, for different sweetness receptors even within the same class of compounds (sulfamates). Although the proposal seems unlikely, it receives further discussion in Section 12.6.

Considerable discussion in Chapter 11 was directed to the notion that there are complex single topological receptor forms to accommodate the sweetness of all substances. Two of them possess multiple active sites to accommodate sweetness ranging from that of ethylene glycol to the extraordinarily sweet substances. They were developed to account primarily for high-potency sweetness, which is presumed to be a fourfold interaction, but that receptor concept is incapable of explaining the varying taste of amino acid enantiomers.

The notion of synergism has also long been offered as evidence for multiple receptors for a given taste, particularly sweetness. While isosteric receptors for sweetness are one possibility to explain the phenomenon, explanations based on a single sweetness receptor also lead to an answer, particularly for the well-known synergism of saccharine and dulcin (Moncrieff, 1967). For example, with an AH,B,γ glycophore, and a commensurate AH,B,γ receptor, several explanations for the notion of synergism, and its nature, are possible. The first is topological, i.e. neither compounds x and y shown below has the appropriate AH,B distance for optimum activity, but x has the proper AH,γ distance and y has the proper B,γ distance. Together, the 'average' geometry, is commensurate with the receptor. Thus, AH,γ fit the receptor better in the case of compound x, but for compound y, B,γ fits better. The net result is then synergism and $1 + 1 \geq 2$.

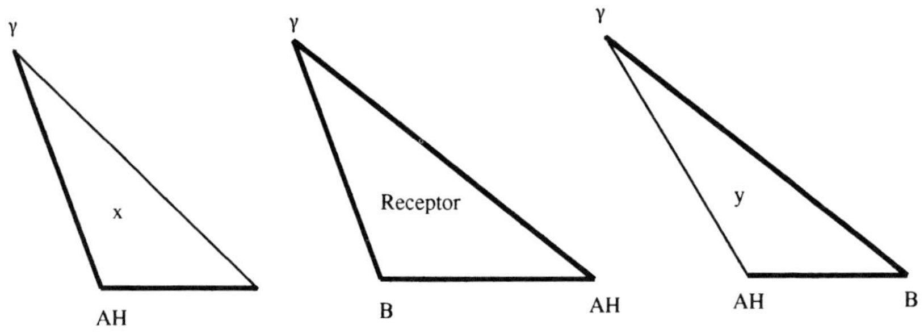

x + y = Synergistic glycophores for the receptor

In a second example, the noncommensurateness between AH,B,γ for the compounds is not completely topological. Each has an element of antipolar dissymmetry with respect to AH versus B. In this case compound x has a greater net positive charge on the AH,B dipole, and compound y has a greater net negative charge on the AH,B dipole. Together, the pair of dipoles represent glycophores with dipoles equally charged. The net effect would be to minimize the bitter note for the dissymmetric dipoles, and accentuate the sweetness note. Parenthetically, both saccharine and dulcin have an intrinsic bitterness attribute.

12.6 A GENERIC RECEPTOR MODEL FOR TASTE

Ultimately, argument for and against different receptors for each of the basic tastes, and for or against different receptors for substances within the same basic taste suggests that there is not a straightforward answer to those arguments. One of the reasons for this is that there are nuance differences within compounds with the same basic taste (taste profile). A second is that it is not always clear what is meant by the term 'different receptor.'

With a protein for the receptor, the number of functional groups available, and to which an AH,B or e/n function can be assigned, is quite limited. For AH, or an electrophilic center, only NH, —NH_2, and —OH is available. For B or a nucleophilic center, only carbonyl and/or hydroxyl groups, and perhaps a peptide bond oxygen atom, are available. For the hydrophobic function, however, not only are hydrophobic side chains available as structural receptor entities, but the lipoid matrix of the membrane surface itself is a possibility. When amino acid sequence is added to the possibilities for the nature of the protein, the potential for different receptors becomes very large.

However, if it is granted that the e/n groups can function in a variety of ways to distinguish among the different taste modalities, then there can be a generic receptor for the general e/n functions. The generic receptor can be considered to be 'different' with respect to the e/n isosteric groups possible, and there is also possibility for 'sweetness best' or 'saltiness best' receptors. Therefore, the belief that there are different receptors, even for the sweetness of the different sulfamates may be true, to a degree. A given receptor may be a multifunctional receptor, and it may also be a 'sweetness best' receptor.

When a hydrophobic function is included for high potency sweetness or bitterness, the number of ways that receptors might function in the same, albeit isosteric, manner becomes even larger. Although all hydrophobic groups are isosteric by definition, their degree of hydrophobicity and lipophilicity, is different. The degree to which electronically isosteric groups are alike or different in their action (i.e., participation in ionic, hydrogen bonding, and other chemical reactions) may also be large.

The above arguments for a single or multiple receptor for the basic tastes is largely resolved by a generic receptor with isosteric functions. It is a 'single' receptor with respect to isosteric functions, and a 'multiple' receptor with respect to the identity of the isosteric functions that have ability to interact with substances that have different chemical properties. Such a generic model was implied for the initial chemistry of taste (Chapter 10), and when the symmetry relations among saporous units were considered (Chapter 11).

A generic receptor is shown in diagrammatic form in Fig. 12.12. The general nucleophilic (N^-) and electrophilic (E^+) groupings of Belitz *et al.* (1981) serve as the functional isosteric components of the basic Kier (1972) triangle. It also has broad directional and distance features for the location of the hydrophobic site (cf. van der Heijden *et al.*, 1985*a*, *b*, and Fig. 9.26). The e/n notion was chosen as the general functions, not to account for the differences among sweeteners, but to account, as a general N/E receptor, for different chemical reactions for the four taste modalities. It can also account for the competition shown among taste substances, either at threshold levels (Section 12.4.2.1), or in mixtures (Section 12.4.2.2). Therefore the competition is derived from N/E functioning in a symmetrical hydrogen bonding manner for sweetness, and in a symmetrical ionic manner for saltiness in sugar/NaCl mixtures. It is also derived from the general competition for a multiply functional common site by the dissymmetric chemistry characteristic of sourness, and the varied dissymmetric chemistries of bitterness. As there are different mechanisms for the four taste modalities, there is no cross-adaptation among taste modalities (Section 12.4.2.3).

Perhaps the most salient feature of the general isosteric receptor is that it is prochiral, and one-sided. The three functional groups are different, and the arrangement is generally skewed 'to the left.' It can distinguish

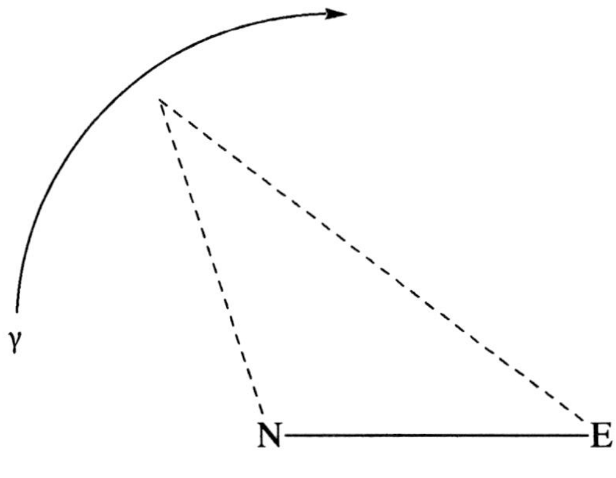

FIG. 12.12. A generic receptor for taste, with variable electronic functions for N and E, and variable positions of γ for intense sweetness.

the respective sweet and bitter tastes of the D- and L-amino acids, and the antipodal intensity of those tastes would be of about the same magnitude. Thus, there is little need for mirror-image receptors, or multiple γ sites (Section 11.3.2). At the same time, the generic receptor can respond about equally to the prosymmetric glycophores of D- and L-glucose, and related sugar enantiomers, and preferentially to fructose enantiomers (Section 11.2.1.2). The greater interaction activity of the higher-potency sweeteners is also taken into account by the generic receptor. That trait is primarily due to the inductive effect that γ and other substituents have on the activity of the AH,B component of the glycophore (Section 10.3.4).

The generic receptor also allows for a 'sweetness best' or 'saltiness best' receptor, etc. (different isosteric N/E groups). One candidate for a sweetness-best receptor is that offered by Suami & Hough (1991) in Section 12.2.5. Indeed, when the amino acid composition and sequence is included, the basis for both common and different features for a generic sweetness receptor are also large.

Epilogue

Although there is a paucity of conventional criteria normally employed to characterize discrete chemical reactions, the first conclusion to be derived from application of chemical principles to taste is that taste is indeed due to chemical reactions, and substances that taste sweet, salty, bitter, or sour have different saporous (sapid, saporophoric) groups that lead to characteristically different chemical reactions. The conclusion is based on the finding that the laws of definite proportion and chemical equivalency apply to taste chemistry and that principles of mass action, chemical equilibria, symmetry point group theory, and stereochemistry (chirality) are either operative, or apply. Moreover, from the thermodynamic viewpoint, the change in free energy is of proper sign (minus) for the interaction of two substances (tastant and receptor), and of proper magnitude for weak nonbonded interactions. Thus, while a given chemical principle alone is insufficient basis for the aforementioned conclusion, they collectively represent substantial evidence for it. The second conclusion is that the different tastes must then be due to weak recognition reactions that occur in solution, and are not the result of binding with, or affinity for a tastant for a particular receptor-type.

For sweetness and bitterness, common saporous and isosteric chemical functions such as glycophores for sweetness (Gr. *glyc*, sweet; *phoros*, to carry) and picrophores for bitterness (Gr. *picros*, bitter) prevail. The halophores for saltiness (Gr. *halos*, salty) are simply metal cations and nonmetal anions. However, only the sodium and chlorine ions jointly elicit a truly salty taste, and those ions have special (albeit antipodal) common properties. The acidophore for sourness is simply the hydrogen (hydronium) ion.

The primary distinction among the different reactions is that sweetness is apparently induced by concerted antiparallel bipartite hydrogen bond interactions. Saltiness is induced by concerted metal and nonmetal antiparallel-ionic interactions particularly characteristic of sodium chloride. Sourness is induced by the ionic activity of a special cation (H^+) and/or by displacement on the hydrogen ion. Bitterness is induced either by ionic interactions with special characteristics and/or joint positive or negative monopolar-hydrophobic interactions.

The different taste intensity for the halophore (Na^+ and Cl^-) for saltiness and the acidophore (H^+) for sourness is a function of concentration (including equivalency) only, and there are no high-potency salt- or sour-tasting substances. On the other hand, the different taste potencies of sweet and bitter substances can be attributed, in part, to an exponential consequence of an interaction between chemical activity and the effective concentration of isosteric glycophores and picrophores, respectively. As recognition reactions in solution serve to increase the effective concentration of tastant through orientation, an additional explanation for the exponential relation between chemical activity and concentration for the sweet and bitter tastes emerges. The explanation serves to begin a marked delineation between taste chemical recognition reactions and affinity (binding) reactions. That there would be an intrinsic exponential response to the inductive effect of a hydrophobic glycophoric component, and to appended groups with high inductive capacity (NO_2, CN, Cl) then logically follows. Furthermore, a compound with multiple saporophore equivalency would also contribute to an exponential increase in orientations that are recognized by the receptor, and produce a response. As a consequence, sweet and bitter taste recognition potency may vary over five log units, even for compounds with a common glycophore, such as an L-aspartyl residue.

While the ability of certain substances to block sweetness, but not other tastes, seems to be due to disruption of the mechanism responsible for sweetness (hydrogen bonding), taste blindness to bitter-tasting compounds seems to have a genetic basis, as an explanation for the phenomenon cannot yet be found in the recognition chemistry mechanisms for taste.

Symmetry attributes can also be assigned to the different chemical recognition mechanisms for sweetness, saltiness, sourness, and bitterness, and serve to classify them. The sweet and salt tastes are characterized by bilaterally symmetrical interactions between the tastant with the receptor, i.e. respective concerted δ^+ and δ^- dipolar or concerted anionic and cationic recognition interactions of opposite sign, but of near-equal magnitude. The latter is the special bipolar property of the sodium and chlorine ions. When the concerted interactions are of unequal magnitude other taste attributes, notably bitterness, are experienced. The sour and bitter tastes, then, are characterized by bilaterally dissymmetric bipolar or dipolar interactions, i.e. the concerted (special ionic or dipolar) interactions are of unequal magnitude. The dissymmetry may result either from unequal numbers of ions or unequal activity. The latter is the special feature of bitter bipolar and dipolar substances, and accounts for the fact that only sodium chloride tastes truly salty. On the other hand, all acids have equal sourness on a hydrogen-ion equivalency basis, and the dissymmetry results from an unequal number of H^+ and OH^- ions, and/or the ionic activity of the hydrogen ion versus the anion.

There is also a stereochemical requirement for the two tastes that are generally initiated by substances that do not ionize in solution, i.e. sweetness and bitterness. The stereospecificity toward right- and left-handed substances, or lack of it in certain cases, is manifested in the two-dimensional prochirality displayed by the receptor. It is also manifested in the occasional prosymmetrical two-dimensional glycophore faces displayed by some enantiomeric substances (e.g., sugars). As a consequence, there is little need to invoke multicomponent topological or even three-dimensional features to the receptor(s) in order to account for observed taste stereochemical phenomena. Finally, it would seem that the aforementioned stereochemical phenomena only become enigmatic in the realm of taste chemistry when the three-dimensional topological notions of chirality are dogmatically invoked. Although the latter application of the notions of chirality and handedness may be intuitive, it is not necessarily correct. The spatial requirements for three-dimensional binding reactions are much more rigid than for two-dimensional recognition reactions.

As an additional consequence of the above conclusions, the extent to which precepts and concepts of enzymology and pharmacology have application to taste chemistry is open to serious question. There are elements of recognition chemistry in the total chemistry of enzyme and drug phenomena, but unlike enzymic and pharmacological action, taste 'substrates' are not transformed to products, and there is no pharmacological effect that is a consequence of a compound's taste. It would appear that taste chemistry is recognition chemistry only, but with corresponding mass action, chemical activity, equilibrium, and stereochemical features. Clearly, then, there is an activity component to taste recognition chemistry phenomena, but no affinity component. That is why conventional enzyme and drug occupancy theory, with attendant notions of competitive, noncompetitive and uncompetitive inhibition do not readily apply to taste chemistry. When taste (particularly sweetness) inhibition occurs, the data obtained display features characteristic of all types of affinity inhibition.

With respect to conventional enzyme and pharmacological 'lock and key' or hand-glove analogies, it is the receptor for taste that is the probe, or 'key,' and the substrate is the 'lock,' or more accurately, a face on the 'lock.' The probe has features that interact strongly or weakly with tastants, and those features depend on tastant electronic characteristics (activity) and concentration, and not the receptor's electronic character or three-dimensional topology.

A number of different receptors for taste are possible, but they probably have many isosteric chemical features. The common features are a prochiral tripartite scalene arrangement of a nucleophilic, electrophilic, and hydrophobic interactive sites, and the receptors are probably contained on the surface of a protein embedded in a lipoid matrix. The fact that proteins are three-dimensionally chiral substances is moot. The multiplicity of the receptor number can then result from the number of different isosteric groups that can function in several different chemical recognition reaction capacities. Therefore, they can interact with tastants with different taste modality by different chemical mechanisms, which accounts for the taste competition displayed by mixtures of substances with different taste. The receptor multiplicity can also be due to the amino acid sequence for the different proteins, different amino acids, or the overall nature of the protein itself (glyco-, lipo-, metallo-protein).

Lastly, it can be stated that many so-called enigmatic taste phenomena that result from interpretation of results by application of principles from different academic disciplines have straightforward explanation upon application of chemical principles. For example, the increased rate and magnitude of adaptation shown by high-potency sweeteners versus sugars is related to glycophore chemical recognition activity alone. On the other hand, the decreasing relative sweetness shown by the high-potency sweeteners with increasing concentration is a manifestation of the psychophysical laws alone, and does not have a chemical recognition basis. The observation that taste is, for the most part, psychologically analytic in nature also supports the conclusion that different tastes are due to different chemical reactions.

Glossary

Inasmuch as stereochemical terms are employed in this text that are not completely defined in the usual dictionary, they have been gathered together, along with other terms, in this glossary for convenient reference. Their principal meaning can be obtained through reference to it, and reference is also made to the Sections in the text where the term is used. Similarly, certain symmetry principles are applied to the chemistry of taste in what may be an unfamiliar manner, e.g. the notions of bilateral symmetry and bilateral dissymmetry. In those cases, an attempt is made at further clarification in the glossary.

Acetal. Whereas a hemiacetal is the product produced by the reaction of a carbonyl group with an alcohol group, reaction of the hemiacetal with another molecule of alcohol (or an additional hydroxyl group) leads to formation of a full acetal (Section 4.4).

Achiral. Symmetrical (Section 2.11.3.5).

Acidophore. A group that elicits acid (sour) taste. Usually the hydrogen ion (Sections 2.9 and 9.1).

Activity. Chemical activity. Chemical activity is the ratio of the effective concentration of a species to its concentration in a standard reference state. Also, the propensity for an interaction to occur (Section 2.4.2.1). The taste chemical recognition activity of a substance can be defined as the response elicited per unit concentration.

Adaptation. The decrease in the perceived intensity of a taste sensation with continuous exposure to the stimulant (Section 1.3.2).

Affinity. Chemical affinity. The attractive force that causes chemical groups or substances to enter into and remain in chemical combination (Section 2.4.2.1), or for a substrate to remain bound to a receptor.

Agonist. A compound that elicits pharmacological or drug activity. A sweetener may be loosely called a 'sweet' agonist, but neither it, nor the sweet taste interaction necessarily elicits a pharmacological response.

***Allo*-isomers**. Diastereoisomers (Section 5.1.1). Examples are L-threonine, and L-*allo*-threonine; D-threonine and D-*allo*-threonine. In carbohydrate chemistry trivial names for diastereoisomers have little in common, and emphasize that many substances with the same molecular formula are, in fact, different compounds, e.g., D-glucose and D-galactose. By the same token, threonine and *allo*-threonine have little in common.

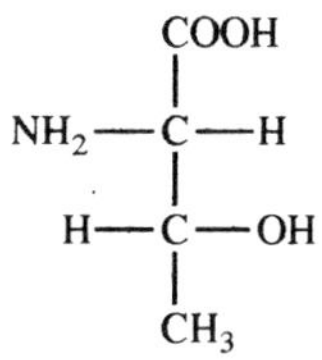

L-Threonine

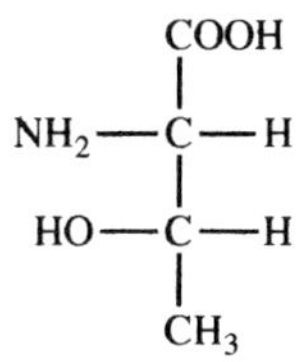

L-*allo*-threonine

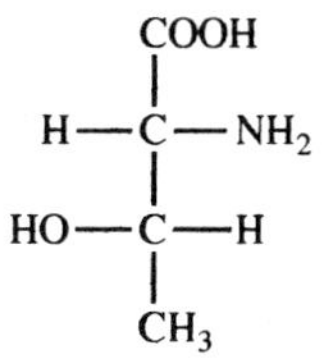

D-Threonine

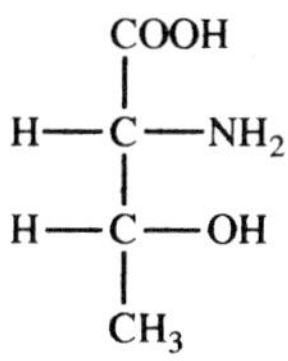

D-*allo*-threonine

Allosteric effect, allosteric inhibition. An effect that is neither competitive, noncompetitive, nor uncompetitive (Section 2.10.4.3). End product inhibition in enzyme chemistry is often given as an example of allosteric inhibition. There probably is little, if any, stereochemistry involved. Taste inhibition is probably allosteric.

Ångstrom unit (**Å**). 0·1 nanometer (nm), or 1×10^{-9} meter. 1 nm = 10 Å.

Anomer. A sugar configurational isomer at the carbon atom rendered asymmetric by virtue of hemiacetal ring formation (Section 5.1.2.1). For aldohexoses it is carbon atom number one. Examples are α-D- and β-D-glucose. For ketohexoses, the anomers result from ring closure at carbon atom number two.

Anti-conformations. A term usually used to describe the opposed disposition of two groups on a molecule characterized by a projected angle of 180° (Section 9.3.7.5).

Antipodes, antipodal. Diametrical opposites, which may refer to geometric structures (right and left hands, enantiomers), the specific optical rotation sign for enantiomers, and even to tastes (sweet versus bitter).

Antistere. Antisteric group. A group with taste functionality opposite to that of another (Section 2.9). See also isostere.

Asymmetric carbon atom. A carbon atom with the four tetrahedrally-oriented valencies substituted with four groups of different size (Section 2.11.3.3). The term is established by use, and such an arrangement of atoms is chiral, or perhaps dissymmetric, rather than asymmetric. The asymmetric carbon atom is often called a 'chiral center.'

Asymmetry. Complete lack of symmetry (Section 2.11.2).

Bilateral dissymmetry. The disproportionate occurrence of antipodal forms. It is manifested in disproportionate left and right forms of different size, shape, number, electrostatic charge, or even activity (Section 2.11.4). Examples are the dissymmetric concentration of hydrogen and hydroxyl ions at pH other than 7; the number of positive charges on an anion versus a cation; the disparity between the activity of a positive and negative dipole; the different sizes of anions and cations, etc.

Bilateral symmetry. The proportionate combination of left and right geometric forms. Also, the equal presence of antipodal or reciprocal elements, such as the number of hydrogen (H^+) and hydroxyl (OH^-) ions at neutral pH. In essence, bilateral symmetry is a form of 'proper proportion' (Section 2.11.4).

Bipole, bipolar. Having diametrically opposed ionic charges. Collectively, separate an ionic and cationic species (Section 2.11).

Calorie (**cal**). An energy expression. 1 cal = 4·184 Joule.

Chiral. Possessing handedness. The occurrence of either left- or right-handed geometric forms. Possessing only a simple axis as the one element of symmetry. Hence, the right and left forms can be distinguished by convention (Section 2.11.3).

Chiral center. An abstraction of the idea of the asymmetric carbon atom, but also used for entire molecules (Section 2.11.3.5).

***Cis, trans* isomerism**. A form of configurational isomerism with no reference to the asymmetric carbon atom, but to whether groups lie on the same (*cis*) or opposite sides of a molecule (*trans*). Examples are *cis* and *trans* 1,2 cyclohexane diol (Section 4.3) as shown in Table 9.2.

Configurational isomer. Isomers formed by transposing the configuration about an asymmetric carbon atom (Section 2.11.3.3). To do so, single covalent bonds need to be broken and reformed. When only one asymmetric carbon atom is present in a structure, the configurational isomers are enantiomers. When two or more asymmetric carbon atoms are present, the possibility for diastereoisomerism arises. When all configurations at the asymmetric carbon atoms have been transposed, the compounds are enantiomers.

Conformational isomer. Isomers formed by rotation about single covalent bonds, in contrast to the severance and reformation of bonds for configurational isomers. Conformational isomers are different forms (shapes) of the same compound. For enantiomers, the favored conformations are also nonsuperposable mirror-image shapes.

Cross-adaptation. The decrease in the perceived intensity of a second sensory stimulant upon adaptation to another (Sections 1.3.2 and 12.5.2.1).

D-Amino acid. An amino acid whereby the amine group on the reference carbon atom is written to the right in Fischer projection structure (Section 7.1.1). See also enantiomer.

D-Sugar. A sugar related configurationally, at the highest-numbered asymmetric carbon atom, to D-glyceraldehyde. Therefore, the hydroxyl group for the Fischer projection structure is written to the right. In contrast, the prefix *d*- or (+) refers to the direction (*dextro*-) of the compound's optical rotatory power (Section 5.1.1.2). See also enantiomer.

Diastereoisomer. (See also *allo*-isomer). A geometric isomer that is not an enantiomer, e.g. galactose and glucose are configurational diastereo-isomers (epimers). β-D-glucopyranose $^4C_1$ and β-D-glucose $^1C_4$ are conformational (shape) diastereoisomers. β-D-Glucopyranose $^4C_1$ and β-D-glucopyranose $^1C_4$ are configurational and conformational geometric opposites, i.e. enantiomers (Section 5.1.1.1).

Dipole, Dipolar. The unequal distribution of positive and negative charges in the structure of a molecule or group (Section 2.7.1).

Dispersion bond. Attraction of two like atoms or groups through induced dipolar configurations (Section 2.7.5).

Dissymmetry. Degraded symmetry. Often used on a one-to-one basis with structural chirality (see however, Section 11.1.1). The term dissym-metry is also used in this treatise to denote degraded bilateral symmetry with respect to size, shape and number of antipodal species (Section 2.11.2).

Eclipsed conformations. Basically a celestial analogy and usually used to describe the disposition of two groups on a molecule characterized by a projected angle of 0° (Section 9.3.7.5).

Electrophilic. A group that attracts and can bind, or is capable of recognizing, an electron (Section 2.3.1).

Enantiomer. Enantiomers have the same composition and molecular formula, but have opposite configuration and conformation. An enantiomer is therefore either one or the other of two possible geometrically opposite chemical structures (Sections 1.5.2 and 2.11.3.3 and Chapters 5 and 7). Both D-threonine and D-*allo*-threonine have an enantiomer (See *allo*-isomer). Further examples are:

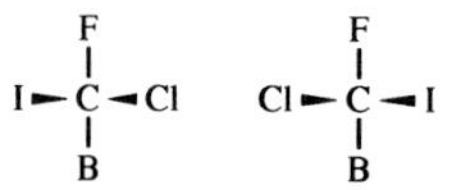

Iodo, fluro, chloro, bromo - methane enantiomers

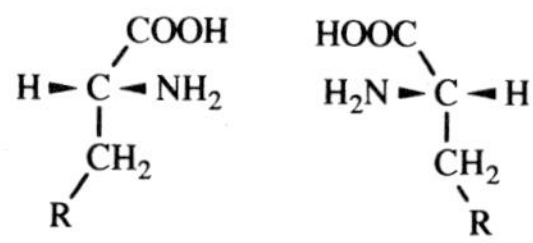

Amino acid enantiomers

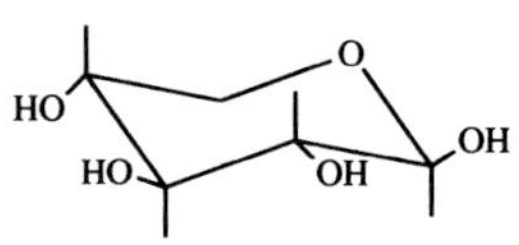

D-Sugar, $^4C_1$ conformation

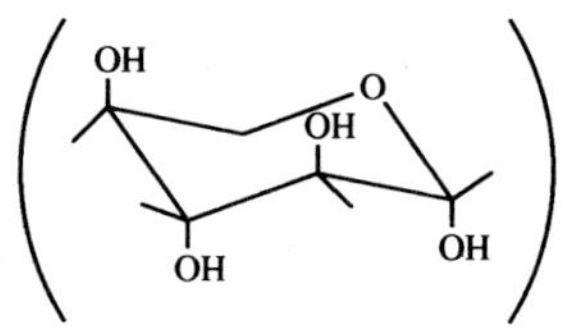

L-Sugar, $^4C_1$ conformation

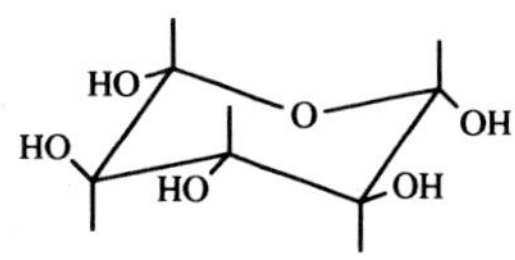

L-Sugar, $^1C_4$ conformation

Epimer. Compounds that differ in configuration at one or more, but not all asymmetric carbon atoms (Section 5.1.1). If they differ in configuration about all asymmetric carbon atoms, then the isomers are enantiomers. In sugar chemistry, if the difference is at the ring-closing carbon atom, the compounds are known as anomers.

Fischer projection structure. The orientation of an asymmetric carbon atom in such a manner that the two horizontal covalent bonds project toward the observer while the two vertical bonds project away.

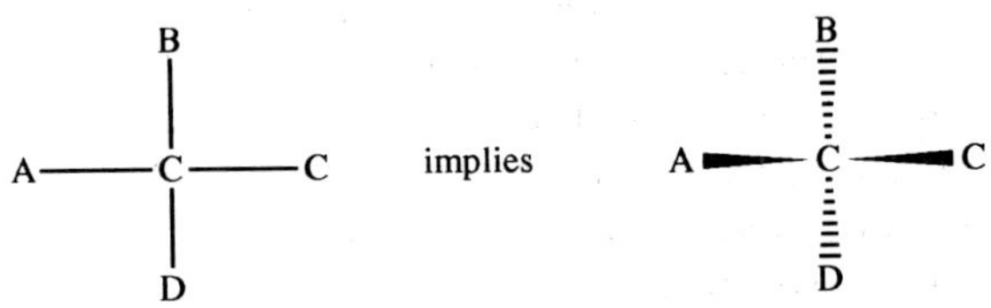

Furanose. A sugar with a five-membered hemiacetal ring structure (Section 5.1.2).
Glycophore. A group that elicits sweet taste (Sections 2.9 and 9.3).
Glycoside. A compound formed by the reaction of a sugar hemiacetal structure with a second alcohol or hydroxyl group. Therefore, a glycoside is a mixed acetal. See also acetal and hemiacetal. Reaction of glucose with methanol yields the mixed glycoside known as methyl α-D-glucopyranoside (Section 5.1.2). See also glycosyl group.

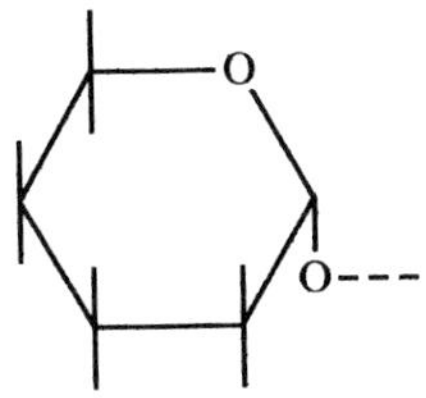

A glycoside radical

Glycosyl group. When the hydroxyl group is dropped from a cyclic form of a sugar, the resulting radical is known as a glycosyl radical (or group).

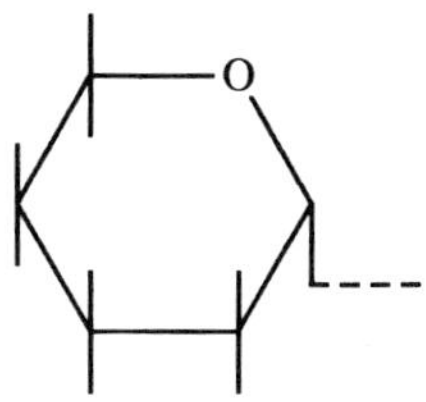

A glycosyl group

Therefore, the generic name for sucrose (Fig. 5.15) is α-D-glucopyranosyl β-D-fructofuranoside.
Halophore. A group that elicits salty taste (Sections 2.9 and 9.2).
Hemiacetal. See also acetal. When a sugar hydroxyl group reacts intramolecularly with a second sugar hydroxyl group, a ring structure is formed by virtue of hemiacetal formation. When a sugar hemiacetal structure is made to react intermolecularly with a second alcohol, or

the hydroxyl group of another sugar, a full acetal, or glycoside is formed. See also acetal.

Homochiral. Forms belonging to only one or the other of the two possible chiral families (Section 2.11.3.1). The D-series of sugars is a homochiral series of substances.

Hydrogen bond. Hydrogen bridge. The attraction or interaction of a hydrogen atom single-covalently bonded to an electronegative atom to/with a second electronegative atom or center (Section 2.7.3).

Hydrophobic bond. Attraction of or interaction between two apolar groups through dispersion forces (Section 2.7.4). See dispersion bond. Hydrophibicity, then, is the potential for participating in a hydrophobic bond, and is distinguished from lipophilicity (Section 9.3.10.2).

Intensity. Taste intensity. The magnitude of a perceived sensory response. It may be minimal or maximal (Section 1.3.1). See also *potency* for a distinction between intensity and potency.

Ionic bond. Electrostatic bond. Transfer of an electron from one atom to another leading to Coulombic attraction between them (Section 2.7.2).

Isostere. Isosteric group. Originally, chemical groups or atoms that are sterically similar. In this treatise the term is used to indicate similar chemical or sensory functionality. An oxygen atom and a sulfur atom may act isosterically in an inductive capacity for sweet taste, but in antisteric taste capacity as components of saporous units (Section 2.9).

j.n.d. A just noticeable difference with respect to a sensory response (Section 1.3.1).

Joule (J). An energy expression. $1\,J = 0{\cdot}2390\,cal$.

k. A specific rate constant (Section 2.5.3). Also, kilo- as in kilocalorie (kcal).

K. An equilibrium constant (Sections 2.3.2, 2.4.2, 2.5.4).

K_a. Activity or association constant (Section 2.6.3).

K_d. A dissociation constant. The reciprocal of K_a (Section 2.6.3).

L-Amino acid. An amino acid whereby the amine group on the reference carbon atom is written to the left in Fischer projection structure (Section 7.1.1). See also enantiomer and D-amino acid.

L-Sugar. A sugar related configurationally, at the highest-numbered asymmetric carbon atom, to L-glyceraldehyde. Therefore, the hydroxyl group for the Fischer projection structure is written to the right. The prefix *l*- or (−) refers to the direction (*levo*-) of the compound's optical rotatory power (Section 5.1.1.2). See also enantiomer and D-sugar.

Lipohilic. Lipophilicity. The affinity of a compound or group, usually a hydrocarbon or hydrocarbon chain, for apolar solvents, i.e. lipid-soluble substances or groups. The term is distinguished from hydrophobicity (Section 9.3.10.2).

***Meso* compound**. A compound with a bilaterally symmetrical combination of right- and left-handed asymmetric carbon atoms (Section 2.11.3.3).

Nonreducing sugar. A sugar without a free (unsubstituted) carbonyl group and therefore incapable of acting as a reducing agent (Section 5.3.1). See also, reducing sugar.

Nucleophilic. A group with a free electron pair that attract and binds, or is capable of recognizing a proton or hydrogen ion (Section 2.3.1).

Oligosaccharide. A compound sugar containing two to about ten monosaccharide units (Section 5.2). Some authors consider disaccharides to be oligosaccharides, others do not.

Picrophore. A group that elicits bitterness (Sections 2.9 and 9.4).

Potency. Taste potency. The relative ability of a substance to exhibit maximum taste response (Section 1.3.1). See also intensity.

Prochiral. A three-component set of groups or sites contained on a surface (face) of a symmetrical molecule that interact in chemical systems as though they were chiral. In fact, the arrangement has elements of two-dimensional chirality, but not necessarily elements of three-dimensional chirality (Sections 2.11.3.6 and 11.2.2).

Prosymmetrical. Reverse prochirality. A three-component set of groups on the face of a three-dimensionally chiral molecule that interact in chemical systems as though the chiral molecule is symmetrical. Three-dimensionally chiral substances (enantiomers) may have a common (superposable) face (Sections 2.11.3.6 and 11.2.2).

Pyranose. A sugar with a six-membered hemiacetal ring structure (Section 5.1.2).

Racemic mixture. An equal mixture of right and left-handed geometric (enantiomeric) structures. A compound may be synthesized as a racemic mixture of D- and L-configurational forms, such as tartaric acid (Section 2.11.3.3). On the other hand, a configurationally symmetrical compound may prevail, in liquid form, as a racemic mixture of geometrically opposite conformational isomers. Examples are ethylene glycol and cyclohexane.

Reducing sugar. A sugar with a free carbonyl group, and therefore capable of acting as a reducing agent in oxidation (gain in positive valence)-reduction (loss of positive valence) reactions (Section 5.3.1).

Saporous. Saporific, sapid, saporophoric. A compound or group (saporophore) capable of eliciting a taste (Section 1.0). See also tastant.

Staggered conformations (gauche conformations). The orientation of adjacent groups on vicinal carbon atoms with a projected angle of 60° (Section 9.3.7.5).

Stoichiometry. Application of the laws of definite proportions. A well-known stoichiometric equation is the combination of carbon dioxide and water by the process of photosynthesis to form carbohydrate, or $CO_2 + H_2O \longrightarrow CH_2O + O_2$ (Sections 2.4.1, 6.6.1).

Superimpose. To merely place one object over another without regard to whether they can be superposed (Section 2.11).

Superpose. To place one object upon another and have them coincide in all geometric respects (Section 2.11).

Tastant. A saporous or sapid substance, i.e. one that elicits a taste. The term is not usually found in dictionaries, although the term *odorant* is found (Section 1.0). Also a saporophore.

Tautomer(s). Easily interconvertible isomers, as obtained, for example, by merely dissolving a crystalline form of a reducing sugar in water (Sections 5.1.2.1 and 5.3.3).

Tridymite. The combination of three pairs (six) of molecules (Section 3.1.2.1).

van der Waals interaction. Attractive and/or repulsive weak chemical interactions (Section 2.7.5).

References

Abbott, E. A. (1884). *Flatland. A Romance of Many Dimensions*, (4th Edition). Seeley & Co., London.

Abderhalden, E. & Weil, A. (1912). Spaltung des racemischen Histidins in seine optisch aktiven Komponenten. *Z. Phys. Chem.* (*Leipzig*) **77,** 435–53.

Acree, T. E., Shallenberger, R. S., Lee, C. Y. & Einset, J. W. (1969). Thermodynamics and kinetics of D-galactose tautomerism during mutarotation. *Carbohydr. Res.* **10,** 355–60.

Acton, E. M., Leaffer, M. A., Oliver, S. M. & Stone, H. (1970*a*). Structure-taste relationships in oximes related to perillartine. *J. Agric. Food Chem.* **18,** 1061–8.

Acton, E. M., Stone, H., Leaffer, M. A. & Oliver, S. M. (1970*b*). Perillartine and some derivatives: clarification of structure. *Experientia* **26,** 473–4.

Adams, M. A. (1985). Substances that modify the perception of sweetness. In *Characterization and Measurement of Flavor Compounds*. ed. D. D. Bills & C. J. Mussinan. ACS Symp. Ser. 289, Am. Chem. Soc., Washington, DC, pp. 11–25.

Adler-Nissen, J. (1986). Relationship of structure to taste of peptides and peptide mixtures. In *Protein Tailoring for Food and Medical Uses*. ed. R. E. Feeney & J. R. Whitaker. Marcel Dekker, New York, pp. 97–122.

Adler-Nissen, J. (1988). Bitterness intensity of protein hydrolyzates — chemical and organoleptic characterization. In *Frontiers of Flavor*. ed. G. Charalambous. Elsevier, Amsterdam, pp. 63–77.

Ajduković, D. (1990). Electrical taste stimulus: Current intensity or current density? *Chem. Senses* **15,** 341–7.

Allerhand, A. & Schleyer, P. v. R. (1963). A survey of C—H groups as proton donors in hydrogen bonding. *J. Am. Chem. Soc.* **85,** 1715–23.

Amerine, M. A., Pangborn, R. M. & Roessler, E. B. (1965). *Principles of Sensory Evaluation of Food.* Academic Press, New York.

Amoore, J. E. (1963*a*). Stereochemical theory of olfaction. *Nature* **198,** 271–2.

Amoore, J. E. (1963*b*). Stereochemical theory of olfaction. *Nature* **199,** 912–3.

Angyal, S. J. (1968). Conformational analysis in carbohydrate chemistry. I. Conformational free energies. The conformations and $\alpha:\beta$ ratios of aldopyranoses in aqueous solution. *Aust. J. Chem.* **21,** 2737–46.

Angyal, S. J. & Bethell, G. S. (1976). Conformational analysis in carbohydrate chemistry. III. The ^{13}C N.M.R. spectra of the hexuloses. *Aust. J. Chem.* **29,** 1249–65.

Angyal, S. J. & Dawes, K. (1968). Conformational analysis in carbohydrate chemistry. II. Equilibria between reducing sugars and their glycosidic anhydrides. *Aust. J. Chem.* **21,** 2747–60.

Ariens, E. J., Beld, A. J., Rodrigues de Miranda, J. F. & Simonis, A. M. (1979). The pharmacon–receptor–effector concept. A basis for understanding the transmission of information in biological systems. In *The Receptors: A Comprehensive Treatise I. General Principles and Procedures.* ed. R. D. O'Brien. Plenum Press, New York, pp. 33–91.

Ariyoshi, Y. (1976). The structure-taste relationships of aspartyl dipeptide esters. *Agric. Biol. Chem.* **40,** 983–92.

Ariyoshi, Y. (1984). Structure-taste relationships of aspartyl tripeptide esters. *Nippon Kagakukai Bull.* **57,** 3197–202.

Ariyoshi, Y., Kohmura, M., Hasegawa, Y., Ota, M. & Nio, N. (1991). Sweet peptides and proteins: Synthetic Studies. In *Sweeteners, Discovery, Molecular Design, and Chemoreception.* ed. D. E. Walters, F. T. Orthoefer & G. E. DuBois, ACS Symposium Series 450, Am. Chem. Soc., Washington, DC, pp. 41–56.

Arnoldi, A., Camarada, L. & Merlini, L. (1986). Synthesis and sweet taste of some 2-phenylbenzodioxanes. *J. Agric. Food Chem.* **34,** 339–44.

Arnoldi, A., Bassoli, A., Merlini, L. & Ragg, E. (1991). Isovanillyl sweeteners. Synthesis, conformational analysis, and structure-activity relationship of some sweet oxygen heterocycles. *J. Chem. Soc., Perkin Trans. II,* **1991,** 1399–406.

Arnoldi, A., Bassoli, A., Merlini, L. & Ragg, E. (1993). Progress in isovanillyl sweeteners. In *Sweet Taste in Chemoreception.* ed. M. Mathlouthi, J. Kanters & G. G. Birch. Elsevier Applied Science, London, 175–184.

Arvidson, K. & Friberg, U. (1980). Human taste: response and taste bud number in fungiform papillae. *Science* **209,** 807–8.

Asao, M., Iwamura, H., Akamatsu, M. & Fujita, T. (1987). Quantitative structure-activity relationships of the bitter thresholds of amino acids, peptides, and their derivatives. *J. Med. Chem.* **30,** 1873–9.

Atkins, P. W. (1978). *Physical Chemistry.* Oxford University Press, Oxford, England.

Audrieth, L. F. & Sveda, M. (1944). Preparation and properties of some N-substituted sulfamic acids. *J. Org. Chem.* **9,** 89–101.

Austin, W. C. & Humoller, F. L. (1934). The preparation of two new crystalline aldohexoses, *l*-allose and *l*-altrose, from *l*-ribose by the cyanohydrin reaction. *J. Am. Chem. Soc.* **56,** 1153–5.

Avenet, P., Hofmann, F. & Lindemann, B (1988). Transduction in taste receptor cells requires cAMP-dependent protein kinase. *Nature* **331,** 351–4.

Baer, E. & Fischer, H. O. L. (1939). Studies on acetone-glyceraldehyde. VII. Preparation of *l*-glyceraldehyde and *l*(−) acetone glycerol. *J. Am. Chem. Soc.* **61,** 761–5.

Baker, B. R. (1967). *Design of Active-Site-Directed Irreversible Enzyme Inhibitors.* Wiley & Sons, New York.

Barker, S. A., Brimacombe, J. S., Foster, A. B., Whiffen, D. H. & Zweifel, G.

(1959). Aspects of stereochemistry. II. Intramolecular hydrogen bonding in some monohydroxy derivatives of tetrahydrofuran, tetrahydropyran, and 1 : 3 dioxan. *Tetrahedron* **7,** 10–8.

Bartoshuk, L. M. (1968). Water taste in man. *Percept. Psychophys.* **3,** 69–72.

Bartoshuk, L. M. (1975). Taste mixtures: Is mixture suppression related to compression? *Physiol. Behav.* **14,** 643–9.

Bartoshuk, L. M. (1977). Modification of taste quality. In *Sensory Properties of Foods*. ed. G. G. Birch, J. G. Brennan & K. J. Parker. Elsevier Applied Science, London, pp. 5–26.

Bartoshuk, L. M. (1987). Is sweetness unitary? An evaluation of the evidence for multiple sweets. In *Sweetness*. ed. J. Dobbing. Springer-Verlag, London, pp. 33–47.

Bartoshuk, L. M. & Cleveland, C. T. (1977). Mixtures of substances with similar tastes: a test of a psychophysical model of taste mixture interactions. *Sensory Processes* **1,** 177–86.

Bartoshuk, L. M. & Gent, J. F. (1985). Taste mixtures: an analysis of synthesis. In *Taste, Olfaction, and the Central Nervous System*. ed. D. W. Pfaff. Rockefeller Univ. Press, New York, pp. 210–32.

Bartoshuk, L. M., McBurney, D. H. & Pfaffmann, C. (1964). Taste of sodium chloride solutions after adaptation to sodium chloride: implications for the "water taste." *Science* **143,** 967–8.

Bartoshuk, L. M., Dateo, G. P., Vandenbelt, D. J., Buttrick, R. L. & Long, Jr., L. (1969). Effects of *Gymnema sylvestre* and *Synespalum dulcificum* on taste in man. In *Olfaction and Taste III*. ed. C. Pfaffmann. Rockefeller Univ. Press, New York, pp. 436–44.

Bartoshuk, L. M., Murphy, C. & Cleveland, C. T. (1978). Sweet taste of dilute NaCl: psychophysical evidence for a sweet stimulus. *Physiol. Behav.* **21,** 609–13.

Bartoshuk, L. M., Rennert, K., Rodin, J. & Stevens, J. C. (1982). Effects of temperature on the perceived sweetness of sucrose. *Physiol. Behav.* **28,** 905–10.

Bates, F. J. & Associates. (1942). Polarimetry, Saccharimetry and the Sugars. *Circular C440, National Bureau of Standards*. U.S. Government Printing Office, Washington, DC.

Bealer, S. L. & Smith, D. V. (1975). Multiple sensitivity to chemical stimuli in single human taste papillae. *Physiol. Behav.* **14,** 795–9.

Beebe-Center, J. G. & Waddell, D. (1948). A general psychological scale of taste. *J. Psychol.* **26,** 517–24.

Beebe-Center, J. G., Rogers, M. S., Atkinson, W. H. & O'Connell, D. N. (1959). Sweetness and saltiness of compound solutions of sucrose and NaCl as a function of concentration of solutes. *J. Exp. Psychol.* **57,** 231–4.

Beets, M. G. J. (1978). *Structure-activity Relationships in Human Chemoreception*. Elsevier Applied Science, London.

Beidler, L. M. (1954). A theory of taste stimulation. *J. Gen. Physiol.* **38,** 133–9.

Beidler, L. M. (1962). Taste receptor stimulation. In *Progress in Biophysics and Biophysical Chemistry*, **12**. ed. J. A. V. Butler, H. E. Huxley & R. E. Zirkle, Pergamon Press, New York, pp. 107–51.

Beidler, L. M. (1963). Dynamics of taste cells. In *Olfaction and Taste*. ed. Y. Zotterman. Pergamon Press, Oxford, pp. 133–44.

Beidler, L. M. (1966). A physiological basis of taste sensation. *J. Food Sci.* **31,** 275–81.

Beidler, L. M. (1967). Anion influences on taste receptor response. In *Olfaction and Taste II.* Pergamon Press, Oxford, pp. 509–34.

Beidler, L. M. & Tonosaki, K. (1985). Multiple sweet receptor sites and taste theory. In *Taste, Olfaction, and the Central Nervous System.* ed. D. W. Pfaff. Rockefeller Univ. Press, New York, pp. 47–64.

Beidler, L M., Fishman, I. Y. & Hardiman, C. W. (1955). Species differences in taste responses. *Am. J. Physiol.* **181,** 235–9.

Békésy, G. von. (1964). Rythmical variations accompanying gustatory stimulation observed by means of localization phenomena. *J. Gen. Physiol.* **47,** 809–25.

Békésy, G. von. (1966). Taste theories and the chemical stimulation of single papillae. *J. Appl. Physiol.* **21,** 1–9.

Belitz, H.-D. & Grosch, W. (1987). *Food Chemistry.* Springer-Verlag, Berlin.

Belitz, H.-D. & Wieser, H. (1985). Bitter compounds: Occurrence and structure-activity relationships. *Food Rev. Int.* **1,** 271–354.

Belitz, H.-D., Chen, W., Jugel, H., Treleano, R., Wieser, H., Gasteiger, J. & Marsili, M. (1979). Sweet and bitter compounds: structure and taste relationship. In *Food Taste Chemistry.* ed. J. C. Boudreau. ACS Symposium Series No. 115, Am. Chem. Soc., Washington, DC, pp. 93–131.

Belitz, H.-D., Chen, W., Jugel, H., Stempfl, W., Treleano, R. & Wieser, H. (1981). Structural requirements for sweet and bitter taste. In *Flavour '81.* ed. P. Schreier, Walter de Gruyter, Berlin, pp. 741–55.

Belitz, H.-D., Rohse, H., Stempfl, W., Wieser, H., Gasteiger, J. & Hiller, C. (1987). Schematic sweet and bitter receptors. In *Olfaction and Taste IX.* ed. S. D. Roper & J. Atema. *Ann. N.Y. Acad. Sci.* **510,** 173–5.

Belitz, H.-D., Rohse, H., Stempfl, W., Wieser, H., Gasteiger, J. & Hiller, C. (1988). Schematic sweet and bitter receptors — a computer approach. In *Frontiers of Flavor.* Proceedings of the 5th International Flavor Conference. ed. G. Charalambous. Elsevier Science Pub., Amsterdam, pp. 49–62.

Belitz, H.-D., Rohse, H., Stempfl, W. & Gries, H. (1990). Trihalogenated benzamides: structure taste relationships. *Z. Lebensm.-Unters.-Forsch.* **190,** 319–24.

Benedetti, E., DiBlasio, B., Pavone, V., Pedone, C., Fuller, W. D., Mierke, D. F. & Goodman, M. (1990). Crystal structure of two retro-inverso sweeteners. *J. Am. Chem. Soc.* **112,** 8909–12.

Benson, G. A., & Spillane, W. J. (1976). Structure-activity studies on sulfamate sweeteners. *J. Med. Chem.* **19,** 869–72.

Bentley, R. (1969). *Molecular Asymmetry in Biology.* Academic Press, New York.

Bentley, R. (1978). Ogsten and the development of prochirality theory. *Nature* **276,** 673–6.

Berg, C. P. (1953). Physiology of the D-amino acids. *Physiol. Rev.* **33,** 145–89.

Bernal, J. D. & Fowler, R. H. (1933). A theory of water and ionic solution, with particular reference to hydrogen and hydroxyl ions. *J. Chem. Phys.* **1,** 515–48.

Bertin, F. & Derouault, J. (1979). Etude structurale des chlorure et sulfate de béryllium hydratés par spectrométrie infrarouge et Raman. *Can. J. Chem.* **57,** 913–9.

Biester, A., Wood, M. W. & Wahlin, C. S. (1925). Carbohydrate studies. I. The relative sweetness of pure sugars. *Am. J. Physiol.* **73,** 387–96.

Bigelow, C. C. & Channon, M. (1976). Hydrophobicities of amino acids and proteins. In *Handbook of Biochemistry and Molecular Biology*, 3rd Ed., Proteins, Vol. 1. ed. G. D. Fasman. CRC Press, Cleveland, pp. 209–43.

Bigiani, A. R. & Soper, S. D. (1991). Mediation of responses to calcium in taste cells by modulation of a potassium conductance. *Science* **252,** 126–8.

Bijvoet, J. M., Peerdeman, A. F. & van Bommel, A. J. (1951). Determination of the absolute configuration of optically active compounds by means of X-rays. *Nature* **168,** 271–2.

Binkley, R. W., Binkley, W. W. & Grey, A. A. (1973). Conformational analysis of di-D-fructose dianhydrides by p.m.r. spectroscopy. *Carbohydr. Res.* **28,** 365–70.

Birch, G. G. (1976). Structural relations of sugars to taste. *CRC Crit. Rev. Food Sci. Nutr.* **8,** 57–95.

Birch, G. G. (1986). Structure, chirality and solution properties of glutamates in relation to taste. In *Umami: Physiology of Its Taste*. ed. Y. Kawamura & M. R. Kare, Marcel Dekker, New York, pp. 3–13.

Birch, G. G. & Kemp, S. E. (1989). Apparent specific volumes and tastes of amino acids. *Chem. Senses* **14,** 249–58.

Birch, G. G. & Lee, C. K. (1971). The chemical basis of sweetness in model sugars. In *Sweetness and Sweeteners*, ed. G. G. Birch, L. F. Green & C. B. Coulson. Elsevier Applied Science, London, pp. 95–111.

Birch, G. G. & Lee, C. K. (1974). Structural functions of taste in the sugar series: Sensory properties of deoxy sugars. *J. Food Sci.* **39,** 947–9.

Birch, G. G. & Lindley, M. G. (1973). Structural functions of taste in the sugar series: Effects of aglycones on the sensory properties of simple glycoside structures. *J. Food. Sci.* **38,** 665–7.

Birch, G. G. & Shamil, S. (1988). Structure, sweetness and solution properties of small carbohydrate molecules. *J. Chem. Soc., Faraday Trans. I.* **84,** 2635–40.

Birch, G. G., Lee, C. K. & Lindley, M. G. (1975). Stereochemistry of sweetness. *Staerke* **27**(**2**), 51–6.

Birch, G. G., Latymer, Z. & Hollaway, M. (1980). Intensity/time relationships in sweetness: evidence for a queue hypothesis in taste chemoreception. *Chem. Senses* **5,** 63–78.

Birch, G. G., Karim, R., Chavez, A. L. & Morini, G. (1993). Sweetness, structure, and specific volume. In *Sweet Taste in Chemoreception*. ed. M. Mathlouthi, J. Kanters & G. G. Birch. Elsevier Applied Science, London, 129–140.

Blakeslee, A. F. & Fox, A. L. (1932). Our different taste worlds. P.T.C. as a demonstration of genetic differences in taste. *J. Hered.* **23,** 97–110.

Blanksma, J. J. & Hoegen, D. (1946). The sweet taste of 4-nitro-2-aminotoluene, 4-nitro-2-aminobenzoic acid, and 2-nitro-4-aminobenzoic acid. *Recl. Trav. Chim.* **65,** 333–7.

Blanksma, J. J. & van der Weyden, P. W. M. (1940). Relationship between taste and structure in some derivatives of meta-nitraniline. *Recl. Trav. Chim.* **59,** 629–32.

Blanksma, J. J., van den Broek, W. J. & Hoegen, D. (1946). On the sweet taste of the 1-halogeno-2-amino-4-nitrobenzenes. *Recl. Trav. Chim.* **65,** 329–32.

Boch, K. & Lemieux, R. U. (1982). The conformational properties of sucrose in aqueous solution: intramolecular hydrogen-bonding. *Carbohydr. Res.* **100,** 63–74.

Bohak, Z. & Li, S.-L. (1976). The structure of monellin and its relation to the sweetness of the protein. *Biochim. Biophys. Acta* **427,** 153–70.

Boudreau, J. C., Oravec, J., Anderson, W., Collings, V. & Nelson, T. E. (1976). Cat neural taste responses to nitrogen compounds. In *Phenolic, Sulfur, and Nitrogen Compounds in Food Flavors.* ed. G. Charalambous, & I. Katz. ACS Symposium Series No. 26, Am. Chem. Soc., Washington, DC, pp. 194–206.

Boyd, W. C. & Matsubara, S. (1962). Different tastes of enantiomorphic hexoses. *Science* **137,** 669.

Brady, J. W. (1990). Molecular dynamic simulations of carbohydrate molecules. *Adv. Biophys. Chem.* **1,** 155–202.

Bragg, R. W., Chow, Y., Dennis, L., Ferguson, L. N., Howell, S., Morga, G., Ogino, C., Pugh, H. & Winters, M. (1978). Sweet organic chemistry. *J. Chem. Educ.* **55,** 281–5.

Brand, J. C. D., Speakman, J. C. & Tyler, J. K. (1975). *Molecular Structure, The Physical Approach.* Wiley & Sons, New York.

Brewster, J. H. (1959). A useful model of optical activity. I. Open chain compounds. *J. Am. Chem. Soc.* **81,** 5475–83.

Brimacombe, J. S., Foster, A. B., Stacey, M. & Whiffen, D. H. (1958). Aspects of stereochemistry. I. Properties and reactions of some diols. *Tetrahedron* **4,** 351–60.

Britton, H. T. S. & Meek, F. H. (1931). Physico-chemical studies of complex formation involving weak acids. Part II. The constitution of acetates in solution. *J. Chem. Soc.* **1931,** 2831–40.

Brouwer, J. N., van der Wel, H., Francke, A. & Henning, G. J. (1968). Miraculin, the sweetness-inducing protein from Miracle fruit. *Nature* (London) **220,** 373–4.

Browne, C. A. (1922). Moisture absorptive power of different sugars and carbohydrates under varying conditions of atmospheric humidity. *Ind. Eng. Chem.* **14,** 712–4.

Browne, C. A. & Zerban, F. W. (1941). *Physical and Chemical Methods of Sugar Analysis.* Wiley & Sons, New York.

Brussel, L. B. P., Peer, H. G. & van der Heijden, A. (1975). Structure-taste relationship of some sweet-tasting dipeptide esters. *Z. Lebensm.-Unters. Forsch.* **159,** 337–43.

Bryant, B. P., Brand, J. G. & Kalinoski, D. L. (1989). Receptor site specificity in taste. In *Chemical Senses: Receptor Events and Transduction in Taste and Olfaction.* ed. J. G. Brand, J. H. Teeter, R. H. Cagan and M. R. Kare. Marcel Dekker, New York, pp. 35–53.

Buitenhuis, J. & Kanters, J. A. (1991). Comments on Shallenberger's chiral principles contained in structure-sweetness relations. *Food Chem.* **40,** 109–12.

Bujas, Z. (1971). Electrical taste. In *Handbook of Sensory Physiology. IV. Chemical Senses. 2. Taste.* ed. L. M. Beidler. Springer-Verlag, Berlin, pp. 180–99.

Cahn, R. S., Ingold, C. & Prelog, V. (1966). Specification of molecular chirality. *Angew. Chem. Intern. Ed. Engl.* **5,** 385–415.

Cameron, A. T. (1947). The taste sense and the relative sweetness of sugars and other sweet substances. *Sugar Research Foundation Sci. Report. Ser. No. 9*, New York.

Carr, C. J., Beck, F. F. & Krantz, Jr., J. C. (1936). Sugar alcohols. V. Chemical constitution and sweet taste. *J. Am. Chem. Soc.* **58,** 1394–5.

Castiglione-Morelli, M. A., Lelj, F., Naider, F., Tallon, M., Tancredi, T. & Temussi, P. A. (1990). Conformation-activity relationship of sweet molecules. Comparison of aspartame and naphthimidazolesulfonic acids. *J. Med. Chem.* **33,** 514–20.

Charton, M. (1965). Application of the Hammett equation to nonaromatic unsaturated systems. III. Dipole moments of *trans*-vinylene and vinylidene sets. *J. Org. Chem.* **30,** 552.

Christofides, J. C. & Davies, D. B. (1985). Co-operative and competitive hydrogen bonding in sucrose determined by SIMPLE ^{1}H NMR Spectroscopy. *J. Chem. Soc., Chem. Commun.,* **1985,** 1533–4.

Clauss, K. & Jensen, H. (1973). Oxathiazinon dioxides — a new group of sweetening agents. *Angew. Chem.* **12,** 869–942.

Cohn, G. (1914). *Die Organischen Geschmackstoffe*. F. Siemenroth, Berlin.

Cole, A. R. H. & Jefferies, P. R. (1956). Stereochemistry of cyclohexane derivatives. V. Infrared spectra and conformations of steroisomeric 1 : 2-diols. *J. Chem. Soc.* **1956,** 4391–7.

Collander, R. (1950). The distribution of organic compounds between *iso*-butanol and water. *Acta Chem. Scand.* **4,** 1085–98.

Compadre, C. M., Hussain, R. A., Lopez de Compadre, R. L., Pezzuto, J. M. & Kinghorn, A. D. (1987). The intensely sweet sesquiterpene hernandulcin: Isolation, synthesis, characterization, and preliminary safety evaluation. *J. Agric. Food Chem.* **35,** 273–9.

Compadre, C. M., Hussain, R. A., Lopez de Compadre, R. L., Pezzuto, J. M. & Kinghorn, A. D. (1988). Analysis of structural features responsible for the sweetness of the sesquiterpene hernandulcin. *Experientia* **44,** 447–9.

Crosby, G. A., DuBois, G. E. & Wingard, Jr., R. E. (1979). The design of synthetic sweeteners. In *Drug Design VIII*. ed. E. J. Ariens. Academic Press, New York, pp. 215–310.

Culbertson, J. C. & Walters, D. E. (1991). Three-dimensional model for the sweet taste receptor. In *Sweeteners, Discovery, Molecular Design, and Chemoreception*. ed. D. E. Walters, F. T. Orthoefer & G. E. Dubois. ACS Symposium Series 450, Am. Chem. Soc., Washington, DC, pp. 214–23.

Curtis, D. W., Stevens, D. A. & Lawless, H. T. (1984). Perceived intensity of the taste of sugar mixtures and acid mixtures. *Chem. Senses* **9,** 107–20.

Dahlberg, A. C. & Penczek, E. S. (1941). The relative sweetness of sugars as affected by concentration. *N.Y. State Agricultural Experiment Station Technical Bulletin* **258**.

Dallenbach, J. W. & Dallenbach, K. M. (1943). The effects of bitter-adaptation on sensitivity to other taste-qualities. *Am. J. Psychol.* **56,** 21–31.

Daniel, J. R. (1989). Sweeteners: theory and design. In *Frontiers in Carbohydrate Research I. Food Applications*. ed. R. P. Millane, J. N. BeMiller & R. Chandrasekaran. Elsevier Applied Science, London, pp. 34–65.

Daniel, J. R. & Whistler, R. L. (1982). Sweetness-structure correlation in carbohydrates. *Cereal Chem.* **59,** 92–5.
Dastoli, F. R. & Price, S. (1966). Sweet-sensitive protein from bovine taste buds: isolation and assay. *Science* **154,** 905–7.
Dastoli, F. R., Lopiekes, D. V. & Doig, A. R. (1968). Bitter-sensitive protein from porcine taste buds. *Nature* **218,** 884–5.
Davies, D. B. & Christofides, J. C. (1987). Comparison of intramolecular hydrogen bonding conformations of sucrose containing oligosaccharides in solution and the solid state. *Carbohydr. Res.* **163,** 269–74.
Davson, H. & Danielli, J. F. (1952). *The Permeability of Natural Membranes.* Cambridge Univ. Press, London.
DeGraaf, C. & Frijters, J. E. R. (1988). Assessment of the taste interaction between two qualitatively similar taste substances: A comparison between comparison rules. *J. Exp. Psychol.: Hum. Percept. Perform.* **14,** 526–38.
DeGraaf, C. & Frijters, J. E. R. (1989). Interrelationships among sweetness, saltiness and total taste intensity of sucrose, NaCl and sucrose/NaCl mixtures. *Chem. Senses* **14,** 81–102.
DeSimone, J. A., Heck, G. L. & DeSimone, S. K. (1981). Active ion transport in dog tongue: A possible role in taste. *Science* **214,** 1039–41.
Deutsch, E. W. & Hansch, C. (1966). Dependence of relative sweetness on hydrophobic bonding. *Nature* **211,** 75.
DeVos, A. M., Hatada, M., van der Wel, H., Krabbendam, H., Peerdeman, A. F. & Sung-Han, K. (1985). Three-dimensional structure of thaumatin I, an intensely sweet protein. *Proc. Natl. Acad. Sci. U.S.A.* **82,** 1406–9.
Diamant, H., Funakoshi, M., Strom, L. & Zotterman, Y. (1963). Electrophysiological studies on human taste nerves. In *Olfaction and Taste I.* ed. Y. Zotterman. Pergamon Press, Oxford, pp. 193–203.
Dick, Jr., W. E. (1981). Structure-taste correlations for flavans and flavanones conformationally equivalent to phyllodulcin. *J. Agric. Food Chem.* **29,** 305–12.
Dickerson, R. E., Gray, H. B. & Haight, Jr., G. P. (1974). *Chemical Principles. 2nd Edition.* W. A. Benjamin, Menlo Park, CA.
Dietrich, S. W., Jorgensen, E. C., Kollman, P. A. & Rothenberg, S. (1976). A theoretical study of intramolecular hydrogen bonding in ortho-substituted phenols and thiophenols. *J. Am. Chem. Soc.* **98,** 8310–24.
Dill, K. (1987). ^{13}C-nuclear magnetic resonance-spectral studies of labeled glycophorins. *Adv. Carbohyd. Chem. Biochem.* **45,** 169–98.
Dmitriev, I. S. (1979). *Symmetry in the World of Molecules.* Mir. Pub., Moscow.
Donohue, J. (1968). Selected topics in hydrogen bonding. In *Structural Chemistry and Molecular Biology.* ed. A. Rich & N. Davidson. W. H. Freeman and Company, San Francisco, pp. 443–65.
Douglas, A. J. & Goodman, M. (1991). Molecular Basis of Taste. A stereoisomeric approach. In *Sweeteners, Discovery, Molecular Design, and Chemoreception.* ed. D. E. Walters, F. T. Orthoefer & G. E. Dubois. ACS Symposium Series 450, Am. Chem. Soc., Washington, DC, pp. 128–42.
DuBois, G. E. & Lee, J. F. (1983). A simple technique for the evaluation of temporal taste properties. *Chem. Senses* **7,** 237–47.
DuBois, G. E., Crosby, G. A., Stephenson, R. A. & Wingard, Jr., R. E. (1977*a*).

Dihydrochalcone sweeteners. Synthesis and sensory evaluation of sulfonate derivatives. *J. Agric. Food Chem.* **25,** 763–72.

DuBois, G. E., Crosby, G. A. & Saffron, P. (1977*b*). Nonnutritive sweeteners: Taste-structure relationships for some new simple dihydrochalcones. *Science* **195,** 397–9.

DuBois, G. E., Crosby, G. A. & Stephenson, R. A. (1981). Dihydrochalcone sweeteners. A study of the atypical temporal phenomena. *J. Med. Chem.* **24,** 408–28.

DuBois, G. E., Walters, D. E., Schiffman, S. S., Warwick, Z. S., Booth, B. J., Pecore, S. D., Gibes, K., Carr, B. T. & Brands, L. M. (1991). Concentration-response relationships of sweeteners: A systematic study. In *Sweeteners, Discovery, Molecular Design, and Chemoreception.* ed. D. E. Walters, F. T. Orthoefer & G. E. DuBois. Am. Chem. Soc. Symposium Series 450, Washington, DC, pp. 261–76.

DuBois, G. E., Walters, D. E. & Kellogg, M. S. (1993). The rational design of ultra-high potency sweeteners. In *Sweet Taste in Chemoreception.* ed. M. Mathlouthi, J. Kanters & G. G. Birch. Elsevier Applied Science, London, 237–268.

Dzendolet, E. & Meiselman, H. L. (1967). Gustatory quality changes as a function of solution concentration. *Percept. Psychophys.* **2,** 29–33.

Dzendolet, E. & Murphy, C. (1974). Electrical stimulation of human fungiform papillae. *Chem. Senses Flavour* **1,** 9–15.

Dziedzic, S. Z. & Birch, G. G. (1981). Structural functions of taste in the sugar series: Function of the γ-attribute in the sweet glycophore. *J. Sci. Food Agric.* **32,** 283–7.

Edens, L. & van der Wel, H. (1985). Microbial synthesis of the sweet-tasting plant protein thaumatin. *Trends Biotechnol.* **3,** 61–4.

Edward, J. T. (1955). Stability of glycosides to acid hydrolysis. A conformational analysis. *Chem. Ind.* (*London*), **1955,** 1102–4.

Ehrlich, F. (1914). Über asymmetrische und symmetrische Einwirkung von Hefe auf Racemverbindungen natürlich vorkommender Aminosäuren. *Biochem. Z.* **63,** 379–401.

Eliel, E. L. (1971). Recent advances in stereochemical nomenclature. *J. Chem. Educ.* **48,** 163–7.

Erickson, R. P. (1985). Grouping in the chemical senses. *Chem. Senses* **10,** 333–40.

Erickson, R. P. & Schiffman, S. S. (1975). The chemical senses: a systematic approach. In *Handbook of Psychobiology.* ed. M. S. Gazzaniga & C. Blakemore. Academic Press, New York, pp. 393–426.

Esaki, S., Kamiya, S. & Konishi, F. (1975). Synthesis and taste of some dihydrochalcone glycosides. *Agric. Biol. Chem.* **39,** 1385–9.

Fabian, F. W. & Blum, H. B. (1943). Relative taste potency of some basic food constituents and their competitive and compensatory action. *Food Res.* **8,** 179–93.

Faurion, A., Saito, S. & MacLeod, P. (1980). Sweet taste involves several distinct receptor mechanisms. *Chem. Senses* **5,** 107–21.

Feeley, T. M., Hargreaves, M. K. & Marshall, D. L. (1968). The circular

dichroism of some α-hydroxyaldehydes. DL-glycoaldehyde. *Tetrehedron Lett.* **46,** 4831–4.
Feinstein, R. D., Polinsky, A., Douglas, A. J., Magnus, C., Beijer, G. F., Chadha, R. K., Benedetti, E. & Goodman, M. (1991). Conformational analysis of the dipeptide sweetener alitame and two stereoisomers by proton NMR, computer simulations and X-ray crystallography. *J. Am. Chem. Soc.* **113,** 3467–73.
Fenton, H. J. H. & Jackson, H. (1899). Crystalline glycollic aldehyde. *J. Chem. Soc.* **75,** 575–9.
Ferguson, L. N. & Childers, L. G. (1960). Ultraviolet spectroscopic studies of some sweet and nonsweet isomeric *m*-nitroanilines. *J. Org. Chem.* **25,** 1971–5.
Ferguson, L. N. & Lawrence, A. R. (1958). The physicochemical aspects of the sense of taste. *J. Chem. Educ.* **35,** 436–44.
Fieser, L. F. & Fieser, M. (1961). *Advanced Organic Chemistry*. Reinhold Pub. Corp., New York.
Fischer, E. (1890). Ueber die optischen isomerism des traubenzuckers, der gluconsäure und der zuckersäure. *Ber.-Dtsch. Chem. Ges.* **23,** 2611–24.
Fischer, E. (1891). Uber die configuration des traubenzuckers und seiner isomeren. *Ber.-Dtsch. Chem. Ges.* **24,** 1836–45.
Fischer, E. (1906). Spaltung der α-Aminoisovaleriansäure in die optisch activen Componenten. *Ber. Dtsch. Chem. Ges.* **39,** 2320–8.
Fischer, E. & Mouneyrat, A. (1900). Spaltung einiger racemischer Aminosäuren in die optisch activen Componenten IV. *Ber. Dtsch. Chem. Ges.* **33,** 2383–93.
Fischer, E. & Warburg, O. (1905). Spaltung des Leucins in die optisch-activen Componenten mittels der Formylverbindung. *Ber. Dtsch. Chem. Ges.* **38,** 3997–4005.
Fox, A. L. (1932). The relationship between chemical constitution and taste. *Proc. Natl. Acad. Sci.* **18,** 115–20.
Frank, G. & Zuber, H. (1976). The complete amino acid sequences of both subunits of the sweet protein monellin. *Hoppe-Seyler's Z. Physiol. Chem.* **357,** 585–92.
Frank, H. S. (1958). Covalency in the hydrogen bond and the properties of water and ice. *Proc. R. Soc.* **247A,** 481–92.
Frank, H. S. (1970). The structure of ordinary water. *Science* **169,** 635–41.
Frank, M. E. (1985). On the neural code for sweet and salty tastes. In *Taste, Olfaction, and the Central Nervous System*. The Rockefeller Univ. Press, New York, pp. 107–28.
Fricker, A., Prochazka, E. & Gutschmidt, J. (1973). Die Süsskraft von Fructose in Abhängigkeit von Temperatur, Konzentration und Substrat. *Lebensm.-Wiss. Technol.* **6,** 63–5.
Frijters, J. E. R. & DeGraaf, C. (1989). Modeling taste mixture interactions in equiratio mixtures. In *Perception of Complex Smells and Tastes*. ed. D. G. Laing, W. S. Cain, R. L. McBride & B. W. Ache. Academic Press, Sydney, Australia, pp. 245–64.
Frings, H. W. (1954). Gustatory stimulation by ions and the taste spectrum. *Abstr. Papers Am. Chem. Soc. 126th Mtng Div. Agric. and Food Chemistry*, p. 14A.
Frost, A. A. & Pearson, R. G. (1961). *Kinetics and Mechanism, A study of homogeneous chemical reactions*. Wiley & Sons, New York.

Fujino, M., Wakimasu, M., Tanaka, K., Aoki, H. & Nakajima, N. (1973). L-Aspartyl-aminomalonic acid diesters. A new group of compounds with intense sweetness. *Naturwissenschaften* **60,** 351.

Fuller, W. D., Goodman, M. & Verlander, M. S. (1985). A new class of amino acid based sweeteners. *J. Am. Chem. Soc.* **107,** 5821–2.

Fuson, R. C. (1950). *Advanced Organic Chemistry.* Wiley & Sons, Inc., New York.

Galema, S. A., Blandamer, M. J. & Engberts, Jan B. F. N. (1990). Stereochemical aspects of the hydration of carbohydrates. Kinetic medium effects of monosaccharides on a water-catalyzed hydrolysis reaction. *J. Am. Chem. Soc.* **112,** 9665–6.

Gan, L.-X., Whistler, R. L. & Daniel, J. R. (1990). Synthesis of sweet-tasting methylthio derivatives of D-fructose and sucrose. *Carbohydr. Res.* **205,** 45–51.

Ganzevles, P. G. J. & Kroeze, J. H. A. (1987). The sour taste of acids. The hydrogen ion and the undissociated acid as sour agents. *Chem. Senses* **12,** 563–76.

Gardner, R. J. (1980*a*). Correlation of bitterness thresholds of amino acids and peptides with molecular connectivity. *J. Sci. Food Agric.* **31,** 23–30.

Gardner, R. J. (1980*b*). Lipid solubility and the sourness of acids: implications for models of the acid taste receptor. *Chem. Senses Flavour* **5,** 185–92.

Giguère, P. A. (1987). The bifurcated hydrogen-bond model of water and amorphous ice. *J. Chem. Phys.* **87,** 4835–9.

Giguère, P. A. & Pigeon-Gosselin, M. (1988). An electrostatic model for hydrogen bonds in alcohols. *J. Solution Chem.* **17,** 1007–14.

Gilman, H. & Dickey, J. B. (1930). Attempted correlations of constitution with sweet taste in the furan series. The very high sweetening power of 5-benzyl-2-furfuraldoxime. *J. Am. Chem. Soc.* **52,** 2010–3.

Giroux, E. L. & Henkin, R. I. (1971). Oral effects of hydrolytic enzymes on taste acuity in man. *Life Sci.* **10(1),** 361–70.

Giroux, E. L. & Henkin, R. I. (1974). Purification and some properties of miraculin, a glycoprotein from *Synsepalum dulcificum* which provokes sweetness and blocks sourness. *J. Agric. Food. Chem.* **22,** 595–601.

Glowaky, R. C., Hendrick, M. E., Smiles, R. E. & Torres, A. (1991). Development and uses of alitame. A novel dipeptide amide sweetener. In *Sweeteners, Discovery, Molecular Design, and Chemoreception.* ed. E. Walters, F. T. Orthoefer & G. E. DuBois, Am. Chem. Soc. Symposium Series 450, Washington, DC, pp. 57–67.

Goldschmidt, H. (1890). Untersuchungen über die isomeren oxime. *Ber. Dtsch. Chem. Ges.* **23,** 2163–80.

Goodman, M., Bland, J., Tsang, J., Coddington, J., Temussi, P. A., Tancredi, T., Lelj, F., Fuller, W. D. & Verlander, M. S. (1985). Molecular correlations of taste including a new class of amino acid based sweeteners. In *Peptides: Structure and Function.* ed. C. M. Deber, V. J. Hruby & K. D. Kopple. Pierce Chem. Co., Rockford, IL, pp. 725–8.

Goodman, M., Coddington, J. & Mierke, D. F. (1987). A model for the sweet taste of stereoisomeric retro-inverso and dipeptide amides. *J. Am. Chem. Soc.* **109,** 4712–4.

Goodman, M., Mierke, D. F. & Fuller, W. D. (1988). A stereoisomeric approach to the molecular basis of taste of retro-inverso and dipeptide amides. In *Peptide Chemistry — 1987*. ed. T. Shiba & S. Sakakibara. Protein Research Foundation, Osaka, pp. 699–704.

Goodman, M., Feinstein, R. & Polinsky, A. (1991). Taste modeling. *Chem. Design Automation News* **6(5),** 1, 23–7.

Goodwin, J. C., Hodge, J. E., Nelson, E. C. & Warner, K. A. (1981). Structure-taste relationships among cyclic glycols, levoglucosan, and methyl glycopyranosides. *J. Agric. Food Chem.* **29,** 929–35.

Guadagni, D. G., Maier, V. P. & Turnbaugh, J. H. (1974). Some factors affecting sensory thresholds and relative bitterness of limonin and naringin. *J. Sci. Food Agric.* **25,** 1199–205.

Ha, S. N., Madsen, L. J. & Brady, J. W. (1988). Conformational analysis and molecular dynamics simulations of maltose. *Biopolymers* **27,** 1927–52.

Haefeli, R. J. & Glaser, D. (1985). Die Kugel-ein mögliches Modell für die Zusammenhänge der Geschmacksqualitäten. *Lebensm.-Wiss.-Technol.* **18,** 267–70.

Haefeli, R. J. & Glaser, D. (1990). Taste responses and thresholds obtained with the primary amino acids in humans. *Lebensm. Wiss.-Technol.* **23,** 523–7.

Hahn, H. (1934). Die adaption des Geschmacksinnes. *Z. Sinnephysiol.* **65,** 105–45.

Hahn, H., Kuckulies, G. & Bissar, A. (1940). Eine systematische Unterschung der Gesmackschwellen. II. *Z. Sinnephysiol.* **68,** 185–260.

Hall, L. D. (1963). Conformations of some ribofuranosides. *Chem. Ind. (London)*, **1963,** 950–1.

Hall, M. J., Bartoshuk, L. M., Cain, W. S. & Stevens, J. C. (1975). PTC taste blindness and the taste of caffeine. *Nature* **253,** 442–3.

Hamilton, W. C. & Ibers, J. A. (1968). *Hydrogen Bonding in Solids: Methods of Molecular Structure Determination.* W. A. Benjamin, Inc., New York.

Hamor, G. H. (1961). Correlation of chemical structure and taste in the saccharin series. *Science* **134,** 1416–7.

Hänig, D. P. (1901). Zur Psychophysik des Geschmackssinnes. *Philosophische Studien* **17,** 576–623.

Hansch, C. (1970). The use of σ^+ in structure-activity correlations. *J. Med. Chem.* **13,** 964–6.

Hansch, C. & Fujita, T. (1964). ρ-σ-π Analysis. A method for the correlation of biological activity and chemical structure. *J. Am. Chem. Soc.* **86,** 1616–26.

Hanson, K. R. (1966). Application of the sequence rule. I. Naming the paired ligands g,g at a tetrahedral atom Xggij. II. Naming the two faces of a trigonal atom Yghi. *J. Am. Chem. Soc.* **88,** 2731–42.

Harper, H. W., Jay, J. R. & Erickson, R. P. (1966). Chemically evoked sensations from single human taste papillae. *Physiol. Behav.* **1,** 319–25.

Harvey, R. B. (1920). The relation between the total acidity, the concentration of the hydrogen ion, and the taste of acid solutions. *J. Am. Chem. Soc.* **42,** 712–4.

Hashimoto, K., Yoda, N. & Iwata, S. (1987). Theoretical study of hydrated beryllium (2^+) ions. *Chem. Phys.* **116,** 193–202.

Haworth, W. N. (1929). *The Constitution of Sugars.* Edward Arnold & Co., London.

Heiduschka, A. & Komm, E. (1925). Beziehungen zwischen konstitution und geschmack von α-aminosäuren. *Z. angew. Chem.* **38,** 291–4.
Hellekant, G. (1969). Inhibitory processes in gustation. *Acta Physiol. Scand.* **75,** 39–48.
Hellekant, G. & Ninomiya, Y. (1991). On the taste of umami in chimpanzee. *Physiol. Behav.* **49,** 927–34.
Hellekant, G., van der Wel, H., Brouwer, J. N., Glaser, D., Hård af Segerstad, C. & Roberts, T. W. (1986). Chorda tympani proper nerve responses to intra-arterial and surface stimulation of the tongue in rhesus monkey and rat. *Chem. Senses* **11,** 557–66.
Hellekant, G., DuBois, G. E., Roberts, T. W. & van der Wel, H. (1988). On the gustatory effect of amiloride in the monkey (*Macaca mulatta*). *Chem. Senses* **13,** 89–93.
Hellekant, G., Walters, D. E., Culbertson, J. C., DuBois, G. E., Nofre, C. & Tinti, J.-M. (1991). Electrophysiological evaluation of sweeteners. In *Sweeteners, Discovery, Molecular Design, and Chemoreception.* ed. E. Walters, F. T. Orthoefer & G. E. DuBois. Am. Chem. Soc. Symposium Series 450, Washington, DC, pp. 290–300.
Henkin, R. I. & Christiansen, R. L. (1967). Taste localization on the tongue, palate, and pharynx of normal man. *J. Appl. Physiol.* **22,** 316–20.
Henkin, R. I. & Shallenberger, R. S. (1970). Aglycogeusia: the inability to recognize sweetness and its possible molecular basis. *Nature* **227,** 965–6.
Henkin, R. I. & Shallenberger, R. S. (1971). Haloforms: sweet taste or smell. *Experientia* **27,** 154–5.
Henkin, R. I., Graziadei, P. P. G. & Bradley, D. F. (1969). Molecular basis of taste and its disorders. *Ann. Intern. Med.* **71,** 791–821.
Henning, G. J., Brouwer, J. N., van der Wel, H. & Francke, A. (1969). Miraculin, the sweetness-inducing principle from miracle fruit. *Olfaction and Taste III.* ed. C. Pfaffmann. Rockefeller Univ. Press, New York, pp. 445–9.
Henning, H. (1915). Die Qualitätenreihe des Geschmacks. *Z. Psychol. Physiol. Sinnesorgane* **74,** 203–19.
Hervé du Penhoat, C., Imberty, A., Roques, N., Michon, V., Mentech, J., Descotes, G. & Pérez, S. (1991). Conformational behavior of sucrose and its deoxy analogue in water as determined by NMR and molecular modeling. *J. Am. Chem. Soc.* **113,** 3720–7.
Heymans, G. (1899). Untersuchungen über psychische Hemmung. *Z. Psychol. Physiol. Sinnesorgane* **21,** 321–59.
Higginbotham, J. D. (1979). Protein sweeteners. In *Developments in Sweeteners, Vol. I.* ed. C. A. M. Hough, K. J. Parker & A. J. Vlitos. Elsevier Applied Science, London, pp. 87–123.
Hiji, Y. (1975). Selective elimination of taste responses to sugars by proteolytic enzymes. *Nature* **256,** 427–9.
Hiji, Y. & Sato, M. (1973). Isolation of the sugar-binding protein from rat taste buds. *Nature New Biol.* **244,** 91–3.
Hirschmann, H. (1960). The nature of substrate asymmetry in stereoselective reactions. *J. Biol. Chem.* **235,** 2762–7.

Hirschmann, H. & Hanson, K. R. (1971). Elements of stereoisomerism and prostereoisomerism. *J. Org. Chem.* **36,** 3293–306.

Höber, R. & Kiesow, F. (1898). Ueber den gesmack von salzen und laugen. *Z. Phys. Chem.* **27,** 601–16.

Hochstrasser, R. M. (1966). *Molecular Aspects of Symmetry.* W. A. Benjamin, New York.

Hodge, J. E. & Inglett, G. E. (1974). Structural aspects of glycosidic sweeteners containing (1′-2)-linked disaccharides. In *Symposium: Sweeteners.* ed. G. E. Inglett. Avi Pub. Co., Westport, CT, pp. 216–34.

Höltje, H.-D. & Kier, L. B. (1974). Sweet taste receptor studies using model interaction energy calculations. *J. Pharm. Sci.* **63,** 1722–5.

Holway, A. H. & Hurvich, L. M. (1937). Differential gustatory sensitivity to salt. *Am. J. Psychol.* **49,** 37–48.

Hooft, R. W. W., Kanters, J. A. & Kroon, J. (1993). Molecular mechanics and dynamics calculations on disaccharides and some derived artificial sweeteners. In *Sweet Taste in Chemoreception.* ed. M. Mathlouthi, J. Kanters & G. G. Birch. Elsevier Applied Science, London, 11–20.

Horowitz, R. M. & Gentili, B. (1969). Taste and structure in phenolic glycosides. *J. Agric. Food Chem.* **17,** 696–700.

Horowitz, R. M. & Gentili, B. (1971). Dihydrochalcone sweeteners. In *Sweetness and Sweeteners.* ed. G. G. Birch, L. F. Green & C. B. Coulson. Elsevier Applied Science, London, pp. 69–80.

Horowitz, R. M. & Gentili, B. (1986). Dihydrochalcone sweeteners from citrus flavanones. In *Alternative Sweeteners.* ed. L. O. Nabors & R C. Gelardi. Marcel Dekker, Inc., New York., pp. 135–53.

Horton, D. & Walaszek, Z. (1982). Tautomeric equilibria of some sugars by partially relaxed, ^{13}C pulse Fourier-transform, nuclear magnetic resonance spectroscopy. *Carbohydr. Res.* **105,** 145–53.

Hough, C. A. M. (1979). Sweet polyhydric alcohols. In *Developments in Sweeteners I.* ed. C. A. M. Hough, K. J. Parker & A. J. Vlitos. Elsevier Applied Sci., London, pp. 69–86.

Hough, C. A. M. & Edwardson, J. A. (1978). Antibodies to thaumatin as a model of the sweet taste receptor. *Nature* **271,** 381–3.

Hough, L. (1989). Sucrose, sweetness, and sucralose. *Int. Sugar J.* **91,** 23–37.

Hough, L. (1990). Applications of the chemistry of sucrose. *Zuckerind.* **115,** 655–61.

Hough, L. & Khan, R. (1978). Intensification of sweetness. *Trends in Biochemical Science* **3(3),** 61–3.

Hough, L. & Khan, R. (1993*a*). The relationship between sweetness and the molecular structure of sucrose and its derivatives. In *Sweet Taste in Chemoreception.* ed. M. Mathlouthi, J. Kanters & G. G. Birch. Elsevier Applied Science, London, 91–102.

Hough, L. & Khan, R. (1993*b*). Intense sweeteners derived from sucrose. In *Sweet Taste in Chemoreception.* ed. M. Mathlouthi, J. Kanters & G. G. Birch. Elsevier Applied Science, London, 269–282.

Hough, L. & Phadnis, S. P. (1976). Enhancement in the sweetness of sucrose. *Nature* **263,** 800.

Humphriss, E. (1940). The inversion of sucrose. *Food Manuf.* **15,** 161–2.
Hyvönen, L., Varo, P. & Koivistoinen, P. (1977*a*). Tautomeric equilibria of D-glucose and D-fructose: Polarimetric measurements. *J. Food Sci.* **42,** 652–3.
Hyvönen, L., Kurkela, R., Koivistoinen, P. & Merimaa, P. (1977*b*). Effects of temperature and concentration on the relative sweetness of fructose, glucose, and xylitol. *Lebensm. Wiss.-Technol.* **10,** 316–20.
Iida, Y. (1955). Method of increasing the sweetness of beet sugar. *Sugar J.* **18**(**5**), 29.
Inglett, G. E. (1971). Intense sweetness of natural origin. In *Sweetness and Sweeteners.* ed. G. G. Birch, L. F. Green & C. B. Coulson. Elsevier Applied Science, London, pp. 32–41.
Inglett, G. E. & May, J. F. (1968). Tropical plants with unusual taste properties. *Econ. Bot.* **22,** 326–31.
Inglett, G. E. & May, J. F. (1969). Serendipity berries — source of a new intense sweetener. *J. Food Sci.* **34,** 408–11.
Inui, T. & Kaneko, T. (1961). Optical resolution of DL-threonine and DL-allothreonine by the use of the oxazolidone derivatives. *Nippon Kagaku Zasshi,* **82,** 1078–81.
Ishibashi, N., Arita, Y., Kanehisa, H., Kouge, K., Okai, H. & Fukui, S. (1987). Bitterness of leucine-containing peptides. *Agric. Biol. Chem.* **51,** 2389–94.
Ishibashi, N., Kouge, K., Shinoda, I., Kanehisa, H. & Okai, H. (1988). A mechanism for bitter taste sensibility in peptides. *Agric. Biol. Chem.* **52,** 819–27.
Ishii, R. & O'Mahony, M. (1987). Taste sorting and naming: can taste concepts be misrepresented by traditional psychophysical labelling systems? *Chem. Senses* **12,** 37–51.
Iwamura, H. (1980). Structure-taste relationship of perillartine and nitro- and cyanoaniline derivatives. *J. Med. Chem.* **23,** 308–12.
Iwamura, H. (1981). Structure-sweetness relationship of L-aspartyl dipeptide analogues. A receptor site topology. *J. Med. Chem.* **24,** 572–83.
Iyengar, R. B., Smits, P., van der Ouderaa, F., van der Wel, H., van Brouwershaven, J., Ravestein, P., Richters, G. & van Wassenaar, P. D. (1979). The complete amino acid sequence of the sweet protein Thaumatin I. *Eur. J. Biochem.* **96,** 193–204.
Jaeger, F. M. (1930). I. Spatial arrangement of atomic systems and optical activity. In *Optical Activity and High Temperature Measurements.* McGraw-Hill, New York, pp. 1–234.
Jaffé, H. H. & Orchin, M. (1965). *Symmetry in Chemistry.* Wiley & Sons, Inc., New York.
Jakinovich, Jr., W. (1976). Stimulation of the gerbil's gustatory receptors by disaccharides. *Brain Res.* **110,** 481–90.
Jakinovich, Jr., W. (1982). Stimulation of the gerbil's gustatory receptors by saccharin. *J. Neurosci.* **2,** 49–56.
Jakinovich, Jr., W. (1983). Methyl 4,6-dichloro-4,6-dideoxy-α-D-galactopyranoside: an inhibitor of sweet taste responses in gerbils. *Science* **219,** 408–10.
Jakinovich, Jr., W. (1985). Sugar taste reception in the gerbil. In *Taste, Olfaction, and the Central Nervous System.* ed. D. W. Pfaff. Rockefeller Univ. Press, New York, pp. 65–91.

Jakinovich, Jr., W. & Goldstein, I. J. (1976). Stimulation of the gerbil's gustatory receptors by monosaccharides. *Brain Res.* **110,** 491–504.

Jakinovich, Jr., W. & Sugarman, D. (1988). Sugar taste reception in mammals. *Chem. Senses* **13,** 13–31.

James, C. E., Hough, L. & Khan, R. (1989). Sucrose and its derivatives. *Prog. Chem. Org. Nat. Prod.* **55,** 117–84.

Janusz, J. M., Young, P. A., Hiler, G. D., Moese, S. A. & Bunger, J. R. (1991). Time-intensity profiles of dipeptide sweeteners. In *Sweeteners, Discovery, Molecular Design, and Chemoreception.* ed. E. Walters, F. T. Orthoefer & G. E. DuBois. Am. Chem. Soc. Symposium Series 450, Am. Chem. Soc., Washington, DC, pp. 277–89.

Jeffrey, G. A. (1993). Hydrogen bonding with sugars and the role of hydrogen bonding in molecular recognition. In *Sweet Taste in Chemoreception.* ed. M. Mathlouthi, J. Kanters & G. G. Birch. Elsevier Applied Science, London, 1–10.

Jeffrey, G. A. & Saenger, W. (1991). *Hydrogen Bonding in Biological Structures.* Springer-Verlag, Berlin.

Jones, H. C. (1914). *The Elements of Physical Chemistry*, 4th Edition. Macmillan Co., New York.

Kabayama, M. A. & Patterson, D. (1958). The thermodynamics of mutarotation of some sugars. II. Theoretical considerations. *Can. J. Chem.* **36,** 563–73.

Kakihana, H. & Sillén, L. G. (1956). Studies on the hydrolysis of metal ions. XVI. The hydrolysis of the beryllium ion, Be^{2+}. *Acta Chem. Scand.* **10,** 985–1005.

Kamiya, S., Esaki, S. & Konishi, F. (1976*a*). Synthesis of some new α- and β-$(1 \rightarrow 2)$ linked disaccharides containing L-quinovose (6-deoxy L-glucose). *Agric. Biol. Chem.* **40,** 273–6.

Kamiya, S., Esaki, S. & Konishi, F. (1976*b*). Synthesis and taste of some analogs of neohesperidin dihydrochalcone. *Agric. Biol. Chem.* **40,** 1731–41.

Kaneko, T. (1938). On the relation between taste and constitution of α-amino acids. *Nippon Kagaku Zasshi* **59,** 433–9.

Kaneko, T. (1939). Taste and constitution of α-amino acids. II. Stereochemical studies on α-amino acids. *Nippon Kagaku Zasshi* **60,** 531–8.

Kaneko, T. (1940). Stereochemistry of α-amino acids. III. Stereochemical constitution of proline and hydroxyproline. *Nippon Kagaku Zasshi* **61,** 207–19.

Kanters, J. A., Roelofsen, G., Alblas, B. P. & Meinders, I. (1977). The crystal and molecular structure of β-D-fructose, with emphasis on anomeric effect and hydrogen-bond interactions. *Acta Crystallogr.* (*Sect. B*) **33,** 665–72.

Kanters, J. A., Scherrenberg, R. L., Leeflang, B. R., Kroon, J. & Mathlouthi, M. (1988). The crystal and molecular structure of an intensely sweet chlorodeoxy-sucrose, 4,1′,6′-trichloro-4,1′,6′-trideoxy-galacto-sucrose. *Carbohydr. Res.* **180,** 175–82.

Kasperson, R. W. & Primack, N. (1986). Cyclamate. In *Alternative Sweeteners.* ed. L. O. Nabors & R. C. Gelardi. Marcel Dekker, Inc., New York, pp. 71–87.

Kelling, S. T. & Halpern, B. P. (1988). Taste judgements and gustatory stimulus duration: taste quality, taste intensity, and reaction time. *Chem. Senses* **13,** 559–86.

Kennedy, L. M. & Halpern, B. P. (1980). A biphasic model for action of the gymnemic acids and ziziphins on taste receptor cell membranes. *Chem. Senses* **5,** 149–58.

Kier, L. B. (1972). A molecular theory of sweet taste. *J. Pharm. Sci.* **61,** 1394–7.

Kier, L. B. (1980). Molecular structure influencing either a sweet or bitter taste among aldoximes. *J. Pharm. Sci.* **69,** 416–9.

Kier, L. B. & Hall, L. H. (1976). *Molecular Connectivity in Chemistry and Drug Research.* Academic Press, New York.

Kiesow, F. (1896). Beiträge zur physiologischen psychologie des geschmackssinnes. *Philosophische Studien* **12,** 255–78.

Kim, S.-H., Kang, C.-H. & Cho, J.-M. (1991). Sweet proteins: Biochemical studies and genetic engineering. In *Sweeteners, Discovery, Molecular Design, and Chemoreception.* ed. D. E. Walters, F. T. Orthoefer & G. E. DuBois. ACS Symposium Series 450, Am. Chem. Soc., Washington, DC, pp. 28–40.

Kimura, K. & Beidler, L. M. (1961). Microelectrode study of taste receptors of rat and hamster. *J. Cell. Comp. Physiol.* **58,** 131–9.

Kinghorn, A. D. & Soejarto, D. D. (1991). New highly sweet compounds from natural sources. In *Sweeteners, Discovery, Molecular Design, and Chemoreception.* ed. E. Walters, F. T. Orthoefer & G. E. DuBois. Am. Chem. Soc. Symposium Series 450, Am. Chem. Soc., Washington, DC, pp. 14–27.

Kinnamon, S. C. (1988). Taste transduction: A diversity of mechanisms. *Trends Neurosci.* **11,** 491–6.

Kionka, H. & Strätz, F. (1922). (Setzt der Geschmack eines Salzes sich zusammen aus dem Geschmack der einzelnen Ionen oder schmeckt man jedes Salz als Gesamtmolekül?) *Arch. Exp. Pathol.* **95,** 241–57.

Kirimura, J., Shimizu, A., Kimizuka, A., Ninomiya, T. & Katsuya, N. (1969). The contribution of peptides and amino acids to the taste of foodstuffs. *J. Agric. Food Chem.* **17,** 689–95.

Kirkwood, J. G. (1937). On the theory of optical rotatory power. *J. Chem. Phys.* **5,** 479–91.

Kodama, S. (1920). Taste. *J. Tokyo Chem. Soc.* **41,** 495–534.

Kolodny, D. E. & Kennedy, L. M. (1988). A model system for receptor cell studies with the taste modifier, hodulcin. *Chem. Senses* **13,** 545–57.

Korolkovas, A. (1970). *Essentials of Molecular Pharmacology: Background for drug design.* Wiley-Interscience, New York.

Krbechek, L., Inglett, G., Holik, M., Dowling, B., Wagner, R. & Riter, R. (1968). Dihydrochalcones. Synthesis of potential sweetening agents. *J. Agr. Food Chem.* **16,** 108–12.

Kroeze, J. H. A. (1982). After repetitious sucrose stimulation saltiness suppression in NaCl-sucrose mixtures is diminished: implications for a central mixture supression mechanism. *Chem. Senses* **7,** 81–92.

Kubota, T. & Kubo, I. (1969). Bitterness and chemical structure. *Nature* **223,** 97–9.

Kuhn, L. P. (1952). The hydrogen bond. I. Intra- and intermolecular hydrogen bonds in alcohols. *J. Am. Chem. Soc.* **74,** 2492–9.

Kulka, K. (1967). Aspects of functional groups and flavor. *J. Agric. Food Chem.* **15,** 48–57.

Kuninaka, A., Kumagai, M., Fujiyama, K., Ogura, M., Sakata, S. & Yonei, S.

(1980). Flavor activity of sulfur-containing compounds related to flavor nucleotides. *Agric. Biol. Chem.* **44,** 1437–9.

Kurihara, K. (1971). Taste modifiers. In *Handbook of Sensory Physiology. IV. Chemical Senses. 2. Taste.* ed. L. M. Beidler. Springer-Verlag, Berlin, pp. 363–78.

Kurihara, K. & Beidler, L. M. (1968). Taste-modifying protein from miracle fruit. *Science* **161,** 1241–3.

Kurihara, Y. (1969). Antisweet activity of gymnemic acid A_1 and its derivatives. *Life Sci.* **8,** 537–43.

Kuznicki, J. T. & McCutcheon, N. B. (1979). Cross-enhancement of the sour taste on single human taste papillae. *J. Exp. Psychol.* **108,** 68–89.

Kuznicki, J. T. & Turner, L. S. (1988). Temporal dissociation of taste mixture components. *Chem. Senses* **13,** 45–62.

Lancet, D. & Ben-Arie, N. (1991). Sweet taste transduction: A molecular-biological analysis. In *Sweeteners, Discovery, Molecular Design, and Chemoreception.* ed. E. Walters, F. T. Orthoefer & G. E. DuBois. Am. Chem. Soc. Symposium Series 450, Am. Chem. Soc., Washington, DC, pp. 226–36.

Lawless, H. T. & Stevens, D. A. (1983). Cross adaptation of sucrose and intensive sweeteners. *Chem. Senses* **7,** 309–15.

Lazareff, P. (1922). Untersuchen über die Ionentheorir der Reizung. III. Ionentheori der Geschmacksreizung. *Arch. Gesamte Physiol.* **194,** 293–7.

Lee, C. K. (1979). Carbohydrate sweeteners: Structural requirements for taste. *World Rev. Nutr. Diet.* **33,** 142–97.

Lee, C. K. (1987). The chemistry and biochemistry of the sweetness of sugars. *Adv. Carbohydr. Chem. Biochem.* **45,** 199–351.

Lee, C. Y., Shallenberger, R. S. & Vittum, M. T. (1970). Free sugars in fruits and vegetables. *New York's Food & Life Sciences Bulletin No. 1.* Cornell Univ., Geneva, NY.

Lee, W. E., III & Pangborn, R. M. (1986). Time intensity: The temporal aspects of sensory perception. *Food Tech.* **40(11),** 71–8.

Lehmann, P. A. (1978). The correlation of sweetness and bitterness of enantiomeric amino acids. *Life Sci.* **22,** 1631–6.

Lelj, F., Tancredi, T., Temussi, P. A. & Toniolo, C. (1976). Interaction of α-L-aspartyl-L-phenylalanine methyl ester with the receptor site of the sweet taste bud. *J. Am. Chem. Soc.* **98,** 6669–75.

Lemieux, R. U. (1982). The binding of carbohydrate structures with antibodies and lectins. In *IUPAC Frontiers of Chemistry.* ed. K. J. Laidler, Pergamon Press, Oxford, pp. 3–24.

Lemieux, R. U. & Brewer, J. T. (1973). Conformational preferences for solvated hydroxymethyl groups in hexopyranose structures. In *Carbohydrates in Solution. Advances in Chemistry Series 117.* ed. R. F. Gould. American Chemical Society, Washington, DC, pp. 121–46.

Lemieux, R. U. & Nagarajan, R. (1964). The configuration and conformation of "di-D-fructose anhydride I." *Can. J. Chem.* **42,** 1270–8.

Lennard-Jones, J. & Pople, J. A. (1951). Molecular association in liquids. I. Molecular association due to lone-pair electrons. *Proc. R. Soc.* **205A,** 155–62.

Lichtenthaler, F. W., Immel, S. & Kreis, U. (1991). Old routs—new branches:

evolution of the structural representation of sucrose. In *Carbohydrates as Organic Raw Materials*. ed. F. W. Lichtenthaler. VCH Pub., New York, pp. 1–32.

Lindley, M. G. (1991). Phenoxyalkanoic acid sweetness inhibitors. In *Sweeteners, Discovery, Molecular Design, and Chemoreception*. ed. E. Walters, F. T. Orthoefer & G. E. DuBois. Am. Chem. Soc. Symposium Series 450, Am. Chem. Soc., Washington, DC, pp. 251–60.

Lindley, M. G. & Birch, G. G. (1975). Structural functions of taste in the sugar series. *J. Sci. Food Agric.* **26,** 117–24.

Lindley, M. G., Birch, G. G. & Khan, R. (1975). Synthesis of methyl ether derivatives of sucrose. *Carbohydr. Res.* **43,** 360–5.

Lindley, M. G., Birch, G. G. & Khan, R. (1976*a*). Sweetness of sucrose and xylitol. Structural considerations. *J. Sci. Food Agric.* **27,** 140–4.

Lindley, M. G., Shallenberger, R. S. & Whistler, R. L. (1976*b*). Comparison of the sweetness of glucose and fructose with their ring-thio analogs. *J. Food Sci.* **41,** 575–7.

Linko, P. (1982). Lactose and lactitol. In *Nutritive Sweeteners*. ed. G. G. Birch & K. J. Parker. Elsevier Applied Science, London, pp. 109–31.

Liptak, A., Szurmai, Z., Nánási, P. & Neszmélyi, A. (1982). ^{13}C-NMR study of methyl- and benzyl ethers of L-arabinose and oligosaccharides having L-arabinose at the reducing end. Synthesis of 2-*O*-β-D-glucopyranosyl-, 2-*O*-α-L-rhamnopyranosyl-, 3-*O*-β-D-glucopyranosyl-2-*O*-α-L-rhamnopyranosyl- and 4-*O*-β-D-glucopyranosyl-2-*O*-α-L-rhamnopyranosyl-L-arabinose. *Tetrahedron* **38,** 3489–97.

Loginov, N. E. (1941). A new method for measuring the sweetness of sugars. *Pischevya Prom.* **1,** 32.

Lowry, T. M. (1935). *Optical Rotatory Power*. Dover Publications, Inc., New York.

Luu, C., Luu, D. V., Rull, F & Sopron, F. (1982). Etude par effet Raman de la perturbation structurale de l'eau liquide par une substance etrangere: Partie I. Modele d'association de l'eau liquide. *J. Mol. Struct.* **81,** 1–10.

Maes, F. W. (1985). Improved best-stimulus classification of taste neurons. *Chem Senses* **10,** 35–44.

Magidson, O. J. & Gorbatschow, S. W. (1923). Zur Frage der Süßigkeit des saccharins. Das *o*-benzoylsulfimid und seine elektrolytische Dissoziation. *Ber. Dtsch. Chem. Ges.* **56B,** 1810–7.

Mapelli, C., Stammer, C. H., Lok, S., Mierke, D. F. & Goodman, M. (1988). Synthesis, taste properties, and conformational analysis of four stereoisomeric cyclopropane analogs of aspartame. *Int. J. Pept. Protein Res.* **32,** 484–95.

Marin, A. B., Barnard, J., Darlington, R. B. & Acree, T. E. (1991). Sensory thresholds: estimation from dose-response curves. *J. Sensory Studies* **6,** 205–25.

Marks, L. E., Stevens, J. C., Bartoshuk, L. M., Gent, J. F., Rifkin, B. & Stone, V. K. (1988). Magnitude-matching: the measurement of taste and smell. *Chem. Senses* **13,** 63–87.

Martin, O. R., Korppi-Tommola, S.-L. & Szarek, W. A. (1982). Structure-sweetness relationships for fructose analogs. Part I. Synthesis and evaluation of sweetness of 5-deoxy-D-*threo*-hexulose. *Can. J. Chem.* **60,** 1857–62.

Mathlouthi, M. (1981). X-ray diffraction study of the molecular association in aqueous solutions of D-fructose, D-glucose and sucrose. *Carbohydr. Res.* **91,** 113–23.

Mathlouthi, M. & Portmann, M. P. (1990). Hydrogen bonding and the sweet taste mechanism. *J. Mol. Struct.* **237,** 327–38.

Mathlouthi, M. & Seuvre, A.-M. (1988). Solution properties and the sweet taste of small carbohydrates. *J. Chem. Soc., Faraday Trans. I.* **84,** 2641–50.

Matoba, T. & Hata, T. (1972). Relationship between bitterness of peptides and their chemical structures. *Agric. Biol. Chem.* **36,** 1423–31.

Matsuo, R. & Yamamoto, T. (1989). Salivary secretion elicited by taste stimulation with umami substances in human adults. *Chem. Senses* **14,** 47–54.

Mazur, R. H. & Ripper, A. (1979). Peptide-based sweeteners. In *Developments in Sweeteners I.* ed. C. A. M. Hough, K. J. Parker & A. J. Vlitos. Elsevier Applied Science, London, pp. 125–34.

Mazur, R. H., Schlatter, J. M. & Goldkamp, A. H. (1969). Structure-taste relationships of some dipeptides. *J. Am. Chem. Soc.* **91,** 2684–91.

Mazur, R. H., Goldkamp, A. H., James, P. A. & Schlatter, J. M. (1970). Structure-taste relationships of aspartic acid amides. *J. Med. Chem.* **13,** 1217–21.

Mazur, R. H., Reuter, J. A., Swiatek, K. A. & Schlatter, J. M. (1973). Synthetic sweeteners. 3. Aspartyl dipeptide esters from L- and D-alkylglycines. *J. Med. Chem.* **16,** 1284–7.

McBride, R. L. (1987). Taste psychophysics and the Beidler equation. *Chem. Senses* **12,** 323–32.

McBurney, D. H. (1969). Effects of adaption on human taste function. In *Olfaction and Taste III.* ed. C. Pfaffmann. Rockefeller Univ. Press, New York, pp. 407–19.

McBurney, D. H. (1972). Gustatory cross adaption between sweet-tasting compounds. *Percept. Psychophys.* **11,** 225–7.

McBurney, D. H. (1974). Are there primary tastes for man? *Chem. Senses Flavour* **1,** 17–28.

McBurney, D. H. & Bartoshuk, L. M. (1972). Water taste in mammals. In *Olfaction and Taste IV.* ed. D. Schneider. Wissenschaftliche Verlagsgesellschaft MBH, Stuttgart, pp. 329–35.

McBurney, D. H. & Bartoshuk, L. M. (1973). Interactions between stimuli with different taste qualities. *Physiol. Behav.* **10,** 1101–6.

McBurney, D. H. & Lucas, J. A. (1966). Gustatory cross adaptation between salts. *Psychon. Sci.* **4,** 301–2.

McBurney, D. H. & Shick, T. R. (1971). Taste and water taste of twenty-six compounds for man. *Percept. Psychophys.* **10,** 249–52.

McBurney, D. H., Smith, D. V. & Shick, T. R. (1972). Gustatory cross adaption: sourness and bitterness. *Percept. Psychophys.* **11,** 228–32.

McFarland, J. W. (1971). Understanding of drug potency. *Fortschr. Arzneimittelforsch.* **15,** 123–46.

Mead, C. A. (1974). Topics in Current Chemistry, Vol. 49: *Symmetry and Chirality*. Springer, New York, 88pp.

Meiselman, H. L. & Dzendolet, E. (1967). Variability in gustatory quality identification. *Percept. Psychophys.* **2,** 496–8.

Mela, D. J. (1989). Bitter taste intensity: the effect of tastant and thiourea taster status. *Chem. Senses* **14,** 131–5.

Minton, A. P. (1970). Can water polymerize? An inquiry into the possible existence of strong bonds between water molecules. *Nature* **226,** 151–2.

Mislow, K. & Bickart, P. (1976/77). An epistemological note on chirality. *Isr. J. Chem.* **15,** 1–6.

Mislow, K. & Siegel, J. (1984). Stereoisomerism and local chirality. *J. Am. Chem. Soc.* **106,** 3319–28.

Moncrieff, R. W. (1967). *The Chemical Senses*, 3rd Edition. Leonard Hill, London.

Monte, W. C. & Maga, J. A. (1981/82). Flavor chemistry of sucrose. *Sugar Technol. Rev.* **8,** 181–204.

Morris, J. A. & Cagan, R. H. (1972). Purification of monellin, the sweet principle of *Dioscoreophyllum cumminsii. Biochim. Biophys. Acta* **261,** 114–22.

Moskowitz, H. R. (1970). Ratio scales of sugar sweetness. *Percept. Psychophys.* **7,** 315–20.

Moskowitz, H. R. (1971*a*). The sweetness and pleasantness of sugars. *Am. J. Psychol.* **84,** 387–405.

Moskowitz, H. R. (1971*b*). Intensity scales for pure tastes and for taste mixtures. *Percept. Psychophys.* **9,** 51–6.

Moskowitz, H. R. (1972). Perceptual changes in taste mixtures. *Percept. Psychophys.* **11,** 257–62.

Moskowitz, H. R. & Sidel, J. L. (1971). Magnitude and hedonic scales of food acceptability. *J. Food Sci.* **36,** 677–80.

Muller, G. W., Madigan, D. L., Culbertson, J. C., Walters, D. E., Carter, J. S., Klade, C. A., DuBois, G. E. & Kellogg, M. S. (1991). In *Sweeteners, Discovery, Molecular Design, and Chemoreception*. ed. D. E. Walters, F. T. Orthoefer & G. E. DuBois. ACS Symposium Series 450, Am. Chem. Soc., Washington, DC, pp. 113–25.

Murray, R. G. (1971). Ultrastructure of taste receptors. In *Handbook of Sensory Physiology. IV. Chemical Senses. 2. Taste*. ed. L. M. Beidler. Springer-Verlag, Berlin, pp. 31–50.

Narten, A. H. & Levy, H. A. (1969). Observed diffraction pattern and proposed models of liquid water. *Science* **165,** 447–54.

Némethy, G. (1967). Hydrophobe Wechselwirkungen. *Angew. Chem.* **79,** 260–71.

Neuman, F. J. (1944). Studies on the psychophysics of taste. Abst. Doctoral Dissert. The Ohio State University **43,** 115–21.

Ney, K. H. (1971). Voraussage der Bitterkeit von Peptiden aus deren Aminosäurezusammensetzung. *Z. Lebensm.-Unters. Forsch.* **147,** 64–8.

Nieman, C. (1958). Relative sweetness of various sugars. *Zucker Suesswaren Wirtsch.* **11,** 420–1090.

Nishida, Y., Ohrui, H. & Meguro, H. (1984). ^{1}H-NMR studies of (6R)- and (6S)-deuterated D-hexoses: assignment of the preferred rotamers about C5—C6 bond of D-glucose and D-galactose derivatives in solutions. *Tetrahedron Lett.* **25,** 1575–8.

Nishimura, T. & Kato, H. (1988). Taste of free amino acids and peptides. *Food Rev. Int.* **4,** 175–94.

Nofre, C. & Sabadie, J. (1972). A propos de la protéine linguale dite 'sensible aux sucres.' *Compt. Rend. Acad. Sci. Ser. D* 274, 2913–15.

Nofre, C. & Pautet, F. (1975). Sodium N-(cyclopentylmethyl)-sulfamate: a new synthetic sweetener. *Naturwissenschaften* **62,** 97.

Nofre, C. & Tinti, J.-M. (1993). In quest of hyperpotent sweeteners. In *Sweet Taste in Chemoreception*. ed. M. Mathlouthi, J. Kanters & G. G. Birch. Elsevier Applied Science, London, 205–236.

Nofre, C., Sabadie, J. & Bal, D. (1974). Sweetening properties of L-alanine *tert*-butyl ester. *Naturwissenschaften* **61,** 404–5.

Nourse, J. G. (1975). Pseudochirality. *J. Am. Chem. Soc.* **97,** 4594–601.

Nybom, N. (1962). Hurt surt är ett äpple? *Sveriges Pomologiska Forening Arsskrift* **63,** 117–34.

Oertly, E. & Myers, R. G. (1919). A new theory relating constitution to taste. *J. Am. Chem. Soc.* **41,** 344–7.

Ogawa, S., Uematsu, Y., Yoshida, S., Sasaki, N. & Suami, T. (1987). Synthesis and sweetness of pseudo-β-D and L-fructopyranose. *J. Carbohydr. Chem.* **6,** 471–8.

Ogston, A. G. (1948). Interpretation of experiments on metabolic processes, using isotopic tracer elements. *Nature* **162,** 963.

Öhrwall, H. (1891). Untersuchungen über den Geschmackssinn. *Skand. Arch. Physiol.* **2,** 1–69.

O'Loane, J. K. (1980). Optical activity in small molecules, nonenantiomorphous crystals, and nematic liquid crystals. *Chem. Rev.* **80,** 41–61.

O'Mahony, M. Goldenberg, M., Stedmon, J. & Alford, J. (1979). Confusion in the use of the taste adjectives "sour" and "bitter." *Chem. Senses Flavour* **4,** 301–18.

Pangborn, R. M. (1960). Taste interrelationships. *Food Res.* **25,** 245–56.

Pangborn, R. M. (1961). Taste interrelationships. II. Suprathreshold solutions of sucrose and citric acid. *J. Food. Sci.* **26,** 648–55.

Pangborn, R. M. (1962). Taste interrelationships. III. Suprathreshold solutions of sucrose and sodium chloride. *J. Food Sci.* **27,** 495–500.

Pangborn, R. M. (1963). Relative taste intensities of selected sugars and organic acids. *J. Food Sci.* **28,** 726–33.

Pangborn, R. M. & Gee, S. C. (1961). Relative sweetness of α- and β-forms of selected sugars. *Nature* **191,** 810–11.

Pasteur, L. (1915). Researches on the molecular asymmetry of natural organic products. *Edinburgh Alembic Club Reprints*, **No. 14**. University of Chicago Press, Chicago.

Paul, T. (1922). The degree of sweetness of sweet substances. *Z. Nahr. Genussm.* **43,** 137–49.

Pauling, L. (1960). *The Nature of the Chemical Bond.* Cornell University Press, Ithaca, NY. 644 pp.

Pautet, F & Nofre, C. (1978). Correlation of chemical structure and taste in the cyclamate series and the steric nature of the chemoreceptor site. *Z. Lebensm.-Unters. Forsch.* **166,** 167–70.

Pearson, R. G. (1963). Hard and soft acids and basis. *J. Am. Chem. Soc.* **85,** 3533–9.

Pérez, S., Meyer, C., Imberry, A. & French, A. (1993). Molecular features and conformational flexibility of sucrose. In *Sweet Taste in Chemoreception*. ed. M. Mathlouthi, J. Kanters & G. G. Birch. Elsevier Applied Science, London, 55–74.

Petersen, S. & Müller, E. (1948). Über eine neue gruppe von süßtoffen. *Chem. Ber.* **81,** 31–8.

Pfaffmann, C. (1941). Gustatory afferent impulses. *J. Cell. Comp. Physiol.* **17,** 243–58.

Pfaffmann, C. (1951). Taste and smell. In *Handbook of Experimental Psychology*, ed. S. S. Stevens. Wiley & Sons, Inc., New York, pp. 1143–71.

Pfaffmann, C. (1955). Gustatory nerve impulses in rat, cat, and rabbit. *J. Neurophysiol.* **18,** 429–40.

Pfaffmann, C. (1959). The sense of taste. In *Handbook of Physiology-Neurophysiology I.* ed. J. Field. Am. Physiol. Soc., Washington, DC, pp. 507–33.

Pfaffmann, C., Bartoshuk, L. M. & McBurney, D. H. (1971). Taste Psychophysics. In *Handbook of Sensory Physiology. IV. Chemical Senses. 2. Taste*. ed. L. M. Beidler, Springer-Verlag, Berlin, pp. 75–101.

Pigman, W. & Horton, D. (1972). Structure and stereochemistry of the monosaccharides. In *The Carbohydrates IA*. ed. W. Pigman & D. Horton. Academic Press, New York, pp. 1–67.

Pimentel, G. C. & McClellan, A. L. (1960). *The Hydrogen Bond*. W. H. Freeman & Co., San Francisco.

Piutti, M. A. (1886). Sur une nouvelle espece d'asparagine. *C. R. Acad. Sci., Paris* **103,** 134–7.

Plattig, K.-H. & Innitzer, J. (1976). Taste qualities elicited by electric stimulation of single human tongue papilla. *Pfluegers Arch.* **361,** 115–20.

Polański, J. & Ratajczak, A. (1991). Determination of the lipophilicity of arylsulphonylalkanoic and arylsulphonylcycloalkanecarboxylic acids by thin-layer chromatography. *J. Chromatogr.* **543,** 195–200.

Prelog, V. (1976). Chirality in chemistry. *Science* **193,** 17–24.

Presta, L. G. & Rose, G. D. (1988). Helix signals in proteins. *Science* **240,** 1632–41.

Price, S. (1972). "Sweet-sensitive protein" from bovine tongues: Stereospecific interactions with amino acids. In *Olfaction and Taste IV*. ed. D. Schneider. Wissenschaftliche Verlagsgesellschaft MBH, Stuttgart, pp. 214–20.

Price, S. & DeSimone, J. A. (1977). Models of taste receptor cell stimulation. *Chem. Senses Flavour* **2,** 427–56.

Probst, M. M., Limtrakul, J. P. & Rode, B. M. (1986). A study of the beryllium (2^+) ion-water system by means of ab initio calculations. *Chem. Phys. Lett.* **132,** 370–6.

Rader, C. P., Tihanyi, S. G. & Zienty, F. B. (1967). A study of the true taste of saccharin. *J. Food Sci.* **32,** 357–60.

Randić, M. (1975). On characterization of molecular branching. *J. Am. Chem. Soc.* **97,** 6609–15.

Ratajczak, A. & Polański, J. (1991*a*). A new class of sweet sulfones. *Naturwissenschaften* **78,** 69–70.

Ratajczak, A. & Polański, J. (1991*b*). A new, sulfone group containing, class of artificial sweeteners. In *Phosphorous, Sulfur, and Silicone*. Gordon & Breach Science Pub., S.A. 59, pp. 259–62.

Ray, A. & Birch, G. G. (1981). Time-dependent inhibition of sucrose sweetness with gymnemic acid: Mode of action. *Life Sci.* **28,** 2773–81.

Reeves, R. E. (1949). Cuprammonium-glucoside complexes. V. The conformation of the pyranose ring in some D-hexose ⟨1, 5⟩ β ⟨1, 6⟩ anhydrides. *J. Am. Chem. Soc.* **71,** 2116–9.

Reinicke, R. (1943). Ueber die chemischen vorbedingungen fur das zustandekommen der geschmacksempfindung "Süss." *Zukerind.* **1,** 79–82.

Renqvist, Y. (1918). Ueber den geschmack. *Skand. Arch. Physiol.* **38,** 97–201.

Richards, T. W. (1898). The relation of the taste of acids to their degree of dissociation. *Am. Chem. J.* **20,** 121–6.

Richardson, J. S. & Richardson, D. C. (1988). Amino acid preferences for specific locations at the ends of α helices. *Science* **240,** 1648–52.

Richter, C. P. & Campbell, K. H. (1939–40). Sucrose taste thresholds of rats and human subjects. *Am. J. Physiol.* **128,** 291–7.

Riddell, F. G. & Robinson, M. J. T. (1974). J. H. van't Hoff and J. A. LeBel — their historical context. *Tetrahedron* **30,** 2001–7.

Rohse, H. & Belitz, H.-D. (1988). Sweet and bitter taste of oxathiazinone dioxides and benzisothiazolone dioxides. *Z. Lebensm.-Unters. Forsch.* **187,** 425–31.

Rohse, H. & Belitz, H.-D. (1991). Shape of sweet receptors studied by computer modeling. In *Sweeteners, Discovery, Molecular Design, and Chemoreception*. ed. D. E. Walters, F. T. Orthoefer & G. E. DuBois. Am. Chem. Soc. Symposium Series 450, Washington, DC, pp. 176–92.

Rohse, H. & Belitz, H.-D. (1993). Sugars and sugar derivatives: Structure-taste relationships. In *Sweet Taste in Chemoreception*. ed. M. Mathlouthi, J. Kanters & G. G. Birch. Elsevier Applied Science, London, 387–402.

Rosanoff, M. A. (1906). On Fischer's classification of stereo-isomers. *Am. Chem. Soc.* **28,** 114–21.

Rossy, P. A., Hoffmann, W. & Müller, N. (1980). Aromatization of dihydrothiophenes. Thiophenesaccharine: A sweet surprise. *J. Org. Chem.* **45,** 617–20.

Ruch, E. (1972). Algebraic aspects of the chirality phenomenon in chemistry. *Acc. Chem. Res.* **5,** 49–56.

Rymon Lipinski, G.-W. von (1986). Acesulfame-K. In *Alternative Sweeteners*. ed. L. O. Nabors & R. C. Gelardi. Marcel Dekker, Inc., New York, pp. 89–102.

Saito, S. & Miyamoto, M. (1988). Discrimination of sugars in human sweet receptor sites as the result of cross-adaptation. 21st Japanese Symposium on taste and smell. Nagasaki, Japan. 11/10-11/87. Abstract. *Chem. Senses* **13,** 312.

Sale, J. W. & Skinner, W. W. (1922). Relative sweetness of invert sugar. *J. Ind. Eng. Chem.* **14,** 522–5.

Salem. L. (1964). Intermolecular forces in biological systems. In *Electronic Aspects of Biochemistry*. ed. B. Pullman, Academic Press, New York, pp. 293–9.

Saroli, A. (1984). Structure-activity relationship of a bitter compound: Denatonium chloride. *Naturwissenschaften* **71,** 428–9.

Saroli, A. (1985). Interaction of denatonium chloride with the bitter taste receptor. *Z. Lebensm.-Unters. Forsch.* **180,** 227–9.

Sausville, T. J. (1974). Acidulants. In *Encyclopedia of Food Technology*. ed. A. H. Johnson & M. S. Peterson. Avi Pub. Co. Westport, CT, pp. 1–6.

Scatchard, G. (1949). The attraction of proteins for small molecules and ions. *Ann. N.Y. Acad. Sci.* **51,** 660–72.

Schifferstein, H. N. J. & Frijters, J. E. R. (1990). Sensory integration in citric acid/sucrose mixtures. *Chem. Senses* **15,** 87–109.

Schiffman, S. S. (1990). The role of sodium and potassium transport pathways in taste transduction: An overview. *Chem. Senses* **15,** 129–35.

Schiffman, S. S. & Erickson, R. P. (1971). A theoretical review: A psychophysical model for gustatory quality. *Physiol. Behav.* **7,** 617–33.

Schiffman, S. S., Sennewald, K. & Gagnon, J. (1981*a*). Comparison of taste qualities and thresholds of D- and L-amino acids. *Physiol. Behav.* **27,** 51–9.

Schiffman, S. S., Cahn, H. & Lindley, M. G. (1981*b*). Multiple receptor sites mediate sweetness: Evidence from cross adaptation. *Pharmacol. Biochem. Behav.* **15,** 377–88.

Schiffman, S. S., Lockhead, E. & Maes, F. W. (1983). Amiloride reduces the taste intensity of Na^+ and Li^+ salts and sweeteners. *Proc. Natl. Acad. Sci. U.S.A.* **80,** 6136–40.

Schiffman, S. S., Sugarman, D., Jakinovich, W., Jr., Paikin, A. & Crofton, V. (1987). Inhibition of sweet taste in humans by methyl 4,6-dichloro-4,6-dideoxy-α-D-galactopyranoside. *Chem. Senses* **12,** 71–6.

Schmidt, C. L. A. (1938). *The Chemistry of the Amino Acids and Proteins.* Thomas Books, Springfield, IL.

Schmitt, F. O. & Bear, R. S. (1939). The ultrastructure of the nerve axon sheath. *Biol. Rev. Cambridge Philos. Soc.* **14,** 27–50.

Schneider, B., Lichtenthaler, F. W., Steinle, G. & Schiweck, H. (1985). Studies on ketoses. 1. Distribution of furanoid and pyranoid tautomers of D-fructose in water, dimethyl sulfoxide, and pyridine via ^1H NMR intensities of anomeric hydroxy groups in $[D_6]$ DMSO[1]. *Liebigs. Ann. Chem.* **1985,** 2443–53.

Schutz, H. G. & Pilgrim, F. J. (1957*a*). Sweetness of various compounds and its measurement. *Food Res.* **22,** 206–13.

Schutz, H. G. & Pilgrim, F. J. (1957*b*). Differential sensitivity in gustation. *J. Exp. Psychol.* **54,** 41–8.

Seki, T., Tamura, M., Nosho, Y., Ohfuji, T. & Okai, H. (1990). δ-Acetyl-L-ornithyl-β-alanine methyl ester hydrochloride, an intermolecule type sweetener. *Agric. Biol. Chem.* **54,** 1811–8.

Shallenberger, R. S. (1963). Hydrogen bonding and the varying sweetness of the sugars. *J. Food Sci.* **28,** 584–9.

Shallenberger, R. S. (1978). Intrinsic chemistry of fructose. *Pure Appl. Chem.* **50,** 1409–20.

Shallenberger, R. S. (1983*a*). *Advanced Sugar Chemistry.* Avi Pub. Co., Westport, CT.

Shallenberger, R. S. (1983*b*). The chiral principles contained in structure-sweetness relations. *Food Chem.* **12,** 89–107.

Shallenberger, R. S. (1993). Chemical aspects of the glucophore/glycophore. In

Sweet Taste in Chemoreception. ed. M. Mathlouthi, J. Kanters & G. G. Birch. Elsevier Applied Science, London, pp. 75–90.
Shallenberger, R. S. & Acree, T. E. (1967). Molecular theory of sweet taste. *Nature* **216,** 480–2.
Shallenberger, R. S. & Acree, T. E. (1971). Chemical structure of compounds and their sweet and bitter taste. In *Handbook of Sensory Physiology. IV. Chemical Senses. 2. Taste*. ed. L. M. Beidler. Springer-Verlag, Berlin, pp. 221–77.
Shallenberger, R. S. & Lindley, M. G. (1977). A lipophylic-hydrophobic attribute and component in the stereochemistry of sweetness. *Food Chem.* **2,** 145–53.
Shallenberger, R. S. & Wienen, W. J. (1989). Carbohydrate stereochemistry. *J. Chem. Educ.* **66,** 67–73.
Shallenberger, R. S., Acree, T. E. & Lee, C. Y. (1969). Sweet taste of D- and L-sugars and amino acids and the steric nature of their chemo-receptor site. *Nature* **221,** 555–6.
Shallenberger, R. S., Lee, C. Y., Acree, T. E., Barnard, J. & Lindley, M. G. (1977). Specific rotation of α-D- and β-D-fructofuranose. *Carbohydr. Res.* **58,** 209–11.
Shallenberger, R. S., Braverman, S. E. & Guild, Jr., W. E. (1980). The structure of β-D-fructofuranose in aqueous solution. *Food Chem.* **5,** 207–16.
Shallenberger, R. S., Wrolstad, R. E. & Kerschner, L. E. (1981). Calculation and specification of the multiple chirality displayed by sugar pyranoid ring structures. *J. Chem. Educ.* **58,** 599–601.
Shamil, S., Birch, G. G., Mathlouthi, M. & Clifford, M. N. (1987). Apparent molar volumes and tastes of molecules with more than one sapophore. *Chem. Senses* **12,** 397–409.
Shamil, S., Birch, G. G. & Njoroge, S. (1988*a*). Intrinsic viscosities and other solution properties of sugars and their possible relation to sweetness. *Chem. Senses* **13,** 457–61.
Shamil, S., Birch, G. G., Jackson, A. A. S. F. & Meek, S. (1988*b*). Use of intensity-time studies as an aid to interpreting sweet taste chemoreception. *Chem. Senses* **13,** 597–605.
Shamil, S., Birch, G. G., Dinovi, M. & Rafka, R. (1989). Structural functions of taste in 5-membered ring structures. *Food Chem.* **32,** 171–88.
Shinoda, I. & Okai, H. (1985). Sweetness and bitterness contributions of structural units of aspartame and some analogues. *J. Agric. Food Chem.* **33,** 792–5.
Shiraishi, H., Okuda, K., Sato, Y., Yamaoka, N. & Tuzimura, K. (1973). Taste of proline-containing peptides. *Agric. Biol. Chem.* **37,** 2427–8.
Shumaker, J. B. & Buchanan, J. H. (1932). A study of caramel color. *Iowa State J. Sci.* **6,** 367–79.
Sicard, P. J. (1982). Hydrogenated glucose syrups, sorbitol, mannitol and xylitol. In *Nutritive Sweeteners, Ind. — Univ. Co-op. Symposium 1981*. ed. G. G. Birch & K. J. Parker. Elsevier Applied Science, Barking, UK, pp. 145–70.
Sienko, M. J. & Plane, R. A. (1974). *Chemical Principles and Properties*. McGraw-Hill, New York, 788 pp.
Simon, S. A. (1991). Mechanisms of sweet taste transduction. In *Sweeteners, Discovery, Molecular Design, and Chemoreception*. ed. E. Walters, F. T.

Orthoefer & G. E. DuBois. Am. Chem. Soc. Symposium Series 450, Am. Chem. Soc., Washington, DC, pp. 237–50.

Singer, S. J. & Nicholson, G. L. (1972). The fluid mosaic model of the structure of cell membranes. *Science* **175,** 720–31.

Skramlik, E. von (1921). Mischungsgleichungen im Gebiete des Geschmacksinns. *Z. Psychol. Physiol. Sinnesorgane* **53,** 36–78.

Skramlik, E. von (1926). *Handbuch der physiologie der niederen sinne: Band 1. Die physiologie des geruchs-und geschmackssinnes.* G. Thieme, Leipzig.

Skramlik, E. von (1963). The fundamental substrates of taste. In *Olfaction and Taste I.* ed. Y. Zotterman. Pergamon Press, Oxford, pp. 125–32.

Smith, D. V. (1985). Brainstem processing of gustatory information. In *Taste, Olfaction, and the Central Nervous System.* The Rockefeller Univ. Press, New York, pp. 151–77.

Smith, D. V. & Frank, M. E. (1972). Cross-adaption between salts in the chorda tympani nerve of the rat. *Physiol. Behav.* **8,** 2133–20.

Smith, D. V. & McBurney, D. H. (1969). Gustatory cross-adaption: Does a single mechanism code the salty taste? *J. Exp. Psychol.* **80,** 101–5.

Sokolovsky, A. (1937). Effect of humidity on hygroscopic properties of sugars and caramel. *Ind. Eng. Chem.* **29,** 1422–3.

Solms, J., Vuataz, L. & Egli, R. H. (1965). The taste of L- and D-amino acids. *Experientia* **21,** 692–4.

Spillane, W. J. & McGlinchey, G. (1981). Structure-activity studies on sulfamate sweeteners II: Semiquantitative structure-taste relationship for sulfamate ($RNHSO_3^-$) sweeteners — the role of R. *J. Pharm. Sci.* **70,** 933–5.

Spillane, W. J. & Sheahan, M. B. (1989). Semi-quantitative structure-state relationships for carbo- and hetero-sulfamate ($RNHSO_3^-$) sweeteners. *J. Chem. Soc., Perkin Trans. II,* **1989,** 741–6.

Stevens, S. S. (1957). On the psychophysical law. *Psychol. Rev.* **64,** 153–81.

Stevens, S. S. (1960). The psychophysics of sensory function. *Am. Sci.* **48,** 226–53.

Stevenson, D. P. (1968). Molecular species in liquid water. In *Structural Chemistry and Molecular Biology.* ed. A. Rich & N. Davidson. W. H. Freeman & Co., San Francisco, pp. 490–506.

Stewart, R. A., Carrico, C. K., Webster, R. L. & Steinhardt, Jr., R. G. (1971). Physicochemical stereospecificity of α-D-mannose and β-D-mannose. *Nature* **234,** 220.

Stillinger, F. H. (1980). Water revisited. *Science* **209,** 451–7.

Stone, H., Oliver, S. & Kloehn, J. (1969). Temperature and pH effects on the relative sweetness of suprathreshold mixtures of dextrose, fructose. *Percept. Psychophys.* **5,** 257–60.

Sträter, P. J. (1986). Palatinit — technological and processing characteristics. In *Alternative Sweeteners.* ed. L. O. Nabors & R. C. Gelardi. Marcel Dekker, Inc., New York, pp. 217–44.

Streitwieser, Jr., A. & Heathcock, C. H. (1976). *Introduction to Organic Chemistry.* Macmillan, New York.

Suami, T. (1987). Synthetic ventures in pseudo-sugar chemistry. *Pure Appl. Chem.* **59,** 1509–20.

Suami, T. & Hough, L. (1991). Molecular mechanisms of sweet taste I. Sweet and non-sweet tasting amino acids. *J. Carbohyd. Chem.* **10,** 851–60.

Suami, T., Ogawa, S. & Toyokuni, T. (1983). Sweet tasting pseudo-sugars. *Chem. Lett (Japan),* **1983(4),** 611–2.

Suami, T., Ogawa, S., Takata, M., Yasuda, K., Suga, A., Takei, K. & Uematsu, Y. (1985). Synthesis of sweet tasting pseudo-β-fructopyranose. *Chem. Lett. (Japan),* **1985(6),** 719–22.

Summerbell, R. K. & Rochen, L. K. (1941). The structure of glycolaldehyde Dimer. *J. Am. Chem. Soc.* **63,** 3241–4.

Szarek, W. A. & Jones, J. K. N. (1980). *L-sucrose and process for producing same.* U.S. Patent 4 207 413.

Szarek, W. A., Korppi-Tommola, S.-L., Shurvell, H. F., Smith, Jr., V. H. & Martin, O. R. (1984). A Raman and infrared study of crystalline D-fructose, L-sorbose, and related carbohydrates. Hydrogen bonding and sweetness. *Can. J. Chem.* **62,** 1512–8.

Tada, M., Shinoda, I. & Okai, H. (1984). L-Ornithyltaurine, a new salty peptide. *J. Agric. Food Chem.* **32,** 992–6.

Tanford, C. (1962). Contribution of hydrophobic interactions to the stability of the globular conformation of proteins. *J. Am. Chem. Soc.* **84,** 4240–7.

Täufel, K. & Klemm, B. (1925). Untersuchungen über natürliche und künstliche Süsstoffe. I. Studien über den Süssungsgrad von saccharin und dulcin. *Z. Nahr. Genussm.* **50,** 264–73.

Täufel, K. & Wagner, C. (1925). Die konstitution wäßriger lösungen von *o*-benzoesäure-sulfinid (Saccharin) und *p*-phenetylcarbamid (Dulcin). *Ber. Dtsch. Chem. Ges.* **58B,** 909–12.

Taylor, E. W., Wilson, S. & Stammer, C. H. (1991*a*). Molecular mechanics of cyclopropane peptide analogs of Aspartame. Implications for three-dimensional requirements of the sweet taste receptor. In *Sweeteners, Discovery, Molecular Design, and Chemoreception.* ed. D. E. Walters, F. T. Orthoefer & G. E. DuBois. ACS Symposium Series 450, Am. Chem. Soc., Washington, DC, pp. 162–75.

Taylor, R. F., Marenchic, G. & Cook, E. J. (1991*b*). Electrical sensors containing biological receptors immobilized in polymer film. U.S. Patent 5,001,048.

Teeter, J. H. & Cagan, R. H. (1989). Mechanisms of taste transduction. In *Neural Mechanisms in Taste.* ed. R. H. Cagan. CRC Press, New York, pp. 1–20.

Teeter, J. & Gold, G. H. (1988). A taste of things to come. *Nature* **331,** 298–9.

Temussi, P. A., Lelj, F. & Tancredi, T. (1978). Three-dimensional mapping of the sweet taste receptor site. *J. Med. Chem.* **21,** 1154–8.

Temussi, P. A., Lelj, F., Tancredi, T., Morelli, M. A. C. & Pastore, A. (1984). Soft agonist receptor interactions: Theoretical and experimental simulation of the active site of the receptor of sweet molecules. *Int. J. Quantum Chem.* **26,** 889–906.

Temussi, P. A., Lelj, F. & Tancredi, T. (1991). Structure–activity relationship of sweet molecules. In *Sweeteners, Discovery, Molecular Design, and Chemoreception.* ed. D. E. Walters, F. T. Orthoefer & G. E. DuBois. ACS Symposium Series 450, Am. Chem. Soc., Washington, DC, pp. 143–61.

Tinti, J.-M. & Nofre, C. (1991*a*). Design of sweeteners. A rational approach. In *Sweeteners, Discovery, Molecular Design, and Chemoreception.* ed. D. E.

Walters, F. T. Orthoefer & G. E. DuBois. ACS Symposium Series 450, Am. Chem. Soc., Washington, DC, pp. 88–99.

Tinti, J.-M. & Nofre, C. (1991*b*). Why does a sweetener taste sweet? In *Sweeteners, Discovery, Molecular Design, and Chemoreception*. ed. D. E. Walters, F. T. Orthoefer & G. E. DuBois. Am. Chem. Soc. Symposium Series 450, Washington, DC, pp. 206–13.

Tinti, J.-M., Durozard, D. & Nofre, C. (1980). Sweet taste receptor. Evidence of separate specific sites for COO^- and NO_2/CN groups in sweeteners. *Naturwissenschaften* **67,** 193–4.

Tinti, J.-M., Nofre, C. & Durozard, D. (1981). Studies on sweeteners requiring the simultaneous presence of both NO_2/CN and COO^- groups. *Naturwissenschaften* **68,** 143.

Tinti, J.-M., Nofre, C. & Peytavi, A.-M. (1982). Interaction of suosan with the sweet taste receptor. *Z. Lebensm.-Unters. Forsch.* **175,** 266–8.

Tonosaki, K. & Funakoshi, M. (1988). Cyclic nucleotides may mediate taste transduction. *Nature* **331,** 354–6.

Tran, V. H. & Brady, J. W. (1990). Disaccharide conformational flexibility. II. Molecular dynamics simulations of sucrose. *Biopolymers* **29,** 977–97.

Trout, G. M. & Sharp, P. F. (1937). The reliability of flavor judgements with special reference to the oxidized flavor of milk. *Cornell Univ. Agric. Exp. Sta. Mem.* **204,** 60 pp.

Tsang, J. W., Schmied, B., Nyfeler, R. & Goodman, M. (1984). Peptide sweeteners. 6. Structural studies on the C-terminal amino acid of L-aspartyl dipeptide sweeteners. *J. Med. Chem.* **27,** 1663–8.

Tsuzuki, Y. & Yamashita, A. (1968). History of the chemistry of taste in Japan. *Jap. Stud. Hist. Sci., Tokyo* **7,** 1–26.

Tsuzuki, Y. & Yamazaki, J. (1953). On the sweetness of fructose and some other sugars, especially its variation with temperature. *Biochem. Z.* **323,** 525–31.

Tsuzuki, Y., Kato, S. & Okazaki, H. (1954). Sweet flavor and resonance. *Kagaku* **24,** 523–4.

Tunaley, A., Thomson, D. M. H. & McEwan, J. A. (1987). Determination of equi-sweet concentrations of nine sweeteners using a relative rating technique. *Int. J. Food Sci. Technol.* **22,** 627–35.

van der Heijden, A. (1993). Sweet and bitter taste. In *Flavor Workshop*. ed. R. Teranishi, & T. E. Acree. ACS Prof. Ref. Book Ser., Washington, DC. (In press).

van der Heijden, A., Brussel, L. B. P. & Peer, H. G. (1979). Quantitative structure-activity relationships (QSAR) in sweet aspartyl dipeptide methyl esters. *Chemical Senses Flavour* **4,** 141–52.

van der Heijden, A., van der Wel, H. & Peer, H. G. (1985*a*). Structure-activity relationships in sweeteners. I. Nitroanalines, sulfamates, oximes, isocoumarins and dipeptides. *Chem. Senses* **10,** 57–72.

van der Heijden, A., van der Wel, H. & Peer, H. G. (1985*b*). Structure-activity relationships in sweeteners. II. Saccharins, acesulfames, chlorosugars, tryptophans and ureas. *Chem. Senses* **10,** 73–88.

van der Wel, H. (1972). Isolation and characterization of the sweet principle from *Dioscoreophyllum cumminsii* (Stapf) Diels. *FEBS Lett.* **21,** 88–90.

van der Wel, H. (1981). The influence of chemical modification of the sweet-tasting proteins thaumatin and monellin on the sweetness intensity and character. In *The Quality of Foods and Beverages: Chemistry and Technology*. ed. G. Charalambous & G. Inglett. Academic Press, New York, pp. 113–8.

van der Wel, H. & Bel, W. J. (1976). Effect of acetylation and methylation on the sweetness intensity of thaumin I. *Chem. Senses* **2,** 211–8.

van der Wel, H. & Bel, W. J. (1978). Structural investigation of the sweet-tasting proteins thaumatin and monellin by immunological studies. *Chem. Senses Flavour* **3,** 99–104.

van der Wel, H. & Bel, W. J. (1980). Enzymic properties of the sweet-tasting proteins thaumatin and monellin after partial reduction. *Eur. J. Biochem.* **104,** 413–8.

van der Wel, H. & Ledeboer, A. M. (1989). The thaumatins. *Biochem. Plants* **15,** 379–91.

van der Wel, H. & Loeve, K. (1972). Isolation and characterization of thaumatin I and II, the sweet-tasting proteins from *Thaumatococcus daniellii* Benth. *Eur. J. Biochem.* **31,** 221–5.

van der Wel, H., Iyengar, R. B., van Brouwershaven, J., van Wassenaar, P. D., Bel, W. J. & van der Ouderaa, F. J. G. (1984). Assignment of the disulfide bonds in the sweet-tasting protein thaumatin I. *Eur. J. Biochem.* **144,** 41–5.

van der Wel, H., van der Heijden, A. & Peer, H. G. (1987). Sweeteners. *Food Rev. Int.* **3,** 193–268.

van der Wel, H., Larson, G., Hladik, A., Hladik, C. M., Hellekant, G. & Glaser, D. (1989). Isolation and characterization of pentadin, the sweet principle of *Pentadiplandra brazzeana* Baillon. *Chem. Senses* **14,** 75–9.

van Nierkerk, D. M. & Koeppen, B. H. (1972). Synthesis of 2-*O*-α-L-rhamnopyranosyl D galactose, a reported partial hydrolysis product of α-solanine, and some taste-eliciting flavonoid 2-*O*-α-L-rhamnopyranosyl-β-D-galactopyranosides. *Experientia* **28,** 123–4.

van't Hoff, J H. (1874). Sur les formules de structure dans l'espace. [a] *Archiv. Neerl. Sci. Exactes Nat.* **9,** 445–54; [b] (1875). Sur les formules de structure dans l'espace. *Bulletin de la Societe chimique de Paris* **23,** 295–305.

Venanzi, T. J. & Venanzi, C. A. (1988*a*). *Ab initio* molecular electrostatic potentials of perillartine analogs: Implications for sweet-taste receptor recognition. *J. Med. Chem.* **31,** 1879–85.

Venanzi, T. J. & Venanzi, C. A. (1988*b*). A molecular electrostatic-potential study of acesulfame. *Anal. Chim. Acta* **210,** 213–8.

Venanzi, T. J. & Venanzi, C. A. (1989). Sweet taste receptor recognition of the conformationally flexible aldoxime molecule. In *QSAR: Quantitative Structure-Activity Relationships in Drug Design*. ed. J. L. Fauchere. Alan R. Liss, Inc., New York, pp. 321-4.

Venanzi, T. J. & Venanzi, C. A. (1991). Electrostatic recognition patterns of sweet-tasting compounds. In *Sweeteners, Discovery, Molecular Design, and Chemoreception*. ed. E. Walters, F. T. Orthoefer & G. E. DuBois. Am. Chem. Soc. Symposium Series 450, Washington, DC, pp. 193–205.

Verkade, P. E. (1968). Organic compounds with a sweet and(or) a bitter taste. *Farmaco Ed. Sci.* **23**(**3**), 248–91.

Verkade, P. E. & Witjens, P. H. (1946). Researches on the alkoxy-amino-nitrobenzenes. II. Synthesis and taste of the ten possible enthoxy and *n*-propoxy compounds. *Recl. Trav. Chim.* **65,** 361–79.

Verkade, P. E., Van Dijk, C. P. & Meerburg, W. (1946). Researches on the alkoxy-amino-nitrobenzenes. *Recl. Trav. Chim.* **65,** 34660.

Walter, G. J. & Mitchell, M. L. (1986). Saccharin. In *Alternative Sweeteners.* ed. L. O. Nabors & R. G. Gelardi. Marcel Dekker, New York, pp. 15–41.

Walter, W. & Ruess, K.-P. (1971). Ueber die Struktur der Thioamide und ihrer Derivate. XIII. Konfiguration monosubstituierter Thioharnstoffe. *Justus Liebigs Ann. Chem.* **743,** 167–76.

Warfield, R. B. (1954). Taste and molecular structure. *Abstr. Papers Am. Chem. Soc. 126th Meeting Div. Agric. and Food Chem.*, p. 15A.

Weyl, H. (1952). *Symmetry.* Princeton University Press, Princeton, NJ. 168 pp.

Whitaker, J. R. (1972). *Principles of Enzymology for the Food Sciences.* Marcel Dekker, Inc., New York, 636 pp.

Whyte, L. L. (1958). Chirality. *Nature* **182,** 198.

Wienen, W. J. & Shallenberger, R. S. (1988). Influence of acid and temperature on the rate of inversion of sucrose. *Food Chem.* **29,** 51–5.

Wieser, H. & Belitz, H. D. (1975). Zusammenhänge zwischen Struktur und Bittergeschmack bei Aminosäuren und Peptiden I. Aminosäuren und verwandte Verbindungen. *Z. Lebensm.-Unters. Forsch.* **159,** 65–72.

Wieser, H., Jugel, H. & Belitz, H.-D. (1977). Zusammenhänge zwischen Struktur und Süssgeschmack bei Aminosäuren. *Z. Lebensm.-Unters. Forsch.* **164,** 277–82.

Wilkins, W. F. (1964). Isolation and identification of the lachrymogenic compound of onion. *Cornell Univ. Agric. Exp. Sta. Memoir 385.* Ithaca, New York.

Williams, C. & Allerhand, A. (1977). Detection of β-D-glucofuranose in aqueous solutions of D-glucose. Application of carbon-13 Fourier transform n.m.r. spectroscopy. *Carbohydr. Res.* **56,** 173–9.

Witzemann, E. J. (1914). The isolation of crystalline dl-glyceric aldehyde from a syrup obtained by the oxidation of glycerol. *J. Am. Chem. Soc.* **36,** 2223–34.

Wolfrom, M. L. & Thompson, A. (1946). L-Fructose. *J. Am. Chem. Soc.* **68,** 791–3.

Wrolstad, R. E. & Shallenberger, R. S. (1981). Free sugars and sorbitol in fruits — A compilation from the literature. *J. Assoc. Off. Anal. Chem.* **64,** 91–103.

Yamaguchi, S. (1987). Fundamental properties of umami in human taste sensation. In *Umami: A basic taste.* ed. Y. Kawamura & M. R. Kare. Marcel Dekker, New York, pp. 41–73.

Yamaguchi, S., Yoshikawa, T., Ikeda, S. & Ninomiya, T. (1970). Studies on the taste of some sweet substances. I. Measurement of relative sweetness. *Agric. Biol. Chem.* **34,** 181–6.

Yamato, M. & Hashigaki, K. (1979). Chemical structure and sweet taste of isocoumarins and related compounds. *Chem. Senses Flavour* **4,** 35–47.

Yamazaki, T., Huang, Z., Pröbstl, A. & Goodman, M. (1991). Peptidometic structures incorporating 2-aminocyclopentane carboxylic acids. In *Proceedings of the 21st European Peptide Symposium, Spain.* ed. E. Giralt & D. Andreu, Escom Sci. Pub. B.V., pp. 389–92.

Zehner, L. R. (1988). D-Tagatose as a low-calorie carbohydrate sweetener and bulking agent. EP 257, 626. U.S. Patent 4,786,722.

Zotterman, Y. (1971). The recording of the electrical response from human taste nerves. In *Handbook of Sensory Physiology. IV. Chemical Senses. 2. Taste.* ed. L. M. Beidler. Springer-Verlag, Berlin, pp. 102–15.

Author Index

Index